Saas-Fee Advanced Course 46

Series Editor

Swiss Society for Astrophysics and Astronomy, Sauverny, Switzerland

The Saas-Fee Advanced Courses, organized by the Swiss Society for Astrophysics and Astronomy (SSAA), have been held annually since 1971. The courses cover important and timely fields of Astrophysics and Astronomy including related fields like computational methods. They are given by two to four reputable researchers in those fields. Each lecturer covers a different aspect of the field on which he has his own expertise.

The books in this series publish the updated lecture notes of the Saas-Fee Advanced Courses. Many of them have become standard texts in their field. They are not only appreciated by graduate students, but are also used and cited as reference by researchers.

More information about this series at http://www.springer.com/series/4284

Mark Dijkstra · J. Xavier Prochaska ·
Masami Ouchi · Matthew Hayes

Lyman-alpha as an Astrophysical and Cosmological Tool

Saas-Fee Advanced Course 46

Swiss Society for Astrophysics and Astronomy
Edited by Anne Verhamme, Pierre North,
Sebastiano Cantalupo and Hakim Atek

 Springer

Authors
Mark Dijkstra
Institute of Theoretical Astrophysics
University of Oslo
Oslo, Norway

J. Xavier Prochaska
Lick Observatory
University of California Observatories
(UCO)
Santa Cruz, CA, USA

Masami Ouchi
Institute for Cosmic Ray Research (ICRR)
University of Tokyo
Kashiwa, Japan

Matthew Hayes
Department of Astronomy
University of Stockholm
Stockholm, Sweden

Volume Editors
Anne Verhamme
Department of Astronomy
Geneva University
Versoix, Geneva, Switzerland

Pierre North
Laboratory of Astrophysics (LASTRO)
Ecole Polytechnique Fédérale de Lausanne
(EPFL)
Versoix, Geneva, Switzerland

Sebastiano Cantalupo
Institute for Astronomy
ETH Zürich
Zürich, Switzerland

Hakim Atek
Institut d'Astrophysique de Paris (IAP)
Paris, France

This Series is edited on behalf of the Swiss Society for Astrophysics and Astronomy: Société Suisse d'Astrophysique et d'Astronomie, Observatoire de Genève, ch. des Maillettes 51, CH-1290 Sauverny, Switzerland.

ISSN 1861-7980   ISSN 1861-8227 (electronic)
Saas-Fee Advanced Course
ISBN 978-3-662-59625-8   ISBN 978-3-662-59623-4   (eBook)
https://doi.org/10.1007/978-3-662-59623-4

Cover illustration: Lyman-alpha line overlaid on a photo of Matterhorn. Credit: [Vincent Favre (http://www.cristaldegivre.com) & Myriam Burgener]

This Springer imprint is published by the registered company Springer-Verlag GmbH, DE part of Springer Nature.
The registered company address is: Heidelberger Platz 3, 14197 Berlin, Germany

Preface

Hydrogen is the most abundant element in the universe, as first understood by Cecilia Payne-Gaposchkin in 1924. The Lyman α line is the strongest line of this element, making it especially important in many astrophysical topics and particularly in cosmology. Because this line falls in the ultraviolet, it has long remained unused for observational reasons, until astronomical satellites opened the UV window and until galaxies were discovered at high enough redshift to make it accessible to ground-based telescopes.

Since then, the hydrogen Lyα line has acquired ever increasing popularity to probe the deep universe, through the use of either its absorption or its emission. The Lyα forest seen in the spectra of high redshift quasars reveals the presence and distribution of absorbing gas in galaxies that would otherwise go unnoticed, and star forming galaxies can be seen to the edge of the observable universe thanks to the intensity of this line in emission. Spectroscopic confirmation of the highest redshifts known to date is generally based on the Lyα line. However, the very visibility of this resonance line when seen in emission is linked with its extremely high opacity, so that the interpretation of the Lyα profile is far from straightforward. Careful modeling is needed to understand the observations, because transfer effects are important enough to prevent, for instance, any simple translation of the line position and width in terms of mean radial velocity and velocity dispersion.

The 46th Saas-Fee advanced course of the Swiss Society for Astrophysics and Astronomy, entitled *Lyman-α as an astrophysical and cosmological tool*, took place in Les Diablerets, a mountain resort of the Swiss Alps. Exceptionally, as many as four lecturers (instead of three according to tradition) have contributed to this successful course: Mark Dijkstra, J. Xavier Prochaska, Masami Ouchi, and Matthew Hayes. The four contributions to this book review respectively the theoretical aspects of the Lyα line formation, and three aspects of how this line is used in extragalactic astronomy: absorption of intervening hydrogen clouds in the line of sight of quasars, Lyα emitting galaxies at high redshift, and detailed emission and absorption processes in local galaxies. Such topics cover much of today's endeavours in observational cosmology, so this book should prove useful and timely for many Ph.D. students as well as more advanced researchers. We are aware of an extragalactic bias, in the sense that this book does not adress more local applications of the Lyα line, such as the study of exoplanet atmospheres. Even so, students in the latter field may still find interest in at least Mark Dijkstra's theoretical contribution.

We thank Mrs. Myriam Burgener, the Secretary of the Département d'Astronomie de l'Université de Genève, for her very efficient help in the organization of the course and for her presence during the whole event. This course, attended by 65 participants from many countries, was sponsored by the Swiss Society for Astrophysics and Astronomy, the Swiss Academy of Sciences, the Ecole Polytechnique Fédérale de Lausanne (EPFL), and the University of Geneva.

Versoix, Switzerland   Anne Verhamme
March 2018   Pierre North
                                                                                        Hakim Atek
                                                                          Sebastiano Cantalupo

Contents

4 Lyman Alpha Emission and Absorption in Local Galaxies 319
Matthew Hayes

Contributors

Mark Dijkstra Institute of Theoretical Astrophysics, University of Oslo, Oslo, Norway, e-mail: astromark77@gmail.com

Matthew Hayes Department of Astronomy & OKC, AlbaNova University Centre, University of Stockholm, Stockholm, Sweden, e-mail: matthew@astro.su.se

Masami Ouchi Institute for Cosmic Ray Research (ICRR), University of Tokyo, Kashiwa, Japan, e-mail: ouchims@icrr.u-tokyo.ac.jp

J. Xavier Prochaska Lick Observatory, University of California Observatories (UCO), Santa Cruz, CA, USA, e-mail: xavier@ucolick.org

Chapter 1
Physics of Lyα Radiative Transfer

Mark Dijkstra

1.1 Introduction

Half a century ago, [205] predicted that the Lyα line should be a good tracer of star forming galaxies at large cosmological distances. This statement was based on the assumption that ionizing photons that are emitted by young, newly formed stars are efficiently reprocessed into recombination lines, of which Lyα contains the largest flux. In the past two decades the Lyα line has indeed proven to provide us with a way to both find and identify galaxies out to the highest redshifts (currently as high as $z = 8.7$, see [290]). In addition, we do not only expect Lyα emission from (star forming) galaxies, but from structure formation in general (e.g. [94]). Galaxies are surrounded by vast reservoirs of gas that are capable of both emitting and absorbing Lyα radiation. Observed spatially extended Lyα nebulae (or 'blobs') may indeed provide insight into the formation and evolution of galaxies, in ways that complement direct observations of galaxies.

The original version of this chapter was revised: Figure 1.35 has been updated and in Page "52" an equation has been corrected. The correction to this chapter can be found at https://doi.org/10.1007/978-3-662-59623-4_5

M. Dijkstra (✉)
University of Oslo, Oslo, Norway
e-mail: astromark77@gmail.com

© Springer-Verlag GmbH Germany, part of Springer Nature 2019
M. Dijkstra et al., *Lyman-alpha as an Astrophysical and Cosmological Tool*,
Saas-Fee Advanced Course 46, https://doi.org/10.1007/978-3-662-59623-4_1

Many new instruments and telescopes[1] are either about to be, or have just been, commissioned that are ideal for targeting the redshifted Lyα line. The sheer number of observed Lyα emitting sources is expected to increase by more than two orders of magnitude at all redshifts $z \sim 2$–7. For comparison, this boost is similar to that in the number of known exoplanets as a result of the launch of the Kepler satellite. In addition, sensitive integral field unit spectrographs will allow us to (*i*) detect sources that are more than an order of magnitude fainter than what has been possible so far, (*ii*) take spectra of faint sources, (*iii*) take spatially resolved spectra of the more extended sources, and (*iv*) detect phenomena at surface brightness levels at which diffuse Lyα emission from the environment of galaxies is visible.

In order to interpret this growing body of data, we must understand the radiative transfer of Lyα photons from its emission site to the telescope. Lyα transfer depends sensitively on the gas distribution and kinematics. This complicates interpretation of Lyα observations. On the other hand, the close interaction of the Lyα radiation field and gaseous flows in and around galaxies implies that observations of Lyα contains information on the medium through which the photons were scattering, and may thus present an opportunity to learn more about atomic hydrogen in gaseous flows in and around galaxies.

1.2 The Hydrogen Atom and Introduction to Lyα Emission Mechanisms

1.2.1 Hydrogen in Our Universe

It has been known from almost a century that hydrogen is the most abundant element in our Universe. In 1925 Cecilia Payne demonstrated in her PhD dissertation that the Sun was composed primarily of hydrogen and helium. While this conclusion was controversial[2] at the time, it is currently well established that hydrogen accounts for the majority of baryonic mass in our Universe: the fluctuations in the Microwave Background as measured by the Planck satellite [211] imply that baryons account for 4.6% of the Universal energy density, and that hydrogen accounts for 76% of the

---

[1]Examples of new instruments/telescopes that will revolutionize our ability to target the Lyα emission line: the Hobby-Eberly Telescope Dark Energy Experiment (HETDEX, http://hetdex.org/) will increase the sample of Lyα emitting galaxies by orders of magnitude at $z \sim 2$–4; Subaru's Hyper Suprime-Cam (http://www.naoj.org/Projects/HSC/) is expected to provide a similar boost out to $z \sim 7$. Integral Field Unit Spectrographs such as MUSE (https://www.eso.org/sci/facilities/develop/instruments/muse.html, also see the Keck Cosmic Web Imager http://www.srl.caltech.edu/sal/keck-cosmic-web-imager.html) will allow us to map out spatially extended Lyα emission down to $\sim$10 times lower surface brightness levels, and take spatially resolved spectra. In the (near) future, telescopes such as the James Webb Space Telescope (JWST, http://www.jwst.nasa.gov/) and ground based facilities such as the Giant Magellan Telescope (http://www.gmto.org/) and ESO's E-ELT (http://www.eso.org/public/usa/teles-instr/e-elt/, http://www.tmt.org/ (TMT).

[2]See https://en.wikipedia.org/wiki/Cecilia_Payne-Gaposchkin.

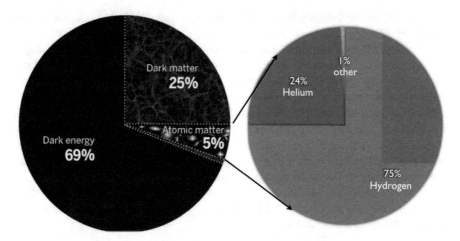

Fig. 1.1 Relative contributions to Universal energy/mass density. Most (∼70%) of the Universal energy content is in the form of dark energy, which is an unknown hypothesized form of energy which permeates all of space, and which is responsible for the inferred acceleration of the expansion of the Universe. In addition to this, ∼25% is in the form of dark matter, which is a pressureless fluid which acts only gravitationally with ordinary matter. Only ∼5% of the Universal energy content is in the form of ordinary matter like baryons, leptons etc. Of this component, ∼75% of all baryonic matter is in Hydrogen, while the remaining 25% is almost entirely Helium. Observing lines associated with atomic hydrogen is therefore an obvious way to go about studying the Universe

total baryonic mass. The remaining 24% is in the form of helium (see Fig. 1.1). The leading constraints on these mass ratios come from Big Bang Nucleosynthesis, which predicts a hydrogen abundance of ∼75% by mass for the inferred Universal baryon density ($\Omega_b h^2 = 0.022$, [211]). Additional constraints come from observations of hydrogen and emission lines of extragalactic, metal poor HII regions [15, 134].

Because of its prevalence throughout the Universe, lines associated with atomic hydrogen have provided us with a powerful window on our Universe. The 21-cm hyperfine transition was observed from our own Milky Way by Ewen and Purcell in [81], shortly after it was predicted to exist by Jan Oort in 1944. Observations of the 21-cm line have allowed us to perform precise measurements of the distribution and kinematics of neutral gas in external galaxies, which provided evidence for dark matter on galactic scales (e.g. [33]). Detecting the redshifted 21-cm emission from galaxies at $z > 0.5$, and from atomic hydrogen in the diffuse (neutral) intergalactic medium represent the main science drivers for many low frequency radio arrays that are currently being developed, including the Murchinson Wide Field Array,[3] the Low Frequency Array,[4] The Hydrogen Epoch of Reionization Array (HERA),[5] the

---

[3]http://www.mwatelescope.org/.

[4]http://www.lofar.org/.

[5]http://reionization.org/.

Precision Array for Probing the Epoch of Reionization (PAPER),[6] and the Square Kilometer Array.[7]

Similarly, the Lyα transition has also revolutionized observational cosmology: observations of the Lyα forest in quasar spectra has allowed us to measure the matter distribution throughout the Universe with unprecedented accuracy. The Lyα forest still provides an extremely useful probe of cosmology on scales that are not accessible with galaxy surveys, and/or the Cosmic microwave background. The Lyα forest will be covered extensively in the lectures by J. X. Prochaska. So far, the most important contributions to our understanding of the Universe from Lyα have come from studies of Lyα absorption. However, with the commissioning of many new instruments and telescopes, there is tremendous potential for Lyα in emission. Because Lyα is a resonance line, and because typical astrophysical environments are optically thick to Lyα, we need to understand the radiative transfer to be able to fully exploit the observations of Lyα emitting sources.

1.2.2 The Hydrogen Atom: The Classical and Quantum Picture

The classical picture of the hydrogen atom is that of an electron orbiting a proton. In this picture, the electrostatic force binds the electron and proton. The equation of motion for the electron is given by

$$\frac{q^2}{r^2} = \frac{m_e v_e^2}{r},$$ (1.1)

where q denotes the charge of the electron and proton, and the subscript 'e' ('p') indicates quantities related to the electron (proton). The acceleration the electron undergoes thus equals $a_e = \frac{v_e^2}{r} = \frac{q^2}{r^2 m_e}$. When a charged particle accelerates, it radiates away its energy in the form of electromagnetic waves. The total energy that is radiated away by the electron per unit time is given by the Larmor formula, which is given by

$$P = \frac{2}{3} \frac{q^2 a_e^2}{c^3}.$$ (1.2)

The total energy of the electron is given by the sum of its kinetic and potential energy, and equals $E_e = \frac{1}{2} m_e v_e^2 - \frac{q^2}{r} = -\frac{q^2}{2r}$. The total time it takes for the electron to radiative away all of its energy is thus given by

$$t = \frac{E_e}{P} = \frac{3c^3}{4r a_e^2} = \frac{3r^3 m_e^2 c^3}{4q^4} \approx 10^{-11} \text{ s,}$$ (1.3)

---

[6]http://eor.berkeley.edu/.

[7]https://www.skatelescope.org/.

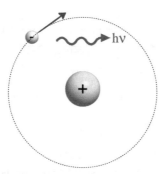

Fig. 1.2 In the classical picture of the hydrogen atom, an electron orbits a central proton at $v \sim \alpha c$. The acceleration that the electron experiences causes it to emit electromagnetic waves and loose energy. This causes the electron to spiral inwards into the proton on a time-scale of $\sim 10^{-11}$ s. In the classical picture, hydrogen atoms are highly unstable, short-lived objects

where we substituted the Bohr radius for r, i.e. $r = a_0 = 5.3 \times 10^{-9}$cm. In the classical picture, hydrogen atoms would be highly unstable objects, which is clearly problematic and led to the development of quantum mechanics (Fig. 1.2).

In quantum mechanics, electron orbits are *quantized*: In Niels Bohr's model of the atom, electrons can only reside in discrete orbitals. While in such an orbital, the electron does not radiate. It is only when an electron transitions from one orbital to another that it emits a photon. Quantitatively, the total angular momentum of the electron $L \equiv m_e v_e r$ can only taken on discrete values $L = n\hbar$, where $n = 1, 2, ...$, and $\hbar$ denotes the reduced Planck constant (Table 1.1). The total energy of the electron is then

$$E_e(n) = -\frac{q^4 m_e}{2n^2 \hbar^2} = -\frac{E_0}{n^2}, \qquad (1.4)$$

where $E_0 = 13.6$ eV denotes 1 Rydberg, which corresponds to the binding energy of the electron in its ground state. In quantum mechanics, the total energy of an electron bound to a proton to form a hydrogen atom can only take on a discrete set of values, set by the 'principal quantum number' n. The quantum mechanical picture of the hydrogen atom differs from the classical one in additional ways: the electron orbital of a given quantum state (an orbital characterized by quantum number n) does not correspond to the classical orbital described above. Instead, the electron is described by a quantum mechanical wavefunction $\psi(\mathbf{r})$, (the square of) which describes the probability of finding the electron at some location $\mathbf{r}$. The functional form of these wavefunctions are determined by the Schrödinger equation. We will not discuss the Schrödinger equation in these lectures, but will simply use that it implies that the quantum mechanical wavefunction $\psi(\mathbf{r})$ of the electron is characterized fully by *two* quantum numbers: the principal quantum number n, and the orbital quantum number l. The orbital quantum number l can only take on the values $l = 0, 1, 2, ..., n - 1$. The electron inside the hydrogen atom is fully characterized by these two numbers. The classical analogue of requiring two numbers to characterize the electron wavefunction

Table 1.1 Symbol dictionary

Symbol	Definition
k_B	Boltzmann constant: $k_B = 1.38 \times 10^{-16}$ erg K^{-1}
h_P	Planck constant: $h_P = 6.67 \times 10^{-27}$ erg s
$\hbar$	Reduced Planck constant: $\hbar = \frac{h_P}{2\pi}$
m_p	Proton mass: $m_p = 1.66 \times 10^{-24}$
m_e	Electron mass: $m_e = 9.1 \times 10^{-28}$ g
q	Electron charge: $q = 4.8 \times 10^{-10}$ esu
c	Speed of light: $c = 2.9979 \times 10^{10}$ cm s^{-1}
ΔE_{ul}	Energy difference between upper level 'u' and lower level 'l' (in ergs)
ν_{ul}	Photon frequency associated with the transition $u \rightarrow l$ (in Hz)
f_{ul}	The oscillator strength associated with the transition $u \rightarrow l$ (dimensionless)
A_{ul}	Einstein A-coefficient of the transition $u \rightarrow l$ (in s^{-1})
B_{ul}	Einstein B-coefficient of the transition $u \rightarrow l$: $B_{ul} = \frac{2h_P \nu_{ul}^3}{c^2} A_{ul}$ (in erg cm^{-2} s^{-1})
B_{lu}	Einstein B-coefficient of the transition $l \rightarrow u$: $B_{lu} = \frac{g_u}{g_l} B_{ul}$
$\alpha_{A/B}$	Case A /B recombination coefficient (in cm^3 s^{-1})
α_{nl}	Recombination coefficient into state (n, l) (in cm^3 s^{-1})
$g_{u/l}$	Statistical weight of upper/lower level of a radiative transition (dimensionless)
ν_α	Photon frequency associated with the Lyα transition: $\nu_\alpha = 2.47 \times 10^{15}$ Hz
ω_α	Angular frequency associated with the Lyα transition: $\omega_\alpha = 2\pi \nu_\alpha$
λ_α	Wavelength associated with the Lyα transition: $\lambda_\alpha = 1215.67$ Å
A_α	Einstein A-coefficient of the Lyα transition: $A_\alpha = 6.25 \times 10^8$ s^{-1}
T	gas temperature (in K)
v_{th}	Velocity dispersion (times $\sqrt{2}$): $v_{th} = \sqrt{\frac{2k_B T}{m_p}}$
v_{turb}	Turbulent velocity dispersion
b	Doppler broadening parameter : $b = \sqrt{v_{th}^2 + v_{turb}^2}$
$\Delta \nu_\alpha$	Doppler induced photon frequency dispersion: $\Delta \nu_\alpha = \nu_\alpha \frac{b}{c}$ (in Hz)
x	'Normalized' photon frequency: $x = (\nu - \nu_\alpha)/\Delta \nu_\alpha$ (dimensionless)
$\sigma_\alpha(x)$	Lyα Absorption cross-section at frequency x (in cm^2), $\sigma_\alpha(x) = \sigma_{\alpha,0}\phi(x)$
$\sigma_{\alpha,0}$	Lyα Absorption cross-section at line center, $\sigma_{\alpha,0} = 5.9 \times 10^{-14} \left(\frac{T}{10^4 \text{ K}}\right)^{-1/2}$ cm^2
$\phi(x)$	Voigt profile (dimensionless)
a_v	Voigt parameter: $a_v = A_\alpha/[4\pi \Delta \nu_\alpha] = 4.7 \times 10^{-4}(T/10^4 \text{ K})^{-1/2}$
I_ν	Specific intensity (in erg s^{-1} Hz^{-1} cm^{-2} sr^{-1})
J_ν	Angle averaged specific intensity (in erg s^{-1} Hz^{-1} cm^{-2} sr^{-1})

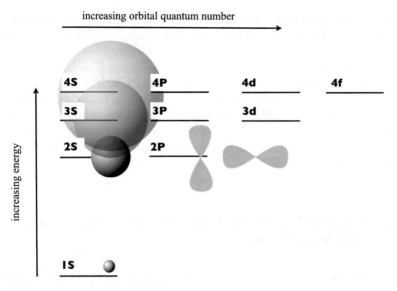

Fig. 1.3 The total energy of the quantum states of the hydrogen atom, and a simplified representation of the associated quantum mechanical wavefunction describing the electron. The level denoted with '1s' denotes the ground state and has a total energy $E = -13.6$ eV. The wavefunction is spherically symmetric and compact. The extent/size of the wavefunction increases with quantum number n. The eccentricity/elongation of the wavefunction increases with quantum number l. The orientation of non-spherical wavefunction can be represented by a third quantum number m

is that we need two numbers to characterize the classical orbit of the electron around the proton, namely energy E and total angular momentum L.

The diagram in Fig. 1.3 shows the total energy of different quantum states in the hydrogen, and a sketch of the associated wavefunctions. This Figure indicates that

- The lowest energy state corresponds to the $n = 1$ state, with an energy of $E = -13.6$ eV. For the state $n = 1$, the orbital quantum number l can only take on the value $l = 0$. This state with $(n, l) = (1, 0)$ is referred to as the '1s'-state. The '1' refers to the value of n, while the 's' is a historical way (the 'spectroscopic notation') of labelling the '$l = 0$'-state. This Figure also indicates (schematically) that the wavefunction that describes the 1s-state is spherically symmetric. The 'size' or extent of this wavefunction relates to the classical atom size in that the expectation value of the radial position of the electron corresponds to the Bohr radius a_0, i.e. $\int dV r |\psi_{1s}(\mathbf{r})|^2 = a_0$.
- The second lowest energy state, $n = 2$, has a total energy $E = E_0/n^2 = -3.4$ eV. For this state there exist two quantum states with $l = 0$ and $l = 1$. The '2s'-state is again characterized by a spherically symmetric wavefunction, but which is more extended. This larger physical extent reflects that in this higher energy state, the electron is more likely to be further away from the proton, completely in line with classical expectations. On the other hand, the wavefunction that describes the '2p'-state ($n = 2, l = 1$) is not spherically symmetric, and consists of two 'lobes'.

The elongation that is introduced by these lobes can be interpreted as the electron being on an eccentric orbit, which reflects the increase in the electron's orbital angular momentum.

- The third lowest energy state $n = 3$ has a total energy of $E = E_0/n^2 = -1.5$ eV. The size/extent of the orbital/wavefunction increases further, and the complexity of the shape of the orbitals increases with n (see e.g. https://en.wikipedia.org/wiki/Atomic_orbital for illustrations). Loosely speaking, the quantum number n denotes the extent/size of the wavefunction, l denotes its eccentricity/elongation. The orientation of non-spherical wavefunction can be represented by a third quantum number m.

1.2.3 Radiative Transitions in the Hydrogen Atom: Lyman, Balmer, ..., Pfund, ... Series

We discussed how in the classical picture of the hydrogen atom, the electron ends up inside the proton after $\sim 10^{-11}$ s. In quantum mechanics, the electron is only stable in the ground state (1s). The life-time of an atom in any excited state is very short, analogous to the instability of the atom in the classical picture. Transitions between different quantum states have been historically grouped into series, and named after the discoverer of these series. The series include

- **The Lyman series**. A series of radiative transitions in the hydrogen atom which arise when the electron goes from $n \geq 2$ to $n = 1$. The first line in the spectrum of the Lyman series—named Lyman α (hereafter, Lyα)—was discovered in 1906 by Theodore Lyman, who was studying the ultraviolet spectrum of electrically excited hydrogen gas. The rest of the lines of the spectrum were discovered by Lyman in subsequent years.
- **The Balmer series**. The series of radiative transitions from $n \geq 3$ to $n = 2$. The series is named after Johann Balmer, who discovered an empirical formula for the wavelengths of the Balmer lines in 1885. The Balmer-α (hereafter Hα) transition is in the red, and is responsible for the reddish glow that can be seen in the famous Orion nebula.
- Following the Balmer series, we have the **Paschen series** ($n \geq 4 \rightarrow n = 3$), the **Brackett series** ($n \geq 5 \rightarrow 4$), the **Pfund series** ($n \geq 6 \rightarrow 5$), ... Especially Pfund-δ is potentially an interesting probe ([[199]-2016 private communication).

Quantum mechanics does not allow radiative transitions between just any two quantum states: these radiative transitions must obey the 'selection rules'. The simplest version of the selection rules—which we will use in these lectures—is that only transitions of the form $|\Delta l| = 1$ are allowed. A simple interpretation of this is that photons carry a (spin) angular momentum given by $\hbar$, which is why the angular momentum of the electron orbital must change by $\pm\hbar$ as well. Figure 1.4 indicates allowed transitions, either as *green solid lines* or as *red dashed lines*. Note that the Lyman-β, γ, ... transitions ($3p \rightarrow 1s$, $4p \rightarrow 1s$, ...) are not shown on purpose. As

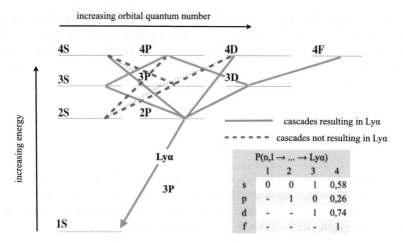

Fig. 1.4 Atoms in any state with $n > 1$ radiatively cascade back down to the ground (1s) state. The quantum mechanical selection rules only permit transitions where $|\Delta l| = 1$. These transitions are indicated with colored lines connecting the different quantum states. The *green solid lines* indicate radiative cascades that result in the emission of a Lyα photon, while *red dotted lines* indicate transitions that do not. We have omitted all direct radiative transitions $np \to 1s$: this corresponds to the 'case-B' approximation, which assumes that the recombining gas is optically thick to all Lyman-series photons, and that these photons would be re-absorbed immediately. The table in the lower right corner indicates Lyα production probabilities from various states: e.g. the probability that at atom in the $4s$ state produces a Lyα photon is ~ 0.58. *Credit from Fig. 1 of* [75], *Lyman Alpha Emitting Galaxies as a Probe of Reionization, PASA, 31, 40D*

we will see later in the lectures, while these transitions are certainly allowed, in realistic astrophysical environments it is better to simply ignore them.

Consider an electron in some arbitrary quantum state (n, l). The electron does not spend much time in this state, and radiatively decays down to a lower energy state (n', l'). This lower energy state is again unstable [unless $(n', l') = (1, 0)$], and the electron again radiatively decays to an even lower energy state (n'', l''). Ultimately, all paths lead to the ground state, even those paths that go through the $2s$ state. While the selection rules do not permit transitions of the form $2s \to 1s$, these transitions can occur, if the atom emits *two* photons (rather than one). Because these two-photon transitions are forbidden, the life-time of the electron in the $2s$ state is many orders of magnitude larger than almost all other quantum states (it is ~ 8 orders of magnitude longer than that of the $2p$-state), and this quantum state is called 'meta-stable'. The path from an arbitrary quantum state (n, l) to the ground state via a sequence of radiative decays is called a 'radiative cascade'.

The *green solid lines* in Fig. 1.4 show radiative cascades that result in the emission of a Lyα photon. The *red dashed lines* show the other radiative cascades. The table in the *lower right corner* shows the probability that a radiative cascade from quantum state (n, l) produces a Lyα photon. This probability is denoted with $P(n, l \to \dots \to \text{Ly}\alpha)$. For example, the probability that an electron in the $2s$ orbital gives rise to Lyα is zero. The probability that an electron in the $3s$ orbital gives Lyα is 1. This is because the only allowed radiative cascade to the ground state from $3s$ is $3s \rightarrow 2p \rightarrow 1s$. This last transition corresponds to the Lyα transition. For $n \geq 4$ the probabilities become non-trivial, as we have to compute the likelihood of different radiative cascades. We discuss this in more detail in the next section.

1.2.4   Lyα Emission Mechanisms

A hydrogen atom emits Lyα once its electron is in the $2p$ state and decays to the ground state. We mentioned qualitatively how radiative cascades from a higher energy state can give rise to Lyα production. Electrons can end up these higher energy quantum states (any state with $n > 1$) in two different ways:

1. **Collisions.** The 'collision' between an electron and a hydrogen atom can leave the atom in an excited state, at the expense of kinetic energy of the free electron. This process is illustrated in Fig. 1.5. This process converts thermal energy of the electrons, and therefore of the gas as whole, into radiation. This process is also referred to as Lyα production via 'cooling' radiation. We discuss this process in more detail in Sect. 1.3.1, and in which astrophysical environments it may occur in Sect. 1.4.

2. **Recombination.** Recombination of a free proton and electron can leave the electron in any quantum state (n, l). Radiative cascades to the ground state can then produce a Lyα photon. As we discussed in Sect. 1.2.3, we can compute the probability that each quantum state (n, l) produces a Lyα photon during the radiative cascade down to the ground-state. If we sum over all these quantum states, and properly weigh by the probability that the freshly combined electron-proton pair ended up in state (n, l), then we can compute the probability that a recombination event gives us a Lyα photon. We discuss the details of this calculation in Sect. 1.3.2. Here, we simply discuss the main results.

   The *upper panel* of Fig. 1.6 shows the total probability $P(\text{Ly}\alpha)$ that a Lyα photon is emitted per recombination event as a function of gas temperature T. This plot contains two lines. The *solid black line* represents 'Case-A', which refers to the most general case where we allow the electron and proton to recombine into *any* state (n, l), and where we allow for all radiative transitions permitted by the selection rules. The *dashed black line* shows 'Case-B', which refers to the case where we do not allow for (*i*) direct recombination into the ground state, which produces an ionizing photon, and (*ii*) radiative transitions of the higher order Lyman series, i.e. Lyβ, Lyγ, Lyδ,.... Case-B represents that most astrophysical gases efficiently re-absorb higher order Lyman series and ionizing photons, which effectively 'cancels out' these transitions (see Sect. 1.3.2 for more discussion on this). This Figure shows that for gas at $T = 10^4$ K and case-B recombination, we have $P(\text{Ly}\alpha) = 0.68$. This value '0.68' is often encountered during discussions on Lyα emitting galaxies. It is worth keeping in mind that

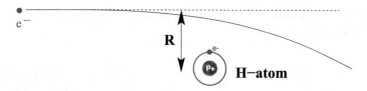

Fig. 1.5 Cooling radiation at the atomic level: an interaction between an electron and a hydrogen atom can leave the hydrogen atom in an excited state, which can produce a Lyα photon. The energy carried by the Lyα photon comes at the expense of the kinetic energy of the electron. Lyα emission by the hydrogen atom thus cools the gas

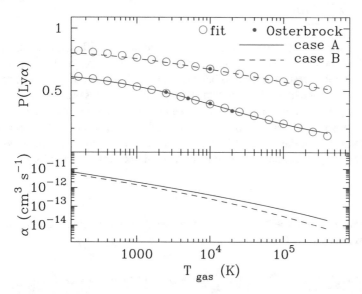

Fig. 1.6 The *top panel* shows the probability $P(\text{Ly}\alpha)$ that a recombination event leads to the production of a Lyα photon, as a function of gas temperature T. The *upper dashed line* (*lower solid line*) corresponds to 'case B' ('case A'). The *lower panel* shows the recombination rate $\alpha(T)$ (in cm^3 s^{-1}) at which electrons and protons recombine. The *solid line* (*dashed line*) represents case-B (case-A). The *red open circles* represent fitting formulae (Eq. 1.5). *Credit from Fig. 2 of* [75], *Lyman Alpha Emitting Galaxies as a Probe of Reionization, PASA, 31, 40D*

the probability $P(\text{Ly}\alpha)$ increases with decreasing gas temperature and can be as high as $P(\text{Ly}\alpha) = 0.77$ for $T = 10^3$ K (also see [42]). The *red open circles* represent the following two fitting formulae

$$P_A(\text{Ly}\alpha) = 0.41 - 0.165 \log T_4 - 0.015(T_4)^{-0.44}$$
$$P_B(\text{Ly}\alpha) = 0.686 - 0.106 \log T_4 - 0.009(T_4)^{-0.44}, \qquad (1.5)$$

where $T_4 \equiv T/10^4$ K. The fitting formula for case-B is taken from [42].

1.3 A Closer Look at Lyα Emission Mechanisms and Sources

The previous section provided a brief description of physical processes that give rise to Lyα emission. Here, we discuss these in more detail, and also link them to astrophysical sources of Lyα.

1.3.1 Collisions

Collisions involve an electron and a hydrogen atom. The efficiency of this process depends on the relative velocity of the two particles. The Lyα production rate therefore includes the product of the number density of both species, and the rate coefficient $q_{1s2p}(P[v_e])$ which quantifies the velocity dependence of this process ($P[v_e]$ denotes the velocity distribution of electrons). If we assume that the velocity distribution of electrons is given by a Maxwellian distribution, then $P[v_e]$ is uniquely determined by temperature T, and the rate coefficient becomes a function of temperature, $q_{1s2p}(T)$. The total Lyα production rate through collisional excitation is therefore

$$R_{\text{coll}}^{\text{Ly}\alpha} = n_e n_{\text{H}} q_{1s2p} \text{ cm}^{-3} \text{ s}^{-1}. \tag{1.6}$$

In general, the rate coefficient q_{lu} is expressed[8] in terms of a 'velocity averaged collision strength' $\langle \Omega_{lu} \rangle$ as

$$q_{lu} = \frac{h_P^2}{(2\pi m_e)^{3/2}(k_B T)^{1/2}} \frac{\langle \Omega_{lu} \rangle}{g_l} \exp\left(-\frac{\Delta E_{lu}}{k_B T}\right) = 8.63 \times 10^{-6} T^{-1/2} \frac{\langle \Omega_{lu} \rangle}{g_l} \exp\left(-\frac{\Delta E_{lu}}{k_B T}\right) \text{ cm}^3 \text{ s}^{-1}. \tag{1.7}$$

Calculating the collision strength is a very complex problem, because for the free-electron energies of interest, the free electron spends a relatively long time near the target atom, which causes distortions in the bound electron's wavefunctions. Complex quantum mechanical interactions may occur, and especially for collisional excitation into higher-n states, multiple scattering events become important (see [26] and references therein). The most reliable collision strengths in the literature are for the $1s \rightarrow nL$, with $n < 4$ and $L < d$ [9, 198, 236]. The *left panel* of Fig. 1.7 shows the velocity averaged collision strength for several transitions. There exist some differences in the calculations between these different groups. Collisional excitation rates still appear uncertain at the 10–20% level.

As we mentioned above, radiation is produced in collisions at the expense of the gas' thermal energy. The total rate at which the gas looses thermal energy, i.e. cools, per unit volume is

---

[8]The subscripts 'l' and 'u' refer to the 'lower' and 'upper' energy states, respectively.

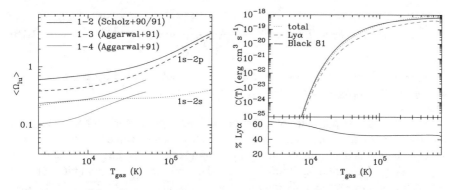

Fig. 1.7 *Left*: Velocity averaged collision strengths $\langle\Omega_{lu}\rangle$ are plotted as a function of temperature for the transitions $1s \rightarrow 2s$ (*dotted line*), $1s \rightarrow 2p$ (*dashed line*), and for their sum $1s \rightarrow 2$ (*black solid line*) as given by [236, 237]. Also shown are velocity averaged collision strengths for the $1s \rightarrow 3$ (*red solid line*, obtained by summing over all transitions $3s$, $3p$ and $3d$), and $1s \rightarrow 4$ (*blue solid line*, obtained by summing over all transitions $4s$, $4p$, $4d$ and $4f$) as given by [10]. Evaluating the collision strengths becomes increasingly difficult towards higher n (see text). *Right*: The *blue-dotted line* in the *top panel* shows the total cooling rate per H-nucleus that one obtaines by collisionally exciting H atoms into all states $n \leq 4$. For comparison, the *red dashed* line shows the total cooling rate as a result of collisional excitation of the $2p$ state, which is followed by a downward transition through emission of a Lyα photon. All cooling rates increase rapidly around $T \sim 10^4$ K. The *lower panel* shows the ratio (in%) of these two cooling rates. This plot shows that $\sim$60% of the total gas cooling rate is in the form of Lyα photons at $T \sim 10^4$ K, and that this ratio decreases to $\sim$45–50% towards higher gas temperatures. At the gas temperatures at which cooling via line excitation is important, $T \lesssim 10^5$ K (see text), $\sim$45–60% of this cooling emerges as Lyα photons. Also shown for comparison as the *black solid line* is the often used fitting formula of [28], and modified following [47]

$$\frac{dE_{\text{th}}}{dVdt} = n_e n_{\text{H}} C(T),\tag{1.8}$$

where

$$C(T) = \sum_u q_{1s \rightarrow u} \Delta E_{1s \rightarrow u} \text{ erg cm}^3 \text{ s}^{-1}.\tag{1.9}$$

Here, the sum is over all excited states 'u'. The *blue-dotted line* in the *top right panel* of Fig. 1.7 shows $C(T)$ including collisional excitation into all states $n \leq 4$. The cooling rate rises by orders of magnitude around $T \sim 10^4$ K, and reflects the strong temperature-dependence of the number density of electrons that are moving fast enough to excite the hydrogen atom. For comparison, the *red dashed* line shows the contribution to $C(T)$ from only collisional excitation into the $2p$ state, which is followed by a downward transition through emission of a Lyα photon. The *lower panel* shows the ratio (in%) of these two rates. This plot shows that $\sim$60% of the total gas cooling rate is in the form of Lyα photons at $T \sim 10^4$ K, and that this ratio decreases to $\sim$45–50% towards higher gas temperatures. The *black solid line* is an often-used analytic fitting formula by [28]

$$C(T) = 7.5 \times 10^{-19} \frac{\exp\left(-\frac{118348}{T}\right)}{(1 + T_5^{1/2})} \text{ erg cm}^3 \text{ s}^{-1}. \tag{1.10}$$

It is good to keep in mind that over the past few decades, the hydrogen collision strengths have changed quite substantially, which can explain the difference between these curves.

The cooling rate per unit volume depends on the product of $C(T)$, n_e, and n_H, and therefore on the ionization state of the gas. If we also assume that the ionization state of the gas is determined entirely by its temperature (the gas is then said to be in 'collisional ionization equilibrium'), then the total cooling rate per unit volume is a function of temperature only (and *overall* gas density squared). Figure 1.8 shows that the cooling curve increases dramatically around $T \sim 10^4$ K, which is due to the corresponding increase in $C(T)$ (see Fig. 1.7). The cooling curve reaches a maximum at $\log T \sim 4.2$, which is because at higher T collisional ionization of hydrogen removes neutral hydrogen, which eliminates the collisional excitation cooling channel. For a cosmological mixture of H and He, collisional excitation of singly ionized Helium starts dominating at $\log T \sim 4.6$ (see Fig. 1.8).

1.3.2 Recombination

The capture of an electron by a proton generally results in a hydrogen atom in an excited state (n, l). Once an atom is in a quantum state (n, l) it radiatively cascades to the ground state $n = 1, l = 0$ via intermediate states (n', l'). The probability that a radiative cascade from the state (n, l) results in a Lyα photon is given by

$$P(n, l \to \text{Ly}\alpha) = \sum_{n', l'} P(n, l \to n', l') P(n', l' \to \text{Ly}\alpha). \tag{1.11}$$

This may not feel satisfactory, as we still need to compute $P(n', l' \to \text{Ly}\alpha)$, which is the same quantity but for $n' < n$. In practice, we can compute $P(n', l' \to \text{Ly}\alpha)$ by starting at low values for n', and then work towards increasingly high n. For example, the probability that a radiative cascade from the $(n, l) = (3, 1)$ state (i.e. the 3p state) produces a Lyα photon is 0, because the selection rules only permit,[9] the transitions $(3, 1) \to (2, 0)$ and $(3, 1) \to (1, 0)$. The first transition leaves the H-atom in the 2s state, from which it can only transition to the ground state by emitting two photons [39]. On the other hand, a radiative cascade from the $(n, l) = (3, 2)$ state (i.e. the 3d state) will certainly produce a Lyα photon, since the only permitted cascade is $(3, 2) \to (2, 1) \overset{\text{Ly}\alpha}{\to} (1, 0)$. Similarly, the only permitted cascade from the 3s state is

---

[9]As we referred to in Sect. 1.2.4. Fig. 1.4 schematically depicts permitted radiative cascades in a four-level H atom. *Green solid lines* depict radiative cascades that result in a Lyα photon, while *red dotted lines* depict radiative cascades that do not yield a Lyα photon.

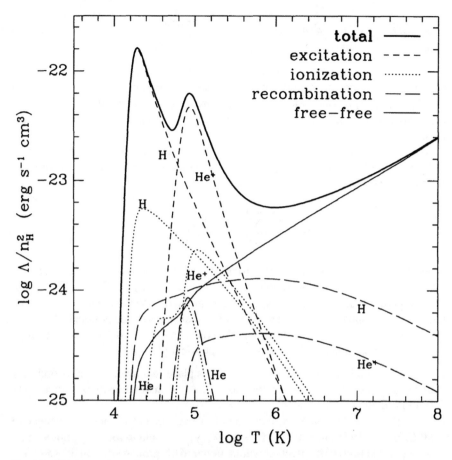

Fig. 1.8 This figure shows the temperature dependence of the cooling rate of primordial gas, under the assumption of collisional ionization equilibrium (i.e. the ionization state of the gas is set entirely by its temperature). Gas cooling due to collisional excitation of atomic hydrogen becomes important at $\log T/K \sim 4$, and reaches a maximum at $\log T/K \sim 4.2$, beyond which collisions can ionize away atomic hydrogen. At $\log T/K \gtrsim 4.6$ cooling is dominated by collisional excitation of singly ionized helium. Figure taken from [258]

$(3, 0) \rightarrow (2, 1) \overset{\text{Lyα}}{\rightarrow} (1, 0)$, and $P(3, 0 \rightarrow \text{Lyα}) = 1$. For $n > 3$, multiple radiative cascades down to the ground state are generally possible, and $P(n, l \rightarrow \text{Lyα})$ takes on values other than 0 or 1 (see e.g. [250] for numerical values). Figure 1.4 also contains a table that shows the probability $P(n', l' \rightarrow \text{Lyα})$ for $n \leq 5$.

When the selection rules permit radiative cascades from a quantum state (n, l) into multiple states (n', l'), then this probability is given by the 'branching ratio' which represents the ratio of the decay rate into state (n', l') into all permitted states, i.e.

$$P(n, l \rightarrow n', l') = \frac{A_{n,l,n',l'}}{\sum_{n'',l''} A_{n,l,n'',l''}}, \quad (1.12)$$

in which $A_{n,l,n',l'}$ denotes the Einstein A-coefficient[10] for the $nl \rightarrow n'l'$ transition, where the quantum mechanical selection rules only permit transitions for which $|l - l'| = 1$. The probability that a radiative cascade from an arbitrary quantum state (n, l) gives rise to a Lyα photon can be computed once we know the Einstein-coefficients $A_{n,l,n',l'}$.

The probability that an arbitrary recombination event results in a Lyα photon follows naturally, if we know the probability that recombination leaves the atom in state (n, l). That is

$$P(\text{Ly}\alpha) = \sum_{n_{\min}}^{\infty} \sum_{l=0}^{n-1} \frac{\alpha_{nl}(T)}{\alpha_{\text{tot}}(T)} P(n, l \rightarrow \text{Ly}\alpha), \quad (1.15)$$

where the first term denotes the fraction of recombination events into the (n, l) state, in which α_{tot} denotes the total recombination coefficient $\alpha_{\text{tot}}(T) = \sum_{n_{\min}}^{\infty} \sum_{l=0}^{n-1} \alpha_{nl}(T)$. The temperature-dependent state specific recombination coefficients $\alpha_{nl}(T)$ can be found in for example [40, 222]. The value of $n_{\min}$ depends on the physical conditions of the medium in which recombination takes place, and two cases bracket the range of scenarios commonly encountered in astrophysical plasmas:

- 'case-A' recombination: recombination takes place in a medium that is optically thin at all photon frequencies. In this case, direct recombination to the ground state is allowed and $n_{\min} = 1$.
- 'case-B' recombination: recombination takes place in a medium that is opaque to all Lyman series[11] photons (i.e. Lyα, Lyβ, Lyγ, ...), and to ionizing photons that were emitted following direct recombination into the ground state. In the so-called 'on the spot approximation', direct recombination to the ground state produces an

---

[10]This coefficient is given by

$$A_{n,l,n',l'} = \frac{64\pi^4 \nu_{ul}^3}{3h_P c^3} \frac{\max(l', l)}{2l + 1} e^2 a_0^2 [M(n, l, n', l')]^2, \quad (1.13)$$

where fundamental quantities e, c, h_P, and a_0 are given in Table 1.1, $h_P \nu_{ul}$ denotes the energy difference between the upper (n,l) and lower (n',l') state. The matrix $M(n, l, n', l')$ involves an overlap integral that involves the radial wavefunctions of the states (n, l) and (n', l'):

$$M(n, l, n', l') = \int_0^{\infty} P_{n,l}(r) r^3 P_{n',l'}(r) dr. \quad (1.14)$$

Analytic expressions for the matrix $M(n, l, n', l')$ that contain hypergeometric functions were derived by [101]. For the Lyα transition $M(n, l, n', l') = M(2, 1, 1, 0) = \sqrt{6}(128/243)$ [128].

[11]At gas densities that are relevant in most astrophysical plasmas, hydrogen atoms predominantly populate their electronic ground state ($n = 1$), and the opacity in the Balmer lines is generally negligible. In theory one can introduce *case-C/D/E/...* recombination to describe recombination in a medium that is optically thick to Balmer/Paschen/Bracket/... series photons.

ionizing photon that is immediately absorbed by a nearby neutral H atom. Similarly, any Lyman series photon is immediately absorbed by a neighbouring H atom. This case is quantitatively described by setting $n_{min} = 2$, and by setting the Einstein coefficient for all Lyman series transitions to zero, i.e. $A_{np,1s} = 0$.

The probability $P(\text{Ly}\alpha)$ that we obtain from Eq. 1.15 was plotted in Fig. 1.6 assuming case-A (*solid line*) and case-B (*dashed line*) recombination. The temperature dependence comes in entirely through the temperature dependence of the state-specific recombination coefficients $\alpha_{nl}(T)$. As we mentioned earlier, for case-B recombination, we have $P(\text{Ly}\alpha) = 0.68$ at $T = 10^4$ K. It is worth keeping in mind that our calculations technically only apply in a low density medium. For 'high' densities, collisions can 'mix' different l-levels at a fixed n. In the limit of infinitely large densities, collisional mixing should cause different l−levels to be populated following their statistical weigths [i.e $n_{nl} \propto (2l − 1)$]. Collisions can be important in realistic astrophysical conditions, as we discuss in more detail in Sect. 1.4.

1.4 Astrophysical Lyα Sources

Now that we have specified different physical mechanisms that give rise to the production of a Lyα photon, we discuss various astrophysical sites of Lyα production.

1.4.1 Interstellar HII Regions

Interstellar HII regions are the most prominent sources of Lyα emission in the Universe. Hot, (mostly) massive and young stars produce ionizing photons in their atmospheres which are efficiently absorbed in the interstellar medium, and thus create ionized HII regions. Recombining protons and electrons give rise to Lyα, Hα, etc lines. These lines are called 'nebular' lines. One of the most famous nebulae is the Orion nebula, which is visible with the naked eye in the constellation of Orion. The reddish glow is due to the Hα line, which at $\lambda = 6536$ Å falls in the middle of the red part of the visual spectrum, that is produced as recombination emission. We showed previously that there is a $P(\text{Ly}\alpha) = 0.68$ probability that a Lyα photon is produced per case-B recombination event at $T = 10^4$ K. A similar analysis can yield the probability that an Hα photon is produced is: $P(\text{H}\alpha) \sim 0.45$. The total ratio of the Lyα to Hα flux is therefore ~ 8. It is interesting to realize that the total Lyα luminosity that is produced in the Orion nebula is almost an order of magnitude larger[12] than the flux contained in the Hα line, which is prominently visible in Fig. 1.9.

---

[12] So next time you look up and see Orion's belt, try and remember that if it were not for the Ozone layer, you would be blasted with Lyα flux strong enough to give you a first-degree burn.

Fig. 1.9 The Orion nebula is one of the closest nearby nebulae, and visible with the naked eye. The reddish glow is due to Hα that was produced as recombination radiation. These same recombination events produce a Lyα flux that is about $8\times$ larger. Recombination line emission is often referred to as 'nebular' emission. *Credit ESA/Hubble*

Recombinations in HII regions in the ISM balance photoionization by ionizing photons produced by the hot stars. The total recombination rate in an equilibrium[13] HII region therefore equals the total photoionization rate in the nebulae, i.e. the total rate at which ionizing photons are absorbed in the HII region. If a fraction $f_{\rm esc}^{\rm ion}$ of ionizing photons is *not* absorbed in the HII region (and hence escapes), then the total Lyα production rate in recombinations is

$$\dot{N}_{\rm Ly\alpha}^{\rm rec} = P({\rm Ly\alpha})(1 - f_{\rm esc}^{\rm ion})\dot{N}_{\rm ion} \approx 0.68(1 - f_{\rm esc}^{\rm ion})\dot{N}_{\rm ion}, \quad {\rm case - B}, \; {\rm T} = 10^4 \; {\rm K},$$
(1.16)

where $\dot{N}_{\rm ion}$ ($\dot{N}_{\rm Ly\alpha}^{\rm rec}$) denotes the rate at which ionizing (Lyα recombination) photons are emitted. The production rate of ionizing photons, $\dot{N}_{\rm ion}$, relates to the abundance

---

[13]The condition of equilibrium is generally satisfied in ordinary interstellar HII regions. In expanding HII regions, e.g. those that exist in the intergalactic medium during cosmic reionization (which is discussed later), the total recombination rate is less than the total rate at which ionising photons are absorbed.

of short-lived massive stars, and therefore closely tracks the star formation rate. For a [226] initial mass function (IMF) with mass limits $M_{low} = 0.1M_\odot$ and $M_{high} = 100M_\odot$ we have [144]

$$\dot{N}_{ion} = 9.3 \times 10^{52} \times SFR(M_\odot/yr) \; s^{-1}$$
$$\Rightarrow L_\alpha = 1.0 \times 10^{42} \times SFR(M_\odot/yr) \; erg \; s^{-1} \quad (Salpeter, \; Z = Z_\odot). \quad (1.17)$$

Equation 1.17 strictly applies to galaxies with continuous star formation over timescales of 10^8 years or longer. This equation is commonly adopted in the literature. A Kroupa IMF gives us a slightly higher Lyα luminosity for a given SFR:

$$L_\alpha = 1.7 \times 10^{42} \times SFR(M_\odot/yr) \; erg \; s^{-1} \quad (Kroupa, \; Z = Z_\odot). \quad (1.18)$$

Finally, for a fixed IMF the Lyα production rate increases towards lower metallicities: stellar evolution models combined with stellar atmosphere models show that the effective temperature of stars of fixed mass become hotter with decreasing gas metallicity [230, 265]. The increased effective temperature of stars causes a larger fraction of their bolometric luminosity to be emitted as ionizing radiation. We therefore expect galaxies that formed stars from metal poor (or even metal free) gas, to be strong sources of nebular emission. Schaerer [231] provides the following fitting formula for $\dot{N}_{ion}$ as a function of *absolute* gas metallicity[14] Z_{gas}

$$\log \dot{N}_{ion} = -0.0029 \times (\log Z_{gas} + 9.0)^{2.5} + 53.81 + \log SFR(M_\odot/yr) \quad (Salpeter). \quad (1.19)$$

Warning: note that this fitting formula is valid for a Salpeter IMF in the mass range $M = 1 - 100M_\odot$. If we substitute $Z = Z_\odot = 0.02$ we get $\log \dot{N}_{ion} = 53.39$, which is a factor of ~ 2.6 times larger than that given by Eq. 1.17. This difference is due to the different lower-mass cut-off of the IMF.

The previous discussion always assumed case-B recombination when converting the production rate of ionizing photon into a Lyα luminosity. However, at $Z \lesssim 0.03Z_\odot$ departures from case-B *increases* the Lyα luminosity relative to case-B (e.g. [216]). This increase of the Lyα luminosity towards lower metallicities is due to two effects: (*i*) the increased temperature of the HII region as a result of a suppressed radiative cooling efficiency of metal-poor gas. The enhanced temperature in turn increases the importance of collisional processes, which enhances the rate at which collisions excite the $n = 2$ level, and which can transfer atoms from the $2s$ into the $2p$ states; (*ii*) harder ionizing spectra emitted by metal poor(er) stars. When higher energy photons (say $E_\gamma \gtrsim 50$ eV) photoionize a hydrogen atom, then it releases an electron with a kinetic energy that is $E = E_\gamma - 13.6$ eV. This energetic electron can heat the gas, collisionally excite hydrogen atoms, and collisionally ionize other hydrogen atoms (see e.g. [243]). Ionizations triggered by energetic electrons created after photoionization by high energy photons are called 'secondary' ionizations. The

---

[14]It is useful to recall that solar metallicity $Z_\odot = 0.02$.

efficiency with which high energy photons induce secondary ionizations depends on the energy of the photon, and strongly on the ionization state of the gas (see Fig. 1.3 in [243]): in a highly ionized gas, energetic electrons rapidly loose their energy through Coulomb interactions with other charged particles. A useful figure to remember is that a 1 keV photon can ionize $\sim$25 hydrogen atoms in a fully neutral medium. Raiter et al. [216] provide a simple analytic formula which captures these effects:

$$\dot{N}_{\text{Ly}\alpha}^{\text{rec}} = f_{\text{coll}} P (1 - f_{\text{esc}}^{\text{ion}}) \dot{N}_{\text{ion}} \quad (\text{non} - \text{case B}), \tag{1.20}$$

where $P \equiv \langle E_{\gamma,\text{ion}} \rangle / 13.6$ eV, in which $\langle E_{\gamma,\text{ion}} \rangle$ denotes the mean energy of ionising photons.[15] Furthermore, $f_{\text{coll}} \equiv \frac{1+an_{\text{HI}}}{b+cn_{\text{HI}}}$, in which $a = 1.62 \times 10^{-3}$, $b = 1.56$, $c = 1.78 \times 10^{-3}$, and n_{HI} denotes the number density of hydrogen nuclei. Equation 1.20 resembles the 'standard' equation, but replaces the factor 0.68 with $P f_{\text{coll}}$, which can exceed unity. Equation 1.20 implies that for a fixed IMF, the Lyα luminosity may be boosted by a factor of a few. Incredibly, for certain IMFs the Lyα line may contain 40% of the total bolometric luminosity[16] of a galaxy.

1.4.2   The Circumgalactic/Intergalactic Medium (CGM/IGM)

Not only nebulae are sources of Lyα radiation. Most of our Universe is in fact a giant Lyα source. Observations of spectra of distant quasars reveal a large collection of Lyα absorption lines. This so-called 'Lyα forest' is discussed in more detail in the lecture notes by X. Prochaska. Observations of the Lyα forest imply that the intergalactic medium is highly ionized, and that the temperature of intergalactic gas is $T \sim 10^4$ K. Observations of the Lyα forest can be reproduced very well if we assume that gas is photoionized by the Universal "ionizing background" that permeates the entire Universe, and that is generated by adding the contribution from all ionizing sources.[17] The *residual* neutral fraction of hydrogen atoms in the IGM is

---

[15]That is, $\langle E_{\gamma,\text{ion}} \rangle \equiv h_{\text{P}} \frac{\int_{13.6 \text{ eV}}^{\infty} d\nu f(\nu)}{\int_{13.6 \text{ eV}}^{\infty} d\nu f(\nu)/\nu}$, where $f(\nu)$ denotes the flux density.

[16]Another useful measure for the 'strength' of the Lyα line is the *equivalent width* (EW, which was discussed in much more detail in the lectures by J.X. Prochaska) of the line:

$$\text{EW} \equiv \int d\lambda \, (F(\lambda) - F_0)/F_0, \tag{1.21}$$

which measures the total line flux compared to the continuum flux density just redward (as the blue side can be affected by intergalactic scattering, see Sect. 1.9.2) of the Lyα line, F_0. For a Salpeter IMF in the range $0.1 - 100 M_\odot$, $Z = Z_\odot$, the UV-continuum luminosity density, L_ν^{UV}, relates to SFR as $L_\nu^{\text{UV}} = 8 \times 10^{27} \times \text{SFR}(M_\odot/\text{yr})$ erg s^{-1} Hz^{-1}. The corresponding equivalent width of the Lyα line would be EW$\sim$70 Å [70]. The equivalent width can reach a few thousand Å for Population III stars/galaxies forming stars with a top-heavy IMF (see [216]).

[17]All sources within a radius equal to the mean free path of ionizing photons, λ_{ion}. For more distant sources ($r > \lambda_{\text{ion}}$), the ionizing flux is reduced by an additional factor $\exp(-r/\lambda_{\text{ion}})$.

$x_{HI} \equiv \frac{n_{HI}}{n_{HI}+n_{HII}} = \frac{n_e \alpha_B(T)}{\Gamma_{ion}}$, where Γ_{ion} denotes the photoionization rate by the ionizing background (with units s^{-1}).

From Sect. 1.3.2 we know that each recombination event in the gas produces ~ 0.68 Lyα photons. First, we note that the *recombination time* of a proton and electron in the intergalactic medium is

$$t_{rec} \equiv \frac{1}{\alpha_B(T)n_e} = 9.3 \times 10^9 \left(\frac{1+z}{4}\right)^{-3} \left(\frac{1+\delta}{1}\right)^{-1} T_4^{0.7} \text{ yr,} \qquad (1.22)$$

where we used that fully ionized gas at mean density of the Universe, $\delta = 0$, has $n_e = \bar{n} = \frac{\Omega_m \rho_{crit,0}}{\mu m_p}(1+z)^3 \sim 2 \times 10^{-7}(1+z)^3$ cm^{-3}, in which $\rho_{crit,0} = 1.88 \times 10^{-29} h^2$ ($H_0 \equiv 100h$ km s^{-1} Mpc^{-1}) denotes the critical density of the Universe today. We also approximated the case-B recombination coefficient as $\alpha_B(T) = 2.6 \times 10^{-13} T_4^{-0.7}$ cm^3 s^{-1}, in which we have adopted the notation $T_4 \equiv (T/10^4$ K). We can compare this number to the Hubble time which is $t_{Hub} \equiv \frac{1}{H(z)} \sim 3([1+z]/4)^{-3/2}$ Gyr (where the last approximation is valid strictly for $z \gg 1$). The recombination time for gas, at mean density ($\delta = 0$) thus exceeds the Hubble time for most of the existence of the Universe. Only at $z \gtrsim 8$ does the recombination time become shorter than the Hubble time. However, at lower redshifts recombination (of course) still happens, and we expect the Universe as a whole to emit Lyα recombination radiation.

Lyα emission from the CGM/IGM differs from interstellar (nebular) Lyα emission in two ways: (*i*) Lyα emission from the CGM/IGM occurs over a spatially extended region. Spatially extended Lyα emission is better characterized by its *surface brightness* (flux per unit area on the sky) than by its overall flux. We present a general formalism for computing the Lyα surface brightness below; and (*ii*) Lyα emission is powered by *external* sources. For example, recombination radiation from the CGM/IGM balances photoionization by either the ionizing background (generate by a large numbers of star forming galaxies and AGN), or a nearby source. This conversion of externally—i.e. *not* within the same galaxy (or even cloud)—generated ionizing radiation into Lyα is known as '*fluorescence*'. Fluorescence generally corresponds to emission of radiation by some material following absorption by radiation at some other wavelength. Certain minerals emit radiation in the optical when irradiated by UV radiation. Fluorescent materials cease to glow immediately when the irradiating source is removed. In the case of Lyα, fluorescently produced Lyα is a product of recombinaton cascade, just as the case of nebular Lyα emission.

Here, we present the general formalism for computing the Lyα surface brightness level. The total flux 'Flux' from some redshift z that we receive per unit solid angle $d\Omega$ equals (see Fig. 1.10 for a visual illustration of the adopted geometry)

$$\frac{\text{Flux}}{d\Omega} = \frac{d_A^2(z)\text{Flux}}{dA} = \frac{d_A^2(z)}{dA}\frac{\text{Luminosity}}{4\pi d_L^2(z)} \qquad (1.23)$$

Fig. 1.10 The adopted geometry for calculating the Lyα surface brightness of recombining gas in the IGM at redshift z

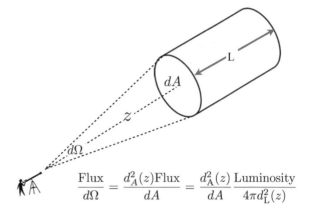

$$\frac{\text{Flux}}{d\Omega} = \frac{d_A^2(z)\text{Flux}}{dA} = \frac{d_A^2(z)}{dA}\frac{\text{Luminosity}}{4\pi d_L^2(z)}$$

where we used the definition of solid angle $d\Omega \equiv dA/d_A^2(z)$, in which $d_A(z)$ denotes the angular diameter distance to redshift z. We also used that Flux $=$ Luminosity$/[4\pi d_L^2(z)]$, in which $d_L(z)$ denotes the luminosity distance to redshift z. We know that $d_L(z) = d_A(z)(1 + z)^2$. Furthermore, the total luminosity that we receive from dA depends on the length of the cylinder L, as Luminosity$=$ $\varepsilon_{\text{Ly}\alpha} \times dA \times L$. Here, $\varepsilon_{\text{Ly}\alpha}$ denotes the Lyα emission per unit volume. Plugging all of this into Eq. (1.23) we get

$$\frac{\text{Flux}}{d\Omega} = \frac{\varepsilon_{\text{Ly}\alpha} L}{4\pi(1+z)^4} \equiv S, \tag{1.24}$$

where we defined S to represent surface brightness. We compute the surface brightness for a number of scenarios next:

Recombination in the diffuse, low density IGM. First, we compute the surface brightness of Lyα from the diffuse, low density, IGM, denoted with S_{IGM}. The expansion of the Universe causes photons that are emitted over a line-of-sight length L to be spread out in frequency by an amount

$$\frac{d\nu_\alpha}{\nu_\alpha} = \frac{d\nu}{c} = \frac{H(z)L}{c} \Rightarrow L = \frac{cd\nu_\alpha}{\nu_\alpha H(z)}. \tag{1.25}$$

When we substitute this into Eq. 1.24 we get

$$S_{\text{IGM}} = \frac{c\varepsilon_{\text{Ly}\alpha}}{4\pi(1+z)^4 H(z)}\frac{d\nu_\alpha}{\nu_\alpha}$$

$$\approx 10^{-21}\left(\frac{1+z}{4}\right)^{0.5}\left(\frac{1+\delta}{1}\right)^2\left(\frac{d\nu_\alpha/\nu_\alpha}{0.1}\right)T_4^{-0.7}\ \text{erg s}^{-1}\ \text{cm}^{-2}\ \text{arcsec}^{-2}\quad\text{for } z \gg 1, \tag{1.26}$$

where we obtained numerical values by substituting that for recombining gas we have $\varepsilon_{\text{Ly}\alpha} = 0.68 h_p\nu_\alpha n_e n_p\alpha_B(T)$ (see Sect. 1.3.2, and that $n_e = n_p \propto (1+z)^3$). We

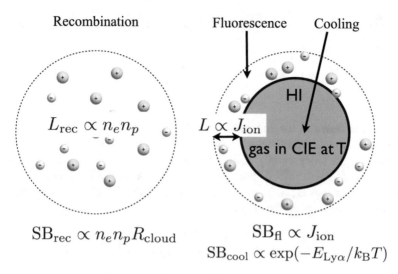

Fig. 1.11 A schematic representation of the dominant processes that give rise to extended Lyα emission in the CGM/IGM. *Left:* the surface brightness from fluorescence from recombination in fully ionized gas scales as $S_{\mathrm{fl,I}} \propto n_e n_p \lambda_J$, where λ_J denotes the Jeans length (see Eq. 1.28). This surface brightness can also be written as $S_{\mathrm{fl,I}} \propto n_e N_{\mathrm{H}}$ where N_{H} denotes the total column of ionized gas (see Eq. 1.27). *Right:* sufficiently dense clouds can self-shield and form a neutral core of gas, surrounded by an ionized 'skin'. The thickness of this skin is set by the mean free path of ionizing photons, λ_{mfp}. The surface brightness of recombination emission that occurs inside this skin is $S_{\mathrm{fl,II}} \propto \Gamma_{\mathrm{ion}}$, where Γ_{ion} denotes the photoionization rate (see Eq. 1.31). The neutral core can also produce Lyα radiation following collisional excitation. The surface brightness of this emission is $S_{\mathrm{cool}} \propto \exp\left(-E_{\mathrm{Ly}\alpha}/k_{\mathrm{B}}T\right)$

adopted $dv \sim 0.1v$, which represents the width of narrow-band surveys, which have been adopted in searches for distant Lyα emitters. For comparison, stacking analyses and deep exposure with e.g. MUSE go down to $SB \sim 10^{-19}$ erg s^{-1} cm^{-2} arcsec^{-2} (e.g. [170, 186, 217, 253, 274]). Directly observing recombination radiation from a representative patch of IGM is still well beyond our capabilities, but would be fantastic as it would allow us to map out the distribution of baryons throughout the Universe.

Fluorescence from recombination in fully ionized dense gas (Fluorescence case I). Equation 1.26 shows that the surface brightness increases as $(1 + \delta)^2$, and the prospects for detection improve dramatically for overdensities of $\delta \gg 1$. The largest overdensities of intergalactic gas are found in close proximity to galaxies. We therefore expect recombining gas in close proximity to galaxies to be potentially visible in Lyα. Note however that this denser gas in close proximity to galaxies is not comoving with the Hubble flow, and Eq. 1.25 cannot be applied. Instead, we need to specify the line-of-sight size of the cloud L. Schaye [233] has shown that the characteristic size of overdense, growing perturbations is the local Jeans length

$\lambda_J \equiv \frac{c_s}{\sqrt{G\rho}} \approx 145([1+z]/4)^{-3/2}(1+\delta)^{-1/2}T_4^{1/2}$ kpc. If we adopt that $L = \lambda_J$, then we obtain[18]

$$S_{\mathrm{fl,I}} = \frac{\varepsilon_{\mathrm{Ly}\alpha}\lambda_J}{4\pi(1+z)^4} \approx 1.5 \times 10^{-21} \left(\frac{1+z}{4}\right)^{1/2} \left(\frac{1+\delta}{100}\right)^{3/2} T_4^{-0.2} \text{ erg s}^{-1} \text{ cm}^{-2} \text{ arcsec}^{-2}. \tag{1.28}$$

For higher densities the enhanced recombination efficiency of the gas gives rise to an enhanced equilibrium neutral fraction, i.e. $x_{\mathrm{HI}} \propto n_e$. With an enhanced neutral fraction, the gas more rapidly 'builds up' a neutral column density $N_{\mathrm{HI}} \gtrsim 10^{-17}$ cm^{-1}, above which the cloud starts self-shielding against the ionizing background. Quantitatively, if gas is photoionized at a rate Γ_{ion}, then the total column density of neutral hydrogen through the cloud is

$$N_{\mathrm{HI}} = \lambda_J(\delta) \times n_{\mathrm{H}}\frac{\alpha_{\mathrm{B}}(T)n_e}{\Gamma_{\mathrm{ion}}} \approx 6 \times 10^{17} \left(\frac{1+\delta}{10^3}\right)^{3/2} \left(\frac{\Gamma_{\mathrm{ion}}}{10^{-12} \text{ s}^{-1}}\right)^{-1} \left(\frac{1+z}{4}\right)^{9/2} T_4^{-0.7} \text{ cm}^{-2}, \tag{1.29}$$

where we used that $n_{\mathrm{HI}} = x_{\mathrm{HI}}n_{\mathrm{H}} = \frac{\alpha_{\mathrm{B}}(T)n_e}{\Gamma_{\mathrm{ion}}}$. The gas becomes self-shielding when $N_{\mathrm{HI}} \gtrsim \sigma_{\mathrm{ion}}^{-1} \sim 10^{17}$ cm^{-2}, which translates to $\delta \gtrsim 500(\Gamma_{\mathrm{ion}}/10^{-12} \text{ s}^{-1})^{2/3}$ $([1+z]/4)^{-3}T_4^{0.47}$ ([233], also see [215] for a much more extended discussion on self-shielding gas). Once gas become denser than this, it can self-shield against an ionizing background, and form a neutral core surrounded by an ionized 'skin' (see Fig. 1.11).

Fluorescence from recombination from the skin of dense clouds (Fluorescence case II). For dense clouds that are capable of self-shielding, only the ionized 'skin' emits recombination radiation. The total surface brightness recombination radiation from this skin depends on its density and thickness, the latter depending directly on the amplitude of the ionizing background. This can be most clearly seen from Eq. 1.24, and replacing $L = \lambda_{\mathrm{mfp}}$, where λ_{mfp} denotes the mean free path of ionizing photons into the cloud. This mean free path is given by

$$\lambda_{\mathrm{mfp}} = \frac{1}{n_{\mathrm{HI}}\sigma_{\mathrm{ion}}} \Rightarrow \lambda_{\mathrm{mfp}} = \frac{1}{\sigma_{\mathrm{ion}}x_{\mathrm{HI}}n_{\mathrm{H}}} = \frac{\Gamma_{\mathrm{ion}}}{\sigma_{\mathrm{ion}}\alpha_{\mathrm{B}}(T)n_e^2}, \tag{1.30}$$

where we used that $x_{\mathrm{HI}} = \alpha_{\mathrm{B}}(T)n_e/\Gamma_{\mathrm{ion}}$. Substituting λ_{mfp} for L into Eq. 1.24 gives for the surface brightness of the skin:

---

[18]Another way to express $S_{\mathrm{fl,I}}$ is by replacing $Ln_p = N_{\mathrm{H}}$, where N_{H} denotes the total column density of hydrogen ions (i.e. protons), which yields (see [125])

$$S_{\mathrm{fl,I}} = \frac{0.68h_{\mathrm{p}}\nu_\alpha n_e \alpha_{\mathrm{B}}(T)N_{\mathrm{H}}}{4\pi(1+z)^4}$$

$$\approx 2.0 \times 10^{-21} \left(\frac{1+z}{4}\right)^{-4} \left(\frac{n_e}{10^{-3} \text{ cm}^{-3}}\right) \left(\frac{N_{\mathrm{H}}}{10^{20} \text{ cm}^{-2}}\right) T_4^{-0.7} \text{ erg s}^{-1} \text{ cm}^{-2} \text{ arcsec}^{-2}. \tag{1.27}$$

.

$$S_{\mathrm{fl,II}} = \frac{0.68 h_p \nu_\alpha \Gamma_{\mathrm{ion}}}{4\pi(1+z)^4 \sigma_{\mathrm{ion}}} \approx 1.3 \times 10^{-20} \left(\frac{\Gamma_{\mathrm{ion}}}{10^{-12}\ s^{-1}}\right) \left(\frac{1+z}{4}\right)^{-4} \mathrm{erg\ s^{-1}\ cm^{-2}\ arcsec^{-2}},$$

$$(1.31)$$

where the T-dependence has cancelled out. A more precise calculation of the surface brightness of fluorescent Lyα emission, which takes into account the spectral shape of the ionizing background as well as the frequency dependence of λ_{mfp}, is presented by [41] (2005, note that this calculation introduces only a minor change to the calculated surface brightness).

'Cooling' by dense, neutral gas. The neutral core of the cloud—the part of the cloud which is truly self-shielded—produces Lyα radiation through collisional excitation. As we mentioned earlier, the rate at which Lyα is produced in collisions depends sensitively on temperature ($\propto \exp(-E_{\mathrm{Ly\alpha}}/k_b T)$, see Eq. 1.7). Note however, that this process is a cooling process, and thus must balance some heating mechanism. Once we know the heating rate of the gas, we can almost immediately compute the Lyα production rate.

The direct environment of galaxies, also known as the 'circum galactic medium' (CGM), represents a complex mixture of hot and cold gas, of metal poor gas that is being accreted from the intergalactic medium and metal enriched gas that is driven out of either the central, massive galaxy or from the surrounding lower mass satellite galaxies. Figure 1.12 shows a snap-shot from a cosmological hydrodynamical simulation [3] which nicely illustrates this complexity. A disk galaxy (total baryonic mass $\sim 2 \times 10^{10} M_\odot$) sits in the center of the snap-shot, taken at $z = 3$. The blue filaments show dense gas that is being accreted. This gas is capable of self-shielding.

Fig. 1.12 A snap-shot from a cosmological hydrodynamical simulation by [3] which illustrates the complexity of the circumgalactic gas distribution. A disk galaxy sits in the center of the snap-shot, taken at $z = 3$. The *blue filaments* show dense gas that is being accreted. The *red gas* has been shock heated to the virial temperature of the dark matter halo hosting this galaxy ($T_{\mathrm{vir}} \sim 10^6$ K). The green clouds show metal rich gas that was stripped from smaller galaxies. This complex mixture of circumgalactic gas produces Lyα radiation through recombination, cooling, and fluorescence. *Credit from Fig. 1 of* [3], *Disc formation and the origin of clumpy galaxies at high redshift, MNRAS, 397L, 64A*

The red gas has been shock heated to the virial temperature ($T_{\rm vir} \sim 10^6$ K) of the dark matter halo hosting this galaxy. The green clouds show metal rich gas that was driven out of smaller galaxies. This complex mixture of gas produces Lyα via all channels described above: there exists fully ionized gas that is emitting recombination radiation with a surface brightness given by Eq. 1.28, the densest gas is capable of self-shielding and will emit both recombination and cooling radiation.

We currently have observations of Lyα emission from the circum-galactic medium at a range of redshifts, covering a range of surface brightness levels:

1. Lyα emission extends further the UV continuum in nearby star forming galaxies. The *left panel* of Fig. 1.13 shows an example of a false-color image of galaxy # 1 from the **L**yman **A**lpha **R**eference **S**ample [79, 107]. In this image, *red* indicates Hα, *green* traces the far-UV continuum, while *blue* traces the Lyα.
2. Stacking analyses have revealed the presence of spatially extended Lyα emission around Lyman Break Galaxies [117, 253] and Lyα emitters at surface brightness levels in the range SB$\sim 10^{-19} - 10^{-18}$ erg s^{-1} cm^{-2} arcsec^{-2} [170, 186].
3. Deep imaging with MUSE has now revealed emission at this level around individual star forming galaxies, which further confirms that this emission is present ubiquitously [274].
4. These previously mentioned faint halos are reminiscent of Lyα 'blobs', which are spatially extended Lyα sources *not* associated with radio galaxies (more on these next, [117, 143, 252]). A famous example of "blob # 1" is shown in the *right panel* of Fig. 1.13 (from [143]). This image shows a 'pseudo-color' image of a Lyα blob. The *red* and *blue* really trace radiation in the red and blue filters, while the *green* traces the Lyα. The *upper right panel* shows how large the Andromeda galaxy would look on the sky if placed at $z = 3$, to put the size of the blob in perspective. The brightest Lyα blobs have line luminosities of $L_\alpha \sim 10^{44}$ erg s^{-1}, though recently two monstrous blobs have been discovered, that are much brighter than this: (*i*) the 'Slug' nebula with a Lyα luminosity of $L_\alpha \sim 10^{45}$ erg s^{-1} [44], (*ii*) the 'Jackpot' nebula, which has a luminosity of $L_\alpha \sim 2 \times 10^{44}$ erg s^{-1}, and contains a quadruple-quasar system [125].
5. The most luminous Lyα nebulae have traditionally been associated (typically) with High-redshift Radio Galaxies (HzRGs, e.g. [172, 219, 267]) with luminosities in excess of $L \sim 10^{45}$ erg s^{-1}.

The origin of extended Lyα emission is generally unclear. To reach surface brightness levels of $\sim 10^{-19} - 10^{-17}$ erg s^{-1} cm^{-2} arcsec^{-2} we need a density exceeding (see Eq. 1.28)

$$\delta \gtrsim \left(1.5 \times 10^3 - 3.5 \times 10^4\right) \times \left(\frac{1+z}{4}\right)^{-1/3} T_4^{-0.13} \tag{1.32}$$

To keep this gas photoionized requires a large $\Gamma_{\rm ion}$:

$$\Gamma_{\rm ion} \gtrsim (5 - 600) \times 10^{-12} \left(\frac{1+z}{4}\right)^{4.5} T_4^{-0.7} \; {\rm s}^{-1}, \tag{1.33}$$

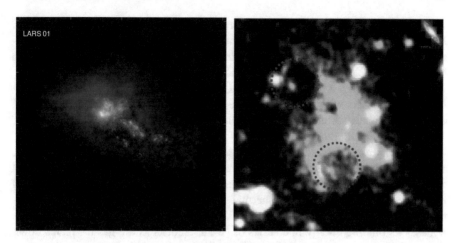

Fig. 1.13 *Left:* A false color image of 'LARS1' (galaxy #1 from the Lyα Reference Sample, *Credit from Fig. 1.1 of* [123] ©*AAS. Reproduced with permission.*). *Red* indicates Hα emission, while *green* traces far-UV continuum. The *blue light* traces the Lyα which extends much further than other radiation. *Right:* A pseudo color image of Lyα blob 1 (LAB1). Here, *red* and *blue* light traces emission from the V and B bands, respectively. The *green light* traces Lyα emission (*Credit from Fig. 2 of* [169], *The Subaru Lyα blob survey: a sample of 100-kpc Ly α blobs at z= 3, MNRAS, 410, L13*). For comparison, the *upper right* shows the Andromeda galaxy to get a sense for the scale of 'giant' Lyα blobs

which is $\sim 10 - 10^3$ times larger than the values inferred from observations of the Lyα forest at this redshift. In case the gas starts to self-shield, then Eq. 1.31 shows that in order to reach $S \sim 10^{-19} - 10^{-17}$ erg s^{-1} cm^{-2} arcsec^{-2} we need an almost identical boost in $\Gamma_{\text{ion}} \sim 10^{-11} - 10^{-9}$ s^{-1}. This enhanced intensity of the ionizing radiation field is expected in close proximity to ionizing sources (e.g. [166]): the photoionization rate at a distance r from a source that emits $\dot{N}$ion ionizing photons per second is

$$\Gamma = \dot{N}_{\text{ion}} \frac{\sigma_{\text{ion}} f_{\text{esc}}^{\text{ion}}}{4\pi r^2} = 5 \times 10^{-10} \left(\frac{f_{\text{esc}}^{\text{ion}}}{1.0}\right) \left(\frac{\dot{N}_{\text{ion}}}{10^{54}\text{s}^{-1}}\right) \left(\frac{r}{10 \text{ kpc}}\right)^{-2}, \qquad (1.34)$$

where $f_{\text{esc}}^{\text{ion}}$ denotes the fraction of ionizing photons that escapes from the central source into the environment. The production rate of ionizing photons can be linked to the star formation rate via Eq. 1.19. Alternatively, ionizing radiation may be powered by an accretion disk surrounding a black hole of mass M_{BH}. Assuming Eddington accretion onto the black hole, and adopting a template spectrum of a radio-quiet quasar, we have

$$\dot{N}_{\text{ion}} = 6.5 \times 10^{53} \left(\frac{M_{\text{BH}}}{10^6 \, M_\odot}\right) \text{s}^{-1} \qquad (1.35)$$

which assumes a broken power-law spectrum of the form $f_\nu \propto \nu^{-0.5}$ for 1050 Å$<$ $\lambda < 1450$ Å, and $f_\nu \propto \nu^{-1.5}$ for $\lambda < 1050$ Å [30]. It therefore seems that fluorescence can explain the observed values of the surface brightness in close proximity to ionizing sources. An impressive recent example of this process is described by [34], who found luminous, spatially extended Lyα halos around *each* of the 17 brightest radio-quiet quasars with MUSE (also see [194] for earlier hints of the presence of extended Lyα halos around a high fraction of radio-quiet quasars, and see [110] for an early theoretical prediction).

Finally, Lyα cooling radiation gives rise to spatially extended Lyα radiation [83, 109], and provides a possible explanation for Lyα 'blobs' [69, 85, 100, 221]). In these models, the Lyα cooling balances 'gravitational heating' in which gravitational binding energy is converted into thermal energy in the gas. Precisely how gravitational heating works is poorly understood. Haiman et al. [109] propose that the gas releases its binding energy in a series of 'weak' shocks as the gas navigates down the gravitational potential well. These weak shocks convert binding energy into thermal energy over a spatially extended region,[19] which is then reradiated primarily as Lyα. We must therefore accurately know and compute all the heating rates in the ISM [43, 85, 221] to make a robust prediction for the Lyα cooling rate. These heating rates include for example photoionization heating, which requires coupled radiation-hydrodynamical simulations (as [221]), or shock heating by supernova ejecta (e.g. [242]).

The previous discussion illustrates that it is possible to produce spatially extended Lyα emission from the CGM at levels consistent with observations, via all mechanisms described in this section.[20] This is one of the main reasons why we have not solved the question of the origin of spatially extended Lyα halos yet. In later lectures, we will discuss how Lyα spectral line profiles (and polarization measurements) contain physical information on the scattering/emitting gas, which can help distinguish between different scenarios.

1.5 Step 1 Towards Understanding Lyα Radiative Transfer: Lyα Scattering Cross-section

The goal of this section is to present a classical derivation of the Lyα absorption cross-section. This classical derivation gives us the proper functional form of the real cross-section, but that differs from the real expression by a factor of order unity, due to a quantum mechanical correction. Once we have evaluated the magnitude of

---

[19]It is possible that a significant fraction of the gravitational binding energy is released very close to the galaxy (e.g. when gas free-falls down into the gravitational potential well, until it is shock heated when it 'hits' the galaxy: [27]). It has been argued that some compact Lyα emitting sources may be powered by cooling radiation (as in [27, 58, 68]).

[20]After these lectures, [167] showed that extended Lyα emission can also be produced by faint satellite galaxies which are too faint to be detected individually (also see [151]).

the cross-section, it is apparent that most astrophysical sources of Lyα emission are optically thick to this radiation, and that we must model the proper Lyα radiative transfer.

The outline of this section is as follows: we first describe the interaction of a free electron with an electromagnetic wave (i.e. radiation) in the classical picture (see Sect. 1.5.1). This discussion provides us with an opportunity to introduce the important concepts of the cross-section and phase function. This discussion also sets us up for discussion of the same interaction in Sect. 1.5.2 for an electron that is *bound* to a proton. The classical picture of this interaction allows us to derive the Lyα absorption cross-section up to a numerical factor of order unity (see Sect. 1.5.3). We introduce the velocity averaged Voigt-profile for the Lyα cross-section in Sect. 1.5.4.

1.5.1 Interaction of a Free Electron with Radiation: Thomson Scattering

Figure 1.14 shows the classical view of the interaction of a free electron with an incoming electro-magnetic wave. The electromagnetic wave consists of an electric field (represented by the *red arrows*) and a magnetic field (represented by the *blue arrows*). The amplitude of the electric field at time t varies as $E(t) = E_0 \sin \omega t$, where ω denotes the angular frequency of the wave. The electron is accelerated by the electric field by an amount $|a_e|(t) = \frac{q|E|(t)}{m_e}$. The total power radiated by this electron is given by the Larmor formula, i.e. $P_{\text{out}}(t) = \frac{2q^2|a_e(t)|^2}{3c^3}$. The *time average*

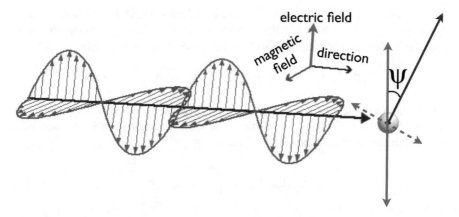

Fig. 1.14 Classical picture of the interaction of radiation with a free electron. The electric field of the incoming wave accelerates the free-electron. The accelerated electron radiates, and effectively scatters the incoming electromagnetic wave. The cross-section for this process is given by the Thomson cross-section. The angle Ψ denotes the angle between the direction of the outgoing electro-magnetic wave, and the oscillation direction of the electron (which corresponds to the electric vector of the incoming electro-magnetic wave)

of this radiated power equals $\langle P_{out} \rangle = \frac{2q^4 E_0^2}{3c^3 m_e^2} \langle \sin^2 \omega t \rangle = \frac{q^4 E_0^2}{3c^3 m_e^2}$. The total power *per unit area* transported by an electromagnetic wave (i.e. the flux) is $F_{in} = \frac{cE_0^2}{8\pi}$. We define the *cross-section* as the ratio of the total radiated power to the total incident flux, i.e.

$$\sigma_T = \frac{P_{out}}{F_{in}} = \frac{\frac{q^2 E_0^2}{3m_e^2 c^3}}{\frac{cE_0^2}{8\pi}} = \frac{8\pi}{3} r_e^2 \approx 6.66 \times 10^{-25} \text{ cm}^2, \qquad (1.36)$$

where $r_e = \frac{q^2}{m_e c^2} = 2.8 \times 10^{-13}$ cm denotes the classical electron radius.

The power of re-emitted radiation is not distributed isotropically across the sky. A useful way to see this is by considering what we see if we observe the oscillating electron along direction $\mathbf{k}_{out}$. The *apparent* acceleration that the electron undergoes is reduced to $\hat{a}(t) \equiv a(t) \sin \Psi$, where Ψ denotes the angle between $\mathbf{k}_{out}$ and the oscillation direction, i.e $\cos \Psi \equiv \mathbf{k}_{out} \cdot \mathbf{e}_E$. Here, the vector $\mathbf{e}_E$ denotes a unit vector pointing in the direction of the E-field. The reduced apparent acceleration translates to a reduced power in this direction, i.e $P_{out}(\mathbf{k}_{out}) \propto \hat{a}^2(t) \propto \sin^2 \Psi$.

The outgoing radiation field therefore has a strong directional dependence with respect to $\mathbf{e}_E$. Note however, that for radiation coming in along some direction $\mathbf{k}_{in}$, the electric vector can point in an arbitrary direction within the plane normal to $\mathbf{k}_{in}$. For unpolarized incoming radiation, $\mathbf{e}_E$ is distributed uniformly throughout this plane. To compute the angular dependence of the outgoing radiation field with respect to $\mathbf{k}_{in}$ we need to integrate over $\mathbf{e}_E$. That is[21]

$$P_{out}(\mathbf{k}_{out}|\mathbf{k}_{in}) \propto \int P_{out}(\mathbf{k}_{out}|\mathbf{e}_E) P(\mathbf{e}_E|\mathbf{k}_{in}) d\mathbf{e}_E. \qquad (1.37)$$

We can solve this integral by switching to a coordinate system in which the x−axis lies along $\mathbf{k}_{in}$, and in which $\mathbf{k}_{out}$ lies in the $x - y$ plane. In this coordinate system $\mathbf{e}_E$ must lie in the $y - z$ plane. We introduce the angles ϕ and θ and write

$$\mathbf{k}_{in} = (1, 0, 0), \quad \mathbf{k}_{out} = (\cos\theta, 0, \sin\theta), \quad \mathbf{e}_E = (0, \cos\phi, \sin\phi). \qquad (1.38)$$

We can then also get that $\cos \Psi \equiv \mathbf{k}_{out} \cdot \mathbf{e}_E = \sin\phi \sin\theta$. This coordinate system is shown in Figure 1.15, which shows that θ denotes the angle between $\mathbf{k}_{in}$ and $\mathbf{k}_{out}$. For this choice of coordinates Eq. 1.37 becomes

$$P_{out}(\mathbf{k}_{out}|\mathbf{k}_{in}) \propto \int_0^{2\pi} d\phi \sin^2 \Psi \qquad (1.39)$$

---

[21] We have adopted the notation of probability theory. In this notation, the function $p(y|b)$ denotes the conditional probability density function (PDF) of y given b. The PDF for y is then given by $p(y) = \int p(y|b)p(b)db$, where $p(b)$ denotes the PDF for b. Furthermore, the joint PDF of y and b is given by $p(y, b) = p(y|b)p(b)$.

Fig. 1.15 The geometry adopted for calculating the phase-function associated with Thomson scattering

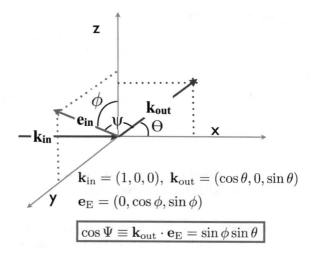

$$\mathbf{k}_{\text{in}} = (1, 0, 0), \ \mathbf{k}_{\text{out}} = (\cos\theta, 0, \sin\theta)$$

$$\mathbf{e}_E = (0, \cos\phi, \sin\phi)$$

$$\boxed{\cos\Psi \equiv \mathbf{k}_{\text{out}} \cdot \mathbf{e}_E = \sin\phi\sin\theta}$$

where we have used that $P_{\text{out}}(\mathbf{k}_{\text{out}}|\mathbf{e}_E) = P_{\text{out}}\sin^2\Psi$, and that $P(\mathbf{e}_E|\mathbf{k}_{\text{in}}) = $ Constant (i.e. $\mathbf{e}_E$ is distributed uniformly in the $y - z$ plane). We have omitted all numerical constants, because we are interested purely in the angular dependence of P_{out}. We will determine the precise constants that should preceed the integral below. If we substitute $\sin^2\Psi = 1 - \cos^2\Psi = 1 - \sin^2\phi\sin^2\theta \equiv 1 - A\sin^2\phi \ (A \equiv \sin^2\theta)$, we get

$$P_{\text{out}}(\mathbf{k}_{\text{out}}|\mathbf{k}_{\text{in}}) \propto \int_0^{2\pi} d\phi[1 - A\sin^2\phi] \propto \int_0^{2\pi} d\phi[1 + A - A - A\sin^2\phi] \propto \ldots \propto (1 + \cos^2\theta).$$

(1.40)

Figure 1.15 shows that our problem of interest is cylindrically symmetric around the x-axis, we therefore have that $P_{\text{out}}(\mathbf{k}_{\text{out}}|\mathbf{k}_{\text{in}})$ only depends on θ, and we will write $P_{\text{out}}(\mathbf{k}_{\text{out}}|\mathbf{k}_{\text{in}}) = P_{\text{out}}(\theta)$ for simplicity.

The angular dependence of the re-emitted radiation is quantified by the so-called *Phase-function* (or the *angular redistribution function*), which is denoted with $P(\theta)$.

$$\frac{P_{\text{out}}(\theta)}{F_{\text{in}}} \equiv \frac{\sigma_T P(\theta)}{4\pi}, \quad \text{with} \quad \int_{-1}^1 d\mu P(\mu) = 2 \ \left(\text{i.e.} \int d\Omega P(\mu) = 4\pi\right),$$

(1.41)

where $\mu \equiv \cos\theta$. Note that the phase-function relates to the *differential cross-section* simply as $\frac{d\sigma}{d\Omega} \equiv \frac{\sigma P(\theta)}{4\pi}$.

There are two important examples of the phase-function we encounter for Lyα transfer. The first is the one we derived above, and describes 'dipole' or 'Rayleigh' scattering:

$$\boxed{P(\mu) = \frac{3}{4}(1 + \mu^2) \quad \text{dipole.}}$$

(1.42)

The other is for isotropic scattering:

$$\boxed{P(\mu) = 1 \quad \text{isotropic.}} \tag{1.43}$$

As we will see further on, the phase function associated with Lyα scattering is either described by pure dipole scattering, or by a superposition of dipole and isotropic scattering.

1.5.2 Interaction of a Bound Electron with Radiation: Lorentzian Cross-Section

We can understand the expression for the Lyα absorption cross-section via an analysis similar to the one described above. The main difference with the previous analysis is that the electron is not free, but instead orbits the proton at a natural (angular) frequency ω_0. We will treat the electron as a harmonic oscillator[22] with natural frequency ω_0. In the classical picture, the electron radiates as it accelerates and spirals inward. To account for this we will assume that the harmonic oscillator is damped. This damped harmonic oscillator with natural frequency ω_0 is 'forced' by the incoming radiation field that again has angular frequency ω. The equation of motion for this forced, damped harmonic oscillator is

$$\ddot{x} + \Gamma \dot{x} + \omega_0^2 x = \frac{q}{m} E(x, t) = \frac{q}{m} E_0 \exp(i\omega t), \tag{1.44}$$

where x can represent one of the Cartesian coordinates that describe the location of the electron in its orbit. The term $\Gamma \dot{x}$ denotes the friction (or damping) term, which reflects that in the classical picture of a hydrogen atom, the electron spirals inwards over a short timescale (see Sect. 1.2.2). The term on the RHS simply represents the electric force ($F = qE$) that is exerted by the electric field. We have represented the electro-magnetic wave as $E(t) = E_0 \exp(i\omega t)$ (which is a more general way of describing a wave than what we used when considering the free electron). It turns out that this simplifies the analysis). We can find solutions to Eq. 1.44 by substituting $x(t) = x \exp(i\omega t)$. We discuss two solutions here:

- In the absence of the electromagnetic field, this yields a quadractic equation for ω, namely $-\omega^2 + i\Gamma\omega + \omega_0^2 = 0$. This equation has solutions of the form $\omega = \frac{i\Gamma}{2} \pm \sqrt{\omega_0^2 - \Gamma^2/4}$. We assume that $\omega_0 \gg \Gamma$ (which is the case for Lyα as we see below), which can be interpreted as meaning that the electron makes multiple orbits around

---

[22] We can justify this picture as follows: we define the $x - y$ plane to be the plane in which the electron orbits the proton. The x-coordinate of the electron varies as $x(t) = x_0 \cos \omega_0 t$. The x-component of the electro-static force on the electron varies as $F_x = F_e \frac{x}{r}$, in which $F_e = \frac{q^2}{r^2}$. That is, the equation of motion for the x-coordinate of the electron equals $\ddot{x} = -kx$, where $k = q^2/r^3$.

the nucleus before there is a 'noticeable' change in its position due to radiative energy losses. The solution for $x(t)$ thus looks like $x(t) \propto \exp(-\Gamma t/2) \cos \omega_0 t$. The solution indicates that the electron keeps orbiting the proton with the same natural frequency ω_0, but that it spirals inwards on a characteristic timescale Γ^{-1} (also see Sect. 1.2.2). We will evaluate Γ later.

- In the presence of an electromagnetic field, substituting $x(t) = x \exp(i\omega t)$ yields the following solution for the amplitude x:

$$x = \frac{qE_0}{m_e} \frac{1}{\omega^2 - \omega_0^2 + i\omega\Gamma} \overset{\omega \approx \omega_0}{\sim} \frac{qE_0}{2m_e\omega_0} \frac{1}{\omega - \omega_0 + i\Gamma/2}, \tag{1.45}$$

where we used that $(\omega^2 - \omega_0^2) = (\omega - \omega_0)(\omega + \omega_0) \approx 2\omega_0(\omega - \omega_0)$. This last approximation assumes that $\omega \approx \omega_0$. It is highly relevant for Lyα scattering, where ω and ω_0 are almost always very close together (meaning that $|\omega - \omega_0|/\omega_0 \ll 10^{-2}$).

With the solution for $x(t)$ in place, we can apply the Larmor formula and compute the time averaged power radiated by the accelerated electron:

$$\langle P_{\text{out}} \rangle = \frac{2q^2 \langle |\ddot{x}|\rangle^2}{3c^3} = \frac{2q^2\omega^4}{3c^3} \frac{q^2 E_0^2}{2 \times 4m_e^2\omega_0^2} \frac{1}{(\omega - \omega_0)^2 + \Gamma^2/4}. \tag{1.46}$$

Where the highlighted factor of 2 comes in from the time-average: $\langle E^2 \rangle = E_0^2/2$. As before, the time average of the total incoming flux $\langle F_{\text{in}} \rangle$ (in erg s^{-1} cm^{-2}) of electromagnetic radiation equals $\langle F_{\text{in}} \rangle = \frac{cE_0^2}{8\pi}$. We therefore obtain an expression for the cross-section as:

$$\sigma(\omega) = \frac{\langle P_{\text{out}} \rangle}{\langle F_{\text{in}} \rangle} = \frac{8\pi \langle P_{\text{out}} \rangle}{cE_0^2} = \frac{16\pi q^4\omega^4}{24m_e^2\omega_0^2c^4} \frac{1}{(\omega - \omega_0)^2 + \Gamma^2/4} = \frac{\sigma_T}{4\omega_0^2} \frac{\omega^4}{(\omega - \omega_0)^2 + \Gamma^2/4}, \tag{1.47}$$

where we substituted the expression for the Thomson cross section $\sigma_T = \frac{8\pi q^4}{3m_e^2c^4}$ (see Eq. 1.36).

The expression for $\sigma(\omega)$ can be recast in a more familiar form when we use the Larmor formula to constrain Γ. The equation of motion shows that 'friction/damping force' on the electron is $F = -m_e\Gamma\dot{x}$. We can also write this force as $F = \frac{dp_e}{dt} = \frac{m_e}{p_e} \frac{d}{dt} \frac{p_e^2}{2m_e} = \frac{m_e}{p_e} \frac{dE_{\text{kin}}}{dt} = -\frac{m_e}{p_e} P_{\text{out}}$, where P_{out} denotes the emitted power. Setting the two equal gives us a relation between Γ and P_{out}: $-\Gamma m_e\dot{x}^2 = P_{\text{out}}$. Using that $\dot{x} = i\omega_0 x$, $\ddot{x} = -\omega_0^2 x$, and the Larmor formula gives us $\Gamma = \frac{2q^2\omega_0^2}{3m_ec^3} = 15 \times 10^8$ s^{-1}. With this the 'classical' expression for the Lyα cross section can be recast as

$$\sigma_{\text{CL}}(\omega) = \frac{3\lambda_0^2}{8\pi} \frac{\Gamma^2(\omega/\omega_0)^4}{(\omega - \omega_0)^2 + \Gamma^2/4} \Rightarrow \sigma_{\text{CL}}(\nu) = \frac{3\lambda_0^2}{8\pi} \frac{(\Gamma/2\pi)^2(\nu/\nu_0)^4}{(\nu - \nu_0)^2 + \Gamma^2/16\pi^2}, \tag{1.48}$$

Fig. 1.16 The Lorentzian
profile $\sigma_{\text{CL}}(\omega)$ for the Lyα
absorption cross-section.
This represents the
absorption cross-section of a
single atom. When averaging
over a collection of atoms
with a Maxwellian velocity
distribution, we obtain the
Voigt profile (shown in
Fig. 1.17)

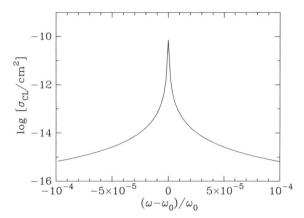

where the subscript 'CL' stresses that we obtained this expression with purely classi-
cal physics. If we ignore the $(\omega/\omega_0)^4$ term, then the functional form for $\sigma(\omega)$ is that
of a *Lorentzian Profile*. In the second equation we re-expressed the cross-section as
a function of frequency $\nu = \omega/(2\pi)$.

Figure 1.16 shows $\sigma_{\text{CL}}(\omega)$ for a narrow range in ω. There are two things to note:

1. The function is sharply peaked on ω_0, at which $\sigma_0 \equiv \sigma_{\text{CL}}(\omega_0) = \frac{3\lambda_0^2}{8\pi} \sim 7 \times 10^{-11}$
   cm^{-2}, where $\lambda_0 = 2\pi c/\omega_0$ corresponds to the wavelength of the electromagnetic
   wave with frequency ω_0. The cross-section falls off by $\gtrsim 5$ orders of magnitude
   with $|\omega - \omega_0|/\omega_0 \sim 10^{-4}$. Note that the cross-section is many, many orders of
   magnitude larger than the Thomson cross-section. This enhanced cross-section
   represents a 'resonance', an '*unusually strong response of a system to an external
   trigger*'.
2. The $\sigma \propto \omega^4$-dependence implies that the atom is slightly more efficient at scattering more energetic radiation. This correspond to the famous Rayleigh scattering
   regime, which refers to elastic scattering of light or other electromagnetic radiation by particles much smaller than the wavelength of the radiation (see https://
   en.wikipedia.org/wiki/Rayleigh_scattering), and which explains why the sky is
   blue,[23] and the setting/rising sun red.

1.5.3 Interaction of a Bound Electron with Radiation: Relation to Lyα Cross-Section

The derivation from the previous section was based purely on classical physics. A
full quantum-mechanical treatment of the Lyα absorption cross-section is beyond the
scope of these lectures, and—interestingly—still subject of ongoing research (e.g.

---

[23]This is not always the case in the Netherlands or Norway.

[11, 16, 17, 161]). A clear recent discussion on this can be found in [188]. There are two main things that change:

1. The parameter Γ reduces by a factor of $f_\alpha = 0.4162$, which is known as the 'oscillator' strength, i.e. $\Gamma \rightarrow f_\alpha \Gamma \equiv A_\alpha$. Here, A_α denotes the Einstein A-coefficient for the Lyα transition.
2. In detail, a simple functional form (Lorentzian, Eq. 1.48, or see Eq. 1.49 below) for the Lyα cross-section does not exist. The main reason for this is that if we want to evaluate the Lyα cross-section far from resonance, we have to take into account the contributions to the cross-section from the higher-order Lyman-series transitions, and even photoionization. When these contributions are included, the expression for the cross-section involves squaring the sum of all these contributions (e.g. [16, 17, 188]). Accurate approximations to this expression are possible close to the resonance(s)—when $|\omega - \omega_0|/\omega_0 \ll 1$, which is the generally the case for practical purposes—and these approximations are in excellent agreement with our derived cross-section (see Eq. 14 in [188], and use that $\omega \approx \omega_0$, and that $\Lambda_{12} = \Gamma/2\pi$).

The *left panel* of Fig. 1.17 compares the different Lyα cross sections. The *black solid line* shows the Lorentzian cross-section. The *red dashed line* shows $\sigma_P(\omega)$, which is the cross section that is given in [207]:

$$\sigma(\omega) = \frac{3\lambda_\alpha^2}{8\pi} \frac{A_\alpha^2 (\omega/\omega_\alpha)^4}{(\omega - \omega_\alpha)^2 + A_\alpha^2 (\omega/\omega_\alpha)^6/4}. \tag{1.49}$$

If we ignore the $(\omega/\omega_\alpha)^6$ term in the denominator, then this corresponds exactly to the cross-section that we derived in our classical analysis (see Eq. 1.48, with $\Gamma \rightarrow A_\alpha$). This cross-section is obtained from a quantum mechanical calculation, and under the assumption that the hydrogen atom has only two quantum levels (the 1s and 2p levels). The *blue dotted line* shows the cross-section obtained from the full quantum mechanical calculation (see [188] for a discussion, which is based on [188]). This Figure shows that these cross-sections differ only in the far wings of the line profile. Close to resonance, the line profiles are practically indistinguishable.

1.5.4 Voigt Profile of Lyα Cross-Section

In the previous section we discussed the Lyα absorption cross section in the frame of a single atom. Because each atom has its own velocity, a photon of a fixed frequency ν will appear Doppler boosted to a slightly different frequency for each atom in the gas. To compute the Lyα absorption cross section for a collection of moving atoms, we must convolve the single-atom cross section with the atom's velocity distribution. That is,

$$\sigma_\alpha(\nu, T) = \int d^3 \mathbf{v} \sigma_\alpha(\nu|\mathbf{v}) f(\mathbf{v}), \tag{1.50}$$

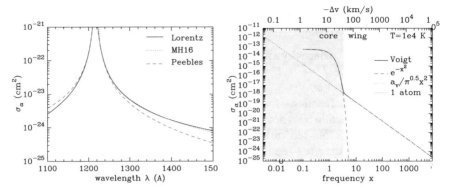

Fig. 1.17 In the *left panel* we compare the different atom frame Lyα cross sections. The *black solid line* shows the Lorentzian cross-section. The *red dashed line* shows the cross-section given in [207], and the *blue dotted line* shows the quantum mechanical cross-section from Mortock and Hirata (in prep, also see [188]). In the *right panel* the *solid line* shows the gas frame (velocity averaged absorption) cross section as given by Eq. 1.55 as a function of the dimensionless frequency x (see text). The *red dashed line* (*green dot-dashed line*) represent the cross section where we approximated the Voigt function as $\exp(-x^2)$ $(a_v/[\sqrt{\pi}x^2])$. Clearly, these approximations work very well in their appropriate regimes. Also shown for comparison as the *blue dotted line* is the symmetric single atom Lorentzian cross section (which was shown in the *left panel* as the *black solid line*). Close to resonance, this single atom cross section provides a poor description of the real cross section, but it does very well in the wings of the line (see text)

where $\sigma_\alpha(\nu|\mathbf{v})f$ denotes the Lyα absorption cross-section at frequency ν for an atom moving at 3D velocity $\mathbf{v}$ (the precise velocity $\mathbf{v}$ changes the frequency in the frame of the atom). Suppose that the photon is propagating in a direction $\mathbf{n}$. We decompose the atoms three-dimensional velocity vectors into directions parallel ($v_{||}$) and orthogonal ($\mathbf{v}_\perp$) to $\mathbf{n}$. These components are independent and $f(\mathbf{v})d^3\mathbf{v} = f(v_{||})g(\mathbf{v}_\perp)dv_{||}d^2\mathbf{v}_\perp$. The absorption cross-section does not depend on $\mathbf{v}_\perp$ because the frequency of the photon that the atoms 'sees' does not depend on $\mathbf{v}_\perp$. We can therefore write

$$\sigma_\alpha(\nu, T) = \underbrace{\int d^2\mathbf{v}_\perp g(\mathbf{v}_\perp)}_{=1} \int_{-\infty}^{\infty} dv_{||}\sigma_\alpha(\nu|v_{||})f(v_{||}). \qquad (1.51)$$

For a Maxwell-Boltzmann distribution $f(v_{||})dv_{||} = \left(\frac{m_p}{2\pi k_B T}\right)^{1/2} \exp\left(-\frac{m_p v_{||}^2}{2kT}\right)$. If we insert this into Eq. 1.51 and substitute Eq. 1.48 for $\sigma_\alpha(\nu|v_{||}) = \sigma(\nu')$, where $\nu' = \nu\left(1 - \frac{v_{||}}{c}\right)$ we find

$$\sigma_\alpha(\nu, T) = \frac{3(\lambda_0 A_\alpha)^2}{32\pi^3}\left(\frac{m_p}{2\pi k_B T}\right)^{1/2} \int_{-\infty}^{\infty} dv_{||} \frac{\exp\left(-\frac{m_p v_{||}^2}{2kT}\right)}{\left[\nu\left(1 - \frac{v_{||}}{c}\right) - \nu_\alpha\right]^2 + \frac{A_\alpha^2}{16\pi^2}}, \qquad (1.52)$$

where we dropped the term $(v/v_0)^4$, which is accurate for $v \approx v_0$ (which is practically always the case). We substitute $y = \sqrt{\frac{m_p v_\parallel^2}{2k_B T}} \equiv v_\parallel/v_{\mathrm{th}}$, and define the *Voigt parameter* $a_v \equiv \frac{A_\alpha}{4\pi \Delta v_\alpha} = 4.7 \times 10^{-4}(T/10^4\ \mathrm{K})^{-1/2}$ in which $\Delta v_\alpha = v_\alpha v_{\mathrm{th}}/c$. We can then recast this expression as

$$\sigma_\alpha(v, T) = \frac{3(\lambda_0 A_\alpha)^2}{32\pi^3}\left(\frac{m_p}{2\pi k_B T}\right)^{1/2}\frac{v_{\mathrm{th}}}{\Delta v_\alpha}\int_{-\infty}^{\infty} dy \frac{\exp(-y^2)}{\left[\frac{v}{\Delta v_\alpha}\left(1 - \frac{y v_{\mathrm{th}}}{c}\right) - \frac{v_\alpha}{\Delta v_\alpha}\right]^2 + a_v^2}.$$
(1.53)

We finally introduce the dimensionless frequency variable $x \equiv (v - v_\alpha)/\Delta v_\alpha$, which we use to express $v = v_\alpha(1 + x v_{\mathrm{th}}/c)$. If we substitute this back into Eq. 1.53 and drop the second order term with $(v_{\mathrm{th}}/c)^2$, then we get

$$\sigma_\alpha(v, T) = \frac{3\lambda_0^2 A_\alpha^2}{32\pi^3 \sqrt{\pi} \Delta v_\alpha^2}\int_{-\infty}^{\infty} dy \frac{\exp(-y^2)}{(x - y)^2 + a_v^2} \equiv \frac{3\lambda_0^2 a_v}{2\sqrt{\pi}} H(a_v, x) \equiv \sigma_{\alpha,0}(T)\phi(x) =$$

$$= 5.9 \times 10^{-14}\left(\frac{T}{10^4\ \mathrm{K}}\right)^{-1/2}\phi(x)\ \mathrm{cm}^2 \qquad (1.54)$$

and the Voigt function as

$$\phi(x) \equiv H(a_v, x) = \frac{a_v}{\pi}\int_{-\infty}^{\infty} \frac{e^{-y^2} dy}{(y - x)^2 + a_v^2} = \begin{cases} \sim e^{-x^2} & \text{core;} \\ \sim \frac{a_v}{\sqrt{\pi}x^2} & \text{wing.} \end{cases} \qquad (1.55)$$

Note that throughout this review we will use both $\phi(x)$ and $H(a_v, x)$ to denote the shape of the Lyα line profile.[24] The transition between core and wing occurs approximately when $\exp(-x^2) = a_v/(\sqrt{\pi}x^2)$.

The *solid line* in the *right panel* of Fig. 1.17 shows the LAB frame cross section— this is also known as the *Voigt-profile*—as given by Eq. 1.55 as a function of the dimensionless frequency x. The *red dashed line* (*green dot-dashed line*) represent the cross section where we approximated the Voigt function as $\exp(-x^2)$ ($a_v/[\sqrt{\pi}x^2]$). Clearly, these approximations work very well in the relevant regimes. Note that this Figure shows that a decent approximation to Voigt function at all frequencies is given simply by the sum of these two terms, i.e. $\phi(x) \approx \exp(-x^2) + a_v/(\sqrt{\pi}x^2)$ (provided that $|x| \gtrsim \sqrt{a_v}$ [148]). This approximation fails in a very narrow frequency regime where the transition from core to wing occurs. A useful fitting function that is accurate at all x is given in [256]. Also shown for comparison as the *blue dotted line* is the symmetric single atom cross section (which was shown in the *left panel* as the *solid line*). Figure 1.17 shows that close to resonance, this single atom cross section provides a poor description of the real cross section. This is because Doppler

---

[24]One has to be a bit careful because in the literature occasionally $\phi(x) = H(a_v, x)/\sqrt{\pi}$, because in this convention the line profile is normalized to 1, i.e. $\int \phi(x)dx = 1$. In our convention $\phi(x = 0) = 1$, while the normalization is $\int \phi(x)dx = \sqrt{\pi}$.

motions 'smear out' the sharply peaked cross-section. Far in the wing however, the single atom cross-section provides an excellent fit to the velocity averaged Voigt profile.

One of the key results from this section is that the Lyα cross-section, evaluated at line center and averaged over the velocity distribution of atoms, is tremendous at $\sigma_{\alpha,0}(T) \sim 5.9 \times 10^{-14}(T/10^4 \text{ K})^{-1/2} \text{ cm}^{-2}$, which is ~ 11 orders of magnitude than the Thomson cross-section. That is, an electron bound to the proton is ~ 11 orders of magnitude more efficient at scattering radiation than a free electron when the frequency of that radiation closely matches the natural frequency of the transition. This further emphasises that the electron 'resonantly scatters' the incoming radiation. To put these numbers in context, it is possible to measure the hydrogen column density, N_{HI}, in nearby galaxies. The observed intensity in the 21-cm line translates to typical HI column densities of order $N_{HI} \sim 10^{19} - 10^{21} \text{ cm}^{-2}$ [33, 53, 140]), which translates to line center optical depths of Lyα photons of order $\tau_0 \sim 10^7 - 10^8$. This estimate highlights the importance of understanding the transport of Lyα photon out of galaxies. They generally are not expected to escape without interacting with hydrogen gas.

1.6   Step 2 Towards Understanding Lyα Radiative Transfer: The Radiative Transfer Equation

The specific intensity $I_\nu(\mathbf{r}, \mathbf{n}, t)$ is defined as the rate at which energy crosses a unit area, per solid angle, per unit time, as carried by photons of energy $h_p\nu$ in the direction $\mathbf{n}$, i.e. $I_\nu(\mathbf{r}, \mathbf{n}, t) = \frac{d^3 E_\nu}{d\Omega dt dA}$. The change in the spectral/specific intensity of radiation at a location $\mathbf{r}$ that is propagating in direction $\mathbf{n}$ at time t is (e.g. [178])

$$\mathbf{n} \cdot \nabla I_\nu(\mathbf{r}, \mathbf{n}, t) + \frac{1}{c}\frac{\partial I(\mathbf{r}, \mathbf{n}, t)}{\partial t} = -\alpha_\nu(\mathbf{r})I(\mathbf{r}, t) + j_\nu(\mathbf{r}, \mathbf{n}, t). \tag{1.56}$$

Here, $\alpha_\nu(\mathbf{r})I(\mathbf{r}, t)$ in the first term on the RHS denotes the 'attenuation coefficient', which accounts for energy loss (gain) due to absorption (stimulated emission). In static media, we generally have that $\alpha_\nu(\mathbf{r})$ is isotropic,[25] which is why we dropped its directional dependence. The emission coefficient $j_\nu(\mathbf{r}, \mathbf{n}, t)$ describes the local specific luminosity per solid angle, per unit volume. The random orientation of atoms/molecules (again) generally causes emission coefficient to be isotropic.

For Lyα radiation, photons are not permanently removed following absorption, but they are generally scattered. To account for this scattering, we must add a third term to the RHS on Eq. 1.56:

---

[25]For radiation at some fixed frequency ν close to the Lyα resonance, the opacity $\alpha_\nu(\mathbf{r}, \mathbf{n})$ depends on $\mathbf{n}$ for non-static media. This directional dependence is taken into account when performing Monte-Carlo Lyα radiative transfer calculations (to be described in Sect. 1.8).

$$\mathbf{n} \cdot \nabla I_\nu(\mathbf{r}, \mathbf{n}, t) + \frac{1}{c}\frac{\partial I(\mathbf{r}, \mathbf{n}, t)}{\partial t} = -\alpha_\nu(\mathbf{r}) I(\mathbf{r}, t) + j_\nu(\mathbf{r}, \mathbf{n}, t)$$

$$+ \int d\nu' \int d^3\mathbf{n}'\, \alpha_{\nu'}(\mathbf{r}) I_{\nu'}(\mathbf{r}, \mathbf{n}, t) R(\nu, \nu', \mathbf{n}, \mathbf{n}'). \quad (1.57)$$

The third term accounts for energy redistribution as a result of scattering. In this term, the so-called 'redistribution function' $R(\nu, \nu', \mathbf{n}, \mathbf{n}')$ describes the probability that radiation that was originally propagating at frequency ν' and in direction $\mathbf{n}'$ is scattered into frequency ν and direction $\mathbf{n}$. We focus on equilibrium solutions and omit the time dependence in Eq. 1.57. Furthermore, we introduce the coordinate s, which measures distance along the direction $\mathbf{n}$. The change of the intensity of radiation that is propagating in direction $\mathbf{n}$ with distance s is then

$$\frac{dI_\nu(s, \mathbf{n})}{ds} = -[\ \underbrace{\alpha_\nu^{\mathrm{HI}}(s)}_{\text{I: absorption}} + \underbrace{\alpha_\nu^{\mathrm{dest}}(s)}_{\text{IV: 'destruction'}}]I_\nu(s, \mathbf{n}) + \underbrace{j_\nu(s)}_{\text{II: emission}}$$

$$+ \underbrace{\int d\nu' \int d^3\hat{\mathbf{n}}'\, \alpha_{\nu'}(s) I_{\nu'}(s, \hat{\mathbf{n}}') R(\nu, \nu', \mathbf{n}, \hat{\mathbf{n}}')}_{\text{III: scattering}}. \quad (1.58)$$

In the following subsections we will discuss each term I–IV on the right hand side in more detail.

1.6.1   I: Absorption Term: Lyα Cross Section

The opacity $\alpha_\nu^{\mathrm{HI}}(s) = \left[n_l(s) - \frac{g_l}{g_u} n_u(s) \right]\sigma(\nu)$. The second term within the square brackets corrects the absorption term for stimulated emission. In most astrophysical conditions all neutral hydrogen atoms are in their electronic ground state, and we can safely ignore the stimulated emission term (see e.g. the Appendix of [23, 66]). That is, in practice we can simply state that $\alpha_\nu(s) = n_l(s)\sigma_\alpha(\nu) = n_{\mathrm{HI}}(s)\sigma_\alpha(\nu)$. We have derived expressions for the Lyα absorption cross-section in Sect. 1.5.

1.6.2   II: Volume Emission Term

The emission term is given by

$$j_\nu(s) = \frac{\phi(\nu)h\nu_\alpha}{4\pi^{3/2}\Delta\nu_\alpha}\left(n_e n_{\mathrm{HI}} q_{1s2p} + n_e n_{\mathrm{HII}}\alpha(T) f_{\mathrm{Ly}\alpha}(T) \right), \quad (1.59)$$

where $\phi(\nu)$ is the Voigt profile (see Sect. 1.5.4), the factor 4π accounts for the fact that Lyα photons are emitted isotropically into 4π steradians, and the factor of

$\sqrt{\pi}\Delta\nu_\alpha$ arises when converting[26] $\phi(x)$ to $\phi(\nu)$. The first term within the brackets denotes the Lyα production rate as a result of collisional excitation of H atoms by (thermal) electrons, and is given by Eq. 1.7, in which the collision strength $\Omega_{1s2p}(T)$ can be read off from Fig. 1.7. The second term within the brackets denotes the Lyα production rate following recombinationn which both the recombination coefficient and the Lyα production probability depend on both temperature and the opacity of the medium to ionizing and Lyman series photons (see Fig. 1.6, and Sect. 1.3.2).

1.6.3 III: Scattering Term

The 'redistribution function' $R(\nu, \nu', \mathbf{n}, \mathbf{n}')$ describes the probability that radiation that was originally propagating at frequency ν' and in direction $\mathbf{n}'$ is scattered into frequency ν and direction $\mathbf{n}$ (see e.g. Zanstra [130, 266, 284] for early discussions). In practise, this probability depends only on the angle between $\mathbf{n}$ and $\mathbf{n}'$, i.e. $R(\nu, \nu', \mathbf{n}, \mathbf{n}') = R(x_{\text{out}}, x_{\text{in}}, \mu)$, in which $\mu \equiv \cos\theta = \mathbf{n} \cdot \mathbf{n}'$. We have also switched to standard dimensionless frequency coordinates (first introduced in Eq. 1.53), and denote with x_{out} (x_{in}) the dimensionless frequency of the photon after (before) scattering. Formally, $R(x_{\text{out}}, x_{\text{in}}, \mu)dx_{\text{out}}d\mu$ denotes the probability that a photon of frequency x_{in} was scattered by an angle in the range $\mu \pm d\mu/2$ into the frequency range $x_{\text{out}} \pm dx_{\text{out}}/2$. Thus, $R(x_{\text{out}}, x_{\text{in}}, \mu)$ is normalized such that $\int_{-1}^{1} d\mu \int_{-\infty}^{\infty} dx_{\text{out}} R(x_{\text{in}}, x_{\text{out}}, \mu) = 1$.

In the remainder of this section, we will compute the redistribution functions $R(x_{\text{in}}, x_{\text{out}}, \mu)$. Following [158] we will employ the notation of probability theory. In this notation, the function $p(y|b)$ denotes the conditional probability[27] density function (PDF) of y given b. The PDF for y is then given by $p(y) = \int p(y|b)p(b)db$, where $p(b)$ denotes the PDF for b. Furthermore, the joint PDF of y and b is given by $p(y, b) = p(y|b)p(b)$. We could just as well have written that $p(y, b) = p(b|y)p(y)$, and by setting $p(y|b)p(b) = p(b|y)p(y)$ we get 'Bayes theorem' which states that

$$p(y|b) = \frac{p(b|y)p(y)}{p(b)}. \tag{1.60}$$

We will use this theorem on several occasions below. We can write the conditional joint PDF for x_{out} and μ given x_{in} as

$$R(x_{\text{out}}, \mu|x_{\text{in}}) = R(x_{\text{out}}|\mu, x_{\text{in}})P(\mu|x_{\text{xin}}). \tag{1.61}$$

---

[26]Recall that $\phi(x)$ was normalized to $\int \phi(x)dx = \sqrt{\pi}$ (Eq. 1.55). Substituting $dx = d\nu/\Delta\nu_\alpha$ (see discussion below Eq. 1.53) gives $\int \phi(\nu)d\nu = \sqrt{\pi}\Delta\nu_\alpha$.

[27]In the lectures I illustrated this with an example in which y denotes my happiness, and in which b denotes the number of snowballs that were thrown in my face in the previous 30 minutes. Clearly, $p(y)$ will be different when $b = 0$ or when $b \gg 1$.

An additional quantity that is of interest in many radiative transfer problems (see for example Sect. 1.6.5) is the conditional PDF for x_{out} given x_{in}:

$$R(x_{out}|x_{in}) = \int_{-1}^{1} d\mu \, R(x_{out}|\mu, x_{in}) P(\mu|x_{xin}). \tag{1.62}$$

To complete our calculation we have to evaluate the PDFs $R(x_{out}|\mu, x_{in})$ and $P(\mu|x_{xin})$. We will do this next.

In most astrophysical conditions, the energy of the Lyα photon before and after scattering is identical in the frame of the absorbing atom. This is because the life-time of the atom in its $2p$ state is only $t = 1/A_\alpha \sim 10^{-9}$ s. In most astrophysical conditions, the hydrogen atom in this state is not 'perturbed' over this short time-interval, and energy conservation forces the energy of the photon to be identical before and after scattering. Because of random thermal motions of the atom, energy conservation in the atom's frame translates to a change in the energy of the incoming and outgoing photon that depends on the velocity of the atom and the scattering direction (see Fig. 1.18 for an illustration). This type of scattering is known as 'partially coherent' scattering.[28]

For notational clarity, we denote the propagation direction and dimensionless frequency of the photon before (after) scattering with $\mathbf{k}_{in}$ and x_{in} ($\mathbf{k}_{out}$ and x_{out}). We assume that the scattering event occurs off an atom with velocity vector $\mathbf{v}$. Doppler boosting between the atom and gas frame corresponds to

$$x_{in}^{atom} = x_{in}^{gas} - \frac{\mathbf{v} \cdot \mathbf{k}_{in}}{v_{th}}, \quad \text{gas} \rightarrow \text{atom}; \quad x_{in}^{gas} = x_{in}^{atom} + \frac{\mathbf{v} \cdot \mathbf{k}_{in}}{v_{th}}, \quad \text{atom} \rightarrow \text{gas}. \tag{1.63}$$

Notice the signs in these equations. When $\mathbf{v}$ and $\mathbf{k}_{in}$ point in the same direction, the atom is moving away from the incoming photon, which reduces the frequency of the photon in the atom's frame. The expressions are the same for the outgoing photon. Partially coherent scattering dictates that $x_{in}^{atom} = x_{out}^{atom}$, which allows us to write down the relation between x_{out}^{gas} and x_{in}^{gas} as

$$x_{out}^{gas} = x_{in}^{gas} - \frac{\mathbf{v} \cdot \mathbf{k}_{in}}{v_{th}} + \frac{\mathbf{v} \cdot \mathbf{k}_{out}}{v_{th}}. \tag{1.64}$$

We always work in the gas frame and will drop the 'gas' superscript from now on. Equation 1.64 still misses the effect of *atomic recoil*, which is illustrated in the drawing on the *left* of Fig. 1.18: in the atom frame, the momentum of the photon prior to scattering is $h_p \nu_{in}/c$ and points to the right. The momentum of the photon

---

[28]It is possible to repeat the analysis of this section under the assumption that (*i*) the energy of the photon before and after scattering is identical, which is relevant when the gas has zero temperature. This corresponds to 'completely coherent' scattering (*ii*) the energy of the re-emitted photon is completely unrelated to the atom of the incoming photon. This can happen in very dense gas where collisions perturb the atom while in the $2p$ state. This case corresponds to 'completely incoherent' scattering.

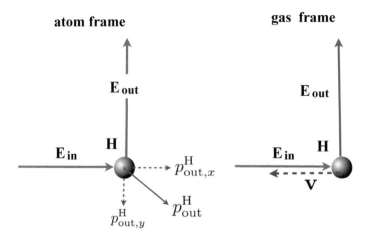

Fig. 1.18 In most astrophysical media, Lyα scattering is *partially coherent*: energy conservation implies that the energy of the photon before and after scattering is the same in the frame of the atom (the 'atom frame' shown on the *left*). In the gas frame however (shown on the *right*) the energy before and after scattering is different. This is because the random thermal motion of the atom induces Doppler shifts to the energy of the photon before and after scattering by an amount that depends on the atoms velocity, and on the scattering angle. Each scattering event induces small changes to the frequency of the Lyα photon. In this case, the atom is moving into the incoming Lyα photon, which causes it to appear at a higher frequency. Scattering by 90° does not induce any additional Doppler boost in this example, and the outgoing photon has a higher frequency (which is why we drew it blue, to indicate its newly acquired blue-shift). The picture on the *left* also illustrates the recoil effect: in the atom frame, momentum conservation requires that the hydrogen atom moves after the scattering event. This newly gained kinetic energy of the H-atom comes at the expense of the energy of the Lyα photon. This recoil effect is generally small

after scattering equals $h_{\rm p}\nu_{\rm out}/c$ and points up. Momentum would therefore not be conserved in the atom's frame, if it were not for the newly acquired momentum of the hydrogen atom itself. This newly acquired momentum corresponds to newly acquired kinetic energy, which in turn came at the expense of the energy of the Lyα photon. The energy of the Lyα photon is therefore (strictly) not conserved exactly, but reduced by a small amount in each scattering event. This effect is not relevant when $\mathbf{k}_{\rm in} = \mathbf{k}_{\rm out}$ (i.e. $\mu = 1$), and maximally important when $\mathbf{k}_{\rm in} = -\mathbf{k}_{\rm out}$ (i.e. $\mu = -1$). Demanding momentum conservation in the atom frame gives additional terms of the form[29]

$$x_{\rm out} = x_{\rm in} - \frac{\mathbf{v} \cdot \mathbf{k}_{\rm in}}{v_{\rm th}} + \frac{\mathbf{v} \cdot \mathbf{k}_{\rm out}}{v_{\rm th}} + \underbrace{g(\mu - 1)}_{\rm recoil} + \mathcal{O}(v_{\rm th}^2/c^2), \qquad (1.65)$$

---

[29]We will not derive this here. The derivation is short. First show that the total momentum of the atom after scattering equals $p_{\rm out}^{\rm H} = \frac{h_{\rm p}\nu_{\rm in}}{c}\sqrt{2 - 2\mu}$, where $\mu = \mathbf{k}_{\rm in} \cdot \mathbf{k}_{\rm out}$. This corresponds to a total kinetic energy $E_e = \frac{[p_{\rm out}^{\rm H}]^2}{2m_{\rm p}}$, which must come at the expense of the Lyα photon. We therefore have $\Delta E = h_{\rm p}\Delta\nu = \frac{1}{m_{\rm p}}\left(\frac{h_{\rm p}\nu_{\rm in}}{c}\right)^2(1 - \mu)$, which corresponds to $\Delta x = \frac{h\nu_\alpha}{m_{\rm p}v_{\rm th}c}$ if we approximate that $\nu_{\rm in} \approx \nu_\alpha$.

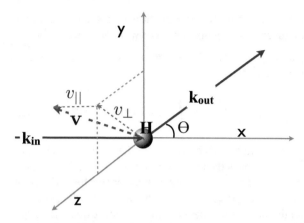

Fig. 1.19 Schematic depiction of the coordinate system that we used to describe the scattering event. The scattering plane is spanned by the wavevectors of the photon before ($\mathbf{k}_{in} = [1, 0, 0]$) and after ($\mathbf{k}_{out} = [\mu, \sqrt{1 - \mu^2}, 0]$) scattering. In other words, the scattering plane corresponds simply to the $x - y$ plane. If we decompose the atom's velocity vector (*the blue dashed vector*) into components parallel ($v_{||}$, *red dotted vector*) and orthogonal ($v_\perp$, *red dotted vector*) to the propagation direction of the incoming photon, then we can derive convenient expressions for the conditional probabilities $R(x_{out}|x_{in}, \mu)$ (see Fig. 1.20) and $R(x_{out}|x_{in})$ (see Fig. 1.21)

where $g = \frac{h\nu_\alpha}{m_p v_{th} c} = 2.6 \times 10^{-4} (T/10^4 \text{ K})^{-1/2}$ is the fractional amount of energy that is transferred per scattering event [86]. Throughout the remainder of this calculation we will ignore recoil. This may be counter-intuitive, as the change in x is of the order $\sim 10^{-4}$ for *each scattering event*, and—as we will see later—Lyα photons can scatter $\gg 10^6$ times. However, (fortunately) this process does not act cumulatively as we will discuss in more detail at the end of this section. Adams [1] first showed that recoil can generally be safely ignored.

For simplicity, but without loss of generality, we define a coordinate system such that $\mathbf{k}_{in} = (1, 0, 0)$, and $\mathbf{k}_{out} = (\mu, \sqrt{1 - \mu^2}, 0)$, i.e. the photon wavevectors lie entirely in the x-y plane (see Fig. 1.19). If we decompose the atom's velocity into components parallel ($v_{||}$) and orthogonal (v_y and v_z) to $\mathbf{k}_{in}$, then $\mathbf{v} = (v_{||}, v_y, v_z)$ and

$$x_{out} = x_{in} - \frac{v_{||}}{v_{th}} + \frac{v_{||}\mu}{v_{th}} + \frac{v_y\sqrt{1 - \mu^2}}{v_{th}} \equiv x_{in} - u + u\mu + w\sqrt{1 - \mu^2}, \quad (1.66)$$

where we have introduced the dimensionless velocity parameters $u = v_{||}/v_{th}$ and $w = v_y/v_{th}$ (see e.g. [4]). Note that the value of v_z is irrelevant in this equation, which is because it does not induce any Doppler boost on either the incoming or outgoing photon.

We were interested in calculating $R(x_{out}|\mu, x_{in})$. We can do this by using Eq. 1.66 and applying probability theory. We first write

$$R(x_{out}|\mu, x_{in}) = \mathcal{N} \int_{-\infty}^{\infty} du \int_{-\infty}^{\infty} dw R(x_{out}|\mu, x_{in}, u, w) P(u|\mu, x_{in}) P(w|\mu, x_{in}),$$

$$(1.67)$$

where $\mathcal{N}$ is a normalization factor. Equation 1.66 states that when x_{out}, x_{in}, μ are fixed, then for a given u, solutions only exist when $w \equiv w_u = \frac{x_{out} - x_{in} + u - u\mu}{\sqrt{1-\mu^2}}$. In other words, $R(x_{out}|\mu, x_{in}, u, w)$ is only non-zero when $w = w_u$. We can therefore drop the integral over w and write

$$R(x_{out}|\mu, x_{in}) = \mathcal{N} \int_{-\infty}^{\infty} du\, P(u|\mu, x_{in}) P(w_u|\mu, x_{in}). \qquad (1.68)$$

The conditional absorption probabilities for both w and u cannot depend on the subsequent emission direction, as the re-emission process is set by quantum mechanics of the wavefunction describing the electron. Therefore we have $P(u|\mu, x_{in}) = P(u|x_{in})$ and $P(w_u|\mu, x_{in}) = P(w_u|x_{in})$. Also note that w denotes the normalized velocity in a direction perpendicular to $\mathbf{k}_{in}$. The incoming photon's frequency, x_{in}—and therefore the absorption probability—cannot depend on w. Therefore, $P(w_u|x_{in}) = P(w_u) = \exp(-w_u^2)/\sqrt{\pi}$, where we assumed a Maxwell-Boltzmann distribution of the atoms' velocities.

The expression for $P(u|x_{in})$ is a bit more complicated. From Bayes Theorem (Eq. 1.60) we know that $P(u|x_{in}) = P(x_{in}|u)P(u)/P(x_{in})$, in which $P(x_{in}|u)$ denotes the absorption probability for a single atom that has a speed u, and $P(x_{in})$ can be interpreted as a normalization factor. The scattering probability of Lyα photons off atoms with velocity component u must scale with the (single-atom) cross-section, i.e. $P(x_{in}|u) \propto \sigma_\alpha(x_{in}|u) = \frac{3\lambda_\alpha^2}{8\pi} \frac{A_\alpha^2}{[\omega_\alpha(x_{in}-u)v_{th}/c]^2 + A_\alpha^2/4}$ (see Eq. 1.49). If we substitute this into Eq. 1.68 and absorb all factors that can be pulled out of the integral into the normalization constant $\mathcal{N}$, then we get

$$R(x_{out}|\mu, x_{in}) = \mathcal{N} \int_{-\infty}^{\infty} du \frac{\exp(-u^2)}{(x_{in} - u)^2 + a_v^2} \exp\left[-\left(\frac{\Delta x - u(\mu - 1)}{\sqrt{1-\mu^2}} \right)^2 \right], \quad (1.69)$$

where we introduced $\Delta x \equiv x_{out} - x_{in}$. It is possible to get an analytic expression[30] for $\mathcal{N}$:

---

[30] This can be seen as follows (note that the second line contains colors to clarify how we got from the L.H.S to the R.H.S):

$$\mathcal{N}^{-1}(\mu) = \int_{-\infty}^{\infty} dx_{out} \int_{-\infty}^{\infty} du \frac{\exp(-u^2)}{(x_{in} - u)^2 + a_v^2} \exp\left[-\left(\frac{\Delta x - u(\mu - 1)}{\sqrt{1-\mu^2}} \right)^2 \right] =$$

$$\int_{-\infty}^{\infty} du \frac{\exp(-u^2)}{(x_{in} - u)^2 + a_v^2} \int_{-\infty}^{\infty} d\Delta x \exp\left[-\left(\frac{\Delta x - u(\mu - 1)}{\sqrt{1-\mu^2}} \right)^2 \right] = \phi(x_{in}) \frac{\pi}{a_v} \sqrt{1-\mu^2} \sqrt{\pi},$$

$$(1.70)$$

where we used that $d\Delta x = dx_{out}$. We rewrote the term in red using the definition of $\phi(x)$ (see Eq. 1.55), and the term in blue is a Gaussian in Δx.

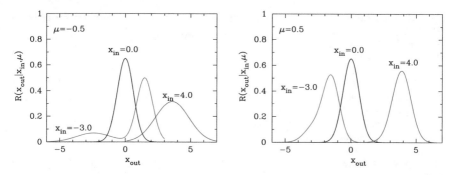

Fig. 1.20 $R(x_{out}|\mu, x_{in})$ is shown as a function of x_{out}. In the *left panel* we plot curves with $\mu = -0.5$, and $x_{in} = -3.0$ (*red line*), $x_{in} = 0.0$ (*black line*), and $x_{in} = 4.0$ (*blue line*). In the *right panel* we changed the sign of μ to $\mu = 0.5$. This Figure shows that for the vast majority of photons $|x_{in} - x_{out}| \lesssim$ a few. Thus the photon frequencies before and after scattering are closely related. Furthermore, $R(x_{out}|\mu, x_{in})$ can depend quite strongly on μ. This is most clearly seen by comparing the curves for $x_{in} = -3.0$ in the *left* and *right panels*

$$R(x_{out}|\mu, x_{in}) = \frac{a_v}{\pi^{3/2}\sqrt{1 - \mu^2}\phi(x_{in})} \int_{-\infty}^{\infty} du \frac{\exp(-u^2)}{(x_{in} - u)^2 + a_v^2} \exp\left[-\left(\frac{\Delta x - u(\mu - 1)}{\sqrt{1 - \mu^2}}\right)^2\right].$$

(1.71)

Examples of $R(x_{out}|\mu, x_{in})$ as a function of x_{out} are plotted in Fig. 1.20. In the *left panel* we plot $R(x_{out}|\mu, x_{in})$ for $\mu = -0.5$, and $x_{in} = -3.0$ (*red line*), $x_{in} = 0.0$ (*black line*), and $x_{in} = 4.0$ (*blue line*). These frequencies were chosen to represent scattering in the wing (for $x_{in} = 4.0$), in the core (for $x_{in} = 0.0$) and in the transition region ($x_{in} = -3.0$, see Fig. 1.17). In the *right panel* we used $\mu = 0.5$. Figure 1.20 shows clearly that (*i*) for the vast majority of photons $|x_{in} - x_{out}| \lesssim$ a few. That is, the frequency after scattering is closely related to the frequency before scattering (*ii*) $R(x_{out}|\mu, x_{in})$ can depend quite strongly on μ. This is most clearly seen by comparing the curves for $x_{in} = -3.0$ in the *left* and *right panels*.

In many radiative transfer problem we are mostly interested in the conditional PDF $R(x_{out}|x_{in})$ (see 1.62) for which one needs to know the 'conditional phase function' $P(\mu|x_{in})$. We introduced the concept of the phase function in Sect. 1.5.1, and discussed cases of (*i*) isotropic scattering, for which $P(\mu) = 1$, and *dipole scattering* for which $P(\mu) = \frac{3}{4}(1 + \mu^2)$. As we will discuss in more detail in Sect. 1.10.1, Lyα scattering represents a superposition of dipole and isotropic scattering. As we will see, there *is* a weak dependence of the phase function $P(\mu)$ to x_{in}, but only over a limited range of frequencies (see Sect. 1.10). It is generally fine to ignore this effect and state that $P(\mu|x_{in}) = P(\mu)$. The conditional PDF $R(x_{out}|x_{in})$ for isotropic and dipole scattering are given by

$$R_A(x_{out}|x_{in}) = \int_{-1}^{1} d\mu R(x_{out}|\mu, x_{in})$$

(1.72)

$$R_B(x_{out}|x_{in}) = \frac{3}{4} \int_{-1}^{1} d\mu R(x_{out}|\mu, x_{in})\left(1 + \mu^2\right),$$

Fig. 1.21 The conditional
PDF $R(x_{out}|x_{in})$, obtained by
marginalizing $R(x_{out}|x_{in}, \mu)$
(shown in Fig. 1.20) over μ,
is shown for isotropic
scattering (*red solid lines*,
$R_A(x_{out}|x_{in})$) and dipole
scattering (*black dotted lines*,
$R_B(x_{out}|x_{in})$) for $T = 10^4$
K. For this gas temperature,
$R(x_{out}|x_{in})$ is very similar
for isotropic and dipole
scattering

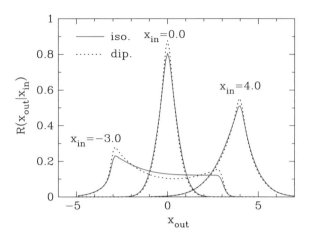

where subscript 'A' ('B') refers to isotropic (dipole) scattering. Figure 1.21 shows
type-A and type-B frequency redistribution functions for $x_{in} = -3.0, 0.0$ and 4.0.
Because Lyα scattering is often a superposition of these two, the actual directionally
averaged redistribution functions are intermediate between the two cases. As these
two cases agree quite closely, they both provide a decent description of actual Lyα
scattering.

There are three important properties of these redistribution functions that play an
essential role in radiative transfer calculations. We discuss these next.

1. Photons that scatter in the wing of the line are pushed back to the line core by an
   amount $-\frac{1}{x_{in}}$ [197], i.e.

$$\boxed{\langle \Delta x|x_{in}\rangle = -\frac{1}{x_{in}}}. \tag{1.73}$$

Demonstrating this requires some calculation. The expectation value for Δx per
scattering event is given by

$$\langle \Delta x|x_{in}\rangle \equiv \int_{-\infty}^{\infty} \Delta x\, R(x_{out}|x_{in})dx_{out}$$

$$= \frac{1}{2}\int_{-\infty}^{\infty} dx_{out} \int_{-1}^{1} d\mu\, \Delta x P(\mu)R(x_{out}|\mu, x_{in}), \tag{1.74}$$

where the factor of $\frac{1}{2}$ reflects that $\int_{-1}^{1} P(\mu)d\mu = 2$ (see Eq. 1.41). For simplicity
we will assume isotropic scattering, for which $P(\mu) = 1$. We previously presented
an expression for $R(x_{out}|\mu, x_{in})$ (see Eq. 1.71). Substituting this expression into
the above equation yields

$$\langle \Delta x | x_{\text{in}} \rangle = \frac{a_v}{2\pi^{3/2}\phi(x_{\text{in}})} \int_{-1}^{1} \frac{d\mu}{\sqrt{1-\mu^2}} \int_{-\infty}^{\infty} \frac{du \exp(-u^2)}{(x_{\text{in}}-u)^2+a_v^2}$$

$$\times \int_{-\infty}^{\infty} d\Delta x \; \Delta x \; \exp\left[- \left(\frac{\Delta x - u(\mu-1)}{\sqrt{1-\mu^2}} \right)^2 \right], \tag{1.75}$$

where we replaced the integral over x_{out} with an integral over Δx. Note that the term in *blue* also represents a 'standard' integral, which equals[31] $u(\mu - 1)\sqrt{\pi(1 - \mu^2)}$. The factor containing μ's can be taken outside of the integral over $d\Delta x$. The integral over μ simplifies to $\int_{-1}^{1} d\mu(\mu - 1) = -2$, and we are left with a single integral over u. We will further assume that $|x_{\text{in}}| \gg 1$, in which case we get

$$\langle \Delta x | x_{\text{in}} \rangle \sim - \frac{a_v}{\pi\phi(x_{\text{in}})} \int_{-\infty}^{\infty} du \frac{u \exp(-u^2)}{(x_{\text{in}} - u)^2} \sim - \frac{a_v}{\pi\phi(x_{\text{in}})x_{\text{in}}^2}$$

$$\times \int_{-\infty}^{\infty} du \; u \exp(-u^2)\Big(\underbrace{1}_{\text{odd} =0} + \underbrace{2\frac{u}{x_{\text{in}}}}_{\text{even} \neq 0} \Big)$$

$$= - \frac{2a_v}{\pi\phi(x_{\text{in}})x_{\text{in}}^3} \int_{-\infty}^{\infty} du \; u^2 \exp(-u^2) = -\frac{2a_v}{\pi\phi(x_{\text{in}})x_{\text{in}}^3} \frac{1}{2}\sqrt{\pi} = -\frac{1}{x_{\text{in}}}, \tag{1.76}$$

where the minus sign appeared after performing the integral over μ. In the last step we used that $\phi(x_{\text{in}}) = a_v/(\sqrt{\pi}x_{\text{in}}^2)$ in the wing of the line profile. This photon is more likely absorbed by atoms that are moving towards the photon, as it would appear closer to resonance for these atoms (i.e. $P(u|x_{\text{in}})$ is larger for those photons with $\mathbf{v} \cdot \mathbf{k}_{\text{in}} < 0$). This photon is likely to experience a Doppler boost to a higher frequency in the frame of the atom. Isotropic (or dipole) re-emission of the photon—on average—conserves this enhancement of the photon's frequency, which pushed it closer to resonance.

This result is very important: it implies that as a Lyα photon is far in the wing at x_{in}, resonant scattering exerts a 'restoring force' which pushes the photon back to line resonance. This restoring force generally overwhelms the energy losses resulting from atomic recoil: Eq. 1.65 indicates that recoil introduces a much smaller average $\Delta x \sim -2.6 \times 10^{-4}(T/10^4 \text{ K})^{-1/2}$, i.e. if a photon finds itself on the red side of line center, then the restoring force pushes the photon back more to line center than recoil pulls it away from it.

2. The r.m.s change in the photon's frequency as it scatters corresponds to 1 Doppler width [197].

$$\boxed{\sqrt{\langle \Delta x^2 | x_{\text{in}} \rangle} = 1}. \tag{1.77}$$

This can be derived with a calculation that is very similar to the above calculation.

3. $R(x_{\text{out}}|x_{\text{in}}) = R(x_{\text{in}}|x_{\text{out}})$. This can be verified by substituting $u = y - \Delta x$ into Eq. 1.71. After some algebra one obtains an expression in which y replaces u, in which x_{out} replaces x_{in}, and in which x_{out} replaces x_{in}.

---

[31]Namely that $\int_{-\infty}^{\infty} dx \; x \exp(-a[x - b]^2) = b\sqrt{\pi/a}$ with $a = (1 - \mu^2)^{-1}$ and $b = u(\mu - 1)$.

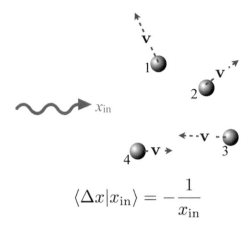

$$\langle \Delta x | x_{\rm in} \rangle = -\frac{1}{x_{\rm in}}$$

Fig. 1.22 Visual explanation of why resonant scattering tends to push Lyα photons far in the wing of the line profile back to the line core. This represents a Lyα photon far in the red wing of the line profile (i.e. $x_{\rm in} \ll 3$). This photon is more likely to be scattered by atoms that are moving towards the photon (in this case atom number 3), as it appears closer to resonance to them. Scattering by these atoms enhances the photon's frequency, and therefore pushes it back towards the core (this is illustrated in Fig. 1.22). The average change in photon frequency is $\langle \Delta x | x_{\rm in} \rangle = -x_{\rm in}^{-1}$

These three properties of the redistribution functions offer key insights into the Lyα radiative transfer problem.

1.6.4 IV: 'Destruction' Term

Absorption of a Lyα photon by a hydrogen atom is generally followed by re-emission of the Lyα photon. Moreover, Lyα photons can be absorbed by something different than hydrogen atoms, which also leads to their destruction. We briefly list the most important processes below:

1. **Dust**. Dust grains can absorb Lyα photons. The dust grain can *scatter* the Lyα photon with a probability which is given by its albedo, $A_{\rm d}$. Dust plays an important role in Lyα radiative transfer, and we will return to this later (see Sect. 1.9.1). Absorption of a Lyα photon by a dust grain increases its temperature, which causes the grain to re-radiate at longer wavelengths, and thus to the destruction of the Lyα photon. This process can be included in the radiative transfer equation by replacing

$$n_{\rm HI}\sigma_{\alpha,0}\phi(x) \rightarrow n_{\rm HI}[\sigma_{\alpha,0}\phi(x) + \sigma_{\rm dust}(x)], \tag{1.78}$$

where $\sigma_{\rm dust}(x)$ denotes the *total* dust cross-section at frequency x, i.e. $\sigma_{\rm dust}(x) = \sigma_{\rm dust,a}(x) + \sigma_{\rm dust,s}(x)$, where the subscript 'a' ('s') stands for 'absorption' ('scattering'). Equation 1.78 indicates that the cross-section $\sigma_{\rm dust}(x)$ is a cross-section *per hydrogen atom*. This definition implies that $\sigma_{dust}(x)$ is not just a property of the dust grain, as it must also depend on the number density of dust grains (if there were no dust grains, then we should not have to add any term). In addition to this, the dust absorption cross section $\sigma_{dust,s}(x)$ (and also the albedo A_d) must depend on the dust properties. For example, [153] shows that $\sigma_{dust} = 4 \times 10^{-22}(Z_{gas}/0.25Z_\odot)$ cm^{-2} for SMC type dust (dust with the same proper-ties as found in the Small Magellanic Cloud), and $\sigma_{dust} = 7 \times 10^{-22}(Z_{gas}/0.5Z_\odot)$ cm^{-2} for LMC (Large Magellanic Cloud) type dust. Here, Z_{gas} denotes the metal-licity of the gas. This parametrization of σ_{dust} therefore assumes that the number density of dust grains scales linearly with the overall gas metallicity, which is a good approximation for $Z \gtrsim 0.3Z_\odot$, but for $Z \lesssim 0.3Z_\odot$ the scatter in dust-to-gas ratio increases (e.g. [78, 218, 235]). Figure 1.23 shows σ_{dust} for dust properties inferred for the LMC (*solid line*), and SMC (*dashed line*). This Figure shows that the frequency dependence of the dust absorption cross section around the Lyα resonance is weak, and in practise it can be safely ignored. Interestingly, as we will discuss in Sect. 1.7.3, dust can have a highly frequency dependent impact on the Lyα radiation field, in spite of the weak frequency dependence of the dust absorption cross-section.

2. **Molecular Hydrogen**. Molecular hydrogen has two transitions that lie close to the Lyα resonance: (*a*) the $v = 1 - 2P(5)$ transition, which lies $\Delta v = 99$ km s^{-1} redward of the Lyα resonance, and (*b*) the $1 - 2R(6)$ transition which lies $\Delta v = 15$ km s^{-1} redward of the Lyα resonance. Vibrationally excited H_2 may therefore convert Lyα photons into photons in the H_2 Lyman bands ([189], and references therein), and thus effectively destroy Lyα. This process can be included in a way that is very similar to that of dust, and by including

$$n_{HI}\sigma_{\alpha,0}\phi(x) \rightarrow n_{HI}[\sigma_{\alpha,0}\phi(x) + f_{H_2}\sigma_{H_2}(x)], \quad (1.79)$$

where $f_{H_2} \equiv n_{H_2}/n_{HI}$ denotes the molecular hydrogen fraction. This destruction process is often overlooked, but it is important to realize that Lyα can be destroyed efficiently by molecular hydrogen. Neufeld [189] provides expressions for frac-tion of Lyα that is allowed to escape as a function of f_{H_2} and HI column density N_{HI}.

3. **Collisional Mixing of the 2s and 2p Levels**. Lyα absorption puts a hydrogen atom in its $2p$ state, which has a life-time of $t = A_\alpha^{-1} \sim 10^{-9}$ s. During this short time, there is a finite probability that the atom interacts with nearby electrons and/or protons. These interactions can induce transitions of the form $2p \rightarrow 2s$. Once in the $2s$-state, the atom decays back to the ground-state by emitting two pho-tons. This process is known as collisional deexcitation of the $2p$ state. Collisional de-excitation from the $2p$ state becomes more probable at high gas densities, and is predominantly driven by free protons. The probability that this process destroys the Lyα photon, p_{dest}, at any scattering event is given by $p_{dest} = \frac{n_p C_{2p2s}}{n_p C_{2p2s} + A_\alpha}$. Here, n_p denotes the number density of free protons, and $C_{2p2s} = 1.8 \times 10^{-4}$ cm^3 s^{-1} (e.g. and references therein [61]) denotes the collisional rate coefficient. This

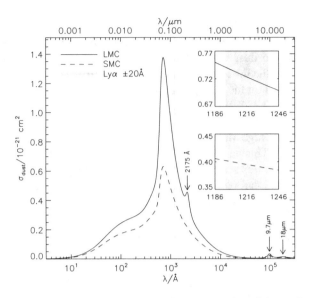

Fig. 1.23 This Figure shows grain averaged absorption cross section of dust grains *per hydrogen atom* for SMC/LMC type dust (*solid/dashed line*, see text). The *inset* shows the cross section in a narrower frequency range centered on Lyα, where the frequency dependence depends practically linearly on wavelength. However, this dependence is so weak that in practise it can be safely ignored (*Credit from Fig. 1 of* [154] ©*AAS. Reproduced with permission.*). We we discuss in Sect. 1.7.3, in spite of the weak frequency dependence of the dust absorption cross-section around the Lyα resonance, dust can have a highly frequency dependent impact in Lyα spectra

process can be included by rescaling the scattering redistribution function (see Sect. 1.6.3) to

$$R(x_{\text{out}}|x_{\text{in}}) \rightarrow R(x_{\text{out}}|x_{\text{in}}) \times (1 - p_{\text{dest}}). \tag{1.80}$$

This ensures that for each scattering event, there is a finite probability (p_{dest}) that a photon is destroyed.

4. **Other**. There are other processes that can destroy Lyα photons, but which are less important. These can trivially be included in Monte-Carlo codes that describe Lyα radiative transfer (see Sect. 1.8). These processes include: (*i*) Lyα photons can photoionize hydrogen atoms *not* in the ground state. The photoionisation cross-section from the $n = 2$ level by Lyα photons is $\sigma_{\text{ion}}^{\text{Ly}\alpha} = 5.8 \times 10^{-19}$ cm^2 (e.g. [57], p 108). This requires a non-negligible populations of atoms in the $n = 2$ state which can occur in very dense media (see e.g. [23]) (*ii*) Lyα photons can detach the electron from the H$^-$ ion. The cross-section for this process is $\sigma = 5.9 \times 10^{-18}$ cm^{-2} (e.g. [240]) for Lyα photons, which is almost an order of magnitude larger than the photoionisation cross-section from the $n = 2$ level at the Lyα frequency. So, unless the H$^-$ number density exceeds $0.1[n_{2\text{p}} + n_{2\text{s}}]$, where $n_{2\text{s}/2\text{p}}$ denotes the number density of H-atoms in the 2s/2p state, this process is not important.

1.6.5 Lyα Propagation Through HI: Scattering as Double Diffusion Process

Scattering of Lyα photons is often compared to a diffusion process, in which the photons undergo a random walk in space and frequency as they scatter off H atoms. Indeed, known analytic solutions to the radiative transfer equation are possible only under certain idealized scenarios (which are discussed below), for which the radiative transfer equation transforms into a diffusion equation. To demonstrate how this transformation works, and to gain some insight into this diffusion process we rewrite the transfer equation (Eq. 1.58) as a diffusion equation. We first simplify Eq. 1.58 in a number of steps:

1. First, we assume that the Lyα radiation field is isotropic, i.e. scattering completely eliminates any directional dependence of $I_\nu(\mathbf{n})$. Under this assumption we can replace the intensity $I_\nu(\mathbf{n})$ with the angle-averaged intensity $J_\nu \equiv \frac{1}{4\pi} \int d\Omega I_\nu(\mathbf{n})$.
2. Second, we replace frequency ν with the dimensionless frequency variable x, introduced in Sect. 1.5.4.
3. Third, we ignore destruction processes and rewrite $\alpha_\nu^{\mathrm{HI}}(s) = n_{\mathrm{HI}}(s)\sigma_\alpha(\nu) = n_{\mathrm{HI}}\sigma_{\alpha,0}\phi(x)$.
4. Fourth, we define $d\tau = n_{\mathrm{HI}}(s)\sigma_{\alpha,0}ds$ and obtain

$$\frac{\partial J(x)}{\partial \tau} = -\phi(x)J(x) + S_x(\tau) + \int dx'\phi(x')J(x')R(x|x'), \qquad (1.81)$$

where $S_x(\tau) \equiv j_x(\tau)/(n_{\mathrm{HI}}\sigma_{\alpha,0})$ denotes the 'source' function. In the integral x denotes the frequency of the photon after scattering, and x' denotes the frequency of the photon before scattering. Equation 1.81 is an integro-differential equation, which is notoriously difficult to solve.

Fortunately, this integro-differential can be transformed into a diffusion equation by Taylor expanding $J(x')\phi(x')$ around x, as we demonstrate next. To keep our notation consistent with Sect. 1.6.3, we denote the photon frequency before (after) scattering with x_{in} (x_{out}). To shorten the notation, we define $f(x) \equiv J(x)\phi(x)$. We would like to rewrite $f(x_{\mathrm{in}})$ as

$$f(x_{\mathrm{in}}) = f(x_{\mathrm{out}}) + (x_{\mathrm{in}} - x_{\mathrm{out}})\frac{\partial f}{\partial x} + \frac{1}{2}(x_{\mathrm{in}} - x_{\mathrm{out}})^2\frac{\partial^2 f}{\partial x^2} + \cdots =$$
$$f(x_{\mathrm{out}}) - \Delta x\frac{\partial f}{\partial x} + \frac{1}{2}\Delta x^2\frac{\partial^2 f}{\partial x^2} + \cdots, \qquad (1.82)$$

where the derivatives are evaluated at x_{out}. Dropping terms that contain Δx^3, Δx^4, ..., we can write the scattering term as

$$\int dx_{\text{in}} f(x_{\text{in}}) R(x_{\text{out}}|x_{\text{in}}) \approx \int_{-\infty}^{\infty} dx_{\text{in}} R(x_{\text{out}}|x_{\text{in}}) \left(f(x_{\text{out}}) - \Delta x \frac{\partial f}{\partial x} + \frac{1}{2} \Delta x^2 \frac{\partial^2 f}{\partial x^2} + \cdots \right) =$$

$$\underset{R(x_{\text{in}}|x_{\text{out}})=R(x_{\text{out}}|x_{\text{in}})}{=} f(x_{\text{out}}) \underbrace{\int_{-\infty}^{\infty} dx_{\text{in}} R(x_{\text{in}}|x_{\text{out}})}_{=1} - \frac{\partial f}{\partial x} \underbrace{\int_{-\infty}^{\infty} dx_{\text{in}} \Delta x R(x_{\text{in}}|x_{\text{out}})}_{\langle \Delta x | x_{\text{in}} \rangle = -\frac{1}{x_{\text{out}}}, \text{ Eq } (1.76)}$$

$$+ \frac{1}{2} \frac{\partial^2 f}{\partial x^2} \underbrace{\int_{-\infty}^{\infty} dx_{\text{in}} \Delta x^2 R(x_{\text{in}}|x_{\text{out}})}_{\langle \Delta x^2 | x_{\text{in}} \rangle = 1, \text{ Eq } (1.77)} = f(x_{\text{out}}) + \frac{1}{x_{\text{out}}} \frac{\partial f}{\partial x} + \frac{1}{2} \frac{\partial^2 f}{\partial x^2} \qquad (1.83)$$

This term can be further simplified by replacing $f(x_{\text{out}})$ with $\phi(x_{\text{out}}) J(x_{\text{out}})$ and taking the derivatives. Note that we derived Eqs. 1.77 and 1.76 assuming that $|x| \gg 1$. This same assumption allows us to further simplify the equation as we see below. For brevity we only write ϕ for $\phi(x_{\text{out}})$ etc. The term then becomes

$$J\phi + \frac{\phi}{x_{\text{out}}} \frac{\partial J}{\partial x} + \frac{J}{x_{\text{out}}} \frac{\partial \phi}{\partial x} + \frac{1}{2} \frac{\partial}{\partial x} \left(\phi \frac{\partial J}{\partial x} + J \frac{\partial \phi}{\partial x} \right) = J\phi - \frac{\phi}{x_{\text{out}}} \frac{\partial J}{\partial x} + \frac{J\phi}{x_{\text{out}}^2} + \frac{\phi}{2} \frac{\partial^2 J}{\partial x^2} = \quad (1.84)$$

$$J\phi \left(1 + \frac{1}{x_{\text{out}}^2} \right) + \frac{1}{2} \frac{\partial}{\partial x} \phi \frac{\partial J}{\partial x} \approx J\phi + \frac{1}{2} \frac{\partial}{\partial x} \phi \frac{\partial J}{\partial x},$$

where we used that $\frac{\partial \phi}{\partial x} = -\frac{2\phi}{x}$, $\frac{\partial^2 \phi}{\partial x^2} = \frac{6\phi}{x^2}$ (see Eq. 1.55), and that $|x_{\text{out}}| \gg 1$ in the wing of the line profile. When we substitute this back into Eq. 1.81, and revert to its original notation, we finally find that [224]

$$\boxed{\frac{\partial J}{\partial \tau} = \frac{1}{2} \frac{\partial}{\partial x} \phi(x) \frac{\partial J}{\partial x} + S_x(\tau)}. \qquad (1.85)$$

This equation corresponds to a diffusion equation with a diffusion coefficient $\phi(x)$ (see https://en.wikipedia.org/wiki/Diffusion_equation), where τ takes the role of the time variable and x denotes the role of space variable. Equation 1.85 therefore states that as photons propagate outwards (as a result of scattering), they diffuse in frequency direction (but with a slight tendency for the photons to be pushed back to the line core). As photons diffuse further into the wings of the absorption line profile, their mean free path increases, which in turn increases their escape probability.

1.7 Basic Insights and Analytic Solutions

In this section, we show that the concept of Lyα transfer as a diffusion process in real and frequency space can offer intuitive insights into some basic aspects of the Lyα transfer process. These aspects include (*i*) how many times a Lyα photon scatters, (*ii*) how long it takes to escape, and (*iii*) the emerging spectrum for a static, uniform medium in Sect. 1.7.1, for an outflowing/contracting uniform medium in Sect. 1.7.2, and for a multiphase medium in Sect. 1.7.3.

1.7.1   *Lyα Transfer Through Uniform, Static Gas Clouds*

We consider a source of Lyα photons in the center of a static, homogeneous sphere, whose line-center optical depth from the center to the edge equals τ_0, where τ_0 is extremely large, say $\tau_0 = 10^7$. This line-centre optical depth corresponds to an HI column density of $N_{HI} = 1.6 \times 10^{20}$ cm^{-2} (see Eq. 1.54). We further assume that the central source emits all Lyα photons at line center (i.e. $x = 0$). As the photons resonantly scatter outwards, they diffuse outward in frequency space. Figure 1.24 illustrates that as the photons diffuse outwards in real space, the spectral energy distribution of Lyα flux, $J(x)$, broadens. If we were to measure the spectrum of Lyα photons crossing some arbitrary radial shell, then we would find that $J(x)$ is constant up to $\sim \pm x_{max}$ beyond which it drops off fast. For Lyα photons in the core of the line profile, the mean free path is negligible compared to the size of the sphere: the mean free path at frequency x is $[\tau_0 \phi(x)]^{-1}$ *in units of the radius of the sphere*. Because each scattering event changes the frequency of the Lyα photon, the mean free path of each photon changes with each scattering event. From the shape of the redistribution function (see Fig. 1.21) we expect that on rare occasions Lyα photons will be scattered further from resonance into the wing of the line (i.e. $|x| \gtrsim 3$). In the wing, the mean free path of the photon increases by orders of magnitude.

From the redistribution function we know that—once in the wing—there is a slight tendency to be scattered back into the core of the line profile. Specifically, we showed that $\langle \Delta x | x_{in} \rangle = -\frac{1}{x_{in}}$). We therefore expect photons that find themselves in the wing of the line profile, at frequency x, to scatter $N_{scat} \sim x^2$ times before returning to

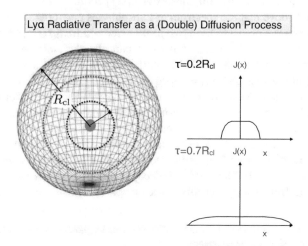

Fig. 1.24 Lyα scattering can be described as a double diffusion process, in which photons diffuse in frequency space as they scatter. In this case photons are emitted in the center of the sphere, at line center. As photons scatter outwards, the spectrum of Lyα photons broadens. Analytic solutions to the Lyα transfer equation indicate that the spectrum remains remarkably constant over a range of frequencies (this range increases as we move outward from the center of the sphere)

the core. During this 'excursion' back to the core, the photon will diffuse a distance $D \sim \sqrt{N_{scat}} \times \lambda_{mfp}(x) \approx \sqrt{N_{scat}}/[\tau_0\phi(x)]$ away from the center of the sphere (recall that this is in units of the radius of the sphere). If we now set this displacement equal to the size of the sphere, i.e. $D = \sqrt{N_{scat}}/[\tau_0\phi(x)] = x/[\tau_0\phi(x)] = 1$, and solve for x using that $\phi(x) = a_v/[\sqrt{\pi}x^2]$, we find [2, 114, 189]

$$\boxed{x_p = \pm[a_v\tau_0/\sqrt{\pi}]^{1/3}}, \qquad (1.86)$$

where x_p denotes the frequency at which the emerging spectrum peaks. Photons that are scattered to frequencies[32] $|x| < |x_p|$ will return to line center before they escape from the sphere (where they have negligible chance to escape). Photons that are scattered to frequencies $|x| > |x_p|$ can escape more easily, but there are fewer of these photons because: (i) it is increasingly unlikely that a single scattering event displaces the photon to a larger $|x|$, and (ii) photons that wish to reach $|x| \gg |x_p|$ through frequency diffusion via a series of scattering events are likely to escape from the sphere before they reach this frequency. We can also express the location of the two spectral peaks at $\pm x_p$ in terms of a velocity off-set and an HI column density as

$$\Delta v_p = |x_p|v_{th} \approx 160\left(\frac{N_{HI}}{10^{20}\ cm^{-2}}\right)^{1/3}\left(\frac{T}{10^4\ K}\right)^{1/6}\ km\ s^{-1}. \qquad (1.87)$$

That is, the full-width at half maximum of the Lyα line can exceed $2\Delta v_p \sim 300$ km s^{-1} for a static medium in which the thermal velocity dispersion of the atoms is only ~ 10 km s^{-1}. Lyα scattering thus broadens spectral lines, which implies that we must exercise caution when interpreting observed Lyα spectra.

We showed above that the spectrum of Lyα photons emerging from the center of an extremely opaque object to have two peaks at $x_p \sim \pm[a_v\tau_0/\sqrt{\pi}]^{1/3}$. More generally, $x_p = \pm k[a_v\tau_0/\sqrt{\pi}]^{1/3}$, where k is a constant of order unity which depends on geometry (i.e. $k = 1.1$ for a slab [114, 189], and $k = 0.92$ for a sphere [63]). This derivation required that photons escaped in a *single excursion*[33]: that is, photons must have been scattered deep enough into the wing (which starts for $|x| > 3$, see Fig. 1.17) to be able to undergo a non-negligible number of wing scattering events before returning to core. So formally our analysis is valid only when $x \gg 3$ when the Lyα photons first start their excursion. Another way of phrasing this requirement is that $x_p \gg 3$, or—when expressed in terms of an optical depth τ_0—when $a_v\tau_0 = \sqrt{\pi}(x_p/k)^3 \gtrsim 1600(x_p/10)^3$. Indeed, analytic solutions of the full spectrum emerging from static optically thick clouds appear in good agreement with

---

[32] Apart from a small recoil effect that can be safely ignored [1], photons are equally likely to scatter to the red and blue sides of the resonance.

[33] Escape in a 'single excursion' can be contrasted with escape in a 'single flight': gases with lower N_{HI} can become optically thin to Lyα photons when they first scattered into the wing of the line profile. For example, gas with $N_{HI} = 10^{17}$ cm^{-2} has a line center optical depth $\tau_0 = 5.9 \times 10^3(T/10^4\ K)^{-1/2}$ (Eq. 1.54). However, Fig. 1.17 shows that the cross-section is $\gtrsim 4$ orders of magnitude smaller when $|x| \gtrsim 3$. A photon that first scattered into the wing would be free to escape from this gas without further scattering.

Key quantities in Lyα transfer.

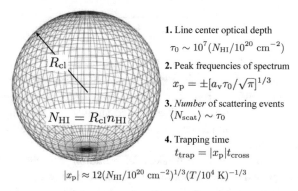

1. Line center optical depth
$$\tau_0 \sim 10^7 (N_{HI}/10^{20}\ cm^{-2})$$

2. Peak frequencies of spectrum
$$x_p = \pm [a_v \tau_0 / \sqrt{\pi}]^{1/3}$$

3. *Number* of scattering events
$$\langle N_{scat} \rangle \sim \tau_0$$

4. Trapping time
$$t_{trap} = |x_p| t_{cross}$$

$$|x_p| \approx 12(N_{HI}/10^{20}\ cm^{-2})^{1/3}(T/10^4\ K)^{-1/3}$$

Fig. 1.25 Lyα radiative transfer through a static, uniform sphere has several interesting features. We denote the line-center optical depth from the center to the edge of the sphere with τ_0. Photons emitted in the center of the sphere diffuse in frequency space as they scatter outward, and emerge with a characteristic double peaked spectrum, for which the peaks occur at $x_p = \pm [a_v \tau_0 / \sqrt{\pi}]^{1/3}$. It takes an average of $N_{scat} \sim \tau_0$ scattering events for a Lyα photon to escape, which *traps* the photon inside the cloud for a time $t_{trap} = |x_p| t_{cross}$ (where $t_{cross} = R/c$, $|x_p| \approx 12(N_{HI}/10^{20}\ cm^{-2})^{1/3}(T/10^4\ K)^{-1/3}$)

full Monte-Carlo calculations (see Sect. 1.8) when $a_v \tau_0 \gtrsim 1000$ (e.g. [63, 189]) (Fig. 1.25).

There are other interesting aspects of Lyα transfer that we can discuss: the first is the mean number of scattering events that each Lyα photon undergoes, N_{scat}. This number of scattering events was calculated by [1], after observing that his numerical results implied that the Lyα spectrum of photons was flat within some frequency range centered on $x = 0$ (as in Fig. 1.24), where the range increases with distance from the center of the sphere. Adams [1] noted that for a flat spectrum, the scattering rate at frequency x is $\propto \phi(x)$. Since partially coherent scattering does not change the frequency much (recall that $\sqrt{\langle \Delta x^2 | x_{in} \rangle} = 1$, see Eq. 1.77), we expect that if we pick a random photon after a scattering event, then the probability that this photon lies in the range $x \pm dx/2$ equals $\phi(x)dx$ (this means that we are more likely to pick a photon close to the core, which simply reflects that these photons scatter more). Adams [1] then noted that photons at x typically scatter x^2 times before returning to the core, and argues that the probability that a photon *first* scattered into the frequency range $x \pm dx/2$ equals $\phi(x)dx/x^2$. The probability that a photon scatters to some frequency which allows for escape in a single excursion then equals

$$P_{esc} = \int_{-\infty}^{-x_p} \frac{\phi(x)dx}{x^2} + \int_{x_p}^{\infty} \frac{\phi(x)dx}{x^2} = 2\int_{x_p}^{\infty} \frac{\phi(x)dx}{x^2} \underset{\text{wing}}{\approx} \frac{2a_v}{3\sqrt{\pi}x_p^3} \underset{\text{Eq (1.86)}}{=} \frac{2}{3\tau_0 k^3}.$$
$$(1.88)$$

We thus expect photons to escape after $N_{\text{scat}} = 1/P_{\text{esc}}$ scattering events, which equals

$$\boxed{N_{\text{scat}} = C\tau_0}, \tag{1.89}$$

where $C \equiv \frac{3k^3}{2\sqrt{\pi}} \approx 1.1$ for a slab, and $C \approx 0.6$ for a sphere. This is an important result: Lyα photons typically scatter $\sim \tau_0$ times before escaping from extremely opaque media. This differs from standard random walks where a photon would scatter $\propto \tau^2$ times before escaping. The reduced number of scattering events is due to frequency diffusion, which forces photons into the wings of the line, where they can escape more easily.

We can also estimate the time it takes for a photon to escape. We now know that photons escape in a single excursion with peak frequency $x_p = \pm[a_\nu\tau_0/\sqrt{\pi}]^{1/3}$. During this excursion the photon scattered x_p^2 times in the wing of the line profile. We also know it took $N_{\text{scat}} = 0.6\tau_0$ scattering events on average before the excursion started. Generally, $N_{\text{scat}} = 0.6\tau_0 \gg x_p^2$, and the vast majority of scattering events occurred in the line core, where the mean free path was very short. The total distance that the photon travelled while scattering in the core is $D_{\text{core}} = N_{\text{scat}}\lambda_{\text{mfp}} \sim \tau_0 \times \tau_0^{-1} \sim 1$, where we used that the mean free path at line center is τ_0^{-1} (in units of the radius of the sphere). The total distance that was travelled in the wing—i.e. during the excursion—equals $D_{\text{wing}} \sim x_p^2 \times [\tau_0\phi(x_p)]^{-1} = x_p^4\sqrt{\pi}/(a_\nu\tau_0)$. If we now substitute the expression for x_p, and add the distance travelled in the core and in the wing we get

$$D = D_{\text{core}} + D_{\text{wing}} \approx 1 + \left(\frac{a_\nu\tau_0}{\sqrt{\pi}}\right)^{1/3} = 1 + |x_p| \approx |x_p|, \tag{1.90}$$

where we used that $|x_p| \gg 1$, i.e. that $D_{\text{wing}} \gg D_{\text{core}}$. If we express this in terms of travel time, then the vast majority of scattering events take up negligible time. The total time it takes for Lyα photons to escape—or the total time for which they are 'trapped'—thus equals

$$\boxed{t_{\text{trap}} = |x_p|t_{\text{cross}}}, \quad |x_p| \approx 12(N_{\text{HI}}/10^{20}\ \text{cm}^{-2})^{1/3}(T/10^4\ \text{K})^{-1/3}. \tag{1.91}$$

This is an interesting, and often overlooked result: for $\tau_0 \sim 10^7$ we now know that Lyα photons scatter 10 million times on average before escaping. Yet, they are only trapped for ~ 15 light crossing times, which is not long.

The last thing I will only mention is that the diffusion equation (Eq. 1.85) can be solved analytically for the angle averaged intensity $J(x, \tau)$ for static, uniform gaseous spheres, slabs, and cubes. These solutions were presented by [114, 189], who showed (among other things) that the angle-averaged Lyα spectrum emerging from a semi-infinite 'slab' has the following analytic solution:

$$J(x) = \frac{\sqrt{6}}{24\sqrt{\pi}a\tau_0}\left(\frac{x^2}{1 + \cosh\left[\sqrt{\frac{\pi^3}{54}\frac{|x^3|}{a\tau_0}}\right]}\right), \tag{1.92}$$

Fig. 1.26 Lyα spectra emerging from a uniform spherical, static gas cloud surrounding a central Lyα source which emits photons at line centre $x = 0$. The total line-center optical depth, τ_0 increases from $\tau_0 = 10^5$ (*narrow histogram*) to $\tau_0 = 10^7$ (*broad histogram*). The *solid lines* represent analytic solutions (*Credit from Figure* **A4** *of* [196], *Can galactic outflows explain the properties of Lyα emitters? MNRAS, 425, 87*)

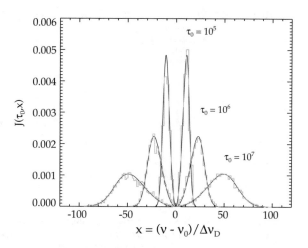

when the photons were emitted in the center of the slab. Here, τ_0 denotes the line center optical depth from the center to the edge of the slab (which extends infinitely in other directions). Similar solutions have been derived for spheres [63] and cubes [257]. The *solid lines* shown in Fig. 1.26 shows analytic solutions for $J(x)$ emerging from *spheres* for $\tau_0 = 10^5$, 10^6 and $\tau_0 = 10^7$. *Histograms* show the spectra emerging from Monte-Carlo simulations of the Lyα radiative transfer process (see Sect. 1.8), which should provide us with the most accurate solution. Figure 1.26 illustrates that analytic solutions derived by [63, 114, 189] approach the true solution remarkably well. The peak flux density for the flux emerging from a slab [sphere] occurs at $x_{max} = \pm 1.1 (a_v \tau_0)^{1/3} \sim x_p$ [$x_{max} = \pm 0.9 (a_v \tau_0)^{1/3}$], where x_p is the characteristic frequency estimated above (see Eq. 1.86).

1.7.2   Lyα Transfer Through Uniform, Expanding and Contracting Gas Clouds

Our previous analysis focussed on static gas clouds. Once we allow the clouds to contract or expand, no analytic solutions are known (except when $T = 0$, see below). We can qualitatively describe what happens when the gas clouds are not static.

Consider an expanding sphere: the predicted spectral line shape must also depends on the outflow velocity profile $v_{out}(r)$. Qualitatively, photons are less likely to escape on the blue side (higher energy) than photons on the red side of the line resonance because they appear closer to resonance in the frame of the outflowing gas. More-over, as the Lyα photons are diffusing outward through an expanding medium, they loose energy because the do 'work' on the outflowing gas [285]. Both these effects combined enhance the red peak, and suppress the blue peak, as illustrated in Fig. 1.27 (taken from [154]). In detail, how much the red peak is enhanced, and the blue peak

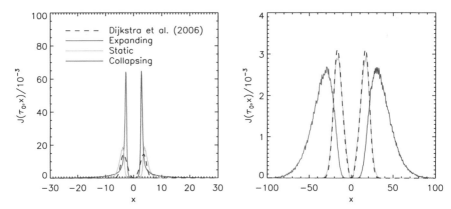

Fig. 1.27 This Figure illustrates the impact of bulk motion of optically thick gas to the emerging Lyα spectrum of Lyα: The *green lines* show the spectrum emerging from a static sphere (as in Fig. 1.26). In the *left/right panel* the HI column density from the centre to the edge of the sphere is $N_{HI} = 2 \times 10^{18}/2 \times 10^{20}$ cm^{-2}. The *red/blue lines* show the spectra emerging from an expanding/a contracting cloud. Expansion/contraction gives rise to an overall redshift/blueshift of the Lyα spectral line (*Credit from Fig. 7 of* [153] ©*AAS. Reproduced with permission*)

is suppressed (and shifted in frequency directions) depends on the outflow velocity and the HI column density of gas.[34] Not unexpectedly, the same arguments outlined above can be applied to a collapsing sphere: here we expect the blue peak to be enhanced and the red peak to be suppressed (e.g. [63, 285]). It is therefore thought that the Lyα line shape carries information on the gas kinematics through which it is scattering. As we discuss in Sect. 1.9.1, the shape and shift of the Lyα spectral line profile has been used to infer properties of the medium through which they are scattering.

There exists one analytic solution to radiative transfer equation through an expanding medium: [162] derived analytic expressions for the radial dependence of the angle-averaged intensity $J(v, r)$ of Lyα radiation as a function of distance r from a source embedded within a neutral intergalactic medium undergoing Hubble expansion, (i.e. $v_{out}(r) = H(z)r$, where $H(z)$ is the Hubble parameter at redshift z). Note that this solution was obtained assuming completely coherent scattering (i.e. $x_{out} = x_{in}$ which corresponds to the special case of $T = 0$), and that the photons frequencies change during flight as a result of Hubble expansion. Formally it describes a somewhat different scattering process than what we discussed before (see for a more detailed discussion [63]). However, [66] have compared this analytic solution to that of a Monte-Carlo code that uses the proper frequency redistribution functions and find good agreement between the Monte-Carlo and analytic solutions.

Finally, it is worth pointing out that in expanding/contracting media, the number of scattering events N_{scat} and the total trapping time both decrease. The main reason

---

[34]Max Gronke has developed an online tool which allows users to vary column density, outflow/inflow velocity of the scattering medium, and directly see the impact on the emerging Lyα spectrum: see http://bit.ly/man-alpha.

for this is that in the presence of bulk motions in the gas, it becomes easier to scatter Lyα photons into the wings of the line profiles where they can escape more easily. The reduction in trapping time has been quantified by [32], and in shell models (which will be discussed in Sect. 1.9.1 and Fig. 1.38) by Dijkstra and Loeb ([66], see their Fig. 1.6).

1.7.3 Lyα Transfer Through Dusty, Uniform and Multiphase Media

The interstellar medium (ISM) contains dust. A key difference between a dusty and dust-free medium is that in the presence of dust, Lyα photons can be destroyed during the scattering process when the albedo (also known as the reflection coefficient) of the dust grains (see Sect. 1.6.4) $A_d < 1$. Dust therefore causes the 'escape fraction' (f_{esc}^{α}), which denotes the fraction Lyα photons that escape from the dusty medium, to fall below unity, i.e. $f_{esc}^{\alpha} < 1$. For a medium with a uniform distribution of gas and dust, the probability that a Lyα photon is destroyed by a dust grain increases with the distance travelled by the Lyα photon. We learned before that as photons diffuse spatially, they also diffuse in frequency. This implies that *dust affects the emerging Ly α spectral line profile in a highly frequency-dependent way, even though the absorption cross-section for this process is practically a constant over the same range of frequencies.* Specifically, if we compare Lyα spectra emerging from two scattering media which differ only in their dust content, then we find that the impact of the dust is largest at large frequency/velocity off-sets from line center (see Fig. 1.28).

Dust also destroys UV-continuum photons, but because Lyα photons scatter and diffuse spatially through the dusty medium, the impact of dust on Lyα and UV-continuum is generally different. This can affect the 'strength' (i.e. the equivalent width) of the Lyα line compared to the underlying continuum emission. In a uniform mixture of HI gas and dust, Lyα photons have to traverse a larger distance before escaping, which increases the probability to be destroyed by dust. In these cases we expect dust to reduce the EW of the Lyα line. The ISM is not smooth and uniform however, which can drastically affect Lyα radiative transfer. The interstellar medium is generally thought to consist of the 'cold neutral medium' (CNM), the 'warm neutral/ionized medium' (WNM/WIM), and the 'hot ionized medium' (HIM, see e.g. the classical paper by [173]). In reality, the cold gas is not in 'clumps' but rather in a complex network of filaments and sheets. Lyα transfer calculations through realistic ISM models have only just begun, partly because modeling the multiphase nature of the ISM with simulations is a difficult task which requires extremely high spatial resolution. There is substantially more work on Lyα transfer through 'clumpy' media that consists of cold clumps containing neutral hydrogen gas and dust, embedded within a (hot) ionized, dust free medium [77, 113, 136, 190], and which represent simplified descriptions of the multi-phase ISM.

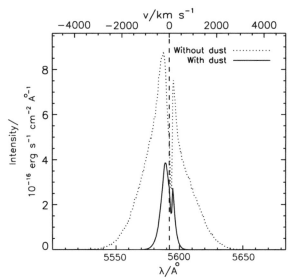

Fig. 1.28 This figure illustrates that dust can suppress the Lyα spectral line profile in a highly frequency dependent way, even though the dust absorption cross-section barely varies over the same frequency range (*Credit from Fig. 8 of* [154] ©*AAS. Reproduced with permission*). The main reason for this is that dust limits the distance Lyα photons can travel before they are destroyed, which limits how much they can diffuse in frequency space. Dust therefore has the largest impact at frequencies furthest from line center

Clumpy media facilitate Lyα escape from dusty interstellar media, and can help explain the detection of Lyα emission from dusty galaxies such as submm galaxies (e.g. [50]), and (U)LIRGs (e.g [126, 193]). In a clumpy medium dust can even *increase* the EW of the Lyα line, i.e. preferentially destroy (non-ionizing) UV continuum photons over Lyα photons: Lyα photons can propagate freely through the 'interclump' medium. Once a Lyα photon enters a neutral clump, its mean free path can be substantially smaller than the clump size. Because scattering is partially coherent, and the frequency of the photon in the clump frame changes only by an r.m.s amount of $\sqrt{\langle \Delta x^2 | x_{in} \rangle} = 1$ (see Eq. 1.77), the mean free path remains roughly the same. Since the photon penetrated the clump by a distance corresponding to $\tau \sim 1$, the photon is able to escape after each scattering event with a significant probability. A Lyα photon that penetrates a clump is therefore likely to escape after ∼5 scattering events [113], rather than penetrate deeper into the clump via a much larger number of scattering events. The Lyα photons effectively scatter off the clump surface (this is illustrated in Fig. 1.29), thus avoiding exposure to dust grains. In contrast, UV continuum photons will penetrate the dusty clumps unobscured by hydrogen and are exposed to the full dust opacity. Clumpy, dusty media may therefore preferentially let Lyα photons escape over UV-continuum.

Laursen et al. [157] and Duval et al. [79] have recently shown however that—while clumpy media facilitate Lyα escape—EW boosting only occurs under physically unrealistic conditions in which the clumps are very dusty, have a large covering

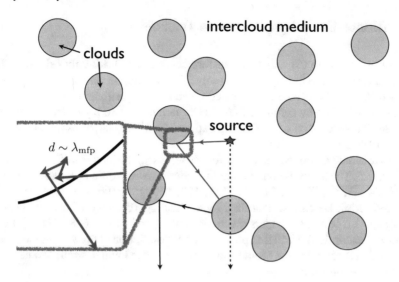

Fig. 1.29 Schematic illustration of how a multiphase medium may favour the escape of Lyα line photons over UV-continuum photons: *Solid/dashed lines* show trajectories of Lyα/UV-continuum photons through clumpy medium. If dust is confined to the cold clumps, then Lyα may more easily escape than the UV-continuum (*Credit from Fig. 1 of* [190] ©*AAS. Reproduced with permission*)

factor, have very low velocity dispersion and outflow/inflow velocities, and in which the density contrast between clumps and interclump medium is maximized. While a multiphase (or clumpy) medium definitely facilitates the escape of Lyα photons from dusty media, EW boosting therefore appears uncommon. We can understand this result as follows: the preferential destruction of UV-continuum photons over Lyα requires at least a significant fraction of Lyα photons to avoid seeing the dust by scattering off the surface of the clumps. How deep the Lyα photons actually penetrate, depends on their mean-free path, which depends on their frequency in the frame of the clumps. If clumps are moving fast, then it is easy for Lyα photons to be Doppler boosted into the wing of the line profile (in the clump frame), and they would not scatter exclusively on the clump surfaces. Finally, we note that the conclusions of [79, 136] apply to the EW-boost averaged over *all* photons emerging from the dusty medium. Gronke and Dijkstra [103] have investigated that for a given model, there can be directional variations in the predicted EW, with large EW boosts occurring in a small fraction of sightlines in directions where the UV-continuum photon escape fraction was suppressed, thus partially restoring the possibility of EW boosting by a multiphase ISM.

1.8   Monte-Carlo Lyα Radiative Transfer

Analytic solutions to the radiative transfer equation (Eq. 1.58) only exist for a few idealised cases. A modern approach to solve this equation is via *Monte-Carlo* methods, which refer to a 'broad class of computational algorithms [...] which change processes described by certain differential equations into an equivalent form interpretable as a succession of random operations' (S. Ulam, see https://en.wikipedia. org/wiki/Monte_Carlo_method).[35]

In Lyα Monte-Carlo radiative transfer, we represent the integro-differential equation (Eq. 1.58) by a succession of random scattering events until Lyα photons escape [5, 22, 24, 41, 63, 75, 85, 91, 146, 151, 153, 160, 162, 196, 210, 239, 256, 269, 279, 285, 286]. Details on how the Monte-Carlo approach works can be found in many papers (see e.g. the papers mentioned above, and Chaps. 6–8 of [155], for an extensive description). I will first provide a brief description of drawing random variables, which is central to the Monte-Carlo method. Then I will describe Monte-Carlo Lyα radiative transfer.

1.8.1   General Comments on Monte-Carlo Methods

Central in Monte-Carlo methods is generating random numbers from probability distributions. If we denote a probability distribution of variable x with $P(x)$, the $P(x)dx$ denotes the probability that x lies in the range $x \pm dx/2$, and we must have $\int_{-\infty}^{\infty} dx P(x) = 1$. The cumulative probability is $C(x) \equiv P(< x) = \int_{-\infty}^{x} P(x')dx'$, where clearly $C(x \to -\infty) = 0$ and $C(x \to \infty) = 1$. We can draw a random x from the distribution $P(x)$ by randomly generating a number (R) between 0 and 1. We then transform $\mathscr{R}$ into x by inverting its cumulative probability distribution $C(x)$, i.e. (Fig. 1.30)

$$\mathscr{R} \equiv C(x) \to \mathscr{R} \equiv \int_{-\infty}^{x} P(x')dx'. \tag{1.93}$$

This implies that we need to (*i*) be able to integrate $P(x)$, and (*ii*) be able to invert the integral equation. Generally, there is no analytic way of inverting Eq. 1.93. One way of randomly drawing x from $P(x)$ is provided by the *rejection method*. This method consists of picking another $T(x)$ which lies above $P(x)$ everywhere, and which we can integrate and invert analytically. We are completely free to pick $T(x)$ however we want. Because $T(x)$ lies above $P(x)$ everywhere, $\int T(x)dx = A$, where $A > 1$. We first generate x_1 from $T(x)$ by generating a random number $\mathscr{R}_1$ between 0 and 1 and inverting

---

[35]The term 'Monte-Carlo' was coined by Ulam and Metropolis as a code-name for their classified work on nuclear weapons (radiation shielding, and the distance that neutrons would likely travel through various materials). 'Monte-Carlo' was the name of the casino where Ulam's uncle had a (also classified) gambling addiction. I was told most of this story over lunch by M. Baes. For questions, please contact him.

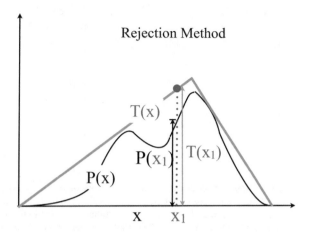

Fig. 1.30 The *rejection method* provides a simple way to randomly draw variables x from arbitrary probability distributions $P(x)$ (see text)

$$\mathcal{R}_1 \equiv \frac{1}{A} \int_{-\infty}^{x_1} T(x')dx'. \tag{1.94}$$

Once we have x_1 we then generate another random number $\mathcal{R}$ between 0 and 1, and accept x_1 as our random pick when

$$\mathcal{R}_2 \leq \frac{P(x_1)}{T(x_1)}. \tag{1.95}$$

If we repeat this procedure a large number of times, and make a histogram for x, we can see that this traces $P(x)$ perfectly (this is illustrated with an example in Fig. 1.31). Equation 1.95 shows that we ideally want $T(x)$ to lie close to $P(x)$ in order not to have to reject the majority of trials (the better the choice for $T(x)$, the smaller the rejected fraction).

1.8.2 Lyα Monte-Carlo Radiative Transfer

Here, we briefly outline the basic procedure that describes the Monte-Carlo method applied to Lyα photons. For each photon in the Monte-Carlo simulation:

1. We first randomly draw a position, **r**, from which the photon is emitted from the emissivity profile[36] $j_\nu(\mathbf{r})$ (see Eq. 1.58). We the assign a random frequency x, which is drawn from the Voigt function $\phi(x)$, and a random propagation direction **k**.

---

[36]For arbitrary gas distributions, the emissivity profile is a 3D-field. We can still apply the rejection method. One way to do this is to discretize the 3D field $j_\nu(x, y, z) \rightarrow j_\nu(i, j, k)$, where we have N_x, N_y, and N_z of cells into these three directions. We can map this 3D-array onto a long 1D array $j_\nu(m)$, where $m = 0, 1, ..., N_x \times N_y \times N_z$, and apply the rejection method to this array.

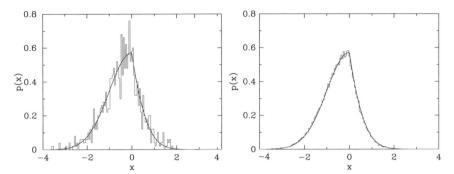

Fig. 1.31 *Red histograms* show the distributions of 10^3 (*left panel*) and 10^5 (*right panel*) randomly drawn values for x from its PDF $P(x)$ by using the rejection method. In this case we randomly truncated a Gaussian PDF with $\sigma = 1$ by suppressing the PDF by an arbitrary function $f(x)$ only for $x > 0$ (note that $f(x) < 1$). This PDF is shown as the *black solid line*

2. We randomly draw the optical depth τ the photon propagates into from the distribution $P(\tau) = \exp(-\tau)$.
3. We convert τ into a physical distance s by (generally numerically) inverting the line integral $\tau = \int_0^s d\lambda \, n_{\mathrm{HI}}(\mathbf{r}')\sigma_\alpha(x'[\mathbf{r}'])$, where $\mathbf{r}' = \mathbf{r} + \lambda\mathbf{k}$ and $x' = x - \mathbf{v}(\mathbf{r}') \cdot \mathbf{k}/(v_{\mathrm{th}})$. Here, $\mathbf{v}(\mathbf{r}')$ denotes the 3D *bulk* velocity vector of the gas at position $\mathbf{r}'$. Note that x' is the dimensionless frequency of the photon in the 'local' frame of the gas at $\mathbf{r}'$.
4. Once we have selected the scattering location, we need to draw the *thermal* velocity components of the atom that is scattering the photon (we only need the thermal velocity components, as we work in the local gas frame). As in Sect. 1.6.3, we decompose the thermal velocity of the atom into a direction parallel to that of the incoming photon, $v_{||}$ (or its dimensionless analogue u, see Eq. 1.66), and a 2D-velocity vector perpendicular to $\mathbf{k}$, namely $\mathbf{v}_\perp$. We discussed in Sect. 1.6.3 what the functional form of the conditional probability $P(u|x)$ (see discussion below Eq. 1.68), and apply the rejection method to draw u from this functional form (see the Appendix [285] for a functional form of $T(u|x)$). The 2 components of $\mathbf{v}_\perp$ can be drawn from a Maxwell-Boltzmann distribution (see e.g. [63]).
5. Once we have determined the velocity vector of the atom that is scattering the photon, we draw an outgoing direction of the photon after scattering, $\mathbf{k}_{\mathrm{out}}$, from the phase-function, $P(\mu)$ (see Eq. 1.41 and Sect. 1.5.1). We will show below that this procedure of generating the atom's velocity components and random new directions generates the proper frequency redistribution functions, as well as their angular dependence.
6. Unless the photon escapes, we replace the photon propagation direction and frequency and go back to (1). Once the photon escapes we record information we are interested in such as the location of last scattering, the frequency of the photon, the thermal velocity components of the atom that last scattered the photon, the number of scattering events the photon underwent, the total distance it travelled through the gas, etc.

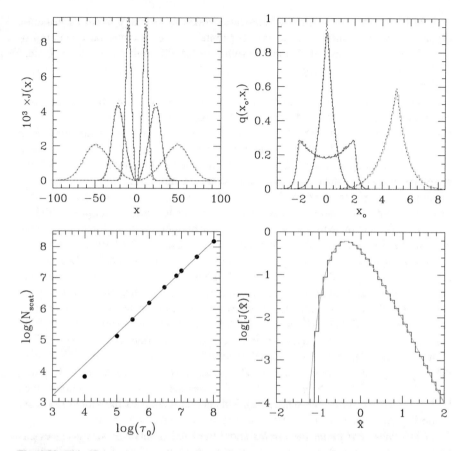

Fig. 1.32 This Figure shows an example set of tests that Monte-Carlo codes must be able pass (*Credit from Fig.* 1 *of* [63] ©*AAS. Reproduced with permission.*). In the *top left panel* show Monte-Carlo calculations of the Lyα spectra emerging from a uniform spherical gas cloud, in which Lyα photons are injected in the center of the sphere, at the line center (i.e. $x = 0$). The total line center optical depth, τ_0 from the center to the edge is $\tau_0 = 10^5$ (*blue*), $\tau_0 = 10^6$ (*red*) and $\tau_0 = 10^7$ (*green*). Overplotted as the *black dotted lines* are the analytic solutions. The agreement is perfect at high optical depth ($a_v\tau \gtrsim 10^3$). *Upper right panel:* The *colored histograms* show the frequency redistribution functions, $R(x_{out}, x_{in})$, for $x_{in} = 0$ (*blue*), $x_{in} = 2$ (*red*) and $x_{in} = 5$ (*green*), as generated by the Monte-Carlo simulation. The *solid lines* are the analytic solutions given by Eq. 1.72. *Lower left panel:* The total number of scattering events that a Lyα photon experiences before it escapes from a slab of optical thickness $2\tau_0$ according to a Monte-Carlo simulation (*circles*). Overplotted as the *red–solid line* is the theoretical prediction by [114]. *Lower right panel:*. The spectrum emerging from an infinitely large object that undergoes Hubble expansion. The histogram is the output from our code, while the *green–solid line* is the (slightly modified) solution obtained by [225] using their Monte Carlo algorithm (see [63] for more details)

It is important to test Lyα Monte-Carlo codes in as many ways as possible. Figure 1.32 shows a minimum set of tests Monte-Carlo codes must be able to reproduce. These comparisons with analytic solutions test different aspects of the code.

The *histograms* in the *top left panel* show Monte-Carlo realizations of the Lyα spectra emerging from a uniform spherical gas cloud, in which Lyα photons are injected in the center of the sphere, at the line center (i.e. $x = 0$). The total line center optical depth, τ_0 from the center to the edge is $\tau_0 = 10^5$ (*blue*), $\tau_0 = 10^6$ (*red*) and $\tau_0 = 10^7$ (*green*). Overplotted as the *black dotted lines* are the corresponding analytic solutions (see Eq. 1.92, but modified for a sphere, see [63]). The agreement is perfect at high optical depth ($a_v \tau \gtrsim 10^3$). At lower optical depth, $a_v \tau_0 \lesssim 10^3$, the analytic solutions are not expected to be accurate any more (see Sect. 1.7.1). Because the analytic solutions were obtained under the assumption that scattering occured in the wing, the agreement between analytic and Monte-Carlo techniques at high τ_0 only confirms that the Monte-Carlo procedure accurately describes scattering in the wing of the line profile. This comparison does not test core scattering, which make up the vast majority of scattering events (see Sect. 1.7.1). The fact that Lyα spectra emerging from optically (extremely) thick media is insensitive to core scattering implies that we need additional tests to test core scattering. However, it also implies we can skip these core-scattering events, which account for the vast majority of all scattering events. That is, Monte-Carlo simulations can be 'accelerated' by skipping core scattering events. We discuss how we can do this in more detail in Sect. 1.8.4.

In the *upper right panel* the colored histograms show Monte-Carlo realization of the frequency redistribution functions, $R(x_{out}, x_{in})$ (see Eq. 1.72), for $x_{in} = 0$ (*blue*), $x_{in} = 2$ (*red*) and $x_{in} = 5$ (*green*). The solid lines are the analytic solutions given by Eq. 1.72 (here for dipole scattering). This comparison tests individual Lyα core scattering events, and thus complements the test we described above.

In the *lower left panel* the *circles* show the total number of scattering events that a Lyα photon experiences before it escapes from a slab of optical thickness $2\tau_0$ in a Monte-Carlo simulation. Overplotted as the *red–solid line* is the theoretical prediction that $N_{scat} = C\tau_0$, with $C = 1.1$ (see Eq. 1.89, [1, 114, 189]). The break down at low τ_0 corresponds to the range of τ_0 where analytic solutions are expected to fail. This test provides another way to test core scattering events, as N_{scat} is set by the probability that a Lyα photon is first scattered sufficiently far into the wing of the line such that it can escape in a single 'excursion'. This test also shows how accurate the analytic prediction is, despite the fact that the derivation presented in Sect. 1.7.1 (following [1]) did not feel like it should be this accurate. This plot also underlines how computationally expensive Lyα transfer can be if we simulate each scattering event.

The *lower right panel* shows Lyα spectrum emerging from a Lyα point source surrounded by an infinitely large sphere that undergoes Hubble expansion. The *black histogram* shows a Monte-Carlo realization. The *green line* represents a 'pseudo-analytic' solution of [162]: as we mentioned earlier in Sect. 1.7.2, [162] provided a fully analytic solution for the angle averaged intensity $J(r, x)$ where r denotes the distance from the galaxy. Unfortunately, their analytic solution does *not* apply to emerging spectrum because as $r \to \infty$, the Lyα photons have redshifted far enough

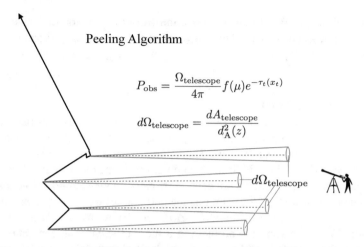

Fig. 1.33 A schematic representation of the *peeling algorithm*. This algorithm allows us to efficiently and accurately extract Lyα observables (spectra, surface brightness profiles, etc) from arbitrary 3D gas distributions. This algorithm overcomes the problem that in Monte-Carlo simulations, the probability that a Lyα photon escapes—*and* into the direction of the telescope such that it lands on the mirror—is practically zero due to the infinitesimally small angular scale of the telescope mirror when viewed from the Lyα source. The peeling algorithm treats each scattering event as a point source with a (scattering induced anisotropic) luminosity $L_\alpha/N_{\rm phot}$, where L_α is the total luminosity of the Lyα source and $N_{\rm phot}$ denotes the number of photons used in the Monte-Carlo run to simulate the source

into the wing that the IGM is optically thin to the Lyα photons, in which case the Lyα radiative transfer problem cannot be reduced to a diffusion problem anymore. Though not shown here, [67] compared the analytic *integrated* $U(r) \propto \int J(r, x)$ to that extracted from Monte-Carlo simulations, and found excellent agreement. The *green line* shown here represents the spectrum[37] obtained from a simplified Monte-Carlo simulation (see [162] for details), and agrees well with the full Monte-Carlo code. This test provides us with a way to test the code when bulk motions in the gas are present (Fig. 1.33).

1.8.3 Extracting Observables from Lyα Monte-Carlo Simulations in 3D Simulations

In Monte-Carlo radiative transfer calculations applied to arbitrary 3D gas distributions, extracting observables requires (a bit) more work than recording the location of last scattering, the photon's frequency etc. This is because formally we are interested only in a tiny subset of Lyα photons that escape, *and* end up in the mir-

---

[37]The frequency $\hat{x}$ here shows another dimensionless frequency variable that was used by [162], and relates to 'standard' dimensionless frequency x as $\hat{x} \sim -3.8 \times 10^3 \, x$ for $T = 10$ K (see [63]).

ror of our telescope. We denote the direction from the location of last scattering towards the telescope with $\mathbf{k}_t$. The mirror of our telescope only subtends a solid angle $d\Omega_{\text{telescope}} = \frac{dA_{\text{telescope}}}{d_A^2(z)}$, where $dA_{\text{telescope}}$ denotes the area of the mirror, and $d_A(z)$ denotes the angular diameter distance to redshift z. The probability that a Lyα photon in our Monte-Carlo simulation escapes from the scattering medium into this tiny solid angle is negligible.

One way to get around this problem is by relaxing the restriction that photons must escape into the solid angle $d\Omega_{\text{telescope}}$ centered on the direction $\mathbf{k}_t$, by increasing $d\Omega_{\text{telescope}}$ to a larger solid angle. This approximation corresponds to averaging over all viewing directions within some angle $\Delta\alpha$ from the real viewing direction. If observable properties of Lyα do not depend strongly on viewing direction, then this approximation is accurate. However, this method implies we are not using information from the vast majority of photons ($\sim \Delta\alpha^2/4\pi$) that we used in the Monte-Carlo simulation. To circumvent this problem in a more efficient (and accurate) way is provided by the so-called 'peeling algorithm'. This algorithm treats each scattering event in the simulation domain as a point source with luminosity L_α/N_{phot}, where L_α is the total Lyα luminosity of the source and N_{phot} denotes the total number of Lyα photons used in the Monte-Carlo simulation to represent this source (see e.g. [256, 282, 285]). The total flux S we expect to get from each point source is

$$S = \frac{L_\alpha}{4\pi d_L^2(z) N_{\text{phot}}} \times \frac{P(\mu_t)}{2} e^{-\tau_t(x_t)}, \tag{1.96}$$

where $d_L(z)$ denotes the luminosity distance to redshift z. Furthermore, $\mu_t \equiv \mathbf{k}_{\text{in}} \cdot \mathbf{k}_t$, in which $\mathbf{k}_{\text{in}}$ (like before) denotes the propagation direction of the photon prior to scattering. The presence of the scattering phase-function $P(\mu)$ reflects that the 'point source' is not emitting isotropically, but that the emission in the direction μ_t is enhanced by a factor of $P(\mu_t)/2$ in direction $\mathbf{k}_t$ (the factor of '2' reflects that $\int d\mu\, P(\mu) = 2$, see Eq. 1.42). The factor $e^{-\tau_t(x_t)}$ denotes the escape fraction in direction $\mathbf{k}_t$, where x_t is the frequency of the photon it *would* have had if it truly had scattered in direction $\mathbf{k}_t$. This frequency x_t can be obtained from Eq. 1.66. Equation 1.96 should be applied for each scattering event: for each scattering event there is a tiny/infinitesimal probablity — $\sim \frac{d\Omega_{\text{telescope}}}{4\pi} e^{-\tau_t(x_t)}$ — that the Lyα scatters into the telescope-mirror. We formally have to reduce the weight of the photon by this probability after each scattering event (we are 'peeling' off the weight of this photon), though in practice this can be ignored because of the tiny probability that a photon scattered into the telescope mirror.

An example of an image generated with the peeling algorithm is shown in Fig. 1.34. Here, a Lyα source is at the origin of a cartesian coordinate system. Each of the 3 coordinate axes has 2 spheres of HI gas at identical distances from the origin. There are no hydrogen atoms outside the sphere, and the Lyα scattering should only occur inside the 6 spheres. Resulting images (taken from Dijkstra and Kramer 2012) from 6 viewing directions are shown in the *left panel* of Fig. 1.34. The 'darkness' of a pixel represents its Lyα surface brightness. The average of these 6 images is

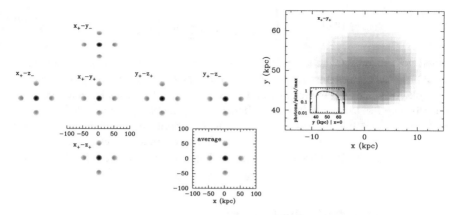

Fig. 1.34 Examples of Lyα images generated with the *peeling algorithm* (see text for a more detailed description, *Credit from Figs.* **A3 and A4** *of* Dijkstra and Kramer 2012, *'Line transfer through clumpy, large-scale outflows: Lyα absorption and haloes around star-forming galaxies', MNRAS, 424, 1672*)

shown in the box. The *right panel* shows a close-up view of the image associated with the sphere on the $+y$-axis. The side of the sphere facing the Lyα source (on the bottom at $y = 0$) is brightest. The *red line* in the *inset* shows an analytic calculation of the expected surface brightness, under the assumption that the sphere as a whole is optically thin (only this assumption allows for analytic solutions). This image shows that the Peeling algorithm gives rise to the sharp features in the surface brightness profiles that should exist. Note that the alternative method we briefly mentioned above, which averages over all viewing directions within some angle $\Delta\alpha$ from $\mathbf{k}_t$, would introduce some blurring to these images.

1.8.4 Accelerating Lyα Monte-Carlo Simulations

As we mentioned above, we do not care about the vast majority of core scattering events for Lyα transfer through extremely optically thick media ($a_\nu\tau_0 \gtrsim 10^3$). We can accelerate Monte-Carlo simulations by 'forcing' photons into the wing of the line profile. Ordinarily, the physical mechanism that puts a Lyα photon from the core into the wing of the line profile is an encounter with a fast moving atom. We can force the scattering atom to have a large velocity when generating its velocity components. A simple way to do this is by forcing the velocity vector of the atom perpendicular to $\mathbf{k}_{in}$, $\mathbf{v}_\perp$, to be large ([6, 63]). We know from Sect. 1.6.3 that $\mathbf{v}_\perp$ follows a 2D Maxwell-Boltzmann distribution $g(\mathbf{v}_\perp)d^2\mathbf{v}_\perp \propto v_\perp \exp(-v_\perp^2)$, where $v_\perp \equiv |\mathbf{v}_\perp|$ is the magnitude of $\mathbf{v}_\perp$. In dimensionless units $u_\perp \equiv v_\perp/v_{th}$, and we have $g(u_\perp)du_\perp = 2\pi u_\perp \exp(-u_\perp^2)/\pi$ (see the discussion under Eq. 1.68). We can force $u_\perp$ to be large by drawing it from a truncated Maxwell-Boltzmann distribution, which states that

$$p(u_\perp)du_\perp = \begin{cases} 0 & |u_\perp| < x_{\text{crit}} \\ \mathcal{N} u_\perp \exp(-u_\perp^2) & |u_\perp| > x_{\text{crit}}. \end{cases}, \qquad (1.97)$$

where $\mathcal{N}$ ensures that the truncated distribution function for $u_\perp$ is normalized. Furthermore, x_{crit} is a parameter that determines how far into the wing we force the Lyα photons. This parameter therefore sets how much Lyα transfer is accelerated. Clearly, one has to be careful when choosing x_{crit}: forcing photons too far into the wing may cause them to escape at frequencies where frequency diffusion would otherwise never take them. Various authors have experimented with choosing x_{crit} based on the local HI-column density of a cell in a simulation, etc [153, 244, 257].

1.9 Lyα Transfer in the Universe

Previous sections discussed the basics of the theory describing Lyα transfer through optically thick media. The goal of this section is to discuss what we know about Lyα transfer in the real Universe. We decompose this problem into several scales: (*i*) Lyα photons have to escape from the interstellar medium (ISM) of galaxies into the circum galactic/intergalactic medium (CGM/IGM). We discuss this in Sect. 1.9.1. We then go to large scales, and describe the subsequent radiative transfer through the CGM/IGM at lower redshift (Sect. 1.9.2) and at higher redshift when reionization is still ongoing (Sect. 1.9.4).

1.9.1 Interstellar Radiative Transfer

[38]Understanding interstellar Lyα radiative transfer requires us to understand gaseous flows in a multiphase ISM, which lies at the heart of understanding star and galaxy formation. Modelling the neutral component of interstellar medium is an extremely challenging task, as it requires resolving the multiphase structure of interstellar medium, and how it is affected by feedback from star-formation (via supernova explosions, radiation pressure, cosmic ray pressure, etc). Instead of taking an 'ab-initio' approach to understanding Lyα transfer, it is illuminating to use a 'top-down' approach in which we try to constrain the broad impact of the ISM on the Lyα radiation field from observations (for this also see the lecture notes by M. Ouchi and M. Hayes for more extended discussions of the observations).

We first focus on observational constraints on the *escape fraction* of Lyα photons, f_{esc}^α. To estimate f_{esc}^α we would need to compare the observed Lyα luminosity to the *intrinsic* Lyα luminosity. The intrinsic Lyα luminosity corresponds to the Lyα luminosity that is actually produced. The best way to estimate the intrinsic Lyα luminosity is from some other non-resonant nebular emission line such as Hα. The observed Hα

---

[38]This discussion represents an extended version of the discussion presented in the review by [75].

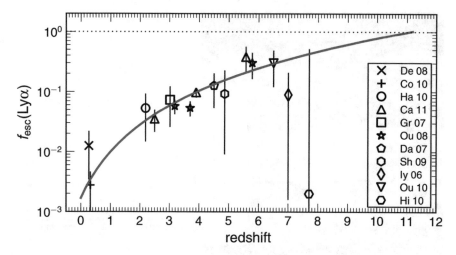

Fig. 1.35 Observational constraints on the redshift-dependence of the volume averaged 'effective' escape fraction, $f_{\mathrm{esc}}^{\mathrm{eff}}$, which contains constraints on the true escape *fraction* $f_{\mathrm{esc}}^{\alpha}$ (*Credit from Fig. 1 of* [121] ©*AAS. Reproduced with permission*)

luminosity can be converted into an intrinsic Hα luminosity once nebular reddening is known (from joint measurements of e.g. the Hα and Hβ lines see lecture notes by M. Hayes). Once the intrinsic Hα luminosity is known, then we can compute the intrinsic Lyα luminosity assuming case-B (or case-A) recombination. This procedure indicates that $f_{\mathrm{esc}}^{\alpha} \sim 1 - 2\%$ at $z \sim 0.3$ [60] and $z \sim 5\%$ at $z \sim 2$ [120].

We do not have access to Balmer series lines at higher redshifts until JWST flies. Instead, it is common to estimate the intrinsic Lyα luminosity from the inferred star formation rate of galaxies (and apply Eq. 1.17 or Eq. 1.18). These star formation rates can be inferred from the dust corrected (non-ionizing) UV-continuum flux density and/or from the IR flux density (e.g. [144]). These analyses have revealed that $f_{\mathrm{esc}}^{\alpha}$ is anti-correlated with the dust-content[39] of galaxies [14, 121, 147]. This correlation may explain why $f_{\mathrm{esc}}^{\alpha}$ increases with redshift from $f_{\mathrm{esc}}^{\alpha} \sim 1$–3 % at $z \sim 0$ [60, 276] to about $f_{\mathrm{esc}}^{\alpha} \sim 30$–50% at $z \sim 6$ ([29, 74, 121], also see Fig. 1.35), as the overall average dust content of galaxies decreases towards higher redshifts (e.g. [35, 88]). It is worth cautioning here that observations are not directly constraining $f_{\mathrm{esc}}^{\alpha}$: Lyα photons that escape from galaxies can scatter frequently in the IGM (or circumgalactic medium) before reaching earth in a low surface brightness glow that cannot be detected yet (see Sect. 1.9.2). These photons would effectively be removed from observations, even though they did escape.

The dependence of $f_{\mathrm{esc}}^{\alpha}$ on dust content of galaxies is an intuitive result, as it is practically the only component of the ISM that is capable of destroying Lyα. However, there is more to Lyα escape. This is probably best illustrated by nearby

---

[39]There is also little observational evidence for EW-boosting by a multiphase medium (e.g. [87, 229]).

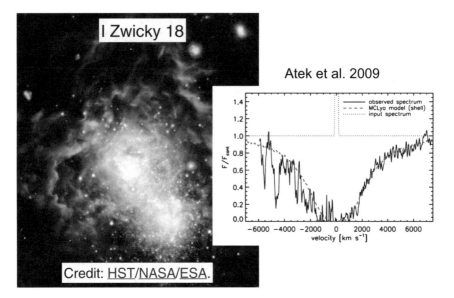

Fig. 1.36 1Zwicky18. A nearby, metal poor, blue young star forming galaxy (*Image Credit HST/NASA/ESA, and A. Aloisi*). While this galaxy is expected to be dust-poor, no Lyα is detected in emission (*Credit Atek et al, A&A, 502, page 791-801, 2009, reproduced with permission ©ESO*). It is thought that this is because there are no (or little) outflows present in this galaxy, which could have facilitated the escape of Lyα. More enriched nearby star forming galaxies that do show Lyα in emission, show evidence for outflows (e.g. [13, 149])

starburst galaxy 1Zwicky18 (shown in Fig. 1.36). This is a metal poor, extremely blue galaxy, and it has even been argued to host Population III stars (i.e. stars that formed our of primordial gas). We would expect this galaxy to have a high $f_{\mathrm{esc}}^{\alpha}$. However, the spectrum of 1Zwicky18 (also shown in Fig. 1.36) shows strong Lyα absorption. In contrast, the more enriched ($Z \sim 0.1 - 0.3 Z_{\odot}$) nearby galaxy ESO 350 does show strong Lyα emission (see Fig. 1.1 of [149]). The main difference between the two galaxies is that ESO 350 shows evidence for the presence of outflowing gas. Kunth et al. [149] observed that for a sample of 8 nearby starburst galaxies, 4 galaxies that showed evidence for outflows showed Lyα in emission, while no Lyα emission was detected for the 4 galaxies that showed no evidence for outflows (irrespective of the gas metallicity of the galaxies). These observations indicate that gas kinematics is a key parameter that regulates Lyα escape [13, 149, 220, 275]. This result is easy to understand qualitatively: in the absence of outflows, the Lyα sources are embedded within a static optically thick scattering medium. The traversed distance of Lyα photons is enhanced compared to that of (non-ionizing) UV continuum photons (see Sect. 1.7.1), which makes them more 'vulnerable' to destruction by dust. In contrast, in the presence of outflows Lyα photons can be efficiently scattered into the wing of the line profile, where they can escape easily.

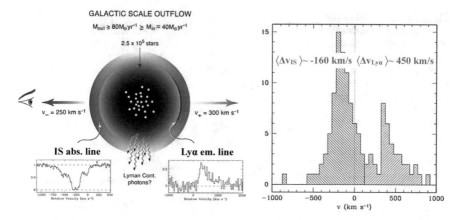

Fig. 1.37 The Figure shows (some) observational evidence for the ubiquitous existence of cold gas in outflows in star-forming galaxies, and that this cold gas affects the Lyα transport: the *right panel* shows the vast majority of low-ionization interstellar (IS) absorption lines are blueshifted relative to the systemic velocity of the galaxy, which is indicative of outflows (as illustrated in the *left panel*. Moreover, the *right panel* illustrates that the Lyα emission line is typically redshifted by an amount that is ∼2–3 times larger than typical blueshift of the IS lines *in the same galaxies*. These observations are consistent with a scenario in which Lyα photons scatter back to the observer from the far-side of the nebular region (indicated schematically in the *left panel*). *Credit figure as a whole corresponds to Fig. 12 of* [75], *Lyman Alpha Emitting Galaxies as a Probe of Reionization, PASA, 31, 40D*

The role of outflows is apparent at all redshifts. Simultaneous observations of Lyα and other non-resonant nebular emission lines indicate that Lyα lines typically are redshifted with respect to these other lines by $\Delta v_{\text{Ly}\alpha}$. This redshift is more prominent for drop-out (Lyman break) galaxies, in which the average $\Delta v_{\text{Ly}\alpha} \sim 460$ km s^{-1} in LBGs [150, 253]), which is larger than the shift observed in LAEs, where the average $\Delta v \sim 200$ km s^{-1} [51, 80, 115, 174, 175, 249, 262].[40] These observations indicate that outflows generally affect Lyα radiation while it is escaping from galaxies. This is not surprising: outflows are detected ubiquitously in absorption in other low-ionization transitions (e.g. [253]). Moreover, the Lyα photons appear to interact with the outflow, as the Lyα line is redshifted by an amount that is correlated with the outflow velocity inferred from low-ionization absorption lines (e.g. [241, 253]). The presence of winds and their impact on Lyα photons is illustrated schematically in Fig. 1.37.

As modelling the outflowing component in interstellar medium is an extremely challenging task (as we mentioned in the beginning of this section), simplified representations, such as the popular 'shell model', have been invoked. In the shell model the outflow is represented by a spherical shell with a thickness that is $0.1 \times$ its inner/outer

---

[40]The different Δv in LBGs and LAEs likely relates to the different physical properties of both samples of galaxies. Shibuya et al. [241] argue that LAEs may contain smaller N_{HI} which facilitates Lyα escape, and results in a smaller shift (also see [249]).

Fig. 1.38 The 'shell model' is a simplified representation of the Lyα transfer process on interstellar scales. The shell model contains six parameters, and generally reproduces observed Lyα spectral line profiles remarkably well (see Fig. 1.39)

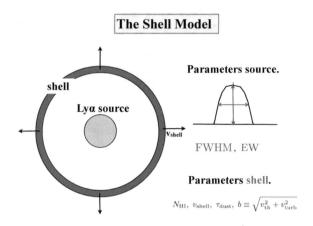

radius. Figure 1.38 summarizes the different ingredients of the shell model. The two parameters that characterize the Lyα sources are (*i*) its equivalent width (EW) which measures the 'strength' of the source compared to the underlying continuum, (*ii*) its full width at half maximum (FWHM) which denotes the width of the spectral line prior to scattering. This width may reflect motions in the Lyα emitting gas. The main properties that characterise the shell are its (*i*) HI-column density, $N_{\rm HI}$, (*ii*) outflow velocity, $v_{\rm sh}$, (*iii*) 'b-parameter' $b^2 \equiv v_{\rm turb}^2 + v_{\rm th}^2$. Here, $v_{\rm th} = \sqrt{2k_{\rm B}T/m_{\rm p}}$ (which we encountered before), and $v_{\rm turb}$ denotes its turbulent velocity dispersion; (*iv*) its dust content (e.g. [7, 269, 270]).

The shell-model can reproduce observed Lyα spectral line shapes remarkably well (e.g. [62, 116, 232, 232, 268, 281]), though not always (see e.g. [21, 51, 150], Forero-Romero et al. in prep.). One example of a good shell model fit to an observed spectrum is shown in Fig. 1.39. Here, the galaxy is a $z \sim 0.3$ green pea galaxy (fit taken from [281] using fitting algorithm from [104]). The 'triangle' diagram (produced using a modified version of https://www.triangle.py, [90]) shows constraints on the 6 shell model parameters, and their correlations. 'Typical' HI column densities in shell models are $N_{\rm HI} = 10^{18} - 10^{21}$ cm^{-2} and $v_{\rm sh} \sim$ a few tens to a few hundreds km s^{-1}. For a limited range of column densities, the Lyα spectrum peaks at $\sim 2v_{\rm sh}$. This peak consists of photons that scatter 'back' to the observer on the far side of the Lyα source, and are then Doppler boosted to twice the outflow velocity,[41] where they are sufficiently far in the wing of the absorption cross section to escape from the medium (the cross section at $\Delta v = 200$ km s^{-1} is only $\sigma_\alpha \sim$ a few times 10^{-20} cm^2, see Eq. 1.54 and Fig. 1.17).

---

[41]This argument implicitly assumes that the scattering is partially coherent (see Sect. 1.6.3): photons experience a Doppler boost $\Delta v/v \sim -v_{\rm out}/c$ when they enter the shell, and an identical Doppler boost $\Delta v/v \sim -v_{\rm out}/c$ when they exit the shell in opposite direction (as is the case for 'back' scattered' radiation). In the case of partially coherent scattering, the frequency of the photon changes only little in the frame of the gas (because $\sqrt{\langle \Delta x_{\rm in}^2 | x_{\rm in} \rangle} = 1$), and the total Doppler boost equals the sum of the two Doppler boosts imparted upon entry and exit from the shell.

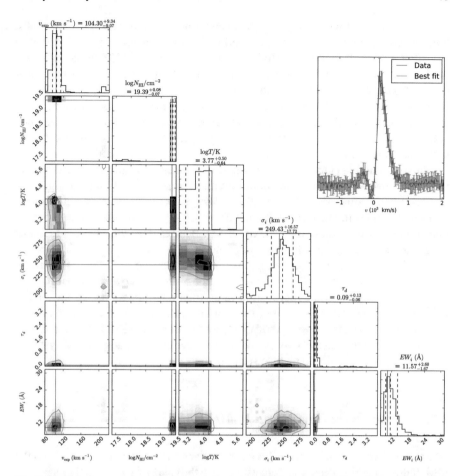

Fig. 1.39 This figure shows an example of an observed Lyα spectral line shape of a $z \sim 0.3$ green pea galaxy [126], which can be reproduced well by the shell model (see [281], using fitting algorithm from [104]). The triangle diagram shows the constraints on the shell model parameters, and degeneracies that exist between them (*Image Credit Max Gronke*)

In spite of its success, there are two issues with the shell-models: (*i*) gas in the shells has a single outflow velocity and a small superimposed velocity dispersion, while observations of low-ionization absorption lines indicate that outflows typically cover a much wider range of velocities (e.g. [126, 150]); and (*ii*) observations of low-ionization absorption lines also suggest that outflows—while ubiquitous—do not completely surround UV-continuum emitting regions of galaxies. Observations by [139] show that the maximum low-ionization covering fraction is 100% in only 2 out of 8 of their $z > 2$ galaxies (also see [124], who find evidence for a low covering factor of optically thick, neutral gas in a small fraction of lower redshift Lyman Break Analogues). There is thus some observational evidence that there exist sight lines that contain no detectable low-ionization (i.e. cold) gas, which may reflect the complex structure associated with outflows which cannot be captured with spherical shells. Two caveats are that (*a*) the inferred covering factors are measured as a function of velocity (and can depend on spectral resolution, see e.g. but [139] discuss why this is likely not an issue in their analysis [214]). Gas at different velocities can cover different parts of the source, and the outflowing gas may still fully cover the UV emitting source. This velocity-dependent covering is nevertheless not captured by the shell-model; (*b*) the low-ionization metal absorption lines only probe enriched cold (outflowing) gas. Especially in younger galaxies it may be possible that there is additional cold (outflowing) gas that is not probed by metal absorption lines.

Shibuya et al. [241] have shown that Lyα line emission is stronger in galaxies in which the covering factor of low-ionization material is smaller (see their Fig. 1.10, also see [262]). Similarly, [138] found the average absorption line strength in low-ionization species to decrease with redshift, which again coincides with an overall increase in Lyα flux from these galaxies [251]. Besides dust, the covering factor of HI gas therefore plays an additional important role in the escape of Lyα photons. These cavities may correspond to regions that have been cleared of gas and dust by feedback processes (see [191, 192] who describe a simple 'blow-out' model).

In short, dusty outflows appear to have an important impact on the interstellar Lyα radiative process, and give rise to redshifted Lyα lines. Low HI-column density holes further facilitate the escape of Lyα photons from the ISM, and can alter the emerging spectrum such that Lyα photons can emerge closer to the galaxies' systemic velocities ([25, 105, 106, 116], also see [288]).

1.9.2 Transfer in the Ionised IGM/CGM

HI gas that exists outside of the galaxy can further scatter Lyα that escaped from the ISM. If this scattering occurs sufficiently close to the galaxy, then this radiation can be detected as a low surface brightness glow (e.g. [286, 287]). As we showed previously in Fig. 1.13, observations indicate that spatially extended Lyα halos appear to exist generally around star-forming galaxies (see e.g. [96, 107, 117–119, 123, 170, 186, 191, 199, 217, 274]). Understanding what fraction of the Lyα flux in these halos consists of scattered Lyα radiation that escaped from the ISM, and what fraction was produced in-situ (as recombination, cooling, and/or fluorescence, see Sect. 1.4) is still an open question,[42] which we can address with integral field spectographs such as MUSE and the Keck Cosmic Web Imager. Polarization measurements (see Sect. 1.10) should also be able to distinguish between in situ-production and scattering [23, 67, 122], although these differences can be subtle [263].

Radiation that scatters 'sufficiently' far out is too faint to be detected with direct observations, and is effectively removed from observations. This applies especially

---

[42]Modeling the distribution of cold, neutral gas in the CGM is also challenging as there is increasing observational support that the CGM of galaxies contains cold, dense clumps (of unknown origin) on scales that are not resolved with current cosmological simulations [44, 125].

to photons that emerge on the blue side of the line resonance (i.e. $v > v_\alpha$ and/or $x > 0$), as these will redshift into the Lyα resonance due to the Hubble expansion of the Universe. Once these photons are at resonance there is a finite probability to be scattered. These photons are clearly not destroyed, but they are removed from the intensity of the radiation pointed at us. For an observer on Earth, these photons are effectively destroyed. Quantitatively, we can take 'sufficiently far' to mean that $r_{IGM} \gtrsim 1.5 r_{vir}$, where r_{vir} denotes the virial radius of halos hosting dark matter halos [156], and r_{IGM} denotes the radius beyond which scattered radiation is effectively removed from obsevations. Clearly, r_{IGM} depends on the adopted surface brightness sensitivity and the total Lyα luminosity of the source, and especially with sensitive MUSE observations, it will be important to verify what 'r_{IGM}' is. At $r > r_{IGM}$ 'intergalactic' radiative transfer consists of simply suppressing the emerging Lyα flux at frequency v_{em} by a factor of $\mathscr{T}_{IGM}(v_{em}) \equiv e^{-\tau_{IGM}(v_{em})}$, where $\tau_{IGM}(v)$ equals

$$\tau_{IGM}(v_{em}) = \int_{r_{IGM}}^{\infty} ds \, n_{HI}(s)\sigma_\alpha(v[s, v_{em}]). \tag{1.98}$$

As photons propagate a proper differential distance ds, the cosmological expansion of the Universe redshifts the photons by an amount $dv = -ds\,H(z)v/c$. Photons that were initially emitted at $v_{em} > v_\alpha$ will thus redshift into the line resonance. Because $\sigma_\alpha(v)$ is peaked sharply around v_α (see Fig. 1.17), we can approximate this integral by taking $n_{HI}(s)$ and $c/v \approx \lambda_\alpha$ outside of the integral. We make an additional approximation and assume that $n_{HI}(s)$ corresponds to $\bar{n}_{HI}(z)$, where $\bar{n}_{HI}(z) = \Omega_b h^2(1 - Y_{He})(1 + z)^3/m_p$ denotes the mean number density of hydrogen atoms in the Universe at redshift z. If we evaluate this expression at $v_{em} > v_\alpha$, i.e. at frequencies blueward of the Lyα resonance, then we obtain the famous Gunn-Peterson optical depth [108]:

$$\tau_{IGM}(v_{em} > v_\alpha) \equiv \tau_{GP} = \frac{\bar{n}_{HI}(z)\lambda_\alpha}{H(z)} \int_0^{\infty} dv \, \sigma_\alpha(v) = \frac{\bar{n}_{HI}(z)\lambda_\alpha}{H(z)} f_\alpha \frac{\pi e^2}{m_e c} \approx 7.0 \times 10^5 \left(\frac{1+z}{10}\right)^{3/2}, \tag{1.99}$$

where we used that $\int dv \, \sigma(v) = f_\alpha \frac{\pi e^2}{m_e c}$ (e.g. p102 [223]). The redshift dependence of τ_{IGM} reflects that $n_{HI}(z) \propto (1 + z)^3$ and that at $z \gg 1$ $H(z) \propto (1 + z)^{3/2}$. Equation 1.99 indicates that if the IGM were 100% neutral, it would be extremely opaque to photons emitted blue-ward of the Lyα resonance. Observations of quasar absorption line spectra indicate that the IGM transmits an *average* fraction $F \sim 85\%$ [$F \sim 40\%$] of Lyα photons at $z = 2$ [$z = 4$] (see lecture notes by X. Prochaska) which imply 'effective' optical depths of $\tau_{eff} \equiv -\ln[F] \sim 0.15$ [$\tau_{eff} \sim 0.9$] (e.g. [84]). The measured values $\tau_{eff} \ll \tau_{GP}$ which is (of course) because the Universe was highly ionized at these redshifts. A common approach to model the impact of the IGM is to reduce the Lyα flux on the blue side of the Lyα resonance by this observed (average) amount, while transmitting all flux on the red side.

The values of F and τ_{eff} mentioned above are averaged over a range of frequencies. In detail, density fluctuations in the IGM give rise to enhanced absorption in overdense regions which is observed as the Lyα forest. It is important to stress that galaxies populate overdense regions of the Universe in which: (*i*) the gas density was likely higher than average (see e.g. Fig. 1.2 of [19]), (*ii*) peculiar motions of gas attracted by the gravitational potential of dark matter halos change the relation between ds and dv, (*iii*) the local ionising background was likely elevated. We thus clearly expect the impact of the IGM[43] at frequencies close to the Lyα emission line to differ from the mean transmission in the Lyα forest: Fig. 1.40 shows the transmitted fraction of Lyα photons averaged over a large number of sight lines to galaxies in a cosmological hydrodynamical simulation [156]. This Figure shows that infall of over dense gas (and/or retarded Hubble flows) around dark matter halos hosting Lyα emitting galaxies can give rise to an increased opacity of the IGM around the Lyα resonance, and even extending somewhat into the red side of the Lyα line [59, 64, 133, 156, 227].

Because these models predict that the IGM can strongly affect frequencies close to the Lyα resonance, the overall impact of the IGM *depends strongly on the Ly* α *spectral line shape* as it emerges from the galaxy (also see [111, 227]). This is illustrated by the *lower three panels* in Fig. 1.41. For Gaussian and/or generally symmetric emission lines centered on the galaxies' systemic velocities, the IGM can transmit as little as[44] $\mathcal{T}_{\mathrm{IGM}} = 10 - 30\%$ even for a fully ionized IGM (e.g. [59, 64, 156, 286]). However, when scattering through outflows shifts the line sufficiently away from line centre, the overall impact of the IGM can be reduced tremendously (e.g. [72, 97, 111, 227]). *That is, not only do we care about how much Lyα escapes from the dusty ISM, we must care as much about how the emerging photons escape in terms of the line profile.* The fraction $\mathcal{T}_{\mathrm{IGM}}$ (also indicated in Fig. 1.41) denotes the 'IGM transmission' and denotes the total fraction of the Lyα radiation emitted by a galaxy that is transmitted by the IGM. The IGM transmission $\mathcal{T}_{\mathrm{IGM}}$ is given by the integral over the frequency-dependent transmission, $e^{-\tau_{\mathrm{IGM}}(\nu)}$. This frequency-dependence can be expressed as a function of the dimensionless frequency variable x, or as a function of the velocity offset $\Delta v \equiv -x v_{\mathrm{th}}$ as:

$$\mathcal{T}_{\mathrm{IGM}} = \int_{-\infty}^{\infty} d\Delta v \, J_\alpha(\Delta v) \exp[-\tau_{\mathrm{IGM}}(z_{\mathrm{g}}, \Delta v)] \qquad (1.100)$$

where $J_\alpha(\Delta v)$ denotes the line profile of Lyα photons as they escape from the galaxy at z_{g}. The IGM opacity discussed above originates in mildly over dense ($\delta = 1 - 20$, see [64]), highly ionized gas. Another source of opacity is provided by Lyman-limit systems (LLSs) and Dampled Lyα absorbers (DLAs). The precise impact of these

---

[43]Early studies defined the IGM to be all gas at $r > 1 - 1.5$ virial radii, which would correspond to the 'circum-galactic' medium by more recent terminology. Regardless of what we call this gas, scattering of Lyα photons would remove photons from a spectrum of a galaxy, and redistribute these photons over faint, spatially extended Lyα halos.

[44]It is worth noting that these models predict that the IGM can reduce the observed Lyα line by as much as $\sim 30\%$ between $z = 5.7$ and $z = 6.5$ [156]. Observations of Lyα halos around star forming galaxies provide hints that scattering in this CGM may be more prevalent at $z = 6.5$ than at $z = 5.7$, although the statistical significance of this claim is weak [186].

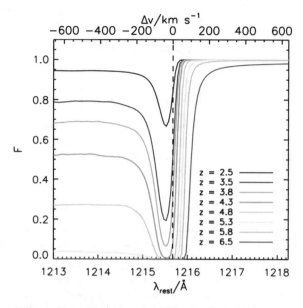

Fig. 1.40 The average fraction of photons that are transmitted though the IGM to the observer as a function of (restframe) wavelength. Overdense gas in close proximity to the galaxy—this gas can be referred to as 'circum-galactic gas'—enhances the opacity in the forest at velocities close to systemic (i.e $v_{sys} = 0$). Inflowing circum-galactic gas gives rise to a large IGM opacity even at a range of velocities redward of the Lyα resonance. Each line represents a different redshift. At wavelengths well on the blue side of the line, we recover the mean transmission measured from the Lyα forest. Overdense gas at close proximity to the galaxy increases the IGM opacity close to the Lyα resonance (and causes a dip in the transmission curve, *Credit from Fig. 2 of* [156] ©*AAS. Reproduced with permission*)

systems on Lyα radiation has only been studied recently [31, 52, 141, 182], and depends most strongly on how they cluster around Lyα emitting galaxies.

1.9.3 Intermezzo: Reionization

Reionization refers to the transformation of the intergalactic medium from fully neutral to fully ionized. For reviews on the Epoch of Reionization (EoR) we refer the reader to e.g. [18, 95, 187], and the recent book by [184]. The EoR is characterized by the existence of patches of diffuse neutral intergalactic gas, which provide an enormous source of opacity to Lyα photons: the Gunn-Peterson optical depth is $\tau_{GP} \sim 10^6$ (see Eq. 1.99) in a fully neutral medium. It is therefore natural to expect that detecting Lyα emitting galaxies from the EoR is hopeless. Fortunately, this is not the case, as we discuss in Sect. 1.9.4.

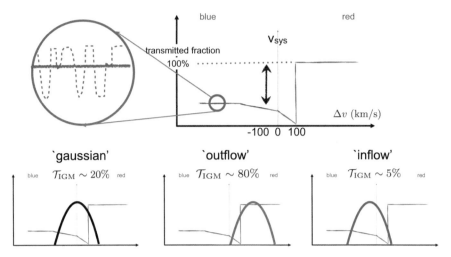

Fig. 1.41 Schematic representation of the impact of a residual amount of intergalactic HI on the fraction of photons that is directly transmitted to the observer, $\mathscr{T}_{\rm IGM}$ for a set of different line profiles. The *top panel* shows that the Lyα forest (shown in the *inset*) suppresses flux on the blue side of the Lyα line. The *lower left figure* shows that for lines centered on $v_{\rm sys} = 0$ (here symmetric around $v_{\rm sys} = 0$ for simplicity), the IGM cuts off a significant fraction of the blue half of the line, and some fraction of the red half of the line. For lines that are redshifted [blueshifted] w.r.t $v_{\rm sys}$, larger [smaller] fraction of emitted Lyα photons falls outside of the range of velocities affected by the IGM. The line profiles set at the interstellar level thus plays a key role in the subsequent intergalactic radiative transfer process. *Credit from 'Understanding the Epoch of Cosmic Reionization: Challenges and Progress', Vol 423, Fig. 3 of Chapter 'Constraining Reionization with Ly α Emitting Galaxies' by Mark Dijkstra (2016), page 145-161, With permission of Springer*

Reionization was likely not a homogeneous process in which the ionization state of intergalactic hydrogen changed everywhere in the Universe at the same time, and at the same rate. Instead, it was likely a temporally extended inhomogeneous process. The first sources of ionizing radiation were highly biased tracers of the underlying mass distribution. As a result, these galaxies were clustered on scales of tens of comoving Mpc (cMpc, [20]). The strong clustering of these first galaxies in overdense regions of the Universe caused these regions to be reionized first (e.g. Fig. 1.1 of [278]), which thus created fluctuations in the ionization field over similarly large scales. As a result a proper description of the reionization process requires simulations that are at least 100 cMpc on the side (e.g. [261]).

Ideally, one would like to simulate reionization by performing full radiative transfer calculations of ionising photons on cosmological hydrodynamical simulations. A number of groups have developed codes that can perform these calculations in 3D (e.g. [55, 89, 98, 99, 132, 206, 245, 248, 260]). These calculations are computationally challenging as one likes to simultaneously capture the large scale distribution of HII bubbles, while resolving the photon sinks (such as Lyman Limit systems) and the lowest mass halos ($M \sim 10^8$ M$_\odot$) which can contribute to the ionising photon budget (see e.g. [261]). Modeling reionization contains many poorly known parameters related to galaxy formation, the ionising emissivity of star-forming galaxies, their spectra etc. Alternative, faster 'semi-numeric' algorithms have been developed which allow for a more efficient exploration of the full parameter space (e.g. [72, 164, 180, 246]). These semi-numeric algorithms utilize excursion-set theory to determine if a cell inside a simulation is ionized or not [93]. Detailed comparisons between full radiation transfer simulations and semi-numeric simulations show both methods produce very similar ionization fields [283].

The picture of reionization that has emerged from analytical consideration and large-scale simulations is one in which the early stages of reionization are characterized by the presence of HII bubbles centered on overdense regions of the Universe, completely separated from each other by a neutral IGM [93, 132, 176]. The ionized bubbles grew in time, driven by a steadily growing number of star-forming galaxies residing inside them. The final stages of reionization are characterized by the presence of large bubbles, whose individual sizes exceeded tens of cMpc (e.g. [164, 283]). Ultimately these bubbles overlapped (percolated), which completed the reionization process. The predicted redshift evolution of the ionization state of the IGM in a realistic reionization model is shown in Fig. 1.42. This Figure illustrates the inhomogeneous, temporally extended nature of the reionization process.

1.9.4 Intergalactic Lyα Radiative Transfer during Reionization

There are indications that we are seeing Lyα emission from galaxies in the reionization epoch: there is increasing observational support for the claim that Lyα emission experiences extra opacity at $z > 6$ compared to the expectation based on extrapolating observations from lower redshift, and which is illustrated in Figu 1.43. The *left panel* shows the drop in the '*Ly α fraction*' in the drop-out (Lyman Break) galaxy

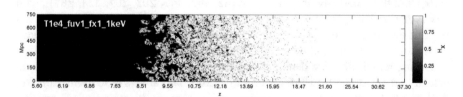

Fig. 1.42 The predicted redshift evolution of the ionization state of the IGM in a realistic reionization model (*Credit Figure kindly provided by Andrei Mesinger, published previously as Fig. 14 in* [75], *Lyman Alpha Emitting Galaxies as a Probe of Reionization, PASA, 31, 40*). The white/black represents fully neutral/ionized intergalactic gas. This Figure demonstrates the inhomogeneous nature of the reionization process which took place over an extended range of redshifts: at $z > 16$ the first ionized regions formed around the most massive galaxies in the Universe (at that time). During the final stages of reionization—here at $z \sim 9$ the IGM contains ionized bubbles several tens of cMpc across

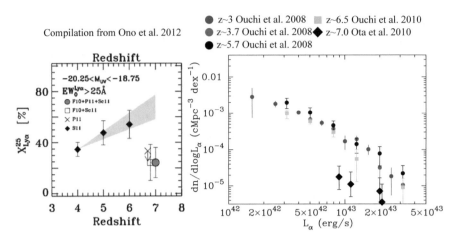

Fig. 1.43 There are two independent observational indications that the Lyα flux from galaxies at $z > 6$ is suppressed compared to extrapolations from lower redshifts. The *left panel* shows the drop in the '*Lyα fraction*' in the drop-out (Lyman Break) galaxy population, the *right panel* shows the sudden evolution in the Lyα luminosity function of Lyα selected galaxies (LAEs). This evolution has been shown to be quantitatively consistent (see [73, 104]). *Credit from 'Understanding the Epoch of Cosmic Reionization: Challenges and Progress', Vol 423, Fig. 1 of Chapter 'Constraining Reionization with Lyα Emitting Galaxies' by Mark Dijkstra (2016) page 145-161, With permission of Springer*

population. The 'Lyα fraction' denotes the fraction of continuum selected galaxies which have a 'strong' Lyα emission line ([251]). The 'strength' of the Lyα emission line is quantified by some arbitrary equivalent width threshold. The Lyα fraction rises out to $z \sim 6$ (this trend continues down to lower redshift), and then suddenly drops at $z > 6$ [45, 46, 195, 208, 209, 234, 264]. The *right panel* shows the sudden evolution in the Lyα luminosity function of Lyα selected galaxies (LAEs). The Lyα luminosity function does not evolve much between $z \sim 3$ and $z \sim 6$ ([203], also see [129]), but then suddenly drops at $z \gtrsim 6$ [82, 142, 143, 202, 204, 228, 289]), with this evolution being most apparent at $z \gtrsim 6.5$ (see e.g. [171, 201]). The observed redshift-evolution of the Lyα luminosity function of LAEs and Lyα fractions have been shown to be quantitatively consistent (see [73, 104]).

During reionization we denote the opacity of the IGM in the *ionized bubbles* at velocity off-set Δv and redshift z with $\tau_{HII}(z, \Delta v)$. This allows us to more explicitly distinguish this source of intergalactic opacity from the 'damping wing' opacity, $\tau_D(z, \Delta v)$, which is the opacity due to the diffuse neutral IGM and which is only relevant during reionization. In other words, τ_{HII} refers to the Lyα opacity in neutral gas that survived in the ionized bubbles. This neutral gas can reside in dense self-shielding clouds,[45] or as residual neutral hydrogen that survived in the ionized bubbles.[46] Equation 1.100 therefore changes to

$$\mathscr{T}_{\mathrm{IGM}} = \int_{-\infty}^{\infty} d\Delta v \, J_{\alpha}(\Delta v) \exp[-\tau_{\mathrm{IGM}}(z_{\mathrm{g}}, \Delta v)], \quad \tau_{\mathrm{IGM}}(z_{\mathrm{g}}, \Delta v) = \underbrace{\tau_{\mathrm{D}}(z_{\mathrm{g}}, \Delta v)}_{\text{reionization only}} + \tau_{\mathrm{HII}}(z_{\mathrm{g}}, \Delta v).$$

(1.101)

Decomposing $\tau_{\mathrm{IGM}}(z_{\mathrm{g}}, \Delta v)$ into $\tau_{\mathrm{D}}(z_{\mathrm{g}}, \Delta v)$ and $\tau_{\mathrm{HII}}(z_{\mathrm{g}}, \Delta v)$ is helpful as they depend differently on Δv: the *left panel* of Fig. 1.44 shows the IGM transmission, $e^{-\tau_{\mathrm{D}}(z_{\mathrm{g}}, \Delta v)}$, as a function of velocity off-set (Δv) for the diffuse neutral IGM for $x_{\mathrm{HI}} = 0.8$. This Figure shows clearly that the neutral IGM affects a range of frequencies that extends much further to the red-side of the Lyα resonance than the ionized IGM (compare with Fig. 1.40). This large opacity $\tau_{\mathrm{D}}(z_{\mathrm{g}}, \Delta v)$ far redward of the Lyα resonance is due to the damping wing of the absorption cross-section (and not due to gas motions at these large velocities), which is why we refer to it as the 'damping wing optical depth'. The *right panel* shows that there is a contribution to the damping wing optical depth from neutral, self-shielding clouds (with $N_{\mathrm{HI}} \gtrsim 10^{17}$ cm^{-2}, also see the high-redshift curves in Fig. 1.40) which can theoretically mimic the impact of a diffuse neutral IGM [31], though this requires a large number density of these clouds (see [31, 52, 141, 182]).

Star-forming galaxies that are luminous enough to be detected with existing telescopes likely populated dark matter halos with masses in excess of $M \gtrsim 10^{10}$ M$_\odot$ (see e.g. [182]). These halos preferentially reside in over dense regions of the Universe, which were reionized earliest. It is therefore likely that (Lyα emitting) galaxies preferentially resided inside these large HII bubbles. This has an immediate implication for the visibility of the Lyα line. Lyα photons emitted by galaxies located inside these HII regions can propagate (to the extent that is permitted by the ionized IGM)—and therefore redshift away from line resonance—through the ionized IGM before encountering the neutral IGM. Because of the strong frequency-dependence of the Lyα absorption cross section, these photons are less likely to be scattered out of the line of sight inside the neutral IGM. A non-negligible fraction of Lyα photons may be transmitted directly to the observer, which is illustrated schematically in Fig. 1.45. *Inhomogeneous reionization thus enhances the prospect for detecting Ly α emission from galaxies inside HII bubbles* (see for a review, and an extensive list of references [75]). It also implies that the impact of diffuse neutral intergalactic gas on the visibility of Lyα flux from galaxy is more subtle than expected in models in which reionization proceeds homogeneously, and that the observed reduction in Lyα flux from galaxies at $z > 6$ requires a significant volume filling fraction of neutral gas (which is indeed the case, as we discuss below). This Figure also illustrates that Lyα

---

[45]To make matters more confusing: self-shielding absorbers inside the ionized bubbles with sufficiently large HI column densities can be optically thick in the Lyα damping wing, and can give rise to damping wing absorption as well. This damping wing absorption is included in $\tau_{\mathrm{HII}}(z, \Delta v)$.

[46]Just as the Lyα forest at lower redshifts—where hydrogen reionization was complete—contains neutral hydrogen gas with different densities, ionization states and column densities.

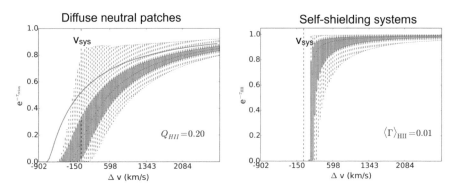

Fig. 1.44 Neutral gas in the intergalactic medium can give rise to a large 'damping' wing opacity $\tau_D(z_g, \Delta v)$ that extends far to the red side of the Lyα resonance, i.e. $\Delta v \gg 1$. The *left panel* shows the IGM transmission, $e^{-\tau_D(z_g, \Delta v)}$, as a function of velocity off-set (Δv) for the diffuse neutral IGM for $x_{HI} = 0.8$. The *right panel* shows the IGM transmission for a (large) number of self-shielding clouds (see text). To obtain the *total* IGM transmission, one should multiply the transmission curves shown here and in Fig. 1.40. *Credit from 'Understanding the Epoch of Cosmic Reionization: Challenges and Progress', Vol 423, Fig. 4 of Chapter 'Constraining Reionization with Ly α Emitting Galaxies' by Mark Dijkstra (2016), page 145–161, With permission of Springer.* Figures adapted from *Figs. 2 and 4* of [182], *'Can the intergalactic medium cause a rapid drop in Ly α emission at z > 6?'*, *MNRAS, 446, 566*

photons emitted by galaxies that lie outside of large HII bubbles, scatter repeatedly in the IGM. These photons diffuse outward, and are visible only as faint extended Lyα halos [135, 145, 162].

We can quantify the impact of neutral intergalactic gas on the Lyα flux from galaxies following our analysis in Sect. 1.9.2. We denote the optical depth in the neutral intergalactic patches with τ_D. We first consider the simplest case in which a Lyα photon encounters one fully neutral patch which spans the line-of-sight coordinate from s_b ('b' stands for beginning) to s_e ('e' stands for end):

$$\tau_D(v) = \int_{s_b}^{s_e} ds\, n_{HI}(s)\sigma_\alpha(v[s]) = \frac{n_{HI}(s)\lambda_\alpha}{H(z)} \int_{v_b(v)}^{v_e(v)} dv'\sigma_\alpha(v'). \quad (1.102)$$

where we followed the analysis of Sect. 1.9.2, and changed to frequency variables, and assumed that $n_{HI}(s)$ is constant across this neutral patch. We eliminate $n_{HI}(s)$ by using the expression for the Gunn-Peterson optical depth in Eq. 1.99, and recast Eq. 1.102 as

$$\tau_D(v) = \tau_{GP} \frac{\int_{v_b(v)}^{v_e(v)} dv'\sigma_\alpha(v')}{\int_0^\infty dv'\sigma_\alpha(v')} = \tau_{GP} \frac{\int_{x_b(v)}^{x_e(v)} dx'\phi(x')}{\int_0^\infty dx'\phi(x')}. \quad (1.103)$$

The denominator can be viewed as a normalisation constant, and we can rewrite Eq. 1.103 as

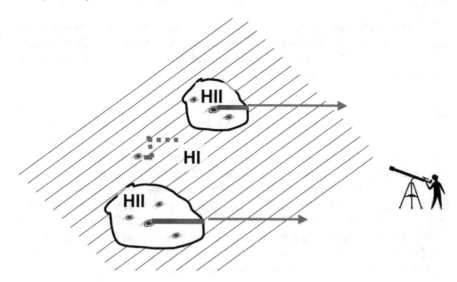

Fig. 1.45 This Figure schematically shows why inhomogeneous reionization boosts the visibility of Lyα emitting galaxies. During the mid and late stages of reionization star-forming—and hence Lyα emitting—galaxies typically reside in large HII bubbles. Lyα photons emitted inside these HII bubbles can propagate—and redshift away from line resonance—through the ionized IGM before encountering the neutral IGM. The resulting reduced opacity of the neutral IGM (Eq. 1.108) to Lyα photons enhances the prospect for detecting Lyα emission from those galaxies inside HII bubbles. *Credit from Fig. 15 of* [75], *Lyman Alpha Emitting Galaxies as a Probe of Reionization, PASA, 31, 40*

$$\tau_D(\nu) = \frac{\tau_{GP}}{\sqrt{\pi}} \int_{x_b(\nu)}^{x_e(\nu)} dx' \, \phi(x'), \qquad (1.104)$$

where the factor of $\sqrt{\pi}$ enters because of our adopted normalisation for the Voigt profile $\phi(x)$.

During the EoR a Lyα photon emitted by a galaxy will generally propagate through regions that are alternating between (partially) neutral and highly ionized. The more general case should therefore contain the sum of the optical depth in separate neutral patches:

$$\tau_D(\nu) = \frac{1}{\sqrt{\pi}} \sum_i \tau_{GP,i} \, x_{HI,i} \int_{x_{b,i}(\nu)}^{x_{e,i}(\nu)} dx \, \phi(x'), \qquad (1.105)$$

where we have placed τ_{GP} within the sum, because τ_{GP} depends on redshift as $\tau_{GP} \propto (1 + z_i)^{3/2}$, and therefore differs slightly for each patch 'i' at redshift z_i, which has a neutral hydrogen fraction $x_{HI,i}$ (Fig. 1.45).

More specifically, the total optical depth of the neutral IGM to Lyα photons emitted by a galaxy at redshift z_g with some velocity off-set Δv is given by Eq. 1.105 with $x_{b,i} = \frac{-1}{v_{th,i}}[\Delta v + H(z_g)R_{b,i}/(1 + z_g)]$, in which $R_{b,i}$ denotes the comoving distance to the beginning of patch 'i' ($x_{e,i}$ is defined similarly). Equation 1.105 must generally be evaluated numerically. However, one can find intuitive approximations: for example, if we assume that (*i*) $x_{\text{HI},i} = 1$ for all 'i' (i.e. all patches are fully neutral), (*ii*) $z_i \sim z_g$, and (*iii*) that Lyα photons have redshifted away from resonance by the time they encounter this first neutral patch,[47] then

$$\tau_D(z_g, \Delta v) = \frac{\tau_{\text{GP}}(z_g)}{\sqrt{\pi}} \sum_i \left(\frac{a_{v,i}}{\sqrt{\pi} x_{e,i}} - \frac{a_{v,i}}{\sqrt{\pi} x_{b,i}} \right) \equiv \frac{\tau_{\text{GP}}(z_g)}{\sqrt{\pi}} \bar{x}_D \left(\frac{a_v}{\sqrt{\pi} x_e} - \frac{a_v}{\sqrt{\pi} x_{b,1}} \right),$$
(1.106)

where $x_{e,i} = x_{e,i}(\Delta v)$ and $x_{b,i} = x_{b,i}(\Delta v)$. It is useful to explicitly highlight the sign-convention here: photons that emerge redward of the Lyα resonance have $\Delta v > 0$, which corresponds to a negative x. Cosmological expansion redshifts photons further, which decreases x further. The $a_v/[\sqrt{\pi} x_{b,i}]$ is therefore less negative, and τ_D is thus smaller. In the last term, we defined the 'patch-averaged' neutral fraction, $\bar{x}_D$, which is related to the volume filling factor of neutral hydrogen $\langle x_{\text{HI}} \rangle$ in a non-trivial way (see [181]).

Following [181], we can ignore the term $a_v/[\sqrt{\pi} x_e]$ and write

$$\tau_D(z_g, \Delta v) \approx \frac{\tau_{\text{GP}}(z_g)}{\pi} \bar{x}_D \frac{a_v}{|x_{b,1}|} = \frac{\tau_{\text{GP}}(z_g)}{\pi} \bar{x}_D \frac{A_\alpha c}{4\pi \nu_\alpha} \frac{1}{\Delta v_{b,1}},$$
(1.107)

where x_e denotes the frequency that photon has redshifted to when it exits from the last neutral patch, while $x_{b,1}$ denotes the photon's frequency when it encounters the first neutral patch. Because typically $|x_e| \gg |x_{b,1}|$ we can drop the term that includes x_e. We further substituted the definition of the Voigt parameter $a_v = A_\alpha/(4\pi \Delta \nu_\alpha)$, to rewrite $x_{b,1}$ as a velocity off-set from line resonance when a photon first enters a neutral patch, $\Delta v_{b,1} = \Delta v + H(z_g) R_{b,i}/(1 + z_g)$.

Substituting numbers gives [70, 185]

$$\tau_D(z_g, \Delta v) \approx 2.3 \bar{x}_D \left(\frac{\Delta v_{b,1}}{600 \text{ km s}^{-1}} \right)^{-1} \left(\frac{1 + z_g}{10} \right)^{3/2}.$$
(1.108)

This equation shows that the opacity of the IGM drops dramatically once photons enter the first patch of neutral IGM with a redshift. This redshift can arise partly at the interstellar level, and partly at the intergalactic level: scattering off outflowing

---

[47]If a photon enters the first neutral patch on the blue side of the line resonance, then the total opacity of the IGM depends on whether the photon redshifted into resonance inside or outside of a neutral patch. If the photon redshifted into resonance inside patch 'i', then $\tau_D(z_g, \Delta v) = \tau_{\text{GP}}(z) x_{\text{HI},i}$. If on the other hand the photon redshifted into resonance in an ionized bubble, then we must compute the optical depth in the ionized patch, $\tau_{\text{HII}}(z, \Delta v = 0)$, plus the opacity due to subsequent neutral patches. Given that the ionized IGM at $z = 6.5$ was opaque enough to completely suppress Lyα flux on the blue-side of the line, the same likely occurs inside ionized HII bubbles during reionization because of (*i*) the higher intergalactic gas density, and (*ii*) the shorter mean free path of ionizing photons and therefore likely reduced ionizing background that permeates ionised HII bubbles at higher redshifts.

material[48] at the interstellar level can efficiently redshift Lyα photons by a few hundred km/s (see Sect. 1.9.1). Because Lyα photons can undergo a larger cosmological subsequent redshift inside larger HII bubbles, Lyα emitting galaxies inside larger HII bubbles may be more easily detected. Equation 1.108 shows that setting $\tau_D = 1$ for $\bar{x}_D = x_{HI} = 1.0$ requires $\Delta v = 1380$ km s^{-1}. This cosmological redshift reduces the damping wing optical depth of the neutral IGM to $\tau_D < 1$ for HII bubbles with radii $R \gtrsim \Delta v/H(z) \sim 1$ Mpc (proper), *independent of z* (because at a fixed R the corresponding cosmological redshift $\Delta v \propto H(z) \propto (1 + z)^{3/2}$, [185]). The presence of large HII bubbles during inhomogeneous reionization may have drastic implications for the prospects of detecting Lyα emission from the epoch.

Current models indicate that if the observed reduction in Lyα flux from galaxies at $z > 6$ is indeed due to reionization—which is plausible (see [184] for recent constraints on the reionization history from a suite of observations)—then this requires a volume filling factor of diffuse neutral gas which exceeds $\langle x_{HI}\rangle \gtrsim 40\%$, which implies that reionization is still ongoing at $z \sim 6$–7 (see [75] for a review). This constraint is still uncertain due to the limited number of Lyα galaxies at $z \sim 6$ and $z \sim 7$, but this situation is expected to change, especially with large surveys for high-z Lyα emitters to be conducted with Hyper Suprime-Cam. These surveys will enable us to measure the variation of IGM opacity on the sky at fixed redshift, and constrain the reionization morphology (see [136, 137, 182]).

1.10 Miscalleneous Topics I: Polarization

Lyα radiative transfer involves scattering. Scattered radiation can be polarized. The polarization of electromagnetic radiation measures whether there is a preferred orientation of its electric and magnetic components. Consider the example that we discussed in Sect. 1.5.1 of a free electron that scatters incoming radiation (shown schematically in Fig. 1.14). If the incoming radiation field were unpolarized, then its electric vector is distributed randomly throughout the plane perpendicular to its propagation direction prior to scattering (denoted with $\mathbf{k}_{in}$).

In Sect. 1.5.1 we discussed how the intensity of scattered radiation $I \propto \sin^2 \Psi$, where $\cos \Psi \equiv \mathbf{k}_{out} \cdot \mathbf{e}_E$ in which $\mathbf{e}_E$ denotes the normalized direction of the electric vector (see Fig. 1.15). Similarly, we can say that the amplitude of the electric-field scales as $E \propto \sin \Psi$ (note at $I \propto |E|^2$), i.e. we project the electric vector onto the plane perpendicular to $\mathbf{k}_{out}$ (see the *left panel* of Fig. 1.46). This same argument can be applied to demonstrate that a free electron can transform unpolarized into a polarized radiation if there is a 'quadrupole anisotropy' in the incoming intensity: the *right panel* of Fig. 1.46 shows a free electron with incident radiation from the left and from the top. If the incident radiation is unpolarized, then the electric field vector

---

[48] Scattering through an extremely opaque *static* medium gives rise to a spectrally broadened double-peaked Lyα spectrum (see Fig. 1.26). Of course, photons in the red peak start with a redshift as well, which boosts their visibility especially for large N_{HI} (see Fig. 1.2 in also see [70, 111]).

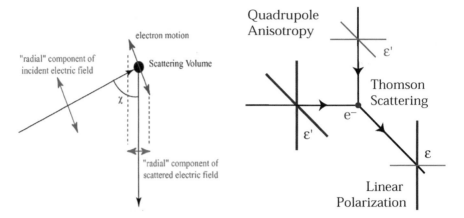

Fig. 1.46 This Figure illustrates how scattering by a free electron can polarize radiation. Both panels illustrate how scattering by a free electron transmits only the electric vector of the incident radiation, projected onto the outgoing radiation direction (also see Fig. 1.15)

points in arbitrary direction in the plane perpendicular to the propagation direction. Consider scattering by 90°. If we apply the projection argument, then for radiation incident from the left we only 'see' the component of the E-field that points upward (shown in *blue*). Similarly, for radiation coming in from the top we only see the E-field that lies horizontally. The polarization of the scattered radiation vanishes if the *blue* and *red* components are identical, which—for unpolarized radiation—requires that the total intensity of radiation coming in from the top must be identical to the that coming in from the left. For this reason, electron scattering can polarize the Cosmic Microwave Background if the intensity varies on angular scales of 90°. If fluctuation exist on these scales, then the CMB is said to have a non-zero quadrupole moment. Similarly, if there were a point source irradiating the electron from the top, then we would also expect only the *red* E-vector to be transmitted.

The first detections of polarization in spatially extended Lyα sources have been reported [23, 122, 131]. The *left panel* in Fig. 1.47 shows Lyα polarization vectors overlaid on a Lyα surface brightness map of 'LAB1' (Lyα Blob 1, see [122]). The *lines* here denote the linear polarization (see a more detailed discussion of this quantity below), which denotes the preferred orientation of the electric vector of the observed Lyα radiation. The longer the lines, the more polarized the radiation. Figure 1.47 shows how the polarization vectors appear to form concentric circles around spots of high Lyα surface brightness. This is consistent with a picture in which most Lyα was emitted in the spots with high surface brightness, and then scattered towards the observer at larger distance from these sites. This naturally gives rise to the observed polarization pattern (also see the *right panel* of Fig. 1.47 for an artistic illustration of this, from [36]).

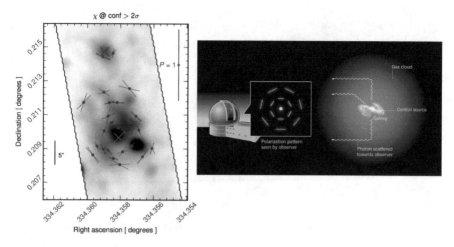

Fig. 1.47 *Left Panel*: Lines indicate the magnitude and orientation of the linear polarization of Lyα radiation in Lyα blob 1, overlaid on the observed Lyα surface brightness (in *gray scale*, (*Reprinted by permission from Macmillan Publishers Ltd:* [122], *Nature 476, 304H, copyright*). Lyα polarization vectors form concentric circles around the most luminous Lyα 'spots' in the map. This is consistent with a picture in which Lyα emission is produced in the locations of high surface brightness, and where lower surface brightness regions correspond to Lyα that was scattered back into the line-of-sight at larger distance from the Lyα source. This process is illustrated visually in the *right panel* (*Reprinted by permission from Macmillan Publishers Ltd:* [36], *Nature 476, 288B, copyright*)

The previous discussion can be condensed into a compact equation if we decompose the intensity of the radiation into a component parallel and perpendicular to the scattering plane, which is spanned by the propagation directions $\mathbf{k}_{in}$ and $\mathbf{k}_{out}$ (see Fig. 1.48). We write $I \equiv I_{\parallel} + I_{\perp}$, with $I_{\parallel} \equiv |\mathbf{e}_{\parallel}|^2 I$, in which $\mathbf{e}_{\parallel}$ denotes the component of $\mathbf{e}_E$ in the scattering plane. Similarly, we have $I_{\perp} \equiv |\mathbf{e}_{\perp}|^2 I$. We define the *scattering matrix*, R, as

$$\begin{pmatrix} I'_{\parallel} \\ I'_{\perp} \end{pmatrix} = \begin{pmatrix} S_1 & S_2 \\ S_3 & S_4 \end{pmatrix} \equiv R \begin{pmatrix} I_{\parallel} \\ I_{\perp} \end{pmatrix}, \tag{1.109}$$

where the total outgoing intensity is given by $I' = I'_{\parallel} + I'_{\perp}$. The scattering matrix quantifies the angular redistribution of both components of a scattered electromagnetic wave. For comparison, the phase-function quantified the angular redistribution of the intensity total I' only. For the case of a free electron, the scattering matrix R_{Ray} is given by

$$R_{dip} = \frac{3}{2} \begin{pmatrix} \cos^2 \theta & 0 \\ 0 & 1 \end{pmatrix}. \tag{1.110}$$

This expression indicates that for scattering by an angle θ, $\mathbf{e}'_{\parallel} = \cos\theta \mathbf{e}_{\parallel}$ while $\mathbf{e}'_{\perp} = \mathbf{e}_{\perp}$ (see Fig. 1.48). The total outgoing intensity $I' \equiv I'_{\parallel} + I'_{\perp} = \frac{3}{2}(\cos^2\theta I_{\parallel} + I_{\perp})$, which for unpolarized incoming radiation ($I_{\parallel} = I_{\perp} = 0.5I$) reduces to $I' = \frac{3}{4}(1 +$

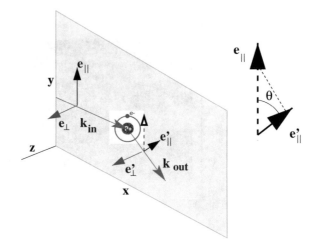

Fig. 1.48 $I_{||}$ ($I'_{||}$) denotes the intensity of the radiation parallel to the scattering plane before (after) scattering (spanned by $\mathbf{k}_{in}$ and $\mathbf{k}_{out}$, indicated in *green*). Furthermore, $I_\perp$ ($I'_\perp$) denotes the intensity perpendicular to this plane. The intensities relate to the polarization vectors $\mathbf{e}_\perp$ and $\mathbf{e}_{||}$ as $I_{||} \equiv (\mathbf{e}_{||} \cdot \mathbf{e}_{||}) I$ etc. Classically, an incoming photon accelerates an electron in the direction of the polarization vector $\mathbf{e}$. The oscillating electron radiates as a classical dipole, and the angular redistribution of the outgoing intensity $I' = I'_{||} + I'_\perp$ scales as $I \propto 1 + \cos^2 \theta$, where $\cos \theta = \mathbf{k}_{in} \cdot \mathbf{k}_{out}$. For scattering by an angle θ, $\mathbf{e}'_{||} = \cos \theta \mathbf{e}_{||}$ while $\mathbf{e}'_\perp = \cos \theta$. For a classical dipole, the intensity $I'_{||}$ is therefore reduced by a factor of $\cos^2 \theta$ relative to $I'_\perp$ (see text)

$\cos^2 \theta) I$. The ratio I'/I corresponds to the phase-function for Rayleigh scattering encountered in Sect. 1.5.1 (see Eq. 1.42). Furthermore, the *linear polarization* of the radiation is defined as

$$P \equiv \frac{I_\perp - I_{||}}{I_{||} + I_\perp}, \tag{1.111}$$

and for the scattered intensity of unpolarized radiation we find

$$P = \frac{1 - \cos^2 \theta}{1 + \cos^2 \theta}. \tag{1.112}$$

Note how this reflects our previous discussion: unpolarized radiation that is scattered by 90° becomes 100% linearly polarized (also see the *right panel* of Fig. 1.46). For comparison, the scattering matrix for an 'isotropic' scatterer is

$$R_{iso} = \frac{1}{2} \begin{pmatrix} 1 & 1 \\ 1 & 1 \end{pmatrix}. \tag{1.113}$$

That is, the outgoing intensity $I'_{||} = I'_\perp = 0.5$ and has no directional dependence. Furthermore, isotropic scattering produces no polarization. An 'isotropic scatterer' scatters photons without caring about the properties of the incoming photon. We introduce this concept because isotropic scattering plays a role in Lyα scattering.

An electron that is bound to a hydrogen atom is confined to orbits defined by quantum physics of the atom (see Fig. 1.3 and Sect. 1.2.2). That is, unlike the case of the free electron discussed above, an electron that is bound to a H nucleus is not completely free to oscillate along the polarization vector of the incoming photon, but is bound to orbits set by quantum mechanics. However, depending on which (sub) quantum state of the $2p$ state (in Sect. 1.10.1 below we discuss how the $2p$ state splits up into several substates) the hydrogen atom is in, the electron may have some memory of the direction and polarization of the incoming photon. In other words, the wavefunction of the electron in the $2p$ state may be aligned along the polarization vector of the incoming photon. It turns out that scattering of Lyα photons by hydrogen atoms can be described as some linear combination of dipole and isotropic scattering, which is described as 'scattering by anisotropic particles' [48]. The scattering matrix for anisotropic particles is given by [48, Eqs. 250–258]

$$R = \frac{3}{2} E_1 \begin{pmatrix} \cos^2 \theta & 0 \\ 0 & 1 \end{pmatrix} + \frac{1}{2} E_2 \begin{pmatrix} 1 & 1 \\ 1 & 1 \end{pmatrix}, \tag{1.114}$$

where $E_1 + E_2 = 1$. Precisely what E_1 and E_2 are is deteremined by the quantum numbers that describe the electron in the $2p$ state. The number E_1 gives the relative importance of dipole scattering, and is sometimes referred to as the '*polarizability*', as it effectively measures how efficiently a scatterer can polarize incoming radiation. Both the angular redistribution—or the phase function—and the polarization of scattered Lyα radiation can therefore be characterized entirely by this single number E_1, which is discussed in more detail next.

1.10.1 Quantum Effects on Lyα Scattering: The Polarizability of the Hydrogen Atom

In order to accurately describe how H atoms scatter Lyα radiation, we must consider the *fine-structure* splitting of the $2p$ level. The spin of the electron causes the $2p$ state quantum state to split into the $2p_{1/2}$ and $2p_{3/2}$ levels, which are separated by ~ 10 Ghz (see Fig. 1.49). The notation that is used here is nL_J, in which $\mathbf{J} = \mathbf{L} + \mathbf{s}$ denotes the total (orbital +spin) angular momentum of the electron. The $1s_{1/2} \rightarrow 2p_{1/2}$ and $1s_{1/2} \rightarrow 2p_{3/2}$ is often referred as the K-line and H-line, respectively.

It turns out that a quantum mechanical calculation yields that $E_1 = \frac{1}{2}$ for the H transition, while $E_1 = 0$ for the K transition (e.g. [6, 37, 112, 159]). When a Lyα scattering event goes through the K-transition, the hydrogen atom behaves like an isotropic scatterer. This is because the wavefunction of the $2p_{1/2}$ state is spherically symmetric (see [273]), and the atom 'forgot' which direction the photon came from or which direction the electric field was pointing to. For the $2p_{3/2}$ state, the wavefunction

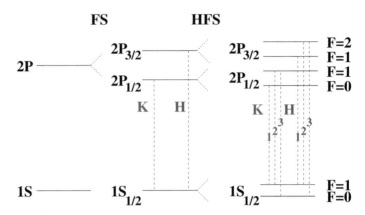

Fig. 1.49 Schematic diagram of the energy levels of a hydrogen atom. The notation for each level is nL_J, where n is the principle quantum number, and L denotes the orbital angular momentum number, and J the total angular momentum ('total' angular momentum means orbital+spin angular momentum, i.e. $\mathbf{J} = \mathbf{L} + \mathbf{S}$). The fine structure splitting of the $2p$ level shifts the $2p_{1/2}$ and $2p_{3/2}$-level by $\Delta E / h_p \sim 11$ Ghz (e.g. [38]). A transition of the form $1s_{1/2} \rightarrow 2p_{1/2}$ ($1s_{1/2} \rightarrow 2p_{3/2}$) is denoted by a K (H) transition. The spin of the nucleus induces further 'hyperfine' splitting of the line. This is illustrated in the *right panel*, where each fine structure level breaks up into two lines which differ only in their quantum number F which measures the total + nuclear spin angular momentum, i.e. $\mathbf{F} = \mathbf{J} + \mathbf{I}$. Hyperfine splitting ultimately breaks up the Lyα transition into six individual transitions. Fine and hyperfine structure splitting plays an important role in polarizing scattered Lyα radiation

is not spherically symmetric and contains the characteristic 'double lobes' shown in Fig. 1.3. The hydrogen in the $2p_{3/2}$ state thus has some memory of the direction of the incoming Lyα photon and its electric vector, and behaves partially as a classical dipole scatterer, and partly as an isotropic scatterer.

However, in reality the situation is more complex and [254] showed that E_1 depends strongly on frequency as

$$E_1 = \frac{\left(\frac{\omega_K}{\omega_H}\right)^2 (b_H^2 + d_H^2) + 2\frac{\omega_K}{\omega_H}(b_H b_K + d_H d_K)}{b_K^2 + d_K^2 + 2\left(\frac{\omega_K}{\omega_H}\right)^2 (b_H^2 + d_H^2)}, \qquad (1.115)$$

where $b_{H,K} = \omega^2 - \omega_{H,K}^2$ and $d_{H,K} = \omega_{H,K}\Gamma_{H,K}$. Here, $\omega_H = 2\pi\nu_H$ ($\omega_K = 2\pi\nu_K$) denotes the resonant angular frequency of the H (K) transition, and $\Gamma_{H,K} = A_\alpha$. The frequency dependence of E_1 is shown in Fig. 1.50, where we have plotted E_1 as a function of wavelength λ. The *black solid line* shows E_1 that is given by Eq. 1.115. This plot shows that

1. $E_1 = 0$ at $\lambda = \lambda_K$ and that $E_1 = \frac{1}{2}$ at $\lambda = \lambda_H$, which agrees with earlier studies (e.g. [37, 112]), and which reflects what we discussed above.
2. E_1 is negative for most wavelengths in the range $\lambda_H < \lambda < \lambda_K$. The classical analogue to this would be that when an atom absorbs a photon at this frequency,

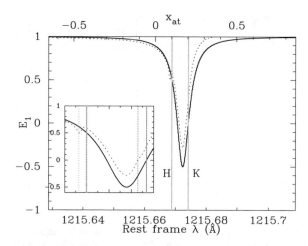

Fig. 1.50 The frequency dependence of the polarizability E_1 of a hydrogen atom is shown. The *black solid line* shows E_1 as a function of λ when fine structure splitting is accounted for (Eq. 1.115, taken from [254]). This plot shows that $E_1 = 0$ at $\lambda = \lambda_K$ and that $E_1 = \frac{1}{2}$ at $\lambda = \lambda_H$. Interestingly, E_1 is negative for most wavelengths in the range $\lambda_H < \lambda < \lambda_K$. However, scattering at these frequencies is very unlikely (see Sect. 1.10.2). More importantly, $E_1 = 1$ when a photon scatters in the wings of the line. That is, for photons that are scattered in the wing of the line profile, the electron in the hydrogen atom behaves like a classical dipole (i.e. as if it were a free electron!). The *red dotted line* shows the wavelength dependence of E_1 when hyperfine splitting is accounted for. Hyperfine splitting introduces an overall slightly higher level of polarization. The box in the *lower left corner* shows a close-up view of E_1 near the resonances (see text)

that then the electron oscillates along the propagation direction of the incoming wave, which is strange because the electron would be oscillating in a direction orthogonal to the direction of the electric vector of the electromagnetic wave. However, scattering at these frequencies is very unlikely (see Sect. 1.10.2).

3. $E_1 = 1$ when a photon scatters in the wings of the line, which is arguably the most bizarre aspect of this plot. Stenflo [254] points out that, when a Lyα photon scatters in the wing of the line profile, it goes simultaneously through the $2p_{1/2}$ and $2p_{3/2}$ states, and as a result, the bound electron is permitted to behave as if it were free.

The impact of 'quantum interference' on the scattering phase function is more than just interesting from a 'fundamental' viewpoint, because it makes a physical distinction between 'core' and 'wing' scattering. This distinction arises because E_1 affects the scattering phase function: the phase function for wing scattering $P_{\text{wing}} \propto (1 + \mu^2)$, while for core scattering the phase function can behave like a (sometimes strange, i.e. when $E_1 < 0$) superposition of isotropic and wing scattering (see Eq. 1.117 for an example of how to compute this phase function from E_1).

For reference: there is hyperfine splitting in the fine structure lines that is induced by the spin of the proton which can couple to the electron spin, which induces further splitting of the line. This is illustrated in Fig. 1.49, where each fine structure level

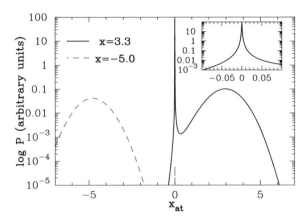

Fig. 1.51 The probability $P(x_{at}, x)$ (Eq. 1.116) that a photon of frequency x is scattered by an atom such that it appears at a frequency x_{at} in the frame of the atom (*Credit from Figure A2 of* [67], *'The polarization of scattered Ly α radiation around high-redshift galaxies', MNRAS, 386, 492D*). The *solid* and *dashed lines* correspond to $x = 3.3$ and $x = -5.0$ respectively. For $x = 3.3$, photons are either scattered by atoms to which they appear *exactly* at resonance (see inset), or to which they appear ∼3 Doppler widths away. For $x = -5$, resonant scattering is less important by orders of magnitude. In combination with Fig. 1.50, this figure shows that if a photon is resonantly scattered then E_1 is either 0 or $\frac{1}{2}$

breaks up into two lines which differ only in their quantum number F which measures the total + nuclear spin angular momentum, i.e. $\mathbf{F} = \mathbf{J} + \mathbf{I}$. The dependence of E_1 on frequency when this hyperfine splitting is accounted for has been calculated by Hirata (2006, his Appendix B. Note however, that E_1 is not computed explicitly. Instead, the angular redistribution functions are given and it is possible to extract E_1 from these). The formula for E_1 is quite lengthy, and the reader is referred to [127] for the full expression. We have overplotted as the *red dotted line* wavelength dependence of E_1 when hyperfine splitting is accounted for. The general frequency dependence of E_1 is not affected. However, there is an overall higher level of polarization. The box in the *lower left corner* of Fig. 1.50 shows a close-up view of E_1 near the resonances. Around the H-resonance, hyperfine splitting introduces new resonances (indicated by the *blue dotted lines*) which affect E_1 somewhat. At wavelengths longward of the H resonance, the E_1 is boosted slightly, which causes scattering through the K resonance to not be perfectly isotropic (another interpretation is that the hyperfine splitting breaks the perfect spherical symmetry of the $2p_{1/2}$ state). An overall boost in the polarizability as a result of hyperfine splitting has also been found by other authors (e.g. [38]).

1.10.2 Lyα Propagation Through HI: Resonant Versus Wing Scattering

We highlighted the distinction between 'core' versus 'wing' scattering previously in Sect. 1.10. As we already mentioned, thd polarizability can be negative $E_1 < 0$ for a range of frequencies between the H and K resonance frequencies (see Fig. 1.50). However, scattering at these frequencies does not occur often enough to leave an observable imprint. The reason for this is that the natural width of the line for both the H and K transitions is much smaller than their separation, i.e $\gamma_{H,K} \equiv \Gamma_{H,K}/[4\pi] \sim 10^8$ Hz $\ll \nu_H - \nu_K = 1.1 \times 10^{10}$ Hz, and the absorption cross-section in the atom's rest-frame scales as, $\sigma(\nu) \propto [(\nu - \nu_{H,K})^2 + \gamma_{H,K}^2]^{-1}$ (see Eq. 1.48). A Lyα photon is therefore much more likely to be absorbed by an atom for which the photon appears exactly at resonance, than by an atom for which the photon has a frequency corresponding to a negative E_1. Quantitatively, the Maxwellian probability P that a photon of frequency x is scattered in the frequency range $x_{at} \pm dx_{at}/2$ in the atom's rest-frame is given by

$$P(x_{at}|x)dx_{at} = \frac{a}{\pi H(a_v, x)} \frac{e^{-(x_{at}-x)^2}}{x_{at}^2 + a_v^2} dx_{at}, \qquad (1.116)$$

where we used that $P(x_{at}|x) = P(u_{at}|x) = P(x|u_{at})P(u_{at})/P(x)$, where $u_{at} = x - x_{at}$ denotes the atom velocity in units of v_{th} (also see the discussion above Eq. 1.69). The probability $P(x_{at}|x)$ is shown in Fig. 1.51 for[49] $x = 3.3$ (*solid line*) and $x = -5.0$ (*dashed line*). Figure 1.51 shows that for $x = 3.3$, photons are scattered either when they are exactly at resonance or when they appear ~ 3 Doppler widths away from resonance. The inset of Fig. 1.51 zooms on the region near $x_{at} = 0.0$. For frequencies $x \lesssim 3$ there are enough atoms moving at velocities such that the Lyα photon appears at exactly at resonance in the frame of the atom. However, this is not the case any more for frequencies $|x| \gtrsim 3$. Instead, the majority of photons is scattered while it is in the wing of the absorption profile. This is illustrated by the *dashed line* which shows the case $x = -5.0$, for which resonant scattering is less likely by orders of magnitude. This discussion illustrates that the transition from 'core' to 'wing' scattering is continuous (though it occurs over a narrow range of frequencies). Photons can 'resonantly' scatter while they are in the wing with a finite probability, and vice versa. While in practise, this is not an important effect, it is worth keeping in mind.

1.10.3 Polarization in Monte-Carlo Radiative Transfer

Incorporating polarization in a Monte-Carlo is complicated if you want to do it correctly. A simple procedure was presented by [12, 162], which is accurate for

---

[49]Note that the transition from core to wing scattering occurs at $x \sim 3$, see Fig. 1.17.

Fig. 1.52 This figure illustrates visually what the 'polarization' angle χ is, which can be used in Monte-Carlo calculations to compute the linear polarization of scattered Lyα as a function of projected distance $r \equiv |\mathbf{r}|$ from the galaxy/Lyα source

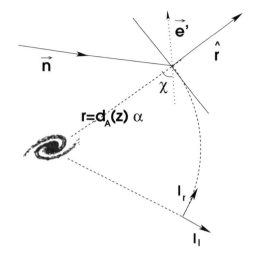

wing-scattering only. In practise this is often sufficient, as Lyα wing photons are the ones that are most likely to escape from a scattering medium, and are thus most likely to be observed.

Rybicki and Loeb [225] assigned 100% linear polarization to each Lyα photon used in the Monte-Carlo simulation by attaching an (normalized) electric vector $\mathbf{e}_E$ to the photon perpendicular to its propagation direction (see discussion above Eq. 1.37). For each scattering event, the outgoing direction ($\mathbf{k}_{out}$) is drawn from the phase-function $P \propto \sin^2 \Psi$ (see the discussion above Eq. 1.37), where $\cos \Psi = \mathbf{k}_{out} \cdot \mathbf{e}$ (see [12, 67, 162] for technical details). This approach thus accounts for the polarization-dependence of the phase function described previously in Sect. 1.5.1.

Consider a Lyα photon in the Monte-Carlo simulation that was propagating in a direction $\mathbf{k}_{in}$, and with an electric vector $\mathbf{e}_E$. It was then scattered towards the observer. The polarization of this photon when it is observed can be obtained as follows: we know that when a photon reaches us, its propagation direction, $\mathbf{k}_{out}$, is perpendicular to the plane of the sky. We therefore know that the polarization vector $\mathbf{e}'_E$ must lie in the plane of the sky. The linear polarization measures the difference in intensity when measured in the two orthogonal directions in the plane of the sky (denoted previously with $I_{||}$ and $I_\perp$). We now define $\mathbf{r}$ to be the vector that connects the location of last scattering to the Lyα source, projected onto the sky. Both $\mathbf{r}$ and $\mathbf{e}'_E$ therefore lie in the plane of the sky, and we let χ denote the angle between them (i.e. $\cos \chi \equiv \mathbf{r} \cdot \mathbf{e}'_E / |\mathbf{r}|$). The photon then contributes $\cos^2 \chi$ to I_l and $\sin^2 \chi$ to I_r. This geometry is depicted in Fig. 1.52. This procedure was tested successfully by [162] against analytic solutions obtained by Schuster [238].

For core scattering the situation is more complicated. We know that scattering through the H-transition corresponds to isotropic scattering, while scattering through the K-transition corresponds to scattering off an anisotropic particle with $E_1 = 0.5$. From their statistical weights we can infer that scattering through the K-transition is twice is likely as scattering through the H-transition ($g_H = 2, g_K = 4$). The weighted average implies that core scattering corresponds to scattering by an anisotropic particle with $E_1 = 1/3$. Equation 1.114 shows that the scattering matrix and phase function take on the following form:

$$R = \begin{pmatrix} \frac{1}{2}\cos^2\theta + \frac{1}{3} & \frac{1}{3} \\ \frac{1}{3} & \frac{5}{6} \end{pmatrix} \quad \Rightarrow \quad P(\mu) = \frac{11}{12} + \frac{3}{12}\mu^2, \qquad (1.117)$$

for unpolarized incident radiation. Formally, the frequency redistribution function for core scattering is therefore neither given by $R_A(x_{out}|x_{in})$ nor $R_B(x_{out}|x_{in})$ (Eq. 1.72), but by some intermediate form. Given the similarity of $R_A(x_{out}|x_{in})$ and $R_B(x_{out}|x_{in})$ (see Fig. 1.21) this difference does not matter in practice. Note that this phase function can be implemented naturally in a Monte-Carlo simulation by treating $1/3$ of all core scattering events as pure dipole scattering events, and the remaining $2/3$ as pure isotropic scattering events [263].

The previous approach assigns electric vectors to each Lyα photon in the Monte-Carlo simulation, and therefore implicitly assumes that each individual Lyα photon is 100% linearly polarized. It would be more realistic if we could assign a fractional polarization to each photon, which would be more representative of the radiation field. Recall that the phase-functions depend on the polarization of the radiation field. An alternative way of incorporating polarization which allows fractional polarization to be assigned to individual Lyα photons is given by the *density-matrix formalism* described in [8]). In this formalism all polarization information is encoded in 2 parameters (the 2 parameters reflect the degrees of freedom for a mass-less spin-0 'particle') within the 'density matrix'. We will not discuss this formalism in this lecture. Both methods should converge for scattering in optically thick gas, but they have not been compared systematically yet (but see [49] for recent work in this direction).

1.11 Applications Beyond Lyα: Wouthuysen-Field Coupling and 21-cm Cosmology/Astrophysics

1.11.1 The 21-cm Transition and its Spin Temperature

Detecting the redshifted 21-cm line from neutral hydrogen gas in the young Universe is one of the main challenges of observational cosmology for the next decades, and serves as the science driver for many low frequency arrays that were listed in Sect. 1.2.1. The 21-cm transition links the two hyperfine levels of the ground state ($1s$) of atomic hydrogen. The energy difference arises due to coupling of the proton to the electron spin: the proton spin S_p gives it a magnetic moment[50] $\mu_p =$

---

[50] A spinning proton can be seen as a rotating charged sphere, which produces a magnetic field.

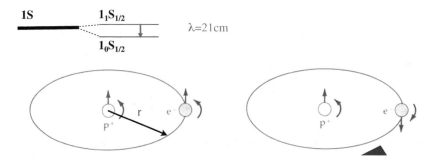

Fig. 1.53 This figure illustrates the classical picture of the origin of the energy difference between the two hyperfine levels of the ground state of atomic hydrogen. The proton spin generates a magnetic dipole moment which in turn generates a magnetic field. This magnetic field introduces an energy dependence to the orientation of the electron spin. In quantum mechanics however, the two states shown here have equal energy, unless the electron finds itself inside the proton (i.e unless $r = \mathbf{0}$)

$\frac{g_p e \hbar}{2 m_p c} S_p$, where the proton's 'g-factor' is $g_p \sim 5.59$. This magnetic dipole generates a magnetic field which interacts with the magnetic moment of the electron (μ_e) due to its spin. Classically, the energy difference between the two opposite electron spin states equals $\Delta E = 2|\mu_e||B_p|$, where B_p denotes the magnetic field generated by the spinning proton. This is illustrated schematically in Fig. 1.53. The 21-cm transition corresponds to the electron's spin 'flipping' in the magnetic field generated by the proton. The 21-cm transition is therefore often referred to as the 'spin-flip' transition. What is interesting about this classical picture is that it fails to convey that quantum mechanically, the energy difference between the two hyperfine levels of the 1s level is actually zero, *unless* $r = \mathbf{0}$. In quantum mechanics, there is a finite probability that the electron finds itself inside the proton. Formally, this is what causes the different hyperfine levels of the ground state of atomic hydrogen to have different energies (special thanks to D. Spiegel for pointing this out to me).

The 21-cm transition is a highly forbidden transition with a natural life-time of $t \equiv A_{21}^{-1} \sim (2.87 \times 10^{-15} \text{ s}^{-1})^{-1} \sim 1.1 \times 10^7$ yr (one way to interpret this long lifetime is to connect it to the low probability of the electron and proton overlapping). The 21-cm line has been observed routinely in nearby galaxies, and has allowed us to map out the distribution and kinematics of HI gas in galaxies. Observations of the kinematics of HI gas have given us galaxy rotation curves, which further confirmed the need for dark matter on galaxy scales. Because of its intrinsic faintness, it is difficult to detect HI gas in emission beyond $z \gtrsim 0.5$ until the Square Kilometer Array (SKA) becomes operational.

1.11.2 The 21-cm Brightness Temperature

It is theoretically possible to detect the 21-cm line from neutral gas during the reionization epoch and even the 'Dark Ages', which refers to the epoch prior to the formation of the first stars, black holes, etc., and when the only source of radiation was the Cosmic Microwave Background which had redshifted out of the visual band and into the infrared. The visibility of HI gas in its 21-cm line can be expressed in terms of a *differential* brightness temperature,[51] $\delta T_b(\nu)$, with respect to the background CMB, and equals (e.g. [70, 95], and references therein):

$$\delta T_b(\nu) \approx 9 x_{\rm HI}(1+\delta)(1+z)^{1/2}\left(1 - \frac{T_{\rm CMB}(z)}{T_S}\right)\left[\frac{H(z)(1+z)}{dv_\parallel/dr}\right] \text{ mK}, \quad (1.118)$$

where $\delta + 1 \equiv \rho/\bar{\rho}$ denotes the overdensity of the gas cloud, z the redshift of the cloud, $T_{\rm CMB}(z) = 2.73(1+z)$ K denotes the temperature of the CMB, the factor in square brackets contains the line-of-sight velocity gradient $dv_\parallel/dr$. Finally, T_S denotes the *spin temperature* (also known as the excitation temperature) of the 21-cm transition, which quantifies the number densities of hydrogen atoms in each of the hyperfine transitions, i.e.

$$\frac{n_1}{n_0} \equiv \frac{g_1}{g_0}\exp\left(\frac{-h\nu_{21\rm cm}}{k_B T_s}\right) = 3\exp\left(\frac{-h\nu_{21\rm cm}}{k_B T_s}\right), \quad (1.119)$$

where n_0 (n_1) denotes the number densities of hydrogen atoms in the ground (excited) level of the 21-cm transition, and where we used that $g_1 = 3$ and $g_0 = 1$. The numerical prefactor of 9 mK in Eq. 1.118 was derived assuming that the gas was undergoing Hubble expansion. For slower expansion rates, we increase the number of hydrogen atoms within a fixed velocity (and therefore frequency) range, which enhances the brightness temperature.

Equation 1.118 states that when $T_{\rm CMB}(z) < T_S$, we have $\delta T_b(\nu) > 0$, and when $T_{\rm CMB}(z) > T_S$ we have $\delta T_b(\nu) < 0$. This means that we see HI in absorption [emission] when $T_S < T_{\rm CMB}(z)$ [$T_S > T_{\rm CMB}(z)$]. The spin temperature thus plays a key role in setting the 21-cm signal, and we briefly discuss what physical processes set T_s below in Sect. 1.11.3. Equation 1.118 highlights why so much effort is going into trying to detect the 21-cm transition: the 21-cm transition corresponds to a line, having a 3D map of the 21-cm line provides us with a unique way of constraining the 3D density, velocity field etc. It provides us with a powerful cosmological and astrophysical probe of the high-redshift Universe.

---

[51]Eq. 1.118 follows from solving the radiative transfer equation, $\frac{dI}{d\tau} = -I + \frac{A_{10}}{B_{10}}\frac{1}{(3n_0/n_1 - 1)}$, in the (appropriate) limit that the neutral IGM is optically thin in the HI 21-cm line, and that it therefore only slightly modifies the intensity I of the background CMB. It is common in radio astronomy to express intensity fluctuations as temperature fluctuations by recasting intensity as a temperature in the Rayleigh-Jeans limit: $I_\nu(\mathbf{x}) \equiv \frac{2k_B T(\mathbf{x})}{\lambda}$.

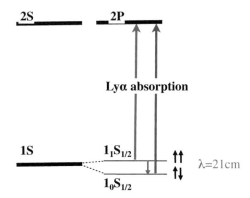

Fig. 1.54 Lyα photons that have a higher energy are (slightly) more likely to be absorbed from the $1_0 S_{1/2}$ level because of the (again slightly) larger energy separation between this level and any $n = 2$ level. The precise spectrum of Lyα photons around the line resonance therefore affects n_0 (number density of hydrogen atoms in the $1_0 S_{1/2}$ level) and n_1 (number density of hydrogen atoms in the $1_1 S_{1/2}$ level). Repeated scattering of Lyα photons in turn modifies the spectrum in such as a way that Lyα scattering drives the spin temperature to the gas temperature. This is known as the Wouthuysen-Field effect

1.11.3 The Spin Temperature and the Wouthuysen-Field Effect

The spin temperature—i.e. how the two hyperfine levels are populated—is set by (i) collisions, which drive $T_S \rightarrow T_{\text{gas}}$, (ii) absorption by CMB photons, which drives $T_S \rightarrow T_{\text{CMB}}$ (and thus that $\delta T_b \rightarrow 0$, see Eq. 1.118), and (iii) Lyα scattering. This is illustrated in Fig. 1.54: absorption of Lyα photons can occur from any of the two hyperfine levels. The subsequent radiative cascade back to the ground state can leave the atom in either hyperfine level. Lyα scattering thus mixes the two hyperfine levels, which drive $T_S \rightarrow T_\alpha$, where T_α is known as the Lyα color temperature, which will be discussed below. Quantitatively, it has been shown that (e.g. [95, 163, 259])

$$\frac{1}{T_S} = \frac{T_{\text{CMB}}^{-1} + x_c T_{\text{gas}}^{-1} + x_\alpha T_\alpha^{-1}}{1 + x_c + x_\alpha}, \tag{1.120}$$

where $x_c = \frac{C_{21} T_*}{A_{21} T_{\text{gas}}}$ denotes the collisional coupling coefficient, in which $k_B T_*$ denotes the energy difference between the hyperfine levels, C_{21} denotes the collisional deexcitation rate coefficient. Furthermore, $x_\alpha = \frac{4 P_\alpha}{27 A_{21} T_\alpha}$, in which P_α denotes the Lyα

scattering rate.[52] The Lyα color temperature provides a measure of the shape of the spectrum near the Lyα resonance [65, 177]:

$$\frac{k_B T_\alpha}{h_P} = \frac{\int J(v)\sigma_\alpha(v)dv}{\int \frac{\partial \sigma_\alpha}{\partial v}J(v)dv} = -\frac{\int J(v)\sigma_\alpha(v)dv}{\int \frac{\partial J}{\partial v}\sigma_\alpha(v)dv}, \tag{1.121}$$

where $\sigma_\alpha(v)$ denotes the Lyα absorption cross-section (see Eq. 1.54), and $J(v)$ denotes the angle-averaged intensity (see Sect. 1.6.5). In the last step, we integrated by parts. The reason the spectral shape near Lyα enters is illustrated in Fig. 1.54, which shows that higher frequency Lyα photons are (slightly) more likely to excite hydrogen atoms from the ground (singlet) state of the 21-cm transition. The relative number of Lyα photons slightly redward and blueward of the resonance therefore affects the 21-cm spin temperature [277]. Equation 1.121 indicates that if $\frac{\partial J(v)}{\partial v} < 0$, then $T_\alpha > 0$. Interestingly, it has been demonstrated that repeated scattering of Lyα photons changes the Lyα spectrum around the resonance such that it drives $T_\alpha \rightarrow T_{\text{gas}}$ [86]. The resulting coupling between T_s and T_{gas} as a result of Lyα scattering is known as the *Wouthuysen-Field* (WF)[53] coupling. We generally expect T_s to be some weighted average of the gas and CMB temperature, where the precise weight depends on various quantities such as gas temperature, density and the WF-coupling strength.

There are two key ingredients in WF-coupling: The Lyα color temperature T_α and the Lyα scattering rate. Theoretically, Lyα scattering rates are boosted in close proximity to star forming galaxies because of the locally enhanced Lyα background [54]. However, this local boost in the WF-coupling strength can be further affected by the assumed spectrum of photons emerging from the galaxy (galaxies themselves are optically thick to Lyα and higher Lyman series radiation), which has not been explored yet.

1.11.4 The Global 21-cm Signal

Equation 1.120 implies that the spin temperature T_s is a weighted average of the gas and CMB temperature. The *left panel* of Fig. 1.55 shows the universally (or globally) averaged temperatures of the gas (*red solid line*) and the CMB (*blue solid line*), and the corresponding 'global' 21-cm signature in the *Right panel*. We discuss these in more detail below:

---

[52]Note that this equation uses that for each Lyα scattering event, the probability that it induces a scattering event is $P_{\text{flip}} = \frac{4}{27}$ (see e.g. [65, 177]). This probability reduces by many orders of magnitude for wing scattering as $P_{\text{flip}} \propto x^{-2}$ (see [65, 127]). In practise wing scattering contributes little to the overall scattering rate, but it is good to keep this in mind.

[53]See [95] for an explanation of how to best pronounce *Wouthuysen*. *Hint:* it helps if you hold your breath 12 s before trying.

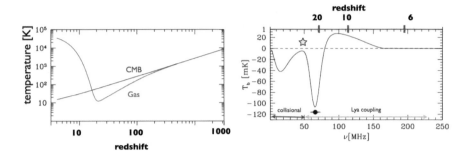

Fig. 1.55 *Left panel:* Globally, averaged redshift evolution of T_{gas} (*red solid line*) and the T_{CMB} (*blue solid line*). Right panel: The 'Global' 21-signal, represents the sky-averaged 21-cm brightness temperature $\delta T_b(\nu)$. See the main text for a description of each of these curves. The global 21-cm signal constrains when the first stars, black holes, and galaxies formed and depends on their spectra. *Credit adapted from slides created by J. Pritchard (based on [212, 213])*

- Adiabatic expansion of the Universe causes $T_{CMB} \propto (1 + z)$ at all redshifts. At $z \gtrsim 100$, the gas temperature remains coupled to the CMB temperature because of the interaction between CMB photons and the small fraction of free electrons that exist because recombination is never 'complete' (i.e. there is a residual fraction of protons that never capture an electron to form a hydrogen atom). When $T_{CMB} = T_{gas}$, we must have $T_s = T_{CMB}$, and therefore that $\delta T_b(\nu) = 0$ mK, which corresponds to the high-z limit in the *right panel*.

- At $z \lesssim 100$, electron scattering can no longer couple the CMB and gas temperatures, and the (non-relativistic) gas adiabatically cools faster than the CMB as $T_{gas} \propto (1 + z)^2$. Because $T_{gas} < T_{CMB}$, we must have that $T_s < T_{CMB}$ and we see the 21-cm line in absorption. When T_{gas} first decouples from T_{CMB} the gas densities are high enough for collisions to keep T_s locked to T_{gas}. However, at $z \sim 70$ ($\nu \sim 20$ MHz) collisions can no longer couple T_s to T_{gas}, and T_s crawls back to T_{CMB}, which reduces $\delta T_b(\nu)$ (at $\nu \sim 20$–50 MHz, i.e. $z \sim$70–30).

- The first stars, galaxies, and accreting black holes emitted UV photons in the energy range $E = 10.2 - 13.6$ eV. These photons can travel freely through the neutral IGM, until they redshift into one of the Lyman series resonances, at which point a radiative cascade can produce Lyα. The formation of the first stars thus generates a Lyα background, which initiates the WF-coupling, which pushes T_s back down to T_{gas}. The onset of Lyα scattering—and thus the WF coupling—causes $\delta T_b(\nu)$ to drop sharply at $\nu \gtrsim 50$ Mhz ($z \lesssim 30$).

- At some point the radiation of stars, galaxies, and black holes starts heating the gas. Especially X-rays produced by accreting black holes can easily penetrate deep into the cold, neutral IGM and contain a lot of energy which can be converted into heat after they are absorbed. The *left panel* thus has T_{gas} increase at $z \sim 20$, which corresponds to onset of X-ray heating. In the *right panel* this onset occurs a bit earlier. This difference reflects that the redshift of all the features (minima and maxima) in $\delta T_b(\nu)$ are model dependent, and not well-known (more on this below). With the onset of X-ray heating (combined with increasingly efficient WF coupling to the build-up of the Lyα background) drives $\delta T_b(\nu)$ up until it becomes positive when $T_{gas} > T_{CMB}$.

- Finally, $\delta T_b(\nu)$ reaches yet another maximum, which reflects that neutral, X-ray heated gas is reionized away by the ionizing UV-photons emitted by star forming galaxies and quasars. When reionization is complete, there is no diffuse intergalactic neutral hydrogen left, and $\delta T_b(\nu) \to 0$.

In detail, the onset and redshift evolution of Lyα coupling, X-ray heating, and reionization depend on the redshift evolution of the number densities of galaxies, and their spectral characteristics. All these are uncertain, and it is not possible to make robust predictions for the precise shape of the global 21-cm signature. Instead, one of the main challenges for observational cosmology is to measure the global 21-cm signal, and from this constrain the abundances and characteristic of first generations of galaxies in our Universe. Detecting the global 21-cm is challenging, but especially the deep absorption trough that is expected to exist just prior to the onset of X-ray heating at $\nu \sim 70$ MHz is something that may be detectable because of its characteristic spectral shape. It is interesting that the presence of this absorption feature relies on the presence of a Lyα background, which must be strong enough to enable WF-coupling.

Acknowledgements I thank Ivy Wong for providing a figure for a section which (unfortunately) was cut as a whole, Daniel Mortlock and Chris Hirata for providing me with tabulated values of their calculations which I used to create Fig. 1.17, Jonathan Pritchard for permission to use one of his slides for these notes, Andrei Mesinger for providing Fig. 1.42, and other colleagues for their permission to re-use Figures from their papers. I thank the astronomy department at UCSB for their kind hospitality when I was working on preparing these lecture notes. I thank the organizers of the school: Anne Verhamme, Pierre North, Hakim Atek, Sebastiano Cantalupo, Myriam Burgener Frick, to Matt Hayes, X. Prochaska, and Masami Ouchi for their inspiring lectures. I thank Pierre North for carefully reading these notes, and for finding and correcting countless typos. Finally, special thanks to the students for their excellent attendance, and for their interest and enthusiastic participation.

References

1. Adams, T.F.: ApJ **168**, 575 (1971)
2. Adams, T.F.: ApJ **174**, 439 (1972)
3. Agertz, O., Teyssier, R., Moore, B.: MNRAS **397**, L64 (2009)
4. Ahn, S.-H., Lee, H.-W., Lee, H.M.: J. Korean Astron. Soc. **33**, 29 (2000)
5. Ahn, S.-H., Lee, H.-W., Lee, H.M.: ApJ **554**, 604 (2001)
6. Ahn, S.-H., Lee, H.-W., Lee, H.M.: ApJ **567**, 922 (2002)
7. Ahn, S.-H., Lee, H.-W., Lee, H.M.: MNRAS **340**, 863 (2003)
8. Ahn, S.-H., Lee, H.-W.: J. Korean Astronomical Soc. **48**, 195 (2015)
9. Aggarwal, K.M.: MNRAS **202**, 15P (1983)
10. Aggarwal, K.M., Berrington, K.A., Burke, P.G., Kingston, A.E., Pathak, A.: J. Phys. B At. Mol. Phys. **24**, 1385 (1991)
11. Alipour, E., Sigurdson, K., Hirata, C.M.: Phys. Rev. D **91**, 083520 (2015)

12. Angel, J.R.P.: ApJ **158**, 219 (1969)
13. Atek, H., Kunth, D., Hayes, M., Östlin, G., Mas-Hesse, J.M.: A&A **488**, 491 (2008)
14. Atek, H., Kunth, D., Schaerer, D., et al.: A&A **506**, L1 (2009)
15. Aver, E., Olive, K.A., Skillman, E.D.: JCAP **7**, 011 (2015)
16. Bach, K., Lee, H.-W.: J. Korean Astronomical Soc. **47**, 187 (2014)
17. Bach, K., Lee, H.-W.: MNRAS **446**, 264 (2015)
18. Barkana, R., Loeb, A.: Phys. Rep. **349**, 125 (2001)
19. Barkana, R.: MNRAS **347**, 59 (2004)
20. Barkana, R., Loeb, A.: ApJ **609**, 474 (2004)
21. Barnes, L.A., Haehnelt, M.G.: MNRAS **403**, 870 (2010)
22. Barnes, L.A., Haehnelt, M.G., Tescari, E., Viel, M.: MNRAS **416**, 1723 (2011)
23. Beck, M., Scarlata, C., Hayes, M., Dijkstra, M., Jones, T.J.: ApJ **818**, 138 (2016)
24. Behrens, C., Niemeyer, J.: A&A **556**, A5 (2013)
25. Behrens, C., Dijkstra, M., Niemeyer, J.C.: A&A **563**, A77 (2014)
26. Bely, O., van Regemorter, H.: ARA&A **8**, 329 (1970)
27. Birnboim, Y., Dekel, A.: MNRAS **345**, 349 (2003)
28. Black, J.H.: MNRAS **197**, 553 (1981)
29. Blanc, G.A., Adams, J.J., Gebhardt, K., et al.: ApJ **736**, 31 (2011)
30. Bolton, J.S., Haehnelt, M.G., Warren, S.J., et al.: MNRAS **416**, L70 (2011)
31. Bolton, J.S., Haehnelt, M.G.: MNRAS **429**, 1695 (2013)
32. Bonilha, J.R.M., Ferch, R., Salpeter, E.E., Slater, G., Noerdlinger, P.D.: ApJ **233**, 649 (1979)
33. Bosma, A.: Ph.D. thesis (1978)
34. Borisova, E., Cantalupo, S., Lilly, S.J., et al.: ApJ **831**, 39 (2016)
35. Bouwens, R.J., Illingworth, G.D., Oesch, P.A., et al.: ApJ **754**, 83 (2012)
36. Bower, R.: Nature **476**, 288 (2011)
37. Brandt, J.C., Chamberlain, J.W.: ApJ **130**, 670 (1959)
38. Brasken, M., Kyrola, E.: A&A **332**, 732 (1998)
39. Breit, G., Teller, E.: ApJ **91**, 215 (1940)
40. Burgess, A.: MNRAS **69**, 1 (1965)
41. Cantalupo, S., Porciani, C., Lilly, S.J., Miniati, F.: ApJ **628**, 61 (2005)
42. Cantalupo, S., Porciani, C., Lilly, S.J.: ApJ **672**, 48 (2008)
43. Cantalupo, S., Lilly, S.J., Haehnelt, M.G.: MNRAS **425**, 1992 (2012)
44. Cantalupo, S., Arrigoni-Battaia, F., Prochaska, J.X., Hennawi, J.F., Madau, P.: Nature **506**, 63 (2014)
45. Caruana, J., Bunker, A.J., Wilkins, S.M., et al.: MNRAS **427**, 3055 (2012)
46. Caruana, J., Bunker, A.J., Wilkins, S.M., et al.: MNRAS **443**, 2831 (2014)
47. Cen, R.: ApJS **78**, 341 (1992)
48. Chandrasekhar, S.: 1960. Dover, New York (1960)
49. Chang, S.-J., Lee, H.-W., Yang, Y.: MNRAS **464**, 5018 (2017)
50. Chapman, S.C., Blain, A.W., Smail, I., Ivison, R.J.: ApJ **622**, 772 (2005)
51. Chonis, T.S., Blanc, G.A., Hill, G.J., et al.: ApJ **775**, 99 (2013)
52. Choudhury, T.R., Puchwein, E., Haehnelt, M.G., Bolton, J.S.: MNRAS **452**, 261 (2015)
53. Chung, A., van Gorkom, J.H., Kenney, J.D.P., Crowl, H., Vollmer, B.: AJ **138**, 1741 (2009)
54. Chuzhoy, L., Zheng, Z.: ApJ **670**, 912 (2007)
55. Ciardi, B., Stoehr, F., White, S.D.M.: MNRAS **343**, 1101 (2003)
56. Cooper, J.L., Bicknell, G.V., Sutherland, R.S., Bland-Hawthorn, J.: ApJ **674**, 157 (2008)
57. Cox, A.N.: Allen's Astrophysical Quantities, 1 (2000)
58. Dayal, P., Ferrara, A., Saro, A.: MNRAS **402**, 1449 (2010)
59. Dayal, P., Maselli, A., Ferrara, A.: MNRAS **410**, 830 (2011)
60. Deharveng, J.-M., Small, T., Barlow, T.A., et al.: ApJ **680**, 1072 (2008)
61. Dennison, B., Turner, B.E., Minter, A.H.: ApJ **633**, 309 (2005)
62. Dessauges-Zavadsky, M., D'Odorico, S., Schaerer, D., Modigliani, A., Tapken, C., Vernet, J.: A&A **510**, A26 (2010)
63. Dijkstra, M., Haiman, Z., Spaans, M.: ApJ **649**, 14 (2006)

64. Dijkstra, M., Lidz, A., Wyithe, J.S.B.: MNRAS **377**, 1175 (2007)
65. Dijkstra, M., Loeb, A.: NA **13**, 395 (2008)
66. Dijkstra, M., Loeb, A.: MNRAS **391**, 457 (2008)
67. Dijkstra, M., Loeb, A.: MNRAS **386**, 492 (2008)
68. Dijkstra, M.: ApJ **690**, 82 (2009)
69. Dijkstra, M., Loeb, A.: MNRAS **400**, 1109 (2009)
70. Dijkstra, M., Wyithe, J.S.B.: MNRAS **408**, 352 (2010)
71. Dijkstra, M., Westra, E.: MNRAS **401**, 2343 (2010)
72. Dijkstra, M., Mesinger, A., Wyithe, J.S.B.: MNRAS **414**, 2139 (2011)
73. Dijkstra, M., Wyithe, J.S.B.: MNRAS **419**, 3181 (2012)
74. Dijkstra, M., Jeeson-Daniel, A.: MNRAS **435**, 3333 (2013)
75. Dijkstra, M.: PASA **31**, e040 (2014)
76. Dijkstra, M., Sethi, S., Loeb, A.: ApJ **820**, 10 (2016)
77. Dijkstra, M., Gronke, M.: ApJ **828**, 71 (2016)
78. Draine, B.T., Dale, D.A., Bendo, G., et al.: ApJ **663**, 866 (2007)
79. Duval, F., Schaerer, D., Östlin, G., Laursen, P.: A&A **562**, A52 (2014)
80. Erb, D.K., Steidel, C.C., Trainor, R.F., et al.: ApJ **795**, 33 (2014)
81. Ewen, H.I., Purcell, E.M.: Nature **168**, 356 (1951)
82. Faisst, A.L., Capak, P., Carollo, C.M., Scarlata, C., Scoville, N.: ApJ **788**, 87 (2014)
83. Fardal, M.A., Katz, N., Gardner, J.P., et al.: ApJ **562**, 605 (2001)
84. Faucher-Giguère, C.-A., Prochaska, J.X., Lidz, A., Hernquist, L., Zaldarriaga, M.: ApJ **681**, 831 (2008)
85. Faucher-Giguère, C.-A., Kereš, D., Dijkstra, M., Hernquist, L., Zaldarriaga, M.: ApJ **725**, 633 (2010)
86. Field, G.B.: ApJ **129**, 551 (1959)
87. Finkelstein, S.L., Rhoads, J.E., Malhotra, S., Grogin, N., Wang, J.: ApJ **678**, 655 (2008)
88. Finkelstein, S.L., Papovich, C., Salmon, B., et al.: ApJ **756**, 164 (2012)
89. Finlator, K., Özel, F., Davé, R.: MNRAS **393**, 1090 (2009)
90. Foreman-Mackey, D., Hogg, D.W., Lang, D., Goodman, J.: PASP **125**, 306 (2013)
91. Forero-Romero, J.E., Yepes, G., Gottlöber, S., et al.: MNRAS **415**, 3666 (2011)
92. Fujita, A., Martin, C.L., Mac Low, M.-M., New, K.C.B., Weaver, R.: ApJ **698**, 693 (2009)
93. Furlanetto, S.R., Zaldarriaga, M., Hernquist, L.: ApJ **613**, 1 (2004)
94. Furlanetto, S.R., Schaye, J., Springel, V., Hernquist, L.: ApJ **622**, 7 (2005)
95. Furlanetto, S.R., Oh, S.P., Briggs, F.H.: Phys. Rep. **433**, 181 (2006)
96. Fynbo, J.U., Møller, P., Warren, S.J.: MNRAS **305**, 849 (1999)
97. Garel, T., Blaizot, J., Guiderdoni, B., et al.: MNRAS **422**, 310 (2012)
98. Gnedin, N.Y.: ApJ **535**, 530 (2000)
99. Gnedin, N.Y.: ApJ **821**, 50 (2016)
100. Goerdt, T., Dekel, A., Sternberg, A., et al.: MNRAS **407**, 613 (2010)
101. Gordon, W.: Annalen der Physik **394**, 1031 (1929)
102. Greig, B., Mesinger, A.: MNRAS **465**, 4838 (2017)
103. Gronke, M., Dijkstra, M.: MNRAS **444**, 1095 (2014)
104. Gronke, M., Bull, P., Dijkstra, M.: ApJ **812**, 123 (2015)
105. Gronke, M., Dijkstra, M.: ApJ **826**, 14 (2016)
106. Gronke, M., Dijkstra, M., McCourt, M., Oh, S.P.: ApJ **833**, L26 (2016)
107. Guaita, L., Melinder, J., Hayes, M., et al.: A&A **576**, A51 (2015)
108. Gunn, J.E., Peterson, B.A.: ApJ **142**, 1633 (1965)
109. Haiman, Z., Spaans, M., Quataert, E.: ApJ **537**, L5 (2000)
110. Haiman, Z., Rees, M.J.: ApJ **556**, 87 (2001)
111. Haiman, Z.: ApJ **576**, L1 (2002)
112. Hamilton, D.R.: ApJ **106**, 457 (1947)
113. Hansen, M., Oh, S.P.: MNRAS **367**, 979 (2006)
114. Harrington, J.P.: MNRAS **162**, 43 (1973)
115. Hashimoto, T., Ouchi, M., Shimasaku, K., et al.: ApJ **765**, 70 (2013)

116. Hashimoto, T., Verhamme, A., Ouchi, M., et al.: ApJ **812**, 157 (2015)
117. Hayashino, T., Matsuda, Y., Tamura, H., et al.: AJ **128**, 2073 (2004)
118. Hayes, M., Östlin, G., Mas-Hesse, J.M., et al.: A&A **438**, 71 (2005)
119. Hayes, M., Östlin, G., Atek, H., et al.: MNRAS **382**, 1465 (2007)
120. Hayes, M., Östlin, G., Schaerer, D., et al.: Nature **464**, 562 (2010)
121. Hayes, M., Schaerer, D., Östlin, G., et al.: ApJ **730**, 8 (2011)
122. Hayes, M., Scarlata, C., Siana, B.: Nature **476**, 304 (2011)
123. Hayes, M., Östlin, G., Schaerer, D., et al.: ApJ **765**, L27 (2013)
124. Heckman, T.M., Borthakur, S., Overzier, R., et al.: ApJ **730**, 5 (2011)
125. Hennawi, J.F., Prochaska, J.X., Cantalupo, S., Arrigoni-Battaia, F.: Science **348**, 779 (2015)
126. Henry, A., Scarlata, C., Martin, C.L., Erb, D.: ApJ **809**, 19 (2015)
127. Hirata, C.M.: MNRAS **367**, 259 (2006)
128. Hoang-Binh, D.: A&A **238**, 449 (1990)
129. Hu, E.M., Cowie, L.L., McMahon, R.G.: ApJ **502**, L99 (1998)
130. Rt, D.G.: MNRAS **125**, 21 (1962)
131. Humphrey, A., Vernet, J., Villar-Martín, M., et al.: ApJ **768**, L3 (2013)
132. Iliev, I.T., Mellema, G., Pen, U.-L., et al.: MNRAS **369**, 1625 (2006)
133. Iliev, I.T., Shapiro, P.R., McDonald, P., Mellema, G., Pen, U.-L.: MNRAS **391**, 63 (2008)
134. Izotov, Y.I., Thuan, T.X., Guseva, N.G.: MNRAS **445**, 778 (2014)
135. Jeeson-Daniel, A., Ciardi, B., Maio, U., et al.: MNRAS **424**, 2193 (2012)
136. Jensen, H., Laursen, P., Mellema, G., et al.: MNRAS **428**, 1366 (2013)
137. Jensen, H., Hayes, M., Iliev, I.T., et al.: MNRAS **444**, 2114 (2014)
138. Jones, T., Stark, D.P., Ellis, R.S.: ApJ **751**, 51 (2012)
139. Jones, T.A., Ellis, R.S., Schenker, M.A., Stark, D.P.: ApJ **779**, 52 (2013)
140. Kalberla, P.M.W., Burton, W.B., Hartmann, D., et al.: A&A **440**, 775 (2005)
141. Kakiichi, K., Dijkstra, M., Ciardi, B., Graziani, L.: MNRAS **463**, 4019 (2016)
142. Kashikawa, N., Shimasaku, K., Malkan, M.A., et al.: ApJ **648**, 7 (2006)
143. Kashikawa, N., Shimasaku, K., Matsuda, Y., et al.: ApJ **734**, 119 (2011)
144. Kennicutt Jr., R.C.: ARA&A **36**, 189 (1998)
145. Kobayashi, M.A.R., Kamaya, H., Yonehara, A.: ApJ **636**, 1 (2006)
146. Kollmeier, J.A., Zheng, Z., Davé, R., et al.: ApJ **708**, 1048 (2010)
147. Kornei, K.A., Shapley, A.E., Erb, D.K., et al.: ApJ **711**, 693 (2010)
148. Krolik, J.H.: ApJ **338**, 594 (1989)
149. Kunth, D., Mas-Hesse, J.M., Terlevich, E., Terlevich, R., Lequeux, J., Fall, S.M.: A&A **334**, 11 (1998)
150. Kulas, K.R., Shapley, A.E., Kollmeier, J.A., et al.: ApJ **745**, 33 (2012)
151. Lake, E., Zheng, Z., Cen, R., et al.: ApJ **806**, 46 (2015)
152. Laursen, P., Sommer-Larsen, J.: ApJ **657**, L69 (2007)
153. Laursen, P., Razoumov, A.O., Sommer-Larsen, J.: ApJ **696**, 853 (2009a)
154. Laursen, P., Sommer-Larsen, J., Andersen, A.C.: ApJ **704**, 1640 (2009b)
155. Laursen, P.: Ph.D. thesis, Niels Bohr Institute, University of Copenhagen (2010). arXiv:1012.3175
156. Laursen, P., Sommer-Larsen, J., Razoumov, A.O.: ApJ **728**, 52 (2011)
157. Laursen, P., Duval, F., Östlin, G.: ApJ **766**, 124 (2013)
158. Lee, J.-S.: ApJ **192**, 465 (1974)
159. Lee, H.-W., Blandford, R.D., Western, L.: MNRAS **267**, 303 (1994)
160. Lee, H.-W., Ahn, S.-H.: ApJ **504**, L61 (1998)
161. Lee, H.-W.: ApJ **594**, 637 (2003)
162. Loeb, A., Rybicki, G.B.: ApJ **524**, 527 (1999)
163. Madau, P., Meiksin, A., Rees, M.J.: ApJ **475**, 429 (1997)
164. Majumdar, S., Mellema, G., Datta, K.K., et al.: MNRAS **443**, 2843 (2014)
165. Martin, C.L., Dijkstra, M., Henry, A., et al.: ApJ **803**, 6 (2015)
166. Mas-Ribas, L., Dijkstra, M.: ApJ **822**, 84 (2016)
167. Mas-Ribas, L., Dijkstra, M., Hennawi, J.F., et al.: ApJ **841**, 19 (2017)

168. Matsuda, Y., et al.: AJ **128**, 569 (2004)
169. Matsuda, Y., Yamada, T., Hayashino, T., et al.: MNRAS **410**, L13 (2011)
170. Matsuda, Y., Yamada, T., Hayashino, T., Yamauchi, R., Nakamura, Y., Morimoto, N., Ouchi, M., Ono, Y., Umemura, M., Mori, M.: MNRAS **425**, 878 (2012)
171. Matthee, J., Sobral, D., Santos, S., et al.: MNRAS **451**, 400 (2015)
172. McCarthy, P.J.: ARA&A **31**, 639 (1993)
173. McKee, C.F., Ostriker, J.P.: ApJ **218**, 148 (1977)
174. McLinden, E.M., Finkelstein, S.L., Rhoads, J.E., et al.: ApJ **730**, 136 (2011)
175. McLinden, E.M., Rhoads, J.E., Malhotra, S., et al.: MNRAS **439**, 446 (2014)
176. McQuinn, M., Hernquist, L., Zaldarriaga, M., Dutta, S.: MNRAS **381**, 75 (2007)
177. Meiksin, A.: MNRAS **370**, 2025 (2006)
178. Meiksin, A.A.: Rev. Mod. Phys. **81**, 1405 (2009)
179. Mellema, G., Iliev, I.T., Alvarez, M.A., Shapiro, P.R.: NA **11**, 374 (2006)
180. Mesinger, A., Furlanetto, S.: ApJ **669**, 663 (2007)
181. Mesinger, A., Furlanetto, S.R.: MNRAS **386**, 1990 (2008)
182. Mesinger, A., Aykutalp, A., Vanzella, E., et al.: MNRAS **446**, 566 (2015)
183. Mesinger, A., Furlanetto, S., Cen, R.: MNRAS **411**, 955 (2011)
184. Mesinger, A.: Understanding the Epoch of Cosmic Reionization: Challenges and Progress. Ap&SSL Conference, Series, p. 423 (2016)
185. Miralda-Escudé, J.: ApJ **501**, 15–22 (1998)
186. Momose, R., Ouchi, M., Nakajima, K., et al.: MNRAS **442**, 110 (2014)
187. Morales, M.F., Wyithe, J.S.B.: ARA&A **48**, 127 (2010)
188. Mortlock, D.J.: Ap&SSL Conf. Ser. **423**, 187 (2016)
189. Neufeld, D.A.: ApJ **350**, 216 (1990)
190. Neufeld, D.A.: ApJ **370**, L85 (1991)
191. Nestor, D.B., Shapley, A.E., Steidel, C.C., Siana, B.: ApJ **736**, 18 (2011)
192. Nestor, D.B., Shapley, A.E., Kornei, K.A., Steidel, C.C., Siana, B.: ApJ **765**, 47 (2013)
193. Nilsson, K.K., Møller, P.: A&A **508**, L21 (2009)
194. North, P.L., Courbin, F., Eigenbrod, A., Chelouche, D.: A&A **542**, A91 (2012)
195. Ono, Y., Ouchi, M., Mobasher, B., et al.: ApJ **744**, 83 (2012)
196. Orsi, A., Lacey, C.G., Baugh, C.M.: MNRAS **425**, 87 (2012)
197. Osterbrock, D.E.: ApJ **135**, 195 (1962)
198. Osterbrock, D.E., Ferland, G.J.: Astrophysics of gaseous nebulae and active galactic nuclei. In: Osterbrock, D.E., Ferland, G.J. (eds.) University Science Books, Sausalito, CA, 2nd (2006)
199. Östlin, G., Hayes, M., Kunth, D., et al.: AJ **138**, 923 (2009)
200. Östlin, G., Hayes, M., Duval, F., et al.: ApJ **797**, 11 (2014)
201. Ota, K., Iye, M., Kashikawa, N., et al.: ApJ **722**, 803 (2010)
202. Ota, K., Iye, M., Kashikawa, N., et al.: ApJ **844**, 85 (2017)
203. Ouchi, M., Shimasaku, K., Akiyama, M., et al.: ApJS **176**, 301 (2008)
204. Ouchi, M., Shimasaku, K., Furusawa, H., et al.: ApJ **723**, 869 (2010)
205. Partridge, R.B., Peebles, P.J.E.: ApJ **147**, 868 (1967)
206. Pawlik, A.H., Schaye, J.: MNRAS **389**, 651 (2008)
207. Peebles, P.J.E.: Princeton Series in Physics. Princeton University Press, Princeton, NJ (1993)
208. Pentericci, L., Fontana, A., Vanzella, E., et al.: ApJ **743**, 132 (2011)
209. Pentericci, L., Vanzella, E., Fontana, A., et al.: ApJ **793**, 113 (2014)
210. Pierleoni, M., Maselli, A., Ciardi, B.: MNRAS **393**, 872 (2009)
211. Planck collaboration, Ade, P.A.R., Aghanim, N., et al.: A&A **594**, A13 (2016)
212. Pritchard, J.R., Loeb, A.: Phys. Rev. D **82**, 023006 (2010)
213. Pritchard, J.R., Loeb, A.: Rep. Prog. Phys. **75**, 086901 (2012)
214. Prochaska, J.X.: ApJ **650**, 272 (2006)
215. Rahmati, A., Pawlik, A.H., Raicevic, M., Schaye, J.: MNRAS **430**, 2427 (2013)
216. Raiter, A., Schaerer, D., Fosbury, R.A.E.: A&A **523**, 64 (2010)
217. Rauch, M., Haehnelt, M., Bunker, A., et al.: ApJ **681**, 856 (2008)
218. Rémy-Ruyer, A., Madden, S.C., Galliano, F., et al.: A&A **563**, A31 (2014)

219. Reuland, M., van Breugel, W., Röttgering, H., et al.: ApJ **592**, 755 (2003)
220. Rivera-Thorsen, T.E., Hayes, M., Östlin, G., et al.: ApJ **805**, 14 (2015)
221. Rosdahl, J., Blaizot, J.: MNRAS **423**, 344 (2012)
222. Rubiño-Martín, J.A., Chluba, J., Sunyaev, R.A.: MNRAS **371**, 1939 (2006)
223. Rybicki, G.B., Lightman, A.P.: Wiley-Interscience, New York (1979)
224. Rybicki, G.B., dell'Antonio, I.P.: ApJ **427**, 603 (1994)
225. Rybicki, G.B., Loeb, A.: ApJ **520**, L79 (1999)
226. Salpeter, E.E.: ApJ **121**, 161 (1955)
227. Santos, M.R.: MNRAS **349**, 1137 (2004)
228. Santos, S., Sobral, D., Matthee, J.: MNRAS **463**, 1678 (2016)
229. Scarlata, C., Colbert, J., Teplitz, H.I., et al.: ApJ **704**, L98 (2009)
230. Schaerer, D.: A&A **382**, 28 (2002)
231. Schaerer, D.: A&A **397**, 527 (2003)
232. Schaerer, D., Verhamme, A.: A&A **480**, 369 (2008)
233. Schaye, J.: ApJ **559**, 507 (2001)
234. Schenker, M.A., Stark, D.P., Ellis, R.S., et al.: ApJ **744**, 179 (2012)
235. Schneider, R., Hunt, L., Valiante, R.: MNRAS **457**, 1842 (2016)
236. Scholz, T.T., Walters, H.R.J., Burke, P.J., Scott, M.P.: MNRAS **242**, 692 (1990)
237. Scholz, T.T., Walters, H.R.J.: ApJ **380**, 302 (1991)
238. Schuster, A.: MNRAS **40**, 35 (1879)
239. Semelin, B., Combes, F., Baek, S.: A&A **474**, 365 (2007)
240. Shapiro, P.R., Kang, H.: ApJ **318**, 32 (1987)
241. Shibuya, T., Ouchi, M., Nakajima, K., et al.: ApJ **788**, 74 (2014)
242. Shull, J.M., McKee, C.F.: ApJ **227**, 131 (1979)
243. Shull, J.M., van Steenberg, M.E.: ApJ **298**, 268 (1985)
244. Smith, A., Safranek-Shrader, C., Bromm, V., Milosavljević, M.: MNRAS **449**, 4336 (2015)
245. So, G.C., Norman, M.L., Reynolds, D.R., Wise, J.H.: ApJ **789**, 149 (2014)
246. Sobacchi, E., Mesinger, A.: MNRAS **440**, 1662 (2014)
247. Sobacchi, E., Mesinger, A.: MNRAS **453**, 1843 (2015)
248. Sokasian, A., Abel, T., Hernquist, L.E.: NA **6**, 359 (2001)
249. Song, M., Finkelstein, S.L., Gebhardt, K., et al.: ApJ **791**, 3 (2014)
250. Spitzer Jr., L., Greenstein, J.L.: ApJ **114**, 407 (1951)
251. Stark, D.P., Ellis, R.S., Chiu, K., Ouchi, M., Bunker, A.: MNRAS **408**, 1628 (2010)
252. Steidel, C.C., Adelberger, K.L., Shapley, A.E., Pettini, M., Dickinson, M., Giavalisco, M.: ApJ **532**, 170 (2000)
253. Steidel, C.C., Erb, D.K., Shapley, A.E., et al.: ApJ **717**, 289 (2010)
254. Stenflo, J.O.: A&A **84**, 68 (1980)
255. Steidel, C.C., Bogosavljević, M., Shapley, A.E., Kollmeier, J.A., Reddy, N.A., Erb, D.K., Pettini, M.: ApJ **736**, 160 (2011)
256. Tasitsiomi, A.: ApJ **645**, 792 (2006)
257. Tasitsiomi, A.: ApJ **648**, 762 (2006)
258. Thoul, A.A., Weinberg, D.H.: ApJ **442**, 480 (1995)
259. Tozzi, P., Madau, P., Meiksin, A., Rees, M.J.: ApJ **528**, 597 (2000)
260. Trac, H., Cen, R.: ApJ **671**, 1 (2007)
261. Trac, H.Y., Gnedin, N.Y.: Advanced Science Letters **4**, 228 (2011)
262. Trainor, R.F., Steidel, C.C., Strom, A.L., Rudie, G.C.: ApJ **809**, 89 (2015)
263. Trebitsch, M., Verhamme, A., Blaizot, J., Rosdahl, J.: A&A **593**, A122 (2016)
264. Treu, T., Schmidt, K.B., Trenti, M., Bradley, L.D., Stiavelli, M.: ApJ **775**, L29 (2013)
265. Tumlinson, J., Shull, J.M.: ApJ **528**, L65 (2000)
266. Unno, W.: PASJ **4**, 100 (1952)
267. van Breugel, W., de Vries, W., Croft, S., et al.: Astronomische Nachrichten **327**, 175 (2006)
268. Vanzella, E., Grazian, A., Hayes, M., et al.: A&A **513**, A20 (2010)
269. Verhamme, A., Schaerer, D., Maselli, A.: A&A **460**, 397 (2006)
270. Verhamme, A., Schaerer, D., Atek, H., Tapken, C.: A&A **491**, 89 (2008)

271. Verhamme, A., Dubois, Y., Blaizot, J., et al.: A&A **546**, A111 (2012)
272. Verhamme, A., Orlitová, I., Schaerer, D., Hayes, M.: A&A **578**, A7 (2015)
273. White, H.E.: IEEE Trans. Appl. Supercond. (1934)
274. Wisotzki, L., Bacon, R., Blaizot, J., et al.: A&A **587**, A98 (2016)
275. Wofford, A., Leitherer, C., Salzer, J.: ApJ **765**, 118 (2013)
276. Wold, I.G.B., Barger, A.J., Cowie, L.L.: ApJ **783**, 119 (2014)
277. Wouthuysen, S.A.: AJ **57**, 31 (1952)
278. Wyithe, J.S.B., Loeb, A.: MNRAS **375**, 1034 (2007)
279. Yajima, H., Li, Y., Zhu, Q., Abel, T.: MNRAS **424**, 884 (2012)
280. Yajima, H., Li, Y., Zhu, Q., et al.: ApJ **754**, 118 (2012)
281. Yang, H., Malhotra, S., Gronke, M., et al.: ApJ **820**, 130 (2016)
282. Yusef-Zadeh, F., Morris, M., White, R.L.: ApJ **278**, 186 (1984)
283. Zahn, O., Mesinger, A., McQuinn, M., et al.: MNRAS **414**, p727–738 (2011)
284. Zanstra, H.: BAN **11**, 1 (1949)
285. Zheng, Z., Miralda-Escudé, J.: ApJ **578**, 33 (2002)
286. Zheng, Z., Cen, R., Trac, H., Miralda-Escudé, J.: ApJ **716**, 574 (2010)
287. Zheng, Z., Cen, R., Weinberg, D., Trac, H., Miralda-Escudé, J.: ApJ **739**, 62 (2011)
288. Zheng, Z., Wallace, J.: ApJ **794**, 116 (2014)
289. Zheng, Z.-Y., Wang, J., Rhoads, J., et al.: ApJ **842**, L22 (2017)
290. Zitrin, A., Labbé, I., Belli, S., et al.: ApJ **810**, L12 (2015)

Chapter 2
HI Absorption in the Intergalactic Medium

J. Xavier Prochaska

2.1 Historical Introduction

The discovery of the intergalactic medium (IGM) was, in essence, precipitated by the discovery of quasars in 1963[1] by [83]. It was through spectroscopy of these enigmatic, distant sources that one could resolve the absorption lines from gas—especially HI Lyα–in the foreground universe. Figure 2.1 shows an early example from [13] taken with the prime-focus spectrograph on the Shane 120 in. telescope at Lick Observatory. Even in these early data, one identifies apparently discrete absorption lines of Hydrogen and heavy elements establishing the presence of diffuse yet enriched gas along the sightline. This thicket of absorption blueward of the quasar's Lyα emission is now commonly referred to as the Lyα Forest.

Spectra like these inspired the first models of the IGM as discrete absorption lines [4] and by inference the first physical insight. Gunn and Peterson [31] recognized that a universe with predominantly neutral hydrogen gas should be opaque to these far-UV photons and inferred—correctly—that the gas must have a neutral fraction x_{HI} of less than 1 part in 10^5. That is, the positive detection of flux at rest wavelengths shortward of the quasar Lyα emission line ($\lambda_{rest} < 1215$ Å), i.e. transmission through the Lyα forest, demands a highly ionized IGM. As an introduction to the material presented in this Chapter, we may offer our own rough estimate. The optical depth of HI Lyα through a $\Delta d = 100$ kpc portion of the $z = 3$ universe at the mean hydrogen density $\bar{n}_H$ is simply

$$\tau(\nu) = \Delta d \, \bar{n}_H \, x_{HI} \, \sigma_{Ly\alpha}(\nu) \tag{2.1}$$

---

[1]There were, however, unsuccessful attempts to search for extragalactic gas in 21 cm absorption [24].

J. X. Prochaska (✉)
UCO/Lick Observatory, UC Santa Cruz, 1156 High Street, Santa Cruz, CA 95064, USA
e-mail: xavier@ucolick.org

© Springer-Verlag GmbH Germany, part of Springer Nature 2019
M. Dijkstra et al., *Lyman-alpha as an Astrophysical and Cosmological Tool*,
Saas-Fee Advanced Course 46, https://doi.org/10.1007/978-3-662-59623-4_2

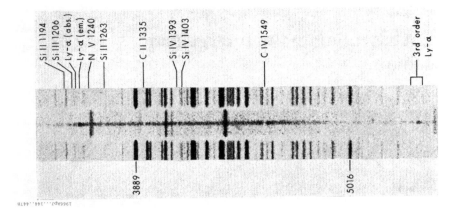

Fig. 2.1 Lick spectrum of 3C 191 obtained in February 1966 with the prime-focus spectrograph on the Shane 120 in. telescope at Lick Observatory. The comparison lamp spectrum shown is that of He+Ar. Taken from [13]. ©AAS. Reproduced with permission

with $\sigma_{\text{Ly}\alpha}$ the Lyα cross-section. We estimate the latter assuming Doppler broadening dominates with a characteristic velocity given by Hubble expansion, $\Delta v \approx H(z) * \Delta d \approx 30$ km/s. Taking a baryonic mass density $\rho_b = 0.0486\rho_c$ at $z = 0$ with ρ_c the critical density and taking 75% of the baryonic mass as Hydrogen, we find $\tau(v) \approx 10^6 x_{\text{HI}}$. Therefore, the positive detection of flux in the Lyα forest demands a highly ionized IGM.

The remainder of the 1960s introduced a series of fundamental papers on the astrophysics of absorption-line analysis, especially by Bahcall and his collaborators. These included the discussion of fundamental diagnostics of the gas [2], the application of absorption from the fine-structure levels of heavy elements [6], and the assertion that the majority of heavy element absorption may be associated to the halos of galaxies [3, 5]. In a number of respects, the theory had outpaced the observations. This held throughout the 1970s, especially for IGM studies with HI Lyα although [12] eventually reported the first intergalactic detection of HI in 21 cm absorption.

In the early 1980s, advances in spectroscopic technology (especially the CCD detector) led to the first high-quality views of the HI Lyα forest (Fig. 2.2; [10, 81, 98]). It was evident from spectra like these that the IGM was characterized by a stochastic forest of HI absorption well-described by discrete lines. In essence, these were the first detailed views of the Lyα forest. This decade also witnessed the first surveys on gas optically thick at the HI Lyman limit (aka Lyman Limit Systems or LLSs; [89]) and on the HI Lyα absorbers with sufficient column density to generate damped Lyα profiles (aka damped Lyα systems or DLAs; [94]), and the first empirical connections between absorption and individual galaxies [9]. The field was suddenly awash with data and theory had now fallen behind. The observers took to developing models of 'spherical' HI clouds and bull's-eye cartoons to describe the gas around

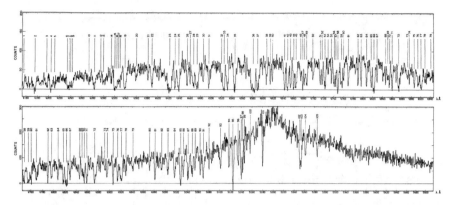

Fig. 2.2 HI Lyα forest spectrum of the quasar PKS 2126–158 obtained by [98]. Data like these provided the first detailed view of the discrete absorption of the Lyα forest. ©AAS. Reproduced with permission

galaxies. At that time, J. Ostriker was the most active theorist on the IGM, publishing a series of papers on applications of the IGM including its first phase diagram [7, 18, 61, 62]. But one of his leading models of the day envisioned "Galaxy formation in an IGM dominated by explosion" [60]. A deeper understanding of the IGM was still to be developed.

In several respects, the 1990s witnessed the true maturation of studies on the IGM. Observationally, the HIRES spectrometer [93] on the 10-m W.M. Keck telescope fully resolved the IGM and at terrific S/N. These spectra represent the pinnacle, analogous to Planck measurements of the CMB. The advance over even the 1980s was profound, as Fig. 2.3 illustrates.

A new paradigm for the IGM emerged from hydrodynamic cosmological simulations [53] and complementary analytic treatments (e.g. [33]). The Lyα clouds were replaced by the Cosmic Web (Fig. 2.4), the filamentary network of dark matter and baryons that describes the large-scale structure of a CDM universe. The HI Lyα forest traces the undulations in this web and this so-called fluctuating Gunn-Peterson approximation offers a terrific description of the IGM with sound analytic underpinnings.

Figure 2.5 compares an early generation model of the IGM from a hydrodynamic simulation against a portion of a Keck/HIRES spectrum. The agreement is remarkable and even the expert reader is challenged to identify which panel is real and which is simulated. The cosmic web paradigm is a true triumph of CDM cosmology and its development ushered in the opportunity to leverage IGM observations for research in cosmology.

For the last decade, observational advances have stemmed largely from the massive spectroscopic surveys of the Sloan Digital Sky Survey (SDSS). These have yielded terrific statistical descriptions of the IGM [43] across large areas of the sky for experiments like measurements of Baryonic Acoustic Oscillations [14]. Large surveys of optically thick HI gas have also been comprised [57, 69, 71] and analysis

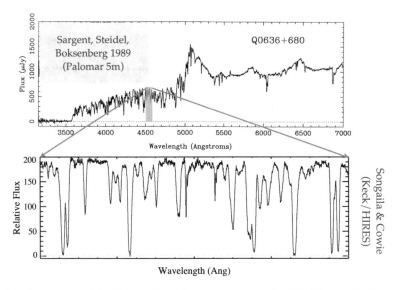

Fig. 2.3 Comparison of the high quality Palomar spectrum of Q0636+680 from [82] against a snippet of spectra obtained with Keck/HIRES by [84]. The latter fully resolves the HI Lyα absorption and, in essence, offers a complete description. ©AAS. Reproduced with permission

Fig. 2.4 Early illustration of the cosmic web from hydrodynamic simulations of the universe [53]. These authors were among the first to associate the Lyα forest with the underlying, dark matter density field. ©AAS. Reproduced with permission

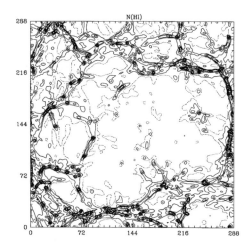

probing the underlying dark matter density field probed by the IGM have emerged [25]. In addition, the ongoing discovery of $z > 6$ quasars and GRBs coupled with high-performance echellette spectrometers have probed the IGM to the epoch of HI reionization. And, a series of increasingly sensitive UV spectrometers on the *Hubble Space Telescope* have anchored the results in the modern universe [16, 64].

This chapter is organized into the following sections: (i) the physics of HI Lyα absorption; (ii) key concepts of spectral-line analysis; (iii) characterizing the HI Lyα

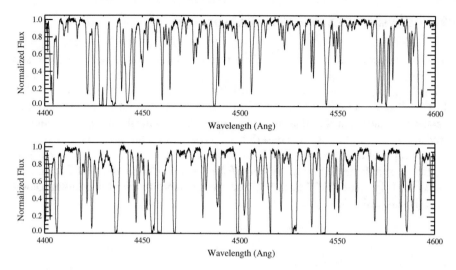

Fig. 2.5 Comparison of a snapshot of Lyα forest spectra (obtained with the Keck/HIRES spectrometer; [38]) with a numerical simulation of the Lyα forest [54]. Even an expert struggles to distinguish one from the other. Can you tell which is which?!

forest as absorption lines; (iv) optically thick HI absorption; and (v) a brief overview of modern analysis and results. The primary goal of these lectures was to provide the foundation and background for a young scientist to engage in empirical analysis of the IGM. The focus throughout is observational and the approaches are largely traditional; excellent reviews with a greater emphasis on theory are given by [50, 52].

This chapter is also supplemented by the lecture notes and slides presented in Saas Fee, and a set of iPython Notebooks illustrating concepts and providing example code for related calculations and modeling. These supplementary materials are publicly available at https://github.com/profxj/SaasFee2016. Python code relevant to HI Lyα absorption and IGM analysis are packaged as `linetools`[2] and `pyigm`[3] on github.

Before continuing, I offer a few caveats on this work: (1) I have not attempted to keep pace with the new literature (since the School); (2) I have only ingested a portion of the DLA lecture. See [96] for a fine review; (3) I have only summarized the slides that ended the series of lectures. The modern literature is evolving rapidly with numerous on-going and upcoming experiments (e.g. eBOSS, CLAMATO, DESI, PFS).

---

[2] https://github.com/linetools/linetools.

[3] https://github.com/pyigm/pyigm.

2.2 Physics of Lyman Series Absorption

The HI Lyα photon may be emitted following one of several processes: (i) the resonant absorption of a Lyα photon by atomic hydrogen; (ii) as the final emission in the recombination cascade of hydrogen; (iii) following the collisional excitation of HI. In this Chapter, we concern ourselves with the first process—HI Lyα resonant-line scattering—although the other processes may play a critical role in Lyα radiative transfer. In this section we will describe the physics of HI line absorption, introduce the concepts of a line-profile, and illustrate the basics of Lyα absorption lines. This section is supplemented by the iPython *HI_Lyman_Series.ipynb* Notebook and the "saasfee16_lymanseries" Lecture. These are available online: https://github.com/profxj/SaasFee2016.

2.2.1 HI Energy Levels

We begin with a derivation of the energy levels for atomic hydrogen. Energies in the classic Rutherford–Bohr model of a Hydrogenic ion with charge Ze are solved from a standard Hamiltonian $H^{(0)}$ with an electrostatic potential[4]

$$H^{(0)} = \frac{-\hbar^2 \nabla^2}{2\mu} - \frac{Ze^2}{r} \quad . \tag{2.2}$$

For energy level n, one recovers

$$E_n = -\frac{1}{2}\mu c^2 \frac{(Z\alpha)^2}{n^2} \quad , \tag{2.3}$$

and the quantum states described by n, ℓ, m, m_s or $|n\ell mm_s>$ are degenerate in ℓ, m, m_s because our Hamiltonian is rotationally invariant. In Eq. 2.3, we have the fine-structure constant $\alpha \equiv e^2/\hbar c \approx 1/137$ and the reduced mass $\mu = \frac{m_e(Zm_p)}{m_e + Zm_p}$. For Hydrogen, $\mu \approx 0.999 m_e$.

From E_n, we may evaluate the wavelengths for the Lyman series (transitions linked to the ground state E_1), as

$$\lambda_{rest,n} = \frac{hc}{E_n - E_1} \quad . \tag{2.4}$$

Table 2.1 lists the calculated values from Eqs. 2.3 and 2.4 for $\approx$20 Lyman series lines, compared against empirical measurements. One identifies a systematic offset of $\delta\lambda \approx 0.015$ Å between the Rutherford–Bohr energies and experiment. These result from perturbations to the standard Hamiltonian that we now consider.

---

[4]We adopt cgs units throughout the chapter.

Table 2.1 HI Lyman series lines

Transition	n	$E_n - E_1$ (eV)	λ_{rest} (Å)	λ_{exp} (Å)
Lyα	2	10.200000	1215.6845	1215.6701
Lyβ	3	12.088889	1025.7338	1025.7223
Lyγ	4	12.750000	972.54759	972.5368
Lyδ	5	13.056000	949.75351	949.7431
Lyε	6	13.222223	937.81375	937.8035
Ly6	7	13.322449	930.75844	930.7483
Ly7	8	13.387500	926.23580	926.2257
Ly8	9	13.432099	923.16041	923.1504
Ly9	10	13.464000	920.97310	920.9631
Ly10	11	13.487604	919.36139	919.3514
Ly11	12	13.505556	918.13933	918.1294
Ly12	13	13.519527	917.19053	917.1806
Ly13	14	13.530613	916.43908	916.429
Ly14	15	13.539556	915.83374	915.824
Ly15	16	13.546875	915.33891	915.329
Ly16	17	13.552942	914.92921	914.919
Ly17	18	13.558025	914.58616	914.576
Ly18	19	13.562327	914.29604	914.286
Ly19	20	13.566000	914.04849	914.039

There are two perturbations with energies that scale as α^4, i.e. the next terms in Eq. 2.3: (1) spin-orbit coupling and (2) the first expansion of the relativistic kinetic energy. Classically, the spin-orbit coupling is described as a magnetic dipole interaction between the spin of the electron and the orbit of the nucleus, i.e. the e^- observes a magnetic field due to the current driven by the nucleus:

$$\mathbf{B} = -\frac{1}{c}\mathbf{v} \times \mathbf{E} = \frac{1}{m_e c r}\mathbf{L}\frac{d\phi}{dr} \quad , \tag{2.5}$$

with ϕ the electric potential and the energy is

$$E = -\boldsymbol{\mu}_s \cdot \mathbf{B} \quad \text{with} \quad \boldsymbol{\mu}_s = -\frac{eg\mathbf{s}}{2m_e c} \quad , \tag{2.6}$$

where $g \approx 2$ for an electron. The Hamiltonian for the perturbation is:

$$H_{SO} = \frac{1}{2m_e^2 c^2}\mathbf{S} \cdot \mathbf{L}\frac{1}{r}\frac{d\phi}{dr} \quad , \tag{2.7}$$

and the standard treatment is to identify an operator that commutes with $H^{(0)}$ and H_{SO}, and also uniquely identifies the degenerate states. We choose $\mathbf{J} \equiv \mathbf{L} + \mathbf{S}$ recognizing

$$\mathbf{L} \cdot \mathbf{S} = \frac{1}{2} \left(|\mathbf{J}|^2 - |\mathbf{L}|^2 - |\mathbf{S}|^2 \right)$$

which yields energies

$$E_{SO} = < H_{SO} > = \frac{1}{2} C \left[j(j+1) - \ell(\ell+1) - s(s+1) \right] \qquad (2.8)$$

with C a constant. For fixed $\mathbf{L}$ and $\mathbf{S}$ (i.e. splitting within a level, e.g. 2P), we have $\Delta E_{SO} = E_{J+1} - E_J = C(j+1)$. For our Hydrogenic ion with $\phi = Ze^2/r$,

$$\frac{d\phi}{dr} = \frac{-Ze^2}{r^2} \qquad (2.9)$$

and

$$H_{SO} = \frac{Ze^2}{2m_e^2 c^2} \frac{1}{r^3} \mathbf{L} \cdot \mathbf{S} \; . \qquad (2.10)$$

Following standard perturbation theory,[5] we find

$$< H_{SO} >= -E_n \frac{Z^2 \alpha^2}{2n} \frac{[j(j+1) - \ell(\ell+1) - s(s+1)]}{\ell(\ell + \frac{1}{2})(\ell+1)} \qquad (2.11)$$

with E_n given by Eq. 2.3. As advertised, spin-orbit coupling is 4th order in α and we now have explicit energy dependence on j, ℓ and s. For the Hydrogen $n = 2$ levels ($Z = 1; n = 2; \ell = 0, 1; j = 1/2, 3/2$), we find

$$< H_{SO} >= \begin{cases} 0 & 2^2S_{\frac{1}{2}} & (\ell = 0; j = 0) \\ \frac{mc^2\alpha^4}{96} & 2^2P_{3/2} & (\ell = 1; j = 3/2) \\ \frac{-mc^2\alpha^4}{48} & 2^2P_{1/2} & (\ell = 1; j = 1/2) \end{cases} \qquad (2.12)$$

giving a $2P_{3/2} - 2P_{1/2}$ splitting of 4.5×10^{-5} eV or $\Delta v \approx \Delta E/cE \approx 1$ km/s.

To derive the Relativistic correction to order α^4, we expand the kinetic energy to the next term in v^2/c^2 from the Lagrangian

$$\text{K.E.} = \frac{p^2}{2m} \left(1 - \frac{1}{4} \frac{v^2}{c^2} \right) \; . \qquad (2.13)$$

---

[5] See the Lecture notes for an expanded derivation.

Table 2.2 Perturbations to the $n = 1, 2$ levels of hydrogen

State	n	j	$< H_{SO} > + < H_{rel} >$
$1\,^2S_{\frac{1}{2}}$	1	$\frac{1}{2}$	-1.8×10^{-4} eV
$2\,^2S_{\frac{1}{2}}, 2\,^2P_{\frac{1}{2}}$	2	$\frac{1}{2}$	-5.7×10^{-5} eV
$2\,^2P_{\frac{3}{2}}$	2	$\frac{3}{2}$	-1.1×10^{-5} eV

This gives a relativistic perturbation

$$H_{rel} = -\frac{1}{2mc^2} \left(\frac{p^2}{2m} \right)^2 \tag{2.14}$$

that has no spin dependence (spherically symmetric) such that $[H_{rel}, L^2] = [H_{rel}, L] = 0$ and the standard $|n\ell m m_s >$ diagonalize H_{rel}. If we recognize that

$$H_{rel} = -\frac{1}{2mc^2} \left(H^{(0)} - V^{(0)} \right)^2$$

with $V^{(0)} = -Ze^2/r$, it is straightforward to compute the energies

$$< H_{rel} > = -E_n \frac{Z^2 \alpha^2}{n} \left(\frac{3}{4n} - \frac{1}{\ell + \frac{1}{2}} \right) \tag{2.15}$$

Combining this result with spin-orbit coupling, we recover a remarkable expression that depends only on j:

$$< H_{SO} > + < H_{rel} > = E_n \frac{Z^2 \alpha^2}{n} \left[\frac{1}{j + \frac{1}{2}} - \frac{3}{4n} \right] \tag{2.16}$$

Altogether, the α^4 term for E_n has no explicit ℓ dependence, nor any s dependence. Furthermore, higher j implies higher energy following the 3rd Hund's rule. The energy shifts from E_n for Hydrogen are then

$$\Delta E_{nj} = -7.25 \times 10^{-4} \text{ eV} \frac{1}{n^3} \left[\frac{1}{j + \frac{1}{2}} - \frac{3}{4n} \right] \tag{2.17}$$

which for the $n = 1, 2$ states of Hydrogen evaluate to the results in Table 2.2.

The splitting of the $n = 2$ level implies that Lyα (1S-2P) is a doublet with ≈ 1 km/s separation which is generally too small to resolve observationally but could be important for radiative transfer treatments.

Returning to Table 2.1, the implied shift from the Rutherford–Bohr energies $\delta\lambda/\lambda \sim \delta E/E$ is calculated using δE from Table 2.2 and accounting for the relative degeneracy of the $2\,^2P_{\frac{1}{2}}, 2\,^2P_{\frac{3}{2}}$ states yields

$$\delta E = \frac{1}{3} \left[2\delta E_{1S \to 2P_{\frac{3}{2}}} + \delta E_{1S \to 2P_{\frac{1}{2}}} \right] \tag{2.18}$$

$$= 1.55 \times 10^{-4} \text{ eV} \tag{2.19}$$

or

$$\Delta \lambda = -\frac{\delta E}{E} \times 1215.68 \text{ Å} = -0.0138 \text{ Å} \quad . \tag{2.20}$$

Voilà!

2.2.2 The Line Profile

We now derive the line profile of HI Lyman series absorption which expresses the energy dependence of the photon cross-section. This results from two physical effects: (1) the quantum mechanical coupling of the energy levels described in the previous sub-section; (2) Doppler broadening from the kinetic motions of the gas. We reserve a discussion of the observed line-profile related to the instrument to the next section.

We express the opacity $\kappa_{jk}(\nu)$ of a gas with number density n_j in state j as:

$$\kappa_{jk}(\nu) = n_j \sigma_{jk}(\nu) \tag{2.21}$$

with σ_ν the photon cross-section at frequency ν for a transition to state k. We separate the frequency dependence by introducing the line-profile $\phi(\nu)$

$$\sigma(\nu) = \sigma_{jk} \phi(\nu) \tag{2.22}$$

with σ_{jk} the integrated cross-section over all frequencies and $\phi_\nu \, d\nu$ reflects the probability an atom will absorb a photon with energy in $\nu, \nu + d\nu$.

A naive guess for $\phi(\nu)$ is the Delta function, i.e. the transition occurs only at the exact energy splitting the energy levels from the previous subsection:

$$\phi_\delta(\nu) = \delta \left(\nu - \nu_{jk} \right) \tag{2.23}$$

Quantum mechanically, however, the excited $n = 2$ state has a half-life $\tau_{\frac{1}{2}}$ given by the Spontaneous emission coefficient, $\tau_{\frac{1}{2}} = 1/A_{jk}$ ($A_{jk} = 6.265 \times 10^8 \text{ s}^{-1}$ for HI Lyα). This finite lifetime implies a finite width ΔE to the energy level which may be estimated from the Heisenberg Uncertainty Principle

$$\Delta E \sim \frac{\hbar}{\Delta t} \sim \hbar A \quad . \tag{2.24}$$

This demands a finite width to the line-profile. Define $W_{jk}(E)$ as the quantum mechanical probability of a transition occurring between states j and k with energy E, and $W_j(E)$ as the probability of state j being characterized by the energy interval $(E_j, E_j + dE_j)$. A standard Quantum mechanical treatment shows that $W_j(E)$ has a Lorentzian shape (aka the Breit–Wigner profile):

$$W_j(E_j)dE_j = \frac{\gamma_j dE_j/h}{(2\pi/h)^2 \left[E_j - < E_j > \right]^2 + (\gamma_j/2)^2} \tag{2.25}$$

with

$$\gamma_j \equiv \sum_{i<j} A_{ij} \tag{2.26}$$

For a coupling between only two states j, k, one derives W_{jk} by convolving W_j and W_k:

$$W_{jk}(E)dE = \frac{\left[\gamma_j + \gamma_k \right] dE/h}{(2\pi/h)^2 \left[E - E_{jk} \right]^2 + ([\gamma_j + \gamma_k]/2)^2} \tag{2.27}$$

We reemphasize that this calculation was restricted to the j and k states whereas a proper calculation needs to consider the coupling of all the energy levels. For our Lyman series lines we note that $\gamma_j = 0$ as, by definition, there are no energy levels below the ground state.

From Eq. 2.27, we introduce the Natural line-profile normalized to have unit integral value in frequency,

$$\phi_N(\nu) = \frac{1}{\pi} \left[\frac{(\gamma_j + \gamma_k)/4\pi}{(\nu - \nu_{jk})^2 + (\gamma_j + \gamma_k)^2/(4\pi)^2} \right] . \tag{2.28}$$

giving (at last)

$$\sigma(\nu) = \sigma_{jk} \phi_N(\nu) \tag{2.29}$$

As illustrated in the Notebook, the "wings" of $\phi_N(\nu)$ are ≈ 10 orders of magnitude down from line-center. Remarkably these can be very important for Lyα.

It is standard practice to express the normalization σ_{jk} in terms of the oscillator strength f_{jk} which is either measured empirically (preferred) or estimated theoretically

$$\sigma_{jk} = \frac{\pi e^2}{m_e c} f_{jk} . \tag{2.30}$$

Our final expression becomes

$$\sigma_\nu = \frac{\pi e^2}{m_e c} f_{jk} \left[\frac{(\gamma_j + \gamma_k)/4\pi^2}{(\nu - \nu_{jk})^2 + (\gamma_j + \gamma_k)^2/(4\pi)^2} \right] \tag{2.31}$$

Expressing the FWHM of the line-profile as the frequency width where $\sigma(v)/\sigma(v)_{max} = 1/2$, we have $\Delta v_{FWHM} = \pm\frac{\gamma_j + \gamma_k}{4\pi}$ which very nearly matches our estimate from the Uncertainty Principle! For HI Lyα with $\gamma_1 = 0$, $\gamma_2 = A_{21}$, we find the FWHM in velocity to be

$$\Delta v_{FWHM} = c\Delta v_{FWHM}/v \approx 1.5 \times 10^{-2}\,\text{km/s} \qquad (2.32)$$

For astrophysical purposes, this nearly is a delta function.

In an astrophysical environment, each atom in a gas has its own motion which spreads the line without changing the total amount of absorption. This Doppler effect, to lowest order in v/c (with v the radial velocity of the atom), is

$$\Delta v = v - v_{jk} = v_{jk}\frac{v}{c} \qquad (2.33)$$

Assuming first that the gas motions are characterized solely by T (i.e. no turbulence), we adopt a Maxwellian distribution for particles of mass m_A giving a profile function:

$$\phi_D(v) = \frac{1}{\Delta v_D\sqrt{\pi}}\exp\left[-\frac{(v - v_{jk})^2}{\Delta v_D^2}\right] \qquad (2.34)$$

$$\text{with } \Delta v_D \equiv \frac{v_{jk}}{c}\sqrt{\frac{2kT}{m_A}} \qquad (2.35)$$

At line-center ($v = v_{jk}$), the cross-section (neglecting stimulated emission) is

$$(\sigma_v^D)_{max} = \sigma_{jk}\phi_D(v_{jk}) = \frac{\pi e^2}{mc}f_{jk}\frac{1}{\Delta v_D\sqrt{\pi}} \qquad (2.36)$$

This cross-section is several orders of magnitude lower than $(\sigma_v^N)_{max}$ (with σ_v^N being the cross-section corresponding to the natural profile of the line) because the absorption has been 'spread' over a velocity interval several orders of magnitude larger than the width estimated by Eq. 2.32.

We can generalize the profile to include random, turbulent motions (characterized by ξ) by modifying the Doppler width,

$$\Delta v_D = \frac{v_{jk}}{c}\left(\frac{2kT}{m_A} + \xi^2\right)^{\frac{1}{2}}. \qquad (2.37)$$

Expressing the line-profile in velocity space, we introduce the Doppler parameter

$$b \equiv \sqrt{\frac{2kT}{m_A} + \xi^2} \qquad (2.38)$$

and the velocity line-profile for Doppler motions is

$$\phi_D(v) = \frac{1}{b\sqrt{\pi}} \exp\left[-\frac{v^2}{b^2}\right] . \tag{2.39}$$

See the Notebook for a series of examples illustrating this profile in comparison to the Natural profile.

Generally, the cross-section has contributions from both Natural and Doppler broadenings, with Doppler broadening dominating the line-center and the Lorentzian of Natural broadening dominates the wings. The overall profile is a convolution of the two terms,

$$\phi_V(v) = \frac{\gamma}{4\pi^2} \int_{-\infty}^{\infty} \frac{\left(\frac{m}{2\pi kT}\right)^{\frac{1}{2}} \exp\left(-\frac{mv^2}{2kT}\right)}{\left(v - v_{jk} - v_{jk}v/c\right)^2 + (\gamma/4\pi)^2} \, dv \tag{2.40}$$

and one is inspired to introduce the Voigt function

$$H(a, u) = \frac{a}{\pi} \int_{-\infty}^{\infty} \frac{e^{-y^2} \, dy}{a^2 + (u - y)^2} \tag{2.41}$$

and we identify a, u as:

$$a \equiv \frac{\gamma}{4\pi \Delta v_D} \tag{2.42}$$

$$u \equiv \frac{v - v_{jk}}{\Delta v_D} \tag{2.43}$$

Altogether, we have

$$\phi_V(v) = \frac{H(a, u)}{\Delta v_D \sqrt{\pi}} \tag{2.44}$$

which has no analytic solution. For speed, one often relies on look-up tables. In Python, the real part of scipy.special.wofz is both accurate and fast (see the Voigt documentation in the linetools package).

The Lorentzian profile from Natural broadening is only an approximation because it ignores the coupling between states other than j, k. Scattering is a second-order quantum mechanical process: annihilation of one photon and the creation of a scattered photon. A proper treatment requires second-order time dependent perturbation theory and requires one to sum over all bound-states and integrate over all continuum-state contributions. Lee [40] have calculated a series expansion of the corrections to the Voigt profile and we refer the reader to his paper for a more detailed presentation. Here, we report the next term in the series for HI Lyα:

$$\sigma_v = \sigma_T \left(\frac{f_{jk}}{2}\right)^2 \left(\frac{v_{jk}}{\delta v}\right)^2 \left[1 - 1.792\frac{\delta v}{v_{jk}}\right] \quad . \tag{2.45}$$

with σ_T the Thomson cross-section. It is evident that the correction is not symmetric about line-center. This leads to an asymmetry which shifts the measured line-center if the gas opacity is very high (e.g. $N_{HI} > 10^{21.7}\,\mathrm{cm}^{-2}$).

2.2.3   Optical Depth (τ_v) and Column Density (N)

We now introduce two quantities central to absorption-line analysis. First, the optical depth τ_v which is defined as the integrated opacity along a sightline. In differential form $d\tau(v) = -\kappa(v)ds$ implying

$$\tau(v) = \sigma(v) \int n_j ds \tag{2.46}$$

which gives an explicit frequency dependence related to the line-profiles of the previous sub-section.

We recognize the second term to be the column density N_j,

$$N_j \equiv \int n_j ds \tag{2.47}$$

which has units of cm^{-2} and is akin to a surface density ($n_j \to \rho$; $N_j \to \Sigma$). As an example, consider the column density of O_2 through 1 m of air. With $\rho_{O_2} = 1.492\,\mathrm{g/L}$, $N_{O_2} = 3 \times 10^{21}\,\mathrm{cm}^{-2}$.

A quantity of particular interest is the optical depth at line-center τ_0 of a gas with N_j and Doppler parameter b. At line-center ($v = v_{jk}$), our line-profile is dominated by Doppler motions $\phi_V(v_{jk}) \approx \phi_D(v_{jk})$ and from Eq. 2.36, we recover

$$\tau_0 = \frac{\sqrt{\pi}e^2}{m_e c} \frac{N_j \lambda_{jk} f_{jk}}{b} \quad . \tag{2.48}$$

For Lyα, with b expressed in km/s:

$$\tau_0^{\mathrm{Ly}\alpha} = 7.6 \times 10^{-13} \frac{N\,[\mathrm{cm}^{-2}]}{b\,[\mathrm{km/s}]} \tag{2.49}$$

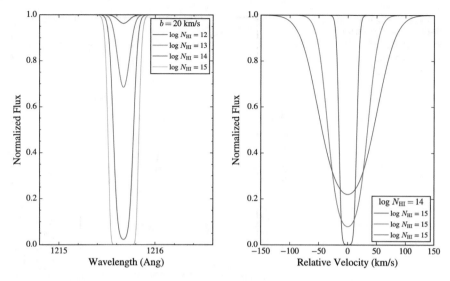

Fig. 2.6 A series of absorption-line profiles for HI Lyα for a gas with varying column density. These are presented as a function of wavelength (left, for $b = 20\,\text{km/s}$) and velocity (right; for $\log N = 14$) with the latter given by $\delta v = c\delta\lambda/\lambda$

2.2.4   Idealized Absorption Lines

Without derivation (see the Lecture notes), we express the radiative transfer for a source with intensity I_ν^* through a medium with optical depth τ_ν as

$$I_\nu = I_\nu^* e^{-\tau_\nu} \tag{2.50}$$

Therefore, we may consider the formation of absorption lines as the simple integration of the optical depth.

In Fig. 2.6, we show a series of idealized HI Lyα lines for a range of HI column densities and Doppler parameters (N_{HI}, b). These illustrate the shapes of the line-profiles, here dominated by Doppler broadening, and the varying optical depth at line-center. These are plotted in velocity space, taking $\delta v = c\delta\lambda/\lambda_0$ with $\delta v = 0\,\text{km/s}$ corresponding to the line-center. In Fig. 2.7 we contrast lines which have $\tau_0 \approx 1$, with an HI absorption line with very high N_{HI} and correspondingly high τ_0. Here the line profile is entirely determined by Natural broadening.

2.2.5   Equivalent Width

The equivalent width for absorption W_λ is a gross measure of the flux absorbed (scattered) by the gas cloud. Strictly, W_λ is the convolution of the optical depth with

Fig. 2.7 Comparison of absorption for gas with optical depth near unit (blue/green lines) to gas exhibiting damped Lyα absorption (red)

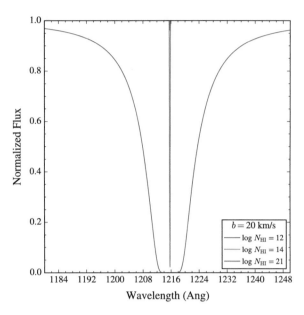

the line profile. And although it is primarily an observational quantity, its value is *independent* of the instrument profile and it does depend on physical properties of the absorbing gas.

Analytically, we define

$$W_\lambda = \int_0^\infty \left[1 - \frac{I_\lambda}{I_\lambda^*} \right] d\lambda \tag{2.51}$$

Substituting our simple radiative transfer equation, this gives

$$W_\lambda = \int_0^\infty \left[1 - \exp(-\tau_\lambda) \right] d\lambda \tag{2.52}$$

One may contrast this equation with analysis of stellar atmospheres which has a very different radiative transfer equation. The W_λ value, expressed in Å, may be visualized as the width of a box-car profile that matches the absorbed flux in a normalized spectrum, e.g. Fig. 2.8.

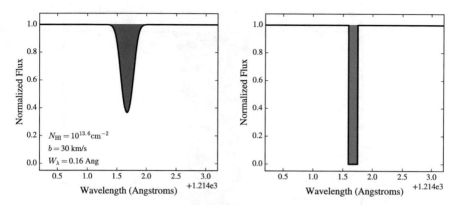

Fig. 2.8 (left) An HI Lyα absorption line with $N_{HI} = 10^{13.6}$ cm^{-2} and $b = 30$ km/s. Integrating the area above the curve (to the normalized continuum), one recovers an equivalent width $W_\lambda = 0.16$ Å. (right) A cartoon representation of the corresponding boxcar profile giving the same equivalent width

2.2.6   Curve of Growth for HI Lyα

It is valuable to develop an intuition of the relation between equivalent width and the physical properties of a gas (N, b). This relationship is generally referred to as the curve-of-growth (COG) and may be inverted to constrain N, b from W_λ measurements. The COG also nicely describes the transition from a Doppler-dominated line to a naturally-broadened line.

One typically considers three regimes for the COG which depend on the central optical depth of the absorption line τ_0 (Eq. 2.48). In the Weak limit ($\tau_0 \ll 1$), Natural broadening is negligible and Doppler broadening dominates,

$$\tau_v = \frac{\pi e^2}{m_e c} f_{jk} N_j \phi_D(v) \ . \tag{2.53}$$

With τ_0 small, Eq. 2.52 reduces to

$$W_\lambda = \frac{\lambda^2}{c} \int \left[1 - \exp(-\tau_v) \right] dv \tag{2.54}$$

$$\approx \frac{\lambda^2}{c} \int_0^\infty \tau_v \, dv \tag{2.55}$$

$$= \frac{\lambda^2}{c} \frac{\pi e^2}{m_e c} f_{jk} N_j \tag{2.56}$$

revealing a linear relationship between W_λ and N_j and no dependence on b. The Weak limit, therefore, is also referred to as the 'linear' portion of the COG. Evaluating for Lyα, we find

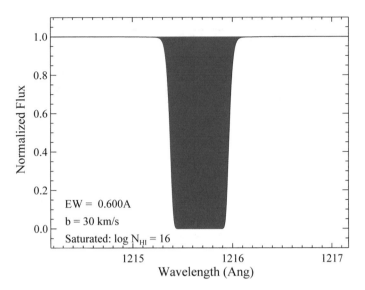

Fig. 2.9 Saturated Lyα line in the Strong regime of the COG. Here the optical depth at line-center τ_0 exceeds unity such that the incident photons are completely absorbed. But, the opacity is insufficient for the Lorentzian wings to appear; the line-profile is instead dominated by the Maxwellian of Doppler broadening (i.e. Gaussian) giving a box-like appearance

$$W_\lambda^{\text{Ly}\alpha} \approx (0.1 \text{ Å}) \, \frac{N_{\text{HI}}}{1.83 \times 10^{13} \text{ cm}^{-2}} \qquad (2.57)$$

Our estimate of τ_0 (Eq. 2.48) requires $N_{\text{HI}} \ll 10^{14} \text{ cm}^{-2}$ for $\tau_0 \ll 1$, which also implies $W_\lambda \ll 1 \text{ Å}$. This column density is many orders of magnitude less than that typical of the Galactic interstellar medium (ISM). It implies a gas that is extremely diffuse and, likely, highly ionized. These are the physical conditions of the IGM.

The Strong line limit of the COG refers to $\tau_0 \gtrsim 1$ and we may still ignore Natural broadening. In this regime, the incident flux is strongly absorbed at line-center and the equivalent width is well described by the width of the line (i.e. it is nearly approximated as a 'box'; Fig. 2.9). Expressing the frequency dependence of the optical depth as $\tau_x = \tau_0 e^{-x^2}$, with $x \equiv \Delta\nu/\Delta\nu_D$, we may estimate the line width by considering the value of x that gives $\tau_x = 1$. Trivially, we find $x_1 = \sqrt{\ln \tau_0}$ and therefore

$$W_\lambda \approx 2x_1 \approx 2\sqrt{\ln \tau_0} \qquad (2.58)$$

in the Strong regime, also known as the saturated limit. To increase W_λ in this saturated limit, we need to increase τ_0 immensely and likewise N, i.e. $W_\lambda \propto (\ln N)^{\frac{1}{2}}$. Whereas W_λ is insensitive to τ_0, it is sensitive to the internal structure of the cloud $W_\lambda \propto x_1 \propto b$.

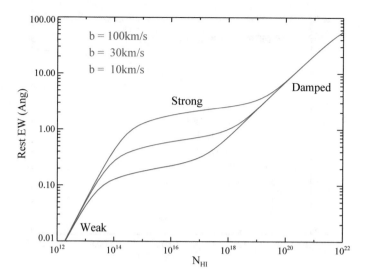

Fig. 2.10 Curve of growth (COG) for the HI Lyα line. Expressed across 10 orders of magnitude, one traverses from the weak ($\tau_0 \ll 1$), to the strong ($\tau_0 \sim 1$), to the damped regimes ($\tau_0 \gg 1$). Note the strong dependence on the Doppler parameter only in the strong regime

Lastly, there is the Damping regime with $\tau_0 \gg 1$ and where Natural broadening dominates. Now the optical depth is given by the Lorentzian profile

$$\tau_x \approx \frac{\tau_0 A}{\sqrt{\pi}} \frac{1}{x^2} \tag{2.59}$$

with $A = \gamma_j + \gamma_k$ and the width (estimated from $\tau_x = 1$) is $x_1 \sim [\tau_0 A]^{\frac{1}{2}}$. Therefore, $W_\lambda \approx 2x_1 \propto N^{\frac{1}{2}}$ or formally

$$\frac{W_\lambda}{\lambda} \approx \frac{2}{c} \left[\lambda^2 N_j \frac{\pi e^2}{m_e c} f_{jk} A \right]^{\frac{1}{2}} \tag{2.60}$$

In the Damping regime, the equivalent width scales as $\sqrt{N}$ with no dependence on the Doppler parameter as the core is fully saturated.

Figure 2.10 shows the COG for a single HI Lyα line with varying N_{HI} and Doppler parameter. The three COG regimes are well-described. We stress that only a small portion of the Weak limit ($W_\lambda \ll 1\,\text{Å}$) permits actual detections with modern spectrometers and that the Damping regime is limited to gas with galactic surface densities. This leaves approximately 6 orders of magnitude in N_{HI} in the Strong (saturated) regime where observations of the equivalent width for Lyα alone offer a weak constraint on the gas column density.

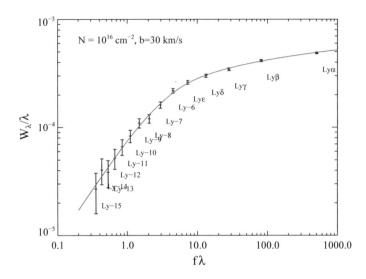

Fig. 2.11 The points are a set of equivalent width measurements for an HI gas for a series of HI Lyman lines. These are plotted against $f\lambda$ which is proportional to τ_0. The curve is a COG fit to the data, parameterized by N and b of the gas

2.2.7 Curve of Growth for the Lyman Series

Another application of the COG is to evaluate the absorption from a series of transitions from as single ion in a single 'cloud' of gas. Such analysis may enable one to derive more precisely the physical parameters (N, b) of the absorbing gas. Recalling the expression for τ_0 (Eq. 2.48)

$$\tau_0 = \frac{\sqrt{\pi} e^2}{m_e c} \frac{\lambda_{jk} f_{jk} N_j}{b} \quad , \tag{2.61}$$

a single hydrogen gas cloud with fixed N, b will show decreasing τ_0 in increasing terms of the Lyman series because both λ_{jk} and f_{jk} decrease along the series. Therefore, the Lyman series absorption generates a COG that may span several of the regimes described above.

The standard analysis is to fit COG curves to a series of W_λ measurements to constrain N, b as illustrated in Fig. 2.11. Another example and related code are provided in the Notebook.

2.3   Basics of Spectral Analysis (for Absorption)

In this section, we provide a brief introduction to several key concepts of spectroscopy and then describe several of the fundamental aspects related to absorption-line analysis with data.

2.3.1   Characteristics of a Spectrum

It is beyond the scope of this chapter to describe in detail the fundamentals of spectrometers in astronomy. For what follows, the reader needs only appreciate that spectrometers' implement a dispersing element (e.g. grating) to redirect incident light as a function of wavelength such that a packet of photons with a distribution of energies are spread (monotonically) across a detector. The number of photons collected at a given detector pixel depends on the flux and the dispersion of the spectrograph. Of course, one cannot achieve infinite resolution and, further, the optics of the instrument affect the resultant spectral image.

One defines the Line Spread Function (LSF) as the functional form a monochromatic source exhibits on the detector. The diffraction of light (e.g. slit + dispersing element) transforms the input PSF of the source (typically a Moffat) into a line profile in wavelength. The precise LSF that results depends on type of spectrometer (e.g. [79]).

In Fig. 2.12, we compare the idealized LSFs for the most common spectrometers in use: a diffraction-limited slit spectrometer and a fiber-fed spectrograph. Although the functional forms are distinct and one recognizes quantitative differences in the wings of the profiles, the inner line-profiles are very similar and one would require high signal data to distinguish the two. Furthermore, system imperfections in the spectrograph, variations in the PSF, and the Central Limit Theorem tends the LSF towards a Gaussian. For this school, we will adopt a Gaussian LSF. In practice, the LSF may be assessed empirically, e.g. through analysis of arc-line emission lamps and/or atmospheric absorption lines imprinted on an external source.

The primary impact of the LSF is to artificially broaden any spectral features through the convolution of the instrument profile with the intrinsic spectrum. While the LSF offers a complete description of this effect, we tend to describe it (and the spectrometer) in terms of the width of the LSF, i.e. the resolution R. The actual metric for R depends on the shape of the LSF (i.e. [79]) and for a Gaussian profile the standard metric is the full-width at half-maximum (FWHM). For an optical slit/fiber spectrograph, R is primarily dependent on the slit/fiber width (inversely proportional) and the grating. It is generally independent of any detector properties. Formally, we define

$$R \equiv \frac{\lambda}{\Delta\lambda_{\mathrm{FWHM}}} \tag{2.62}$$

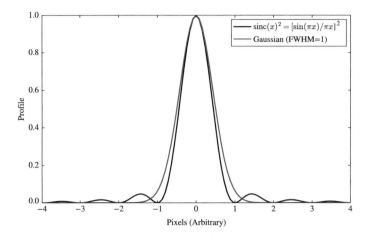

Fig. 2.12 Comparison of the LSF for a slit spectrometer (black) with that for a fiber-fed spectrometer (blue). Despite quantitative differences in the wings of the profiles, the majority of light is similarly described by the inner portion of the profiles

with λ the wavelength where the LSF has a measured FWHM of $\Delta\lambda_{FWHM}$.

While R describes the width of the LSF, the spectral width of a pixel on the detector is given by the dispersion $\delta\lambda$. Obviously, this depends on both R and the characteristics of the detector (i.e. the physical size of the pixel). To further confuse matters, many detectors can be 'binned' in electronics to increase $\delta\lambda$ and reduce the detector noise per effective pixel.

The sampling of the LSF is the number of pixels per resolution element, i.e. $\Delta\lambda_{FWHM}/\delta\lambda$. To achieve the full resolution of the spectrograph in the theoretical limit, one must sample the LSF by 2 or more pixels. This is commonly referred to as the Nyquist limit.

Commonly used spectrometers for HI absorption studies have R ranging from 2,000 (Keck/LRIS; SDSS) to many thousands (echellettes; Keck/ESI, VLT/X-Shooter), to tens of thousands (echelles; Keck/HIRES, VLT/UVES). Figure 2.13 compares spectra of the quasar FJ0812+32 observed with spectrometers having a range of R. It is evident that for the lower resolution spectra (SDSS), the intrinsic width of the lines have been substantially smeared by the instrumental LSF.

Although Eq. 2.62 is the formal definition of R, astronomers frequently refer to only the FWHM of the LSF. In velocity, $\Delta v_{FWHM} = c/R$, e.g. $\Delta v_{FWHM} \approx 10$ km/s for $R = 30,000$. This provides a physical intuition for the best sensitivity one achieves with the spectrometer. Old-timers (and many instrument web-pages) frequently refer to $\Delta\lambda_{FWHM}$, e.g. ≈ 2 Å for SDSS.

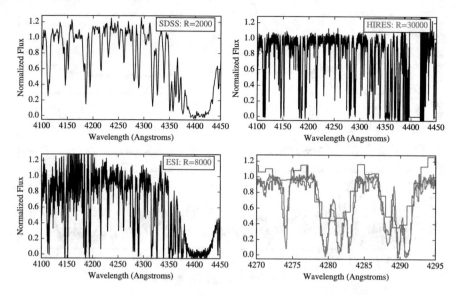

Fig. 2.13 Three spectra of the quasar FJ0812+32 observed with the SDSS, Keck/ESI, and Keck/HIRES spectrometers with resolution ranging from $R \approx 2,000$ to 30,000. The lower-right panel shows a zoom-in where one can easily appreciate the effects of varying spectral resolution

2.3.2 Continuum Normalization

Studies of HI in absorption require a background, light source with emission at FUV wavelengths (and bluer!) to excite HI at Lyα and higher order transitions. For extragalactic research, this has included the O and B stars in star-forming galaxies [42], the bright afterglows of gamma-ray bursts [28], and quasars (the focus of this Chapter). To estimate the opacity, one must estimate the source continuum f_λ^C then normalize the flux,

$$\bar{f}_\lambda = \frac{f_\lambda^{obs}}{f_\lambda^C} \tag{2.63}$$

In essence, we aim to remove any trace of the background source as it is (usually) scientifically irrelevant to the analysis. This process is referred to as continuum normalization and it can be the dominant systematic uncertainty in absorption analysis.

To date, quasars have been the most commonly observed background sources because they are the most common, steadily luminous, distant phenomena. Unfortunately, they may also exhibit the most complexity in their spectral energy distribution (SED). This includes very strong and wide emission lines, an underlying power-law continuum with varying exponent, intrinsic absorption from gas associated with the quasars, and a plethora of unresolved emission features (e.g. Fig. 2.14).

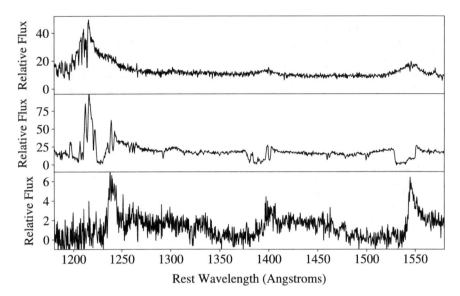

Rest Wavelength (Angstroms)

Fig. 2.14 Three quasar spectra from the SDSS survey (J021740.97-085447.9, J004143.16-005855.0, J003859.34-004252.4) which demonstrate diversity in the degree of associated absorption due to highly ionized gas in the quasar environment

The challenge is compounded by the fact that no well-developed physical model exists for the quasar SED. The phenomenon is well-modeled by a combination of synchrotron emission, optically thick emission lines arising in a wind, a hot accretion disk emitting preferentially at UV wavelengths, and a partially obscuring dust torus that radiates at IR wavelengths (e.g. [46]). But each of these is at best empirically parameterized, and may contribute largely independent of one another with great diversity. Furthermore, the emission is poorly constrained at $\lambda_{rest} < 900\,\text{Å}$ in part because the IGM greatly absorbs the SED making it very difficult to assess the intrinsic flux (see [47]).

Despite all of these vagaries and variabilities, the average quasar spectrum exhibits remarkably little evolution across cosmic time. Figure 2.15 compares composite quasar SED spectra generated by averaging tens of $z \sim 1$ sources [88] and thousands of $z \sim 2$ sources [92]. There are significant differences in the broad emission lines (e.g. Lyα, CIV 1550), but the underlying SED is very similar between the two samples. This commonality holds to the highest redshifts where quasars are observed (Fig. 2.16 [8]). This offers hope that quasar continua can at least empirically be estimated.

Historically, and even today (e.g. [59]), the majority of continuum normalization has been performed manually. Redward of Lyα emission, where absorption from intervening gas is minimal (e.g. Fig. 2.14), one can identify unabsorbed regions (by-eye or by-algorithm) to fit the continuum with a model. A higher order ($n \sim 7$) polynomial suffices over each span of $\approx 100\,\text{Å}$ to capture undulations due to underlying

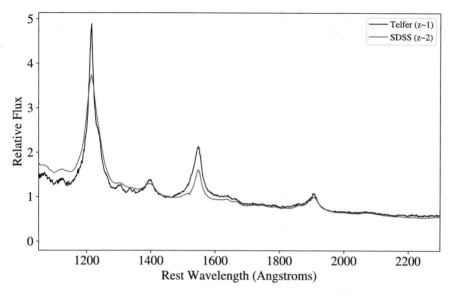

Fig. 2.15 Comparison of the quasar composite spectrum derived from *HST* spectra of $z \sim 1$ quasars [88] with a composite spectrum derived primarily from $z \sim 2$ quasars taken from the SDSS survey [92]. Each composite is normalized to unity at $\lambda_{\text{rest}} \approx 1450$ Å. While there are notable differences in the strength and widths of the emission line features (dominated by the so-called Baldwin effect), the overall SED is remarkably similar

emission features. One can routinely achieve a precision of several percent, depending on the data quality. An example in the *Spectral_Analysis* Notebook shows the call to a GUI used for continuum fitting.

Blueward of Lyα emission, where the IGM opacity is substantial, continuum estimation is very difficult. Human (by-eye) analysis typically involves spline evaluations through spectral regions purportedly free of IGM absorption. But even at echelle resolution, there may be not a single pixel unscathed by the IGM.

Modern analysis has aimed to eliminate the human aspect by empirically predicting the quasar continuum. Suzuki [87] was the first to perform a Principal Component Analysis (PCA) of quasar continua. This mathematical technique analyzes a cohort of individual spectra (they used the same $z \sim 1$ quasar spectra of the Telfer composite) to calculate the eigenvectors that best describe variations in the spectra off the mean. Formally, we have

$$|q_i> \; = \; |\mu> + \sum_{j=1}^{m} c_{ij}|\xi_j> \tag{2.64}$$

where $|q_i>$ is any quasar spectrum, $|\mu>$ is the mean (composite) spectrum, $|\xi_j>$ are the Principal Components, and c_{ij} are the eigenvalues. These eigenvectors are orthogonal and generally have little physical meaning. The eigenvectors derived by

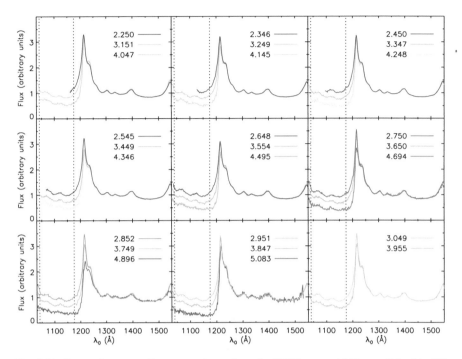

Fig. 2.16 Composite spectra for quasars drawn from the SDSS over redshifts $z \approx 2.2-5$ by [8]. Note the close similarity in the SEDs across these several Gyr. The obvious exception is at $\lambda_{rest} <$ 1215 Å, where the IGM opacity increase with increasing redshift. We will return to an analysis of the spectra in this figure in the following section. Reproduced by permission of the MNRAS

[87] are shown in Fig. 2.17 and they found the first 3 PCA components accounted for over 80% of the observed variance in the quasar spectra.

The modern approach to quasar fitting, especially in lower resolution spectra, is to fit the quasar SED with these PCA eigenvectors outside the Lyα forest ($\lambda_{rest} >$ 1215.67 Å) and then extrapolate the model to wavelengths impacted by the IGM. A more recent example of this technique is described in [63]. We also refer the student to analysis that includes Mean Flux Regulation [41] where the PCA estimate is further refined in the Lyα forest by imposing that the IGM opacity match the average (which we derive in a later section).

2.3.3 Equivalent Width Analysis

Now that we have normalized the spectrum by the intrinsic source continuum, we may proceed to analyze the observed absorption. The fundamental observable of absorption strength is the equivalent width W_λ. Observationally, this metric describes

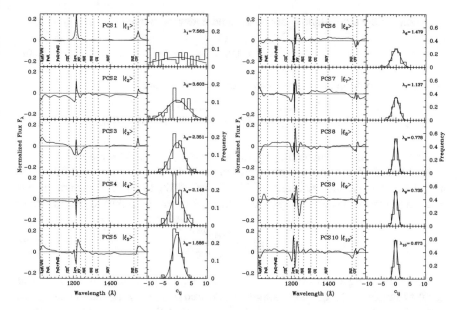

Fig. 2.17 Each of the sub panels shows one PCA eigenvector ξ_j (left) and the distribution of eigenvalues c_{ij} (right) for the $z \sim 1$ quasars analyzed by [87]. The eigenvectors from top-left to lower-right are ordered in decreasing relevance. Taken from [87]. ©AAS. Reproduced with permission

the fraction of incident flux absorbed by the gas. Empirically, there are two standard approaches to estimating W_λ: box-car integration and line-profile analysis.

A non-parametric evaluation may be performed with boxcar integration, i.e., the simple summation over the pixels covering the absorption line. With the flux continuum normalized (and expressed as $\bar{f}$), we have

$$W_\lambda = \sum_i (1 - \bar{f}_i)\delta\lambda_i \tag{2.65}$$

where $\delta\lambda_i$ is the dispersion of each pixel and the sum is performed over 'all' pixels of the profile. The uncertainty follows from propagation of error (neglecting continuum uncertainty),

$$\sigma^2(W_\lambda) = \sum_i \sigma^2(\bar{f}_i)(\delta\lambda_i)^2 \tag{2.66}$$

One may also adopt parametric techniques by fitting a model of the line-profile to the observed absorption; this is referred to as line-fitting. Most HI absorption features (aside from the very strongest) may be well approximated by a (upside-down) Gaussian. Or by considering $(1 - \bar{f})$ one can fit with

$$g(\lambda) = A \exp[-(\lambda - \lambda_0)^2/(2\sigma_\lambda^2)] \tag{2.67}$$

and calculate the equivalent-width as the area under the curve

$$W_\lambda = A\sigma_\lambda \sqrt{2\pi} \ . \tag{2.68}$$

Uncertainty should include the co-variance between A and σ

$$\sigma(W_\lambda) = W_\lambda \sqrt{\sigma^2(A)/A^2 + \sigma^2(\sigma_\lambda)/\sigma_\lambda^2 + 2\sigma(A)\sigma(\sigma_\lambda)/[A\sigma_\lambda]} \tag{2.69}$$

We show examples of calculating W_λ in the *Spectral_Analysis* Notebook.

Analysis of W_λ is generally performed in the observer (i.e. measured) frame, but W_λ^{obs} depends on λ and therefore redshift. To relate W_λ to physical quantities (e.g. N, b via a curve-of-growth analysis; Fig. 2.11), one requires a shift to the rest-frame

$$W_\lambda^{\mathrm{rest}} = \frac{W_\lambda^{\mathrm{obs}}}{1+z} \tag{2.70}$$

When designing an absorption-line observation, one generally must estimate the signal-to-noise (S/N) required to achieve the desired sensitivity. The latter may follow from a desired equivalent width limit W_λ^{lim}. We relate the two quantities by asking (and answering): *What is the limiting equivalent width one can measure from data with a given set of spectral characteristics?* Returning to our boxcar estimate of the uncertainty (Eq. 2.66), we assume a constant dispersion $\delta\lambda$ and recognize that $\sigma(\bar{f})$ is equivalent to the inverse of the S/N per pixel for normalized data. If the S/N is roughly constant across the absorption, then the terms inside the sum in Eq. 2.66 come out and the sum evaluates to M, the number of pixels across the line, i.e.,

$$\sigma(W_\lambda) = \frac{\sqrt{M}\delta\lambda}{S/N} \ . \tag{2.71}$$

For a 3σ detection, we relate $W_\lambda^{\mathrm{lim}} = 3\sigma(W_\lambda)$ and we can invert Eq. 2.71 to calculate S/N:

$$S/N = 3\frac{\sqrt{M}\delta\lambda}{W_\lambda^{\mathrm{lim}}} \tag{2.72}$$

The lingering unknown is the number of pixels in the integration. If the line is unresolved (or barely resolved), M is simply the sampling ($\Delta\lambda_{\mathrm{FWHM}}/\delta\lambda$). Otherwise, M depends on the intrinsic line-width.

We may also characterize the data quality in terms of a limiting equivalent width. As an example, consider HI Lyα in SDSS spectra (see also the Notebook). For $R = 2{,}000$, S/N $= 10$ per pixel, and a sampling of 2 pixels, we estimate $W_{\mathrm{lim}}^{3\sigma} \approx 1$ Å. This falls on the saturated portion of the COG for Lyα (Fig. 2.10). One expects, therefore, to have poor sensitivity to N_{HI} in such spectra. As another example, consider the required S/N for a 30 mÅ detection at 3σ significance in an echelle spectrum with $R = 30,000$ and 3 pixel sampling. Using Eq. 2.72, we estimate that a S/N ≈ 10 is desired.

2.3.4 Line-Profile Analysis

While W_λ offers an intuitive, observationally based measure of the absorption, it is difficult to translate such measurements to physical quantities, i.e. the column density and/or kinematics of the gas. At high spectral resolution ($R > 20,000$), one may fully resolve the line profile and estimate the optical depth directly. In turn, a line-profile analysis yields the line-centroid (i.e. redshift) and the N, b values described in the previous section. A full discussion of line-profile fitting techniques is beyond the scope of this Chapter. We refer the reader to a simple example in the *Spectral_Analysis* Notebook. And we further refer the reader to two of the widely adopted packages for line-profile analysis: VPFIT (developed by R. Carswell): http://www.ast.cam.ac.uk/~rfc/vpfit.html and ALIS (R. Cooke): https://github.com/rcooke-ast/ALIS. Each package performs χ^2 minimization on a set of input absorption lines to derive physical quantities. We will see examples of the results from such analysis in the following sections.

2.4 HI Lines of the Lyα Forest $f(N_{HI}, z)$

The term Lyα forest refers to the thicket of absorption observed in the spectra of a distant source at $\lambda_{rest} < 1215.67$ Å in the source rest-frame. The absorption lies blueward of the source Lyα emission because these photons redshift while traveling to intervening HI gas at $z < z_{source}$ to then scatter at Lyα in the rest-frame of this gas. Figure 2.18 shows Lyα forest spectra for sources ranging from $z \approx 0 - 6$. Two aspects are immediately obvious: (i) the opacity fluctuations are substantial at any given redshift and at least has the appearance of discrete lines; (ii) there is a significant increase in the average opacity with increasing redshift.

In this section, we introduce the standard approaches used to characterize the Lyα forest opacity. These techniques are primarily empirical although the results offer valuable constraints on modern cosmology. Fundamentally, any precise measurement from the IGM offers a valuable constraint on our cosmology because we have a well-developed theory of structure formation (e.g. N-body simulations) and on quasi-linear scales, the baryons track the dark matter. Indeed, one may predict the Lyα opacity provided an estimate of the universe's ambient radiation field (which may also be constrained). This section focuses on results for gas that is optically thin at the HI Lyman limit (i.e. $N_{HI} < 10^{17}$ cm^{-2}, as derived in the following section). We note further that true intergalactic gas likely corresponds to even lower column densities ($N_{HI} < 10^{14}$ cm^{-2}).

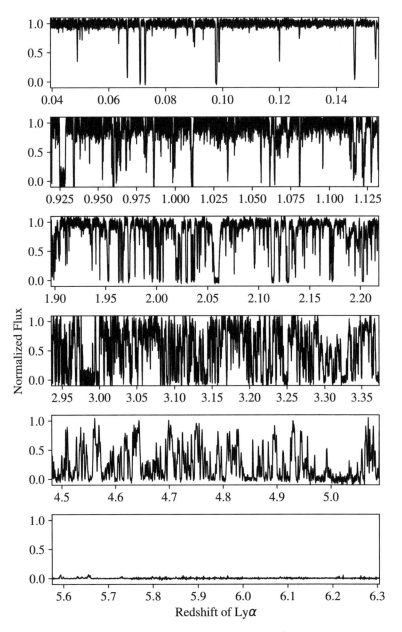

Fig. 2.18 Snapshots of the Lyα forest in a series of spectra with increasing source redshift. The data are plotted against the redshift for Lyα, i.e., $z_{\text{Ly}\alpha} = \lambda_{\text{obs}}/1215.67\,\text{Å} - 1$. Note the evolution from a nearly transparent Lyα forest at $z_{\text{Ly}\alpha} \approx 0$ to essentially zero transmission at $z_{\text{Ly}\alpha} \approx 6$

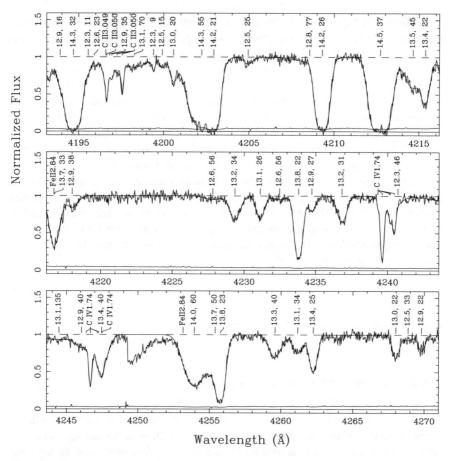

Fig. 2.19 Examples of a series of fits to lines in high S/N, echelle spectra of the Lyα forest. The dashes indicate individual lines with corresponding N, b values from fits to the data. The redshift of each line is given by the fitted line centroid. Taken from [38]. ©AAS. Reproduced with permission

2.4.1 N_{HI} Frequency Distribution $f(N_{HI})$: Concept and Definition

Motivated by the discrete appearance of lines in the Lyα forest (e.g. Fig. 2.18), the first approach developed was to model the gas as a series of absorption lines. Each line has three physical parameters (N, b, z) and the Lyα forest is described by the distributions of N and b across cosmic time. This technique demands echelle resolution spectra and high S/N to precisely constrain the absorption-line parameters.

In practice, one laboriously fits the individual lines in a set of spectra (Sect. 2.3.4). Figure 2.19 shows a snippet of spectra analyzed in this manner by [38]. The approach has the positive features of capturing the stochastic nature of the Lyα Forest and its absorption-line appearance. Furthermore, it is primarily model-independent aside from the underlying ansatz that the gas arises primarily in discrete lines. On the other hand, the N, b, z distributions do not represent a truly physical model and similar quantities are not easily derived from actual models or simulations of the IGM [76]. Furthermore, the analysis tends to be very expensive (laborious), human interaction implies non-reproducible results, and the data required is expensive to obtain (demanding many hours of integration with echelle spectrometers on 10-m class telescopes). Nonetheless, it offers a precise description of the IGM.

Formally, one defines a frequency distribution for the lines of the Lyα forest $f(N_{HI}, b, z) \, dN_{HI} \, db \, dz$, the number of lines on average in the intervals N_{HI}, N_{HI} + dN_{HI}; $b, b + db$; $z, z + dz$. This distribution function is akin to a luminosity function $\phi(L)dL$, which describes the average number of galaxies per volume in a luminosity interval $L, L + dL$. The absorption, however, occurs along sightlines instead of within a volume. A standard assumption adopted in most analyses is that the Doppler parameter (b-value) distribution has minimal N_{HI} (or z) dependence, i.e. the N_{HI} and b-value distributions are separable,

$$f(N_{HI}, b) = f(N_{HI}) \, g(b) \tag{2.73}$$

Indeed, in the following, we discuss these separately.

2.4.2 Binned Evaluations of $f(N_{HI})$

We may estimate $f(N_{HI})$ from a set of absorption-line measurements as follows. Consider first, a single sightline to a source at $z = 3$. A spectrum spanning $\lambda = 3500-5000$ Å will cover Lyα absorption at $z \approx 2 - 3$. One may slice the sightline into multiple redshift intervals, e.g. $\delta z = 0.5$. To bolster the statistics for evaluating $f(N_{HI})$ at a given redshift, we repeat the experiment for many $\mathcal{N}$ sightlines and cover a survey path (see the *characterizing_lya_forest* Slides for an illustration):

$$\Delta z = \sum_{i}^{\mathcal{N}} (\delta z)_i \tag{2.74}$$

Later in this chapter, I will define the ΔX survey path which offers a more physical definition than the observational redshift path.

The simplest estimator of $f(N_{HI})$ is a binned evaluation of lines having N_{HI} in a finite ΔN_{HI} interval, ideally large enough to have statistical significance. We evaluate,

$$f(N_{HI}, z) = \frac{\text{\# lines in } [\Delta N_{HI}, \Delta z]}{\Delta N_{HI} \, \Delta z} \tag{2.75}$$

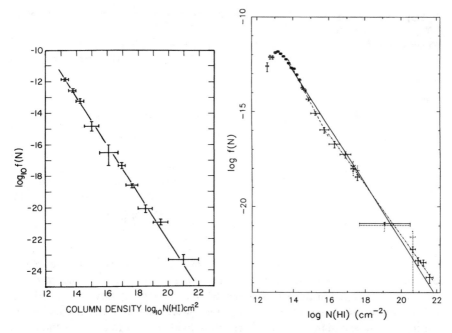

Fig. 2.20 Binned evaluations of $f(N_{HI})$ from early data sets on the Lyα forest (left) [90], (right) [65]. ©AAS and ©MNRAS. Reproduced with permission

where again Δz is the survey path for lines in the ΔN_{HI} interval. Uncertainty is simply assumed to follow Poisson statistics. An important subtlety with this estimator is that evolution in $f(N_{HI})$ with N_{HI} or z skews the detection within a bin (to lower N_{HI} and higher z). Therefore, simple model fitting with χ^2 analysis using the center of the bins will be (marginally) wrong.

Early results adopting this approach are shown in Fig. 2.20. There is a steep and monotonic decrease in $f(N_{HI})$ with increasing N_{HI}, partly resulting from the fact that ΔN_{HI} appears in the denominator of Eq. 2.75. As the figures illustrate, over many orders of magnitude in N_{HI}, the data are reasonably well-described with a power-law. This led [90] to propose a single-population of gas 'clouds'. In contrast, [65] observed significant departures from a single power-law and argued for distinct populations. Indeed, the modern data and interpretation are even more complex.

With the commissioning of an echelle spectrometer on the 10-m Keck I telescope (HIRES [93]), observers had access to exquisite data on multiple sightlines to assess $f(N_{HI})$. Figure 2.21 show early results from [38] derived from line-profile fitting analysis along the sightline to HS 1946 + 7658 observed to very high S/N. Their results show the decline in $f(N_{HI})$ for $N_{HI} > 10^{12}$ cm^{-2} as in previous work, but also an apparent turn-over in $f(N_{HI})$ at lower N_{HI}. This turn-over at low N_{HI} could reflect a lack of sensitivity (i.e. incompleteness); indeed, HI Lyα with $N_{HI} = 10^{12}$ cm^{-2} has $\tau_0 \approx 0.025$. However, the data are exquisite (see Fig. 2.19) and a careful assessment

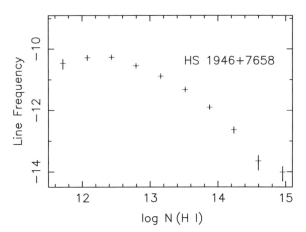

of incompleteness was performed. Another possibility is that the authors have underestimated the quasar continuum; even a few percent error in the continuum would overwhelm the signal at these optical depths. But the data quality (S/N > 100) is such that this is not the most likely explanation. Instead, we are likely witnessing the transition from describing the Lyα forest by individual lines to a more continuous opacity. We return to this concept towards the end of the section.

Modern estimates of $f(N_{HI})$ now include tens of sightlines with exquisitely small statistical error [37, 80]. The results, however, are not in uniform agreement suggesting that systematic error is the limiting factor.

2.4.3 Models for $f(N_{HI})$

The binned evaluations of $f(N_{HI})$ describe the shape of distribution function and have motivated empirical models. These models reduce the binned evaluations to a few, well-constrained parameters. In turn, one can examine redshift evolution in the amplitude and shape of $f(N_{HI})$ and use the model to evaluate $f(N_{HI})$ at any N_{HI} for calculations on the Lyα forest (see below [51]). We emphasize, however, that we have no a priori physical model for $f(N_{HI})$ and therefore there is the freedom to introduce any functional form. We also note that the results will depend on the N_{HI} range considered.

As the observational constrains have improved, the functional forms adopted for $f(N_{HI})$ have evolved from a single power-law [90], $f(N_{HI}) = B N^\beta$ to broken power-laws [65, 71], to broken, disjoint power-laws [80] to the sum of several functional forms [35], to spline evaluations [74].

In the early days, simple χ^2 analysis to binned evaluations of $f(N_{HI})$ was sufficient but these results are sensitive to the choice of bins and did not account for evolution within the bin. A more robust approach is to apply Maximum Likelihood techniques [15, 86].

Consider a sightline sample with a total survey path Δz covering the interval δz at redshift z (refer back to Eq. 2.74). We wish to model $f(N_{\mathrm{HI}})$ at this redshift over a range of N_{HI}. Divide the distribution space into M cells, each with 'volume': $\delta V = \delta z\, \delta N_{\mathrm{HI}}$. Here, we have divided ΔN_{HI} into small intervals and have assumed that $f(N_{\mathrm{HI}})$ does not evolve in the small redshift interval δz considered. The expected number of lines μ_i in ith cell follows from our definition of $f(N_{\mathrm{HI}})$,

$$\mu_i = f(N_{\mathrm{HI}})_i\, \Delta z\, \delta N_{\mathrm{HI}} \tag{2.76}$$

where we consider the full survey path Δz. Therefore, the probability of detecting m absorbers within cell i is given by Poisson statistics,

$$P(m; \mu_i) = e^{-\mu_i} \frac{\mu_i^m}{m!} \tag{2.77}$$

and construct a Likelihood function as the simple product over all cells

$$\mathscr{L} = \prod_i^M P(m; \mu_i) \ . \tag{2.78}$$

Now perform what appears to be a swindle. Reduce the volume of the cell so that each one contains at most one absorber. This is achieved by letting $\delta N_{\mathrm{HI}} \to 0$. In this case, $P(m; \mu_i)$ reduces to $e^{-\mu_i}$ for an empty cell and $e^{-\mu_i}\mu_i$ for a cell with 1 system. We may then evaluate the likelihood by summing over all M cells. Explicitly, let there be p lines detected giving $g = M - p$ empty cells. It follows that

$$\mathscr{L} = \prod_{i=1}^{g} e^{-\mu_i} \prod_{j=1}^{p} e^{-\mu_j} \mu_j \tag{2.79}$$

$$= \prod_{i=1}^{M} e^{-\mu_i} \prod_{j=1}^{p} \mu_j \tag{2.80}$$

and the log-Likelihood is expressed as

$$\ln \mathscr{L} = \sum_{i=1}^{M} -\mu_i + \sum_{i=1}^{p} \ln \mu_i \tag{2.81}$$

$$= \sum_{i=1}^{M} -f(N_{\mathrm{HI}})_i\, \Delta z\, \delta N_{\mathrm{HI}} + \sum_{j=1}^{p} \ln f(N_{\mathrm{HI}})_j\, \Delta z + p \ln[\delta N_{\mathrm{HI}}] \tag{2.82}$$

In our limit with $\delta N_{HI} \rightarrow 0$, we ignore the last term (a constant) and take the integral form of the first term to derive

$$\ln \mathscr{L} = - \int_{N_{min}}^{N_{max}} f(N_{HI}) \Delta z \, dN_{HI} + \sum_{j=1}^{p} \ln f(N_{HI})_j \Delta z \qquad (2.83)$$

If Δz is independent of N_{HI} it may be ignored, but see [15] for a treatment where Δz is dependent on N_{HI}.

Lastly, we maximize $\mathscr{L}$ for the parameterization of $f(N_{HI})$ to derive the "best" values for the parameters. We emphasize that the resultant model need not provide a good description of the data. It is simply the best model for the functional form imposed. The assessment of goodness of fit requires another statistical test (see 2.5.4).

With modern data and the assumption of a simple power-law, [80] report $f(N_{HI}) \propto N_{HI}^{-1.65 \pm 0.02}$ for $N_{HI} = 10^{13.5} - 10^{17}$ cm^{-2} at $z \approx 2.5$. With a similar but distinct dataset [37] derive $f(N_{HI}) \propto N_{HI}^{-1.52 \pm 0.02}$ for $N_{HI} = 10^{12.75} - 10^{18}$ cm^{-2} at $z \approx 2.8$. Formally, these are statistically incompatible. Progress with the $f(N_{HI})$ approach requires resolving systematic errors, including human biases. Or, alternatively, fundamentally different methods to analyze the data.

2.4.4 b-Value Distribution

The results from line-profile fitting also yield a distribution of measured b-values. Data from [38] are shown in Fig. 2.22. The distribution of b-values is dominated by lines with $b \approx 20-30$ km/s. Further, one identifies no strong dependence on HI column density aside from a possible increase in the minimum b-value with increasing N_{HI}.

Hui and Rutledge [34] derived a functional form for the b-value distribution based on theoretical expectations (fluctuations in the IGM optical depth):

$$g(b) = \frac{4b_\sigma^4}{b^5} \exp\left(-\frac{b_\sigma^4}{b^4} \right) \qquad (2.84)$$

described by a single parameter b_σ. Fitting to the results in Fig. 2.22 they achieved a good description of the observations with $b_\sigma = 26.3$ km/s (Fig. 2.23). This yields an average b-value

$$<b> = \frac{\int b g(b) \, db}{\int g(b) \, db} = b_\sigma \Gamma(3/4) \approx 32 \, \text{km/s} \qquad (2.85)$$

It is evident in both of the above figures that the Lyα forest lines exhibits a lower limit to the measured b-values. This is $b_{min} \approx 18$ km/s. What sets this apparent lower bound to the b-value (which is well above the spectral resolution of the instrument)?

Fig. 2.22 Scatter plot of *b* versus N_{HI} measurements for the absorption lines analysed along the sightline to quasar HS 1946 + 7658 by [38]. ©AAS. Reproduced with permission

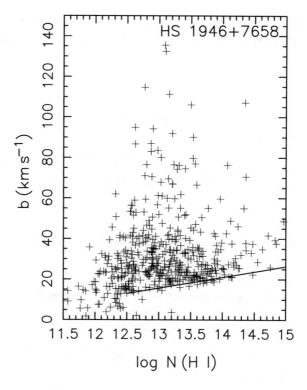

Fig. 2.23 Fit by [34] to the observed *b*-value distribution of lines in the Lyα forest [38]. This one-parameter model has an average *b*-value of 32 km/s. ©AAS. Reproduced with permission

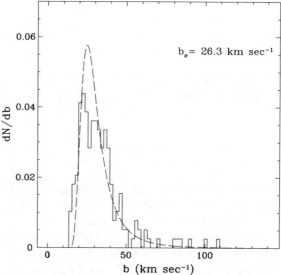

Recalling that $b = \sqrt{\frac{2kT}{m_A} + \xi^2}$ with ξ^2 a characteristic turbulence of the gas, one is tempted to associate b_{min} to the IGM temperature. For purely thermal broadening of Hydrogen, we have

$$T = 10^4 \, \text{K} \left(\frac{b}{13 \, \text{km/s}} \right)^2 \tag{2.86}$$

and may estimate $T \approx 20{,}000$ K. We comment on recent estimations in the final section.

One also notes an upper bound to the distribution of $b_{max} \approx 100$ km/s. It is difficult to establish whether this represents a sensitivity limit (recall $\tau_0 \propto b^{-1}$) or a true physical limit (e.g. collisional ionization of HI).

2.4.5 Line Density (the Incidence of Lines in the Lyα Forest)

Referring back to Fig. 2.18, the average opacity of the Lyα forest clearly evolves with redshift. In our description of the IGM as a series of lines, this implies evolution in the line density (or their incidence). Confusingly, the literature is abound with notation for this quantity: $dN/dz, n(z), N(z)$. I have adopted my own: $\ell(z)dz$ defined as the number of lines detected on average in the interval $z, z + dz$ over an interval of column density $N_{HI} = [N_{min}, N_{max}]$. This is the zeroth moment of our frequency distribution,

$$\ell(z)dz = \int_{N_{min}}^{N_{max}} f(N_{HI}, z) \, dN_{HI} \, dz \tag{2.87}$$

It is akin to the number density of galaxies derived from a luminosity function.

Taking the results for $f(N_{HI})$ from [37] at $z \approx 2.8$, and integrating from $N_{min} = 10^{12.75} \, \text{cm}^{-2}$ to $N_{max} = 10^{17} \, \text{cm}^{-2}$, we estimate

$$\ell(z \approx 2.8) = \frac{10^{9.1}}{0.52} N_{HI}^{-0.52} |_{N_{min}}^{N_{max}} \approx 560 \tag{2.88}$$

This integral is dominated, of course, by the lowest N_{HI} systems. If we consider a 5 Å patch of spectrum at 4600 Å, we estimate $\delta z = (1 + z)(\delta \lambda / \lambda) = 0.004$ and predict $\mathcal{N} = \ell(z) \delta z = 2.4$ lines on average.

Binned evaluations of $\ell(z)$ from [37] are shown in Fig. 2.24 where one observes strong redshift evolution, as evident in individual spectra. Their analysis also indicates differing behavior for lines of differing N_{HI} implying a likely evolution in the shape of $f(N_{HI})$.

We can model the redshift evolution in $\ell(z)$, which is also the redshift evolution in the normalization of $f(N_{HI})$. Here we have some physical guidance on the functional

Fig. 2.24 Evolution in the incidence of Lyα forest lines with redshift for two intervals of N_{HI}. As inferred from a visual inspection of the data (Fig. 2.18), the incidence decreases steeply with decreasing redshift

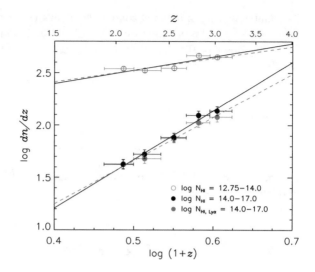

form (see also ?). Imagine a population of absorbers with physical (proper) number density $n_p(z)$ and proper cross-section $A_p(z)$, e.g. a population of spherical absorbing cows. On average, we expect to intersect $\ell(r) = n_p(z)A_p(z)$ absorbers per proper path length dr_p. In our Cosmology,

$$\frac{dr_p}{dz} = \frac{c}{H(z)(1+z)} \quad . \tag{2.89}$$

Recognizing $\ell(r)dr = \ell(z)dz$, we have

$$\ell(z) = n_p(z)A_p(z)\frac{c}{H(z)(1+z)} \tag{2.90}$$

The Hubble Parameter in a flat ΛCDM universe is given by

$$H(z) = H_0\sqrt{\Omega_m(1+z)^3 + \Omega_\Lambda} \tag{2.91}$$

and at $z > 2$, the universe is matter dominated and Ω_Λ may be ignored. This gives $H(z) \approx H_0\Omega_m^{1/2}(1+z)^{3/2}$.

Now ansatz that the absorbing gas (cows) is a constant comoving population, $n_p(z) = n_c(1+z)^3$. This gives

$$\ell(z) \propto n_c(z)A_p(z)(1+z)^{\frac{1}{2}} \tag{2.92}$$

which (on its own) implies a relatively shallow redshift dependence (indeed, much weaker than observed). However, if the gas is undergoing Hubble expansion, then $\bar{\rho} \propto (1+z)^3$ and one predicts an increasing neutral fraction with z provided the emissivity from ionizing sources does not rise more steeply than $(1+z)^3$. In any case, the number density of ionizing sources should also scale as $(1 + z)^\gamma$ and altogether, we have a physical motivation for a $(1 + z)^\gamma$ evolution in $\ell(z)$.

Fitting this model to the data gives the results at $z \sim 2-3$ shown in Fig. 2.24 which correspond to

$$\ell(z) = 100(1 + z)^{1.12 \pm 0.24} \tag{2.93}$$

for $N_{\mathrm{HI}} < 10^{14} \, \mathrm{cm}^{-2}$ and

$$\ell(z) = 0.4 \, (1 + z)^{4.14 \pm 0.6} \tag{2.94}$$

for $N_{\mathrm{HI}} = [10^{14}, 10^{17}] \, \mathrm{cm}^{-2}$. Both of these have $\gamma > 1/2$ implying physical evolution in the Lyα forest.

2.4.6 Mock Spectra of the Lyα Forest

Provided the $f(N_{\mathrm{HI}})$ distribution, one may generate mock spectra of the Lyα forest (with or without an underlying source continuum). This produces an empirical description that basically ignores any underlying physical model and its implications (e.g. clustering of gas in the IGM). However, because it is derived directly from the data it may have a greater realism in many respects.

An example is provided in the *fN* Notebook and the code is available in the *pyigm* package. We provide here the basic recipe:

1. Define a redshift interval δz for the mock Forest
2. Define N_{HI} bounds for $f(N_{\mathrm{HI}})$ (usually the full dynamic range)
3. Calculate the average number of lines ($N_{lines} = \ell(z)\delta z$)
4. Random draw (Poisson) from N_{lines}
5. Draw N_{HI} values from $f(N_{\mathrm{HI}})$
6. Draw b from $g(b)$
7. Draw z from $\ell(z)$
8. Calculate $\tau_{\lambda,i}$, including all Lyman series lines (as applicable)

    - On a wavelength grid fine enough to capture the line profile
    - i.e. a perfect spectrometer

9. Sum $\tau_{\lambda,i}$
10. Calculate $F_\lambda = \exp[-\tau_\lambda]$
11. Include spectral characteristics

    - Convolve with instrument LSF
    - Rebin to final wavelength array
    - Add in Noise

12. Multiply in a source SED.

2.4.7 Effective Lyα Opacity: $\tau_{eff}^{Ly\alpha}$

As emphasized above, the opacity of the Lyα forest is highly stochastic, both within a given sightline and from sightline to sightline. This implies an underlying, stochastic density field associated with the underling large-scale structure of the universe. While there is great scientific value in the undulations (e.g. [43]), one may also derive valuable constraints and inferences from the mean opacity across cosmic time.

Let $< F >_{\text{norm}}$ be the average, normalized flux in the Lyα forest as observed across many sightlines. See the *characterizing_lya_forest* slides for a visualization. One defines an effective Lyα opacity

$$\tau_{\text{eff},\alpha} \equiv - \ln < F >_{\text{norm}} \tag{2.95}$$

as the average opacity of the IGM from Lyα absorption alone (and generally limited to lower density gas, i.e. $N_{\text{HI}} < 10^{17} \text{ cm}^{-2}$). An alternative description used is the Lyα decrement

$$D_A = 1 - < F >_{\text{norm}} \tag{2.96}$$

Cosmologically (as described below), this average opacity is a balance between the baryon density Ω_b and the ambient radiation field. With an accurate measurement of $\tau_{\text{eff},\alpha}$ one can infer either Ω_b or constrain the radiation field.

In principle, one can estimate $\tau_{\text{eff},\alpha}$ from $f(N_{\text{HI}})$ as described by [55, 67]. We describe the technique here for completeness but note that $\tau_{\text{eff},\alpha}$ is measured directly from observations. As such, it can be used as a constraint on $f(N_{\text{HI}})$. Consider the number of lines $\mathcal{N}$ per unit rest wavelength with

$$d\lambda_{\text{rest}} = \lambda_{\text{rest}} dz/(1 + z) \ . \tag{2.97}$$

We can calculate the number of lines per $\mathcal{N} = \ell(z)dz$ from our HI frequency distribution

$$\ell(z) \, dz = \int f(N_{\text{HI}}, b, z) dN db dz \tag{2.98}$$

Translating to a rest wavelength interval,

$$\mathcal{N} = \frac{1+z}{\lambda_{\text{rest}}} \int f(N_{\text{HI}}, b, z) dN db d\lambda_{rest} \tag{2.99}$$

For these lines, the mean equivalent width is

$$\bar{W}_\lambda = \frac{1+z}{\mathcal{N} \lambda_{\text{rest}}} \int f(N_{\text{HI}}, b, z) W_\lambda(N, b) dN db \tag{2.100}$$

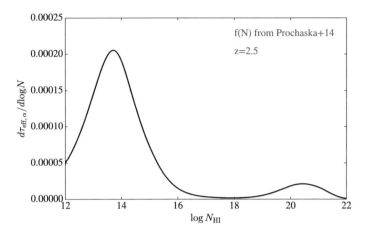

Fig. 2.25 Differential contribution to $\tau_{\text{eff},\alpha}$ for $z = 2.5$ using the $f(N_{\text{HI}})$ function from [74]

If the lines are randomly distributed (i.e. no clustering), then the mean transmission $(1 - D_A)$ is[6]

$$1 - D_A = \exp(-\mathcal{N}\,\bar{W}_\lambda) \tag{2.101}$$

An expression for the effective opacity follows,

$$\tau_{\text{eff},\alpha} = \frac{1 + z}{\lambda_{\text{rest}}} \int f(N_{\text{HI}}, b, z) W_\lambda^{\text{Ly}\alpha}(N, b)\, dN db \ . \tag{2.102}$$

This equation is relatively straightforward, but not analytic. A typical cheat is to assume the average b-value instead of a distribution which gives a one-to-one correspondence between N_{HI} and W_λ. The *fN* Notebook shows example calculations using the *pyigm* software package.

Using $f(N_{\text{HI}})$ from [74], we estimate $\tau_{\text{eff},\alpha} = 0.24$ at $z = 2.5$. Figure 2.25 shows the differential contribution with respect to $\log N_{\text{HI}}$. It is evident that $\tau_{\text{eff},\alpha}$ is dominated by lines with $\tau_0 \approx 1$ (see Eq. 2.49) and that the result depends on the choice of N_{min}.

Returning to actual measurements of $\tau_{\text{eff},\alpha}$, the first results were derived from 24 quasar spectra observed at Lick Observatory [39]. The authors employed an 'army' of undergrad students to fit continua, normalize the data, and assess systematic uncertainty. After masking metal absorption and HI lines with $N_{\text{HI}} > 10^{17}$ cm^{-2}, they reported D_A measurements for $z \approx 2 - 3$ (Fig. 2.26). Modeling D_A as $(1 + z)^\gamma$, they found

$$D_A = 0.0062\,(1 + z)^{2.75} \tag{2.103}$$

---

[6]This equation appears intuitive yet a proper derivation is remarkably complex!

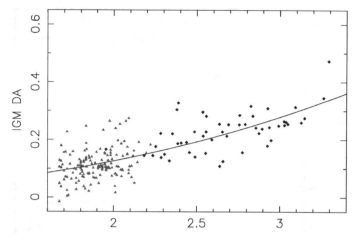

Fig. 2.26 Measurements of the Lyα decrement D_A at $z \approx 2-3$ from a modest sample of quasar spectra, by [39]. The fitted line is given by Eq. 2.103. Reproduced by permission of MNRAS

which gives $\tau_{\text{eff},\alpha} = 0.22$ at $z = 2.5$.

Analyzing a larger dataset ($\mathcal{N} \sim 100$) at higher spectral resolution, [22] extended the results to $z = 4$ and improved the precision of the measurements. These authors corrected statistically for metal absorption, assessed continuum bias on mock spectra, and reported on a disturbing 'wiggle' in their measurements (which has appeared to disappear in later datasets). Their best-fit power law is $\tau_{\text{eff},\alpha} = 0.0018 (1 + z)^{3.92}$.

Most recently, [8] have considered the full dataset of SDSS and leveraged the nearly constant mean continuum of quasar spectra (see Fig. 2.16) to measure $\tau_{\text{eff},\alpha}$ in stacked, composite spectra. Their analysis yields the *relative* evolution in $\tau_{\text{eff},\alpha}$ with redshift (Fig. 2.27). Tying their relative measurement to the absolute value at $z = 2.5$ in [22], they report

$$\tau_{\text{eff},\alpha}(z) = 0.751 \left(\frac{1+z}{1+3.5} \right)^{2.9} - 0.132 \qquad (2.104)$$

which includes statistical corrections for $N_{\text{HI}} > 10^{17} \text{ cm}^{-2}$ absorbers and metal absorption. These measurements provide a blunt yet powerful test for any cosmological model of the IGM.

2.4.8   Fluctuating Gunn–Peterson Approximation (FGPA)

The modern paradigm of the Lyα forest, or IGM, is that the observed opacity results from undulations in the underlying dark matter density. The baryons, which are predominantly ionized with an ionization state dictated by photoionization from the

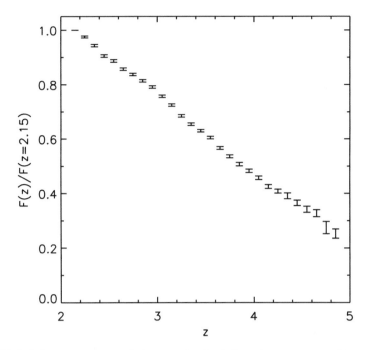

Fig. 2.27 Relative measurements of the mean flux in the Lyα forest as measured in composite quasar spectra by [8]. After tying their measurements to the $z \approx 2.5$ effective optical depth estimate of [22], they constructed a model for the evolution in $\tau_{\mathrm{eff},\alpha}$ given by Eq. 2.104. Reproduced by permission of MNRAS

extragalactic UV background (EUVB), trace these dark matter fluctuations to yield the observed Lyα forest.

This paradigm lends to the formalism now known as the fluctuating Gunn–Peterson approximation (FPGA). The standard formalism is derived as follows. We begin with our result from quantum mechanics that the optical depth of Lyα at line center is

$$\tau_0 = \frac{\pi e^2}{m_e c \sqrt{\pi}} \frac{N_{\mathrm{HI}} f_{\mathrm{Ly}\alpha}}{\Delta v_D} \ . \tag{2.105}$$

Expressing the N_{HI} column density as

$$N_{\mathrm{HI}} = n_{\mathrm{HI}} \Delta r \tag{2.106}$$

with Δr an interval in space and n_{HI} the gas density. From Cosmology (see Eq. 2.89 for dr/dz), we can relate distance to redshift

$$\Delta r = \frac{c \Delta z}{H(z)(1+z)} \ , \tag{2.107}$$

and we have

$$\tau_0 = \frac{\pi e^2}{m_e} \frac{n_{\text{HI}} \, \Delta z f_{\text{Ly}\alpha}}{\Delta v_D H(z)(1+z)} \tag{2.108}$$

Because the absorption occurs over a small redshift interval, we can express $\Delta z/(1+z) = \Delta v_{\text{Ly}\alpha}/v_{\text{Ly}\alpha}$ and identify Δv_D in Eq. 2.108 with $\Delta v_{\text{Ly}\alpha}$ (i.e. the line width is given by cosmic expansion). Ionization balance for n_{HI} with ionization rate Γ and recombination rate $\alpha(T)$ gives

$$n_{\text{HI}} \Gamma = \alpha(T) n_{\text{HII}} n_e \tag{2.109}$$

Altogether now,

$$\tau = \frac{\pi e^2 f_{\text{Ly}\alpha}}{m_e v_{\text{Ly}\alpha}} \frac{1}{H(z)} \frac{\alpha(T) n_{\text{HII}} n_e}{\Gamma} \tag{2.110}$$

To link the gas density with the dark matter density, we introduce the over-density δ where

$$\rho = \bar{\rho}(1+\delta) \tag{2.111}$$

with ρ and $\bar{\rho}$ the dark matter density and its cosmic mean. We then express our number densities in terms of the over-density. Furthermore, we define ionization and mass fractions—X is the Hydrogen mass fraction and x is the ionized Hydrogen fraction—to express

$$n_{\text{HII}} = \frac{\rho_{\text{crit}} \Omega_b}{m_p} X x \, (1+\delta)(1+z)^3 \tag{2.112}$$

A similar (uglier) expression for n_e includes Helium (with Y, y_{II}, y_{III} defined similarly):

$$n_e = \frac{\rho_{\text{crit}} \Omega_b}{m_p} [Xx + 0.25Y(y_{II} + 2y_{III})] \, (1+\delta)(1+z)^3 \tag{2.113}$$

For gas with $T \sim 10^4$ K (as expected for a photoionized medium), $\alpha(T)$ is a power-law of the form,

$$\alpha(T) \approx \alpha_0 T^{-0.7} \quad . \tag{2.114}$$

Lastly, one assumes the IGM gas follows a power-law temperature-density relation, (as derived from hydrodynamic analysis by [33])

$$T = T_0(1+\delta)^\beta \tag{2.115}$$

The FGPA expression becomes

$$\tau = A(z)(1+\delta)^{2-0.7\beta} \tag{2.116}$$

Fig. 2.28 Estimates of the
ionization rate Γ from
measurements of the
effective optical depth in
Lyα and adopting the
fluctuating Gunn–Peterson
approximation. One recovers
a nearly constant value at
$z \approx 2$–4 in contrast with the
estimation from quasars
alone over that same epoch
(which may decline steeply
at high z). Taken from [21]
and HRH07?. ©AAS.
Reproduced with permission

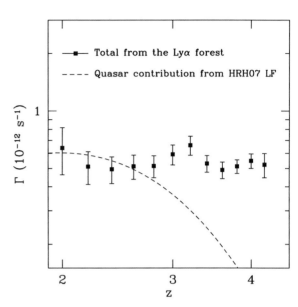

where the opacity "fluctuates" with the local over-density. The $A(z)$ term is proportional to $(1 + z)^6/[H(z)\Gamma]$ or, in its glory,

$$A(z) \equiv \frac{\pi e^2 f_{Ly\alpha}}{m_e \nu_{Ly\alpha}} \left(\frac{\rho_{crit}\Omega_b}{m_p} \right)^2 \frac{1}{H(z)} Xx[Xx + 0.25Y(y_{II} + 2y_{III})]\frac{\alpha_0 T_0^{-0.7}}{\Gamma}(1 + z)^6 \tag{2.117}$$

Given a probability density function for the overdensity $P(\Delta)$ with $\Delta \equiv 1 + \delta$, we
may calculate the mean flux (and opacity)

$$<F>(z) = \int_0^\infty P(\Delta; z)\exp(-\tau)d\Delta \ . \tag{2.118}$$

See [54] for models of $P(\Delta)$ based on numerical simulations. This FGPA provides
an analytic expression, calibrated against cosmological simulations, for the IGM
opacity.

Returning to $\tau_{eff,\alpha}$ measurements, we can relate Eq. 2.118 to the data with the only
significant unknown[7] being the photoionization rate Γ. Figure 2.28 from [21] shows
estimates for Γ based on the $\tau_{eff,\alpha}$ measurements of [22]. The figure also shows their
estimate for Γ from quasars alone and the authors argued that star-forming galaxies
must dominate the radiation field by $z \sim 3$. Recent work, however, has raised the
possibility that faint quasars play as strong a role [30, 49]. This issue awaits deeper,
high-z surveys of faint AGN.

---

[7]There is a weak dependence on T which leads to some degeneracy.

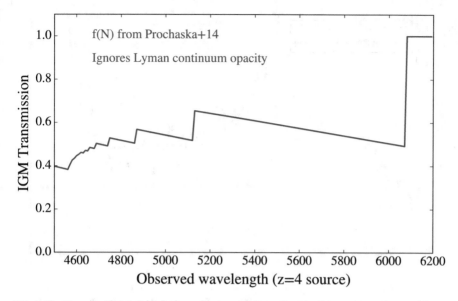

Fig. 2.29 The so-called 'sawtooth' transmission function $\mathscr{T}$ of the HI Lyman series of the IGM for a source at $z = 4$ using the $f(N_{HI})$ model of [74]. Redward of HI Lyα there is no attenuation ($\mathscr{T} = 1$). From $\lambda_{obs} \approx 5100-6000$ Å, only HI Lyα contributes and the increasing transmission with decreasing redshift is due to evolution in the Lyα line density. At shorter wavelengths, additional HI Lyman series lines contribute and reduce the transmission accordingly

2.4.9 Effective Lyman Series Opacity

We finish this section by extending our calculations for the effective opacity of Lyα to the full HI Lyman series. Each line up the HI energy ladder contributes additional opacity at shorter wavelengths with sequentially smaller opacity ($\tau_0 \propto \lambda f N$). A complex 'blending' of the Lyman series from various redshifts occurs.

For example consider the quasar light emitted at $\lambda_{rest} = 920$ Å from a quasar at $z_{em} = 4$ (we observe these photons at $\lambda_{obs} = 4600$ Å). The photons emitted by the quasar will redshift into resonance with HI Lyα at $z_{Ly\alpha} = 2.78$, Lyβ at $z_{Ly\beta} = 3.48$, Lyγ at $z_{Ly\gamma} = 3.73$, Lyδ at $z_{Ly\delta} = 3.84$, etc.

The effective opacity from each Lyman series line is independent and sums simply. An example calculation for the predicted transmission $\mathscr{T}$ through the IGM with $\mathscr{T} = \exp(-\tau_{eff})$ is shown in Fig. 2.29. This 'sawtooth' curve has teeth at each of the additional Lyman series lines. The slope of increasing transmission with decreasing wavelength is due to the decreasing incidence of lines with decreasing redshift. The *fN* Notebook shows an example calculation using software from *pyigm*.

In Fig. 2.30, we show the average relative flux (normalized to unity at $\lambda_{rest} = 1450$ Å) of ≈150, $z \sim 4$ quasars from the SDSS. Overlaid on the data is the quasar SED from [88] attenuated by the effective opacity shown in Fig. 2.29. The good

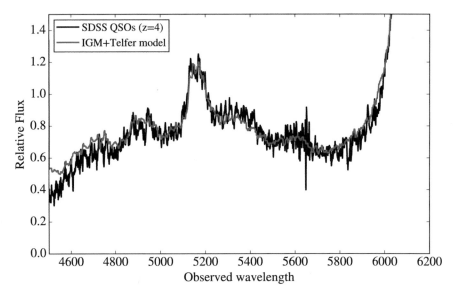

Fig. 2.30 Composite spectrum of 150 quasars at $z \sim 4$ from the SDSS (black). These are shown in the observer frame for $z = 4$ (but were stacked in the rest-frame) and were normalized to unity at $\lambda_{rest} = 1450$ Å. Overlayed on the data is a model using the quasar SED from [88] and our estimate of the IGM transmission from the full HI Lyman series. The excellent agreement (down to at least 4700 Å) is quite remarkable

agreement down to $\lambda \approx 4700$ Å is truly remarkable and demonstrates the quality of our characterization of the IGM.

Of course, IGM attenuation is imprinted in the spectra of *all* distant sources. This includes galaxies and for a $z = 4$ galaxy, the g-band flux suffers a decrement of $\Delta m_g \approx 1$ mag. This insight led to the discovery of the first high-z star-forming galaxies [45, 48, 85]. This attenuation will also affect our Lyα emitters (LAEs). At $z = 5$, $\tau_{eff,\alpha} \approx 1.5$ which likely contributes to the observed asymmetry in Lyα emission.

2.5 Optically Thick HI Absorption

Thus far, our discussion has focused entirely on Lyman series opacity for atomic Hydrogen (i.e. Lyα, Lyβ). These are the bound-bound transitions of HI. The atom also exhibits a bound-free continuum opacity beyond the so-called Lyman limit at energies $h\nu \geq I_H$ with I_H the ionization energy (13.6 eV). This corresponds to wavelengths $\lambda < 911.7$ Å. The HI continuum opacity leads to additional attenuation of the radiation field at high z, both for ambient radiation (frequently referred to as the extragalactic UV background or EUVB) and for individual sources.

Theoretically, one predicts that optically thick gas originates in environments within and around galaxies and, possibly, denser gas within the IGM. Work on the gas around galaxies (also referred to as the circumgalactic medium or CGM) indicates a high incidence of such gas to at least 100 kpc [9, 73, 75]. In turn, the study of optically thick HI bears on the accretion of gas onto galaxies (aka cold flows) and the flow of cool gas away from galaxies (aka feedback).

In this section, we focus on optically thick HI absorption, commonly referred to as Lyman Limit Systems (or LLSs). We also provide a short subsection on the damped Lyα systems (DLAs) which are systems with the highest HI column densities.

2.5.1   Physics of Continuum Opacity

The bound-free absorption process occurs when a photon with energy $\hbar\omega > I_H$ strikes an atom to eject an electron with energy $\hbar\omega - I_H > 0$, i.e. absorption occurs over a continuous range of frequencies. Our approach to describing the physics of this encounter is to combine an expression for the probability of absorption (w_{fi}) with the density of final states and then calculate the probability/time for a transition to a 'cell' of phase space. We will express this probability in terms of a cross-section which will then (when combined with N_{HI}) provide the optical depth.

Consider the final state for the emitted, free electron

$$\hbar \mathbf{k}_f = \mathbf{p}_f \tag{2.119}$$

and assume a wave function normalized to a very large volume L^3

$$|\mathbf{k}_f> \equiv \frac{1}{L^{3/2}}|f> \tag{2.120}$$

with energy

$$E_f = \frac{p_f^2}{2m} = \frac{\hbar^2 k_f^2}{2m} \tag{2.121}$$

The density of final states using a usual phase-space argument (i.e. periodic boundary conditions) is

$$k_x = \frac{2\pi n_x}{L} \quad \text{with } n_x \text{ any integer} \tag{2.122}$$

For final momentum $\hbar k_f$ propagating in direction $d\Omega$, the density of final states is

$$n^2 dn d\Omega \quad \text{with } n^2 = n_x^2 + n_y^2 + n_z^2 = k_f^2 \left(\frac{L}{2\pi}\right)^2 . \tag{2.123}$$

Finally, we derive $g(E)$, the density of final states, by equating

$$n^2 dn d\Omega = g(E) dE d\Omega$$

$$= \left(\frac{L}{2\pi}\right)^3 \frac{m_e}{\hbar^2} k_f \, dE \, d\Omega \tag{2.124}$$

We may define the Differential Cross-section as

$$d\sigma = \frac{\text{(Energy/unit time) absorbed by atom } (i \rightarrow f)}{\text{Energy flux of the radiation field}} \tag{2.125}$$

The transition probability per unit time (without derivation) is given by

$$w_{fi} = \frac{4\pi^2 e^2}{m^2 c} \frac{I(\omega_{fi})}{\omega_{fi}^2} \left| <\phi_f | e^{i\mathbf{k}\cdot\mathbf{r}} \hat{\varepsilon} \cdot \nabla |\phi_i> \right|^2 \tag{2.126}$$

For our photoelectric effect,

$$d\sigma_p = \frac{w_{fi} \, g(E) d\Omega \cdot \hbar\omega}{I(\omega)} \tag{2.127}$$

Some algebraic manipulation gives (introducing the fine structure constant $\alpha = e^2/(\hbar c)$)

$$\frac{d\sigma_p}{d\Omega} = \frac{\alpha}{2\pi} \frac{k_f}{m\hbar\omega} \left| <f | e^{i\mathbf{k}\cdot\mathbf{r}} \hat{\varepsilon} \cdot \mathbf{p} |\phi_i> \right|^2 \tag{2.128}$$

Finally, we adopt the dipole approximation, which is valid for energies near the ionization potential ($\hbar\omega \approx I_H$). In this case, instead of $|f> = e^{i\mathbf{k}_f \cdot \mathbf{r}}$, we let $e^{i\mathbf{k}\cdot\mathbf{r}} = 1$ assuming $k a_0 \ll 1$, which is automatically satisfied near the threshold energy. The differential cross-section becomes

$$\frac{d\sigma_p^D}{d\Omega} = \frac{\alpha}{2\pi} \frac{m\omega k_f}{\hbar} \left(\hat{\varepsilon} \cdot \mathbf{r}\right)_{fi} \tag{2.129}$$

For photoionization out of the ground state (1s),

$$|i> = |n = 1; \ell = 0; m = 0; m_s = \pm 1> \tag{2.130}$$

The total cross-section, integrating over $d\Omega$, is evaluated as,

$$\sigma_{photo}^D = \frac{2^9 \pi^2}{3} \alpha a_0^2 \left(\frac{I_H}{\hbar\omega}\right)^4 f(\eta) \tag{2.131}$$

where

$$f(\eta) \equiv \frac{\exp\left[-4\eta \cot^{-1}\eta\right]}{1 - \exp\left[-2\pi\eta\right]} \tag{2.132}$$

$$\eta \equiv \left(\frac{I_H}{E_f}\right)^{\frac{1}{2}} = \left(\frac{I_H}{\hbar\omega - I_H}\right)^{\frac{1}{2}} \tag{2.133}$$

For general atomic K-shell (n = 1) absorption, replace

$$I_H \rightarrow Z^2 I_H$$
$$a_0 \rightarrow a_0 Z$$
$$\eta \rightarrow \left[Z^2 I_H / (\hbar\omega - Z^2 I_H)\right]^{\frac{1}{2}}$$
$$\alpha \rightarrow 2\alpha \quad \{2 \text{ K-shell electrons}\}$$

Near threshold, $\hbar\omega \approx I_H$ giving $\eta \gg 1$ and we can approximate (using the approximation to $cot^{-1}(x)$ for large x values: $cot^{-1}(x) \sim 1/x - 1/(3x^3) + ...$)

$$f(\eta) \approx e^{-4+4/3\eta^2} \approx e^{-4}\left(1 + \frac{4}{3\eta^2}\right) \tag{2.134}$$

$$\approx e^{-4}\left(1 + \frac{1}{\eta^2}\right)^{4/3} = e^{-4}\left(\frac{\hbar\omega}{I_H}\right)^{4/3} \tag{2.135}$$

Altogether, one finds the following frequency dependence:

$$\sigma \propto \begin{cases} 0 & \hbar\omega < I_H \\ \nu^{-8/3} & \hbar\omega \approx I_H \\ \nu^{-3} & \hbar\omega \gtrsim I_H \\ \nu^{-7/2} & \hbar\omega \gg I_H \end{cases} \tag{2.136}$$

See the *Optically_Thick* Notebook for an illustration. As an aside, one can use the Milne relation to relate the photoionization cross-section to the recombination rate. Evaluating Eq. 2.131 at $h\nu \approx 1$ Ryd, we calculate

$$\sigma_{photo}(1 \text{ Ryd}) = 6.339 \times 10^{-18} \text{ cm}^2 \tag{2.137}$$

and ask: What HI column density is required to give $\tau_{LL} = 1$? Of course, our continuum opacity is simply $\tau_{LL}(\nu) = \sigma_{phot}(\nu) N_{HI}$ which we can invert to give an expression for column density:

$$N_{HI} = \frac{\tau_{LL}}{\sigma_{phot}} \tag{2.138}$$

Evaluating at 1 Ryd, we estimate $N_{HI} = 10^{17.198} \text{ cm}^{-2}$. Therefore, an HI gas with integrated $N_{HI} = 10^{17.2} \text{ cm}^{-2}$ will exhibit a Lyman limit optical depth of unity at

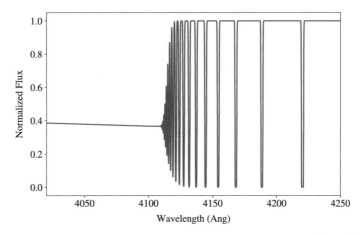

Fig. 2.31 Idealized absorption spectrum of an LLS at $z = 3.5$ with $N_{HI} = 10^{17.2}\,\mathrm{cm}^{-2}$. One sees the Lyman series absorption at $\lambda > 4110\,\text{Å}$ that converges to the continuum opacity at $\lambda_{rest} < 911.7\,\text{Å}$ with optical depth of 1

$\lambda_{LL} = 911.7\,\text{Å}$. This opacity is continuous and declines as $\nu^{-\alpha}$ with $\alpha \approx 3$ (closer to 2.75).

2.5.2 Lyman Limit System (LLS)

In quasar absorption line parlance, we define an HI absorption system that is optically thick at the Lyman limit to be a Lyman Limit System or LLS. The value of τ_{LL} defining an LLS is somewhat arbitrary, but $\tau_{LL} \geq 1$ is probably the most sensible.[8] Given Eq. 2.138, this implies HI systems with $N_{HI} \geq 10^{17.198}\,\mathrm{cm}^{-2}$ or $\log N_{HI} \geq 17.2$. At this column density, the first ≈ 10 lines of the Lyman series are highly saturated ($\tau_0 \gg 1$). Furthermore, The integrated opacity of the highest order Lyman series lines matches τ_{LL} at 1 Ryd, i.e. the effective optical depth is continuous through the Limit. This is illustrated in Fig. 2.31.

Figure 2.32 shows a real-world example, the LLS identified in the VLT/X-Shooter spectrum of J0529–3526 taken from the XQ-100 Survey [44]. The LLS at $z \approx 4.37$ ($\lambda_{LL} \approx 4900\,\text{Å}$) absorbs approx 66% of the quasar flux that is transmitting through the Lyα forest. This implies an optical depth from the continuum opacity of $\tau_{LL} \approx 1$ and therefore $N_{HI} \approx 10^{17.2}\,\mathrm{cm}^{-2}$. Careful inspection reveals a second LLS at $\lambda \approx 4470\,\text{Å}$ which completely absorbs the quasar flux (i.e. $\tau_{LL} > 3$).

In analysis of actual spectra (e.g. Fig. 2.32), one can recover a precise measurement for the optical depth only over a limited range: $\tau_{LL} = 0.2-3$. For $\tau > 3$ the flux is zero

---

[8]With an LLS defined as HI gas with $\tau_{LL} \geq 1$, systems with $\tau_{LL} \lesssim 1$ are termed partial LLS or pLLS.

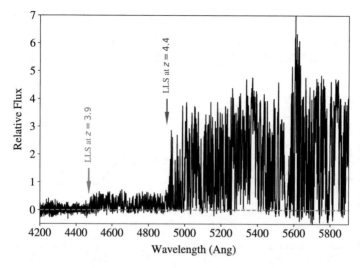

Fig. 2.32 A portion of the quasar spectrum of J0529–3526 recorded by the XQ-100 survey [44]. One observes a sharp break in the flux at $\lambda \approx 4900\,\text{Å}$ corresponding the Lyman limit opacity of an HI absorber at $z \approx 4.37$. The flux decrement corresponds to an optical depth of $\tau_{LL} \approx 1$. Further to the blue ($\lambda \approx 4470\,\text{Å}$), one observes a second and complete break beyond which the quasar flux is completely absorbed. This second LLS at $z \approx 3.9$ has $\tau_{LL} > 3$

and one only establishes a lower limit to the opacity. For $\tau < 0.2$, the flux decrement is difficult to detect amidst the fluctuating IGM and variations in the quasar SED. This limited range of τ_{LLS} implies a limited range of sensitivity for measuring N_{HI} from the Lyman limit of $N_{HI} \approx 10^{16.5} - 10^{17.8}\,\text{cm}^{-2}$.

2.5.3   LLS Surveys

As a first estimate for the universe's average opacity to ionizing photons, one may survey for LLS in a large spectroscopic sample of sources. This requires, of course, spectra blueward of the Lyman Limit and that the source emits photons beyond its Lyman limit. Conveniently, most quasars are transmissive at $h\nu > 1$ Ryd as illustrated by the *HST*/WFC3 quasar spectrum shown in Fig. 2.33. Evidently, Type I Quasars are sufficiently bright to photoionize the sightline through their own host galaxy. This proximity effect extends to 1 Mpc and possibly beyond [72].

A survey for LLS is relatively straightforward: search for and measure a continuum break associated to the Lyman continuum opacity. There is, however, a subtlety. The discovery of one LLS generally precludes the discovery of any others at lower redshift because the quasar flux has been significantly attenuated. This is unique amongst absorption systems or any other astrophysical phenomena. And it inspires a unique approach to analysis of the survey.

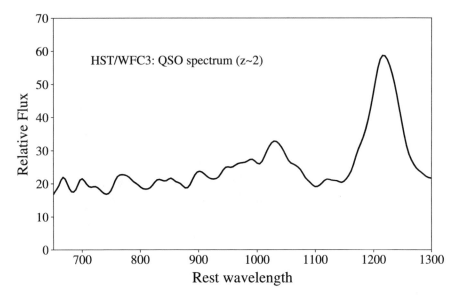

Fig. 2.33 Spectrum of a $z \approx 2$ quasar plotted in its rest-frame taken from the *HST*/WFC3 survey of [58]. This source shows a sustained and largely unattenuated flux blueward of its Lyman limit which enables the search for Lyman limit absorption from intervening gas (none is observed here)

Tytler [89] introduced the concepts and techniques of Survival Statistics to analyze his own survey of LLS. We briefly present his formalism. Let t be the redshift separation of the LLS from the QSO redshift, i.e. $t = z_{em} - z_{LLS}$. For a spectrum with finite spectral coverage (λ_{min} to λ_{max}), the LLS search must truncate at $z_{min} = \lambda_{min}/\lambda_{LL} - 1$. For example, $z_{min} \approx 3.2$ for $\lambda_{min} = 3800\,\text{Å}$ for spectra from the SDSS survey. The maximum value of t for a given sightline is therefore $T = z_{em} - z_{min}$.

Define three functions to describe the distribution of LLS: (1) the Observed density function, the probability that an LLS is observed in a line of sight at $t = x$

$$f(t) = \frac{\text{Prob}(t < x < t + \Delta t)}{\Delta t} \quad \text{as } \Delta t \to 0 \tag{2.139}$$

(2) the Survival function, the probability that *no* LLS occurs in the interval $(0, t)$

$$S(t) = 1 - \int_0^t f(t')dt' \tag{2.140}$$

and (3) the parent population density function, the probability that an LLS is observed in the interval $(t, t + \Delta t)$.

$$\lambda(t) = \frac{\text{Prob}(t < x < t + \Delta t | x > t)}{\Delta t} \quad \text{as } \Delta t \to 0 \tag{2.141}$$

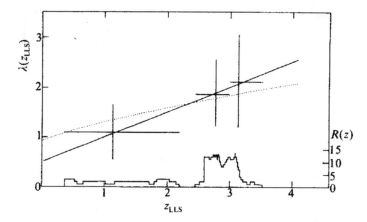

Fig. 2.34 Binned evaluations of the incidence of LLS per unit redshift (here expressed as $\lambda(z_{LLS})$) from [89]. The histogram describes the sensitivity function of their survey, i.e. the number of quasars searched for LLS at a given redshift. Reprinted by permission from Nature

Ansatz that the parent population is constant with t, i.e. $\lambda(t) = \lambda_0$. This is equivalent to a Poisson distribution for the number of LLS as a function of t. With $\lambda(t) = \lambda_0$, our survival function (Eq. 2.140) evaluates trivially to $S(t) = \exp(-\lambda_0 t)$. This is akin to the decay of a particle with decay time ($\tau = 1/\lambda$) and the probability that the particle will survive for a time t. One may also derive $f(t) = \ell \exp(\lambda_0 t)$ and use (simple) statistical methods from biomathematics to derive the maximum likelihood for λ_0 from the observed distribution of LLS. We relate λ to the incidence of LLS at a given redshift, $\ell(z)_{LLS}$. Figure 2.34 shows the first estimates of this quantity from [89].

The lower curve in Fig. 2.34, defined there with notation $R(z)$, is the sensitivity function. In modern literature, it is notated with $g(z)$ where $g(z)\,dz$ is the number of sightlines that one may analyze for absorption in the interval $z, z + dz$. In essence, $g(z)$ expresses the 'volume' of any absorption line survey. In the [89] survey (Fig. 2.34), $R(z = 3) \approx 10$ indicates that the survey included 10 quasars whose spectra permitted the search for an LLS at $z = 3$. Again, in an LLS survey, the search along a given sightline runs from z_{em} to the maximum of (z_{min}, z_{LLS}) where z_{min} is set by the spectral coverage and z_{LLS} is the redshift of the first LLS identified along the sightline (if any).

It is also common practice to ignore systems close to the quasar by beginning the search several thousand km/s blueward of z_{em}. This accounts for quasar redshift error and minimizes any bias related to the 'proximity' of the quasar (e.g. an elevated incidence of LLS due to galaxy–galaxy clustering). An example of the selection function for an LLS survey using the SDSS dataset [71] is shown in Fig. 2.35. The shape of $g(z)$ is determined primarily by the quasar population recovered by the survey.

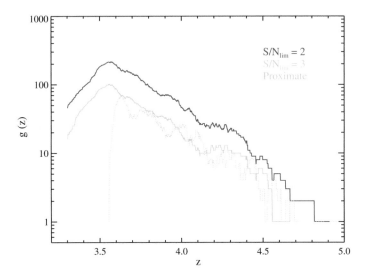

Fig. 2.35 Sensitivity function of [71], cut by S/N. The increase in $g(z)$ at low redshifts is due to their avoiding $z < 3.5$ quasars which are biased in SDSS to exhibit LLS (see Sect. 2.5.5). The monotonic decrease in $g(z)$ beyond $z \approx 3.5$ is due to the decreasing number of quasars with high quality spectra in the survey. ©AAS. Reproduced with permission

2.5.4 Incidence of LLS: $\ell_{LLS}(z)$

Another approach to estimating $\ell_{LLS}(z)$, motivated by survival analysis, is through a binned evaluation of the number of systems discovered relative to the integrated sensitivity function (i.e. survey path). Consider, $\ell_{LLS}(z)$ in a redshift interval $[z_1, z_2]$.

The estimator is

$$\ell_{LLS}(z) = \frac{\text{Number of systems detected in } [z_1, z_2] \ (\mathcal{N})}{\int\limits_{z_1}^{z_2} g(z)\, dz} \tag{2.142}$$

One estimates uncertainty using the variance from Poisson statistics: $\sigma^2(\mathcal{N})$. Figure 2.36 shows an example of results from $z \sim 4$ for LLS with optical depths $\tau_{LL} \geq 2$.

We may model $\ell_{LLS}(z)$ independent of binning, with the Maximum Likelihood technique described in Sect. 2.4.3 for $f(N_{HI})$ (but without the N_{HI} dependence). For the same arguments made on the Lyα Forest, we may expect the redshift evolution to scale as $(1 + z)^\gamma$. Ignoring the N_{HI} dependence in Eq. 2.82, the log-Likelihood probability for a given sightline is given by:

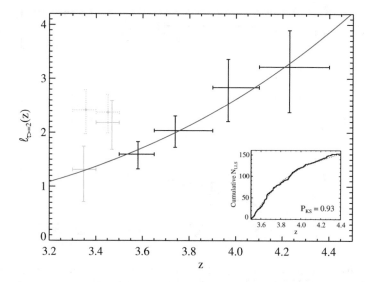

Fig. 2.36 Incidence of LLS with $\tau_{LL} \geq 2$ per unit redshift $\ell_{LLS}(z)$ from $z \approx 3.3-4.4$ derived from quasar spectra in the SDSS DR7 [71]. The black binned evaluations show an increasing incidence with redshift. The inset compares the observed, cumulative distribution of LLS redshifts against the model prediction. The points with dotted error bars show $\ell(X)$ for the full set of SDSS quasars observed in the survey. Because of their color selection (see Sect. 2.5.5), the quasars with $z_{em} < 3.5$ are biased to show LLS. ©AAS. Reproduced with permission

$$\ln \mathscr{L}_j = \sum_{i=1}^{M} -\mu_i + \sum_{k=1}^{p} \ln \mu_k \qquad (2.143)$$

$$= \sum_{i}^{M} -\ell(z_i)dz + \sum_{k}^{p} \ln \ell(z_k) \qquad (2.144)$$

This expression is for the jth quasar and we have related the expected number of LLS in a cell with 'volume' dz to be $\mu_i = \ell(z_i)\delta z$. We also recognize that for an LLS survey $p = 0$ or 1 because we are limited to detecting one LLS per sightline.

Letting $dz \rightarrow 0$,

$$\ln \mathscr{L}_j = -\int_{z_{start}}^{z_{end}} \ell(z)dz + \sum_{k}^{p} \ln \ell(z_k), \qquad (2.145)$$

The full log-likelihood $\ln \mathscr{L}$ is the sum over all quasars. Maximizing $\ln \mathscr{L}$ yields the best parameters for a given model. We may then assess the goodness of the model by comparing the observed cumulative distribution of LLS redshifts against that predicted by the model after integrating over $g(z)dz$. This is frequently assessed

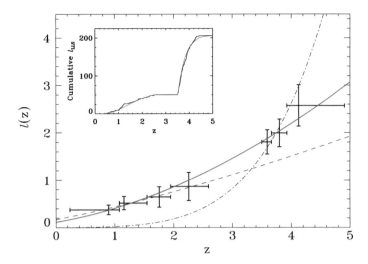

Fig. 2.37 Incidence of LLS with $\tau_{LL} \geq 2$ per unit redshift $\ell_{LLS}(z)$ from $z \approx 0{-}5$ combining surveys with the *HST* [77] and SDSS-DR7 [71]. The continuous curve, derived with Maximum likelihood techniques, is a $(1+z)^{\gamma}$ power-law with $\gamma = 1.83$ which describes the data well across cosmic time. ©AAS. Reproduced with permission

with a one-sided Kolmogorov–Smirnov (KS) test. This is shown in Fig. 2.36 with $P_{KS} \sim 1$ indicating a fully acceptable model.

Results on $\ell_{LLS}(z)$ across a wide redshift interval are shown in Fig. 2.37, again for LLS with $\tau_{LL} \geq 2$ [77]. The data have been modeled with a power-law of the form

$$\ell(z) = \ell_* \left[\frac{1+z}{1+z_*} \right]^{\gamma} \tag{2.146}$$

and the authors report $z_* = 3.23$, $\ell_* = 1.62$, $\gamma = 1.83 \pm 0.21$.

Consider the distance Δr that one travels on average to intersect one LLS at $z = 3.5$. The redshift path is approximately $\Delta z \approx 1/\ell(z = 3.5) \approx 0.56$. From cosmology, the physical distance is

$$\Delta r \approx \frac{c\Delta z}{H(z)(1+z)} \tag{2.147}$$

and we estimate $\Delta r \approx 100\,\text{Mpc}$ (see also the *Optically_Thick* Notebook). This exceeds the mean separation between faint quasars by a factor of $3{-}10$ [23]. Therefore, we expect each volume in the universe sees the radiation from multiple sources, although we caution that the opacity to ionizing photons may be dominated by gas with $\tau_{LL} < 2$. We discuss a more accurate measurement of the mean free path below.

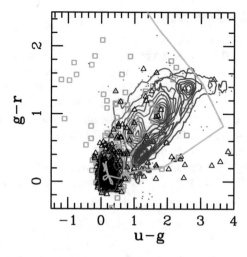

Fig. 2.38 Color-color plot of point sources in SDSS photometry (contours), primarily Galactic stars and white dwarfs. The black triangles are spectroscopically confirmed Type 1 quasars and the cyan curve is the color-color track for quasars from $z = 0$ to 7. In this parameter space, quasars only separate themselves from stars when they have a *red $u - g$* color. This occurs for $z = 3$ quasars when an intervening LLS absorbs the u-band flux. Taken from [78]. ©AAS. Reproduced with permission

2.5.5 Survey Subtleties

Thus far we have ignored any systematic effects in HI absorption analysis related to the selection of background sources. For the Lyα forest, it is possible that these are largely negligible. But for stronger HI absorption, the gas has substantial effect on the observed source flux.

Indeed, LLS attenuate the color of the background source and may bias their selection. For a quasar at $z = 3$, the u-band flux is greatly diminished by the continuum opacity of an LLS. Projects which select by optical (rest-frame UV) color are then sensitive to the presence of an intervening LLS. For quasars, this is particularly an issue at $z \sim 3$ when their colors lie close to the stellar locus (see Fig. 2.38). In general, the two are indistinguishable using only optical photometry. But if the intrinsic color is highly altered, e.g. the u-band flux of a quasar is absorbed by an intervening LLS, the source separates from the stellar locus and may be targeted. Indeed, the SDSS quasars at $z \sim 3$ exhibit a higher incidence of LLS than the true parent population (see [71] and Fig. 2.35).

In short, when one achieves high statistical precision from today's large datasets, systematic biases like these become relevant and must be carefully considered.

2.5.6 Absorption Path $X(z)$

Bahcall and Peebles [3] introduced a pathlength definition as a means to test cosmological models with the measured incidence of absorption-line systems. The basic
concept is to define $X(z)$ such that the incidence of absorbers per dX is *constant*
provided the product of the following two quantities is constant: (a) the comoving
number density of absorbers $n_c(z)$; and (b) the physical cross-section $A_p(z)$. Under
the (brazen) assumption of a non-evolving population, one can then solve for the
cosmological parameters that yield a constant $n_c A$ product.

We nearly derived $X(z)$ in our discussion of redshift evolution for the line density
in the Lyα forest (Sect. 2.4.5). Recall, we noted that the incidence of absorption per
proper distance is

$$\frac{d\mathcal{N}}{dr_p} = n_p(z)A_p(z) \tag{2.148}$$

with n_p the proper density of absorbers. Expressed in terms of dz, we have

$$\frac{d\mathcal{N}}{dz} = \ell(X) = n_p(z)A_p(z)\frac{c}{H(z)(1+z)} \tag{2.149}$$

Now introduce dX such that,

$$\ell(X) = C\,n_c(z)A_p(z) \ , \tag{2.150}$$

with C a constant. In an expanding universe, $n_c(z) = n_p(z)/(1+z)^3$. And, if we
wish dX to be a dimensionless quantity (like redshift), we choose

$$dX = \frac{H_0}{H(z)}(1+z)^2 dz \tag{2.151}$$

to yield

$$\ell(X) = \frac{c}{H_0}n_c(z)A_p(z) \tag{2.152}$$

As desired, this quantity is static with an expanding universe provided the comoving
number density of sources is constant and their physical size does not evolve.

In Fig. 2.39, we show the dX/dz curves versus redshift for a few cosmologies.
Two trends are apparent: (1) there is significant evolution in dX/dz with redshift
and (2) dX/dz is sensitive to the cosmology (but not H_0). Therefore, a non-evolving
population of absorbers (if any such population exists!) offers significant sensitivity
to cosmology. Alternatively, we may adopt our concordance ΛCDM cosmology (e.g.
[66]) and interpret evolution in $\ell(X)$ as intrinsic evolution in the gas.

We can derive $\ell_{\mathrm{LLS}}(X)$ for the LLS by repeating the Maximum likelihood analysis
after replacing dz with dX. All other aspects of the analysis are identical. Figure 2.40
shows the results across cosmic time. We identify a steep decline with decreasing

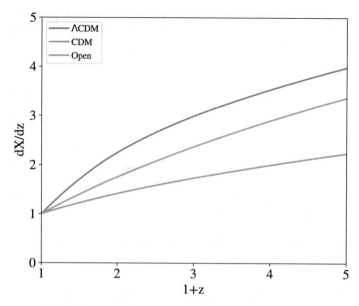

Fig. 2.39 dX/dz curves for a range of cosmologies. The significant differences imply sensitivity to the underlying cosmological parameters [3]

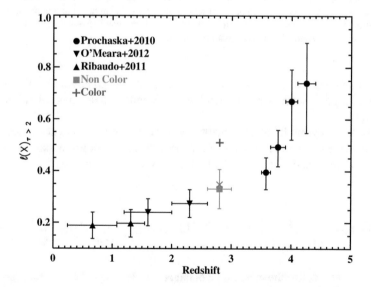

Fig. 2.40 Redshift evolution of the incidence of LLS $\ell_{LLS}(X)$ with $\tau_{LL} \geq 2$ per absorption path length dX. The red and blue points distinguish between quasar samples selected on the basis of color or not (respectively); focus on the latter. If the structures giving rise to LLS had a static comoving number density and physical size, $\ell(X)$ would be constant in time (provided we have adopted the correct cosmology). The steep increase in $\ell(X)$ with z indicates substantial intrinsic evolution in the structures hosting optically thick gas. Taken from [27]. ©AAS. Reproduced with permission

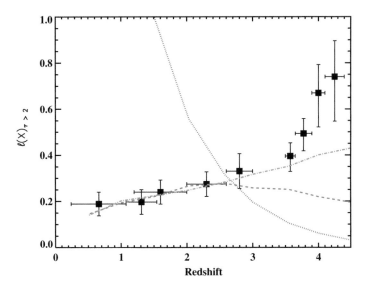

Fig. 2.41 Comparison of the $\ell_{\mathrm{LLS}}(X)$ data from Fig. 2.40 with toy models for the evolution in $\ell(X)$ assuming LLS arise primarily in dark matter halos. The curves correspond to fixed covering fraction of optically thick gas (dotted), a wind model that ejects the gas (dashed) and an accretion scenario that also reduces the covering fraction with decreasing redshift (dotted-dashed). The latter two models offer a good description of the observations for $z < 3$, but fail at higher redshifts. These results argue for an increasing contribution to the Lyman limit opacity from structures in the IGM. Taken from [27]. ©AAS. Reproduced with permission

redshift that demands an evolving population. Either the comoving number density of optically thick sources greatly declines (n_c) and/or their physical size (A_p) is decreasing.

Consider a toy model of dark matter halos within hierarchical cosmology [27]. This model yields a prediction for $\ell(X)$ with redshift according to assumptions made on the covering fraction f_c of optically thick gas in each halo.

$$\ell(X) = \frac{4c\pi}{H_0} \int\limits_{\log M_{\mathrm{low}}}^{\log M_{\mathrm{up}}} R_{vir}^2(M_{vir}, z)\, f_c(M_{vir}, z)\frac{dn_c}{d\log M_{vir}}\, d\log M_{vir} \qquad (2.153)$$

Here M_{low}, M_{up} define the mass range of halos contributing to LLS and we recognize $A_p = f_c \pi R_{vir}^2$ with R_{vir} the virial radius. The last term in Eq. 2.153 expresses the comoving number density. Figure 2.41 compares this toy model against the observations for several assumptions on f_c with each tuned to match at $z \approx 3$ [27]. The dotted line in the Figure assumes constant f_c with redshift. Because of the growth of structure with decreasing redshift in our cosmology, this model predicts a steep *increase* in $\ell(X)$ towards the present-day that strongly violates the observed evolution.

The dashed and dotted-dashed lines are simple "wind" and gas accretion scenarios which are tuned to describe the observations at $z < 3$. These also fail at high-z and one infers that the IGM, and not dark matter halos, contributes greatly to optically thick gas at $z > 3.5$.

2.5.7   Mean Free Path λ_{mfp}^{912}

Optically thick gas attenuates the flux of ionizing photons throughout the Universe. The LLSs act, in essence, as brick walls absorbing all photons with $h\nu > 1$ Ryd. However, gas with $N_{HI} \lesssim 10^{17}\,\mathrm{cm}^{-2}$ will also attenuate the ionizing flux and, if sufficiently common, can dominate the absorption. The integrated effect of all gas yields an HI mean free path to ionizing photons.

Cosmologically, the mean free path is a fundamental input to calculations of the extragalactic UV background (EUVB). It relates the emissivity of sources to the mean intensity [23, 32]:

$$J_\nu(z) \approx \frac{1}{4\pi} \lambda_{mfp}^{912} \varepsilon_\nu(z) \tag{2.154}$$

The mean free path is also relevant to calculations of the topology of HI reionization at $z > 5$, altering the leakage of ionizing sources into the IGM. Currently, λ_{mfp}^{912} cannot be predicted from first principles and one must constrain its value (and evolution) with observation.

Let us define $\lambda_{mfp}^{912}(z_{em})$ as the physical distance that a packet of photons travel from a source at $z = z_{em}$ before suffering an e^{-1} attenuation. In contrast to other physical quantities, this definition is directly coupled to the source redshift. If we define an effective optical depth for Lyman continuum opacity τ_{eff}^{LL}, then λ_{mfp}^{912} is the distance from the source where $\tau_{eff}^{LL} = 1$. There is, however, an important and subtle concept associated with defining and measuring λ_{mfp}^{912} due to the expansion of the universe. Consider a packet of photons with energy of 1 Ryd emitted by the source. These will travel only an infinitesimal distance[9] before redshifting beyond 1 Ryd to become insensitive to the HI continuum opacity. This means that the packet of photons which suffers an e^{-1} attenuation must have rest-frame energy $h\nu > 1$ Ryd. Analysis of λ_{mfp}^{912} therefore requires a proper treatment of cosmic expansion and the frequency dependence of the continuum opacity.

The traditional approach to estimate λ_{mfp}^{912} was to calculate τ_{eff}^{LL} from our $f(N_{HI})$ distribution and relate $\tau_{eff}^{LL} = 1$ to λ_{mfp}^{912}. Here, τ_{eff}^{LL} is the effective optical depth from Lyman continuum opacity that a photon of $\nu \geq \nu_{912}$ experiences, $F_{obs} = F \exp(-\tau_{eff}^{LL})$. In contrast to the effective opacity from Lyman series lines (e.g. $\tau_{eff,\alpha}$), τ_{eff}^{LL} is an integral (cumulative) quantity. That is, it depends on the IGM along the entire path traveled until the photon with frequency ν redshifts to ν_{912} at

---

[9] In reality, peculiar motions of the gas including Doppler broadening imply a greater than infinitesimal distance.

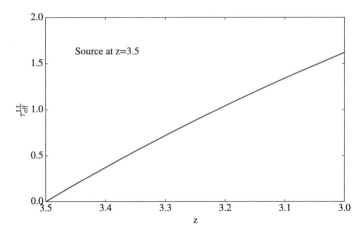

Fig. 2.42 Effective opacity as a function of $z < z_{em}$ with $z_{em} = 3.5$. In contrast to $\tau_{eff}^{Ly\alpha}$ (or the effective opacity of any Lyman series line), τ_{eff}^{LL} is a cumulative quantity and must increase monotonically with decreasing redshift from the source

$$z_{912} \equiv (1 + z_{em})\frac{\nu_{912}}{\nu} - 1 \qquad (2.155)$$

with $\nu_{912} = 1\,\mathrm{Ryd}/h$ and $\nu = \nu_{em}$.

The derivation from $f(N_{HI})$ is as follows. The fractional attenuation from Lyman continuum opacity for gas with N_{HI} is

$$1 - \exp[-N_{HI}\sigma_{photo}(\nu)] \qquad (2.156)$$

i.e. a transparent source gives null attenuation and a 'wall' gives complete attenuation. For a population of absorbers, we weight by $f(N_{HI})$. Lastly, we must account for the frequency dependence of σ_{photo} (using Eq. 2.155) to recover

$$\tau_{eff}^{LL}(z_{912}, z_{em}) = \int_{z_{912}}^{z_{em}} \int_{0}^{\infty} f(N_{HI}, z')\{1 - \exp\left[-N_{HI}\sigma_{ph}(z')\right]\}dN_{HI}dz' \qquad (2.157)$$

For the $f(N_{HI})$ model of [74], we calculate τ_{eff}^{LL} as shown in Fig. 2.42. Of course, τ_{eff}^{LL} rises with decreasing redshift as the photons travel to ever greater distances from the source. We see that $\tau_{eff}^{LL} = 1$ at $z \approx 3.2$ and the proper distance from $z = 3.5$ gives $\lambda_{mfp}^{912} \approx 60\,\mathrm{Mpc}$.

This technique of integrating $f(N_{HI})$ is indirect, i.e. we have predicted the integrated Lyman continuum from analysis of the HI Lyman series opacity of the Lyα forest. But $f(N_{HI})$ at $N_{HI} \approx 10^{17}\,\mathrm{cm}^{-2}$ is highly uncertain because the Lyman series lines lie on the saturated portion of the COG. Furthermore, line-blending poses a significant problem and surveys of $\tau \approx 1$ LLS are difficult to perform. In-

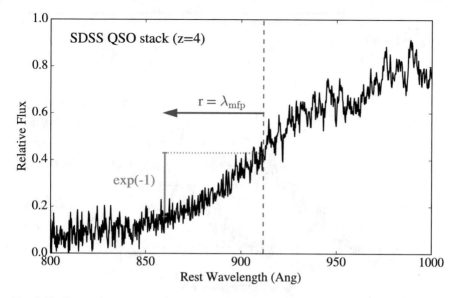

Fig. 2.43 Composite spectrum of $\sim$150 quasars at $z_{em} \approx 4$ (black). The declining flux at $\lambda <$ 911.7 Å (dashed line) is due to the cumulative effective opacity of Lyman limit absorption τ_{eff}^{LL}. By estimating the redshift where the flux has suffered an e^{-1} attenuation, one directly measures λ_{mfp}^{912}

deed, the $f(N_{HI})$ function of [74] used an estimation of λ_{mfp}^{912} as a constraint at $N_{HI} \lesssim 10^{17}$ cm^{-2}.

Prochaska et al. [70] proposed a new method to directly assess the average Lyman continuum opacity τ_{eff}^{LL} and thereby λ_{mfp}^{912}. A similar concept was introduced in Sect. 2.4.7 where $\tau_{eff,\alpha}$ was estimated from a composite spectrum. By averaging the IGM absorption along many sightlines, the composite spectrum (without weighting) provides an optimal means to measure the distance traveled for an e^{-1} attenuation. The concept is illustrated in Fig. 2.43.

In practice, we model the IGM attenuation with an effective opacity κ_{LL} for Lyman continuum absorption

$$\kappa_{LL}(z, \nu) = \tilde{\kappa}_{912}(z) \left(\frac{\nu}{\nu_{912}} \right)^{-2.75} \tag{2.158}$$

where the -2.75 exponent is a better empirical estimate than -3 [70]. The frequency term captures the energy dependence of the photoionization cross-section and κ_{912} is a normalization which captures IGM evolution with redshift. Integrating this opacity over a physical distance and allowing for redshift of the photons

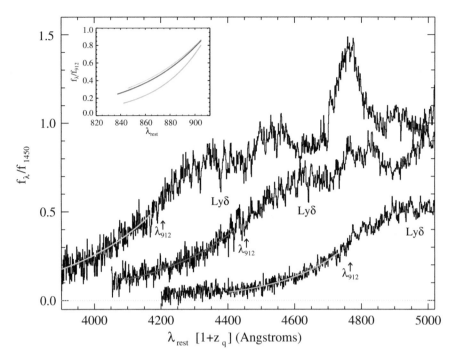

Fig. 2.44 Composite spectra for a series of $z_{em} \approx 3.8-4.5$ quasars from the SDSS (black curves). Overplotted on the data are models of the IGM attenuation from HI continuum opacity. The e^{-1} attenuation measured from these models gives mean free path λ_{mfp}^{912} measurements. Taken from [70]. ©AAS. Reproduced with permission

$$\tau_{eff}^{LL} = \int_{0}^{r} \kappa_{LL}(r') \, dr' \tag{2.159}$$

$$= \int_{z_{912}}^{z_{em}} \kappa_{LL}(z', \nu) \frac{dr}{dz} dz' \tag{2.160}$$

As demonstrated above, at $z > 2$ we have $dr/dz \propto (1+z)^{\beta}$ with $\beta \approx -2.5$. Our model reduces to a simple integral of $(1+z)$ to some power and one additional parameter (κ_{912}). This provides great statistical power (i.e. a one parameter fit to many pixels). The primary uncertainties are our estimate for the quasar SED and sample variance (estimated from a bootstrap approach).

Model fits to composite quasar spectra at $z \sim 4$ are shown in Fig. 2.44 from [70], which also assumes a flat quasar SED below the Lyman limit. The mean free path measured from the direct evaluation is larger than predicted from older $f(N_{HI})$ models. In turn, it implies less attenuation of ionizing sources and therefore fewer ionizing sources required to generate the intensity of the EUVB.

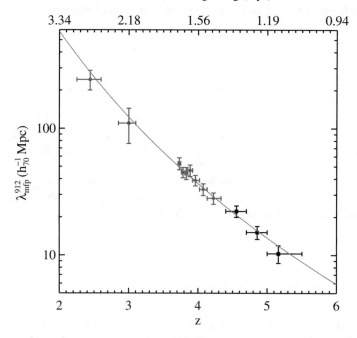

Fig. 2.45 Series of $\lambda_{\mathrm{mfp}}^{912}$ measurements at a series of source redshifts from analysis of composite spectra. The overplotted model is a $(1 + z)^{\gamma}$ fit with $\gamma \approx -5.4$. This very steep evolution in $\lambda_{\mathrm{mfp}}^{912}$ implies ongoing ionization of gas in the IGM down to $z < 3$. Taken from [97]. Reproduced by permission of MNRAS

Similar analysis has been performed at a range of redshifts [27, 58, 97]. Each composite yields an estimate of $\lambda_{\mathrm{mfp}}^{912}$ by estimating the distance where the flux is attenuated by e^{-1}. These are presented in Fig. 2.45. Assuming a $(1 + z)^{\gamma}$ power-law evolution in $\lambda_{\mathrm{mfp}}^{912}$, the authors report

$$\lambda_{\mathrm{mfp}}(z) = (37 \pm 2\, h_{70}^{-1}\mathrm{Mpc}) \left[\frac{1 + z}{5} \right]^{-5.4 \pm 0.4} \tag{2.161}$$

As with the LLS, this is far steeper than cosmological expansion alone would predict. It requires the disappearance of optically thick gas with decreasing redshift. This likely results from the continued ionization of the universe as sources (quasars) rise in number and/or the accretion of optically thick gas onto galaxies.

These new results on $\lambda_{\mathrm{mfp}}^{912}$ also impact estimations for the escape fraction of ionizing photons from galaxies f_{esc}. The observational experiment is to directly observe the flux of a galaxy (e.g. at $z = 3$) at wavelengths just blueward of its Lyman limit. For this, one employs a narrow-band filter (e.g. [56]) or integrates a spectrum near the the Lyman limit: $< f_\nu >_{\mathrm{LL,obs}}$. The escape fraction is then defined as the flux of radiation that escapes versus that which the galaxy produced (from O and B stars):

$$f_{\mathrm{esc}} = \frac{< f_\nu >_{\mathrm{LL,obs}}}{< f_\nu >_{\mathrm{LL,intrinsic}}} \tag{2.162}$$

One is interested here in the flux attenuation by gas in the galaxy, but $< f_\nu >_{\mathrm{LL,obs}}$ is also attenuated by the IGM.

As an example, a medium-band filter (FWHM $\approx 100\,\text{Å}$) placed just blueward of a $z = 3.5$ galaxy's Lyman limit covers $z = 3.4$ to 3.5 at the Lyman limit. We estimate the IGM effective opacity to be $< \tau_{\mathrm{eff}}^{\mathrm{LL}} >= 0.25$ which gives a 30% correction to f_{esc} that increases steeply with redshift.

If we extrapolate the evolution in $\lambda_{\mathrm{mfp}}^{912}$ to $z < 2$, we find that $\lambda_{\mathrm{mfp}}^{912}$ exceeds the Horizon at $z \approx 1.6$. This is referred to as the "breakthrough" redshift where photons are no longer (significantly) attenuated by the IGM. Beyond this epoch, every ionizing source can see every other.

2.5.8 Connecting LLS to Theory

We conclude our discussion of LLS with a few additional comments on the origin of optically thick gas in a cosmological context. Early cosmological simulations woefully underpredicted the incidence of LLS [29]. In hindsight, this was not surprising; the requirements to model LLS on simulations are steep. LLS are related to non-linear structure formation and are sensitive to the radiative transfer of ionizing photons. This is a terrific challenge in terms of resolution (one needs kpc-scale sensitivity over hundreds of Mpc) and as regards the physics one can implement in cosmological, hydrodynamic simulations.

More recently, theory has connected LLS to an emerging paradigm for accreting cold ($T \sim 10^4$) gas into dark matter halos, a.k.a. cold streams or cold flows (e.g. [17, 36]). Such gas may be optically thick (or partially optically thick) and a series of papers have associated at least a fraction of LLS with these streams (e.g. [20, 26, 91]). An illustration of the results is shown in Fig. 2.46. Here one finds extended, optically thick gas throughout the underlying, dark matter halo. One estimates that a modest ($\approx 10-20\%$) fraction of the halo is covered by optically thick gas which is metal-poor but not primordial. This leads to a significant number of LLS, but it remains difficult to reconcile the incidence of LLS with dark matter halos alone (see the $\ell(X)$ measurements above). Again, at $z > 3$ one speculates that the dominant contribution to LLS is from gas in large-scale structure (e.g. filaments). Ultimately, this is a fundamental benchmark for future cosmological simulations with a full treatment of radiative transfer.

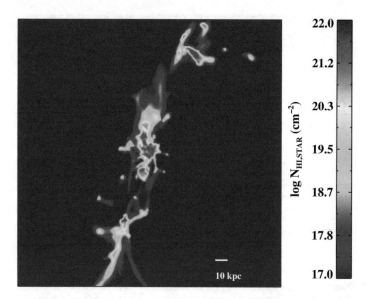

Fig. 2.46 Visualization of the HI column density of streams flowing into a dark matter halo at $z \approx 3$. At the center of the image is a star-forming galaxy and its ISM dominates the HI gas. Beyond ≈ 10 kpc from the galaxy, the gas is primarily related to streams and dwarf galaxies within those streams. These structures are predicted to dominate the cross-section of optically thick gas within dark matter halos. Taken from [26]. Reproduced by permission of MNRAS

2.5.9 Damped Lyα Systems

In the school, I presented a short lecture on the damped Lyα systems (DLAs), HI gas with $N_{\mathrm{HI}} \geq 2 \times 10^{20}$ cm^{-2}. Obviously, these systems are also optically thick at the Lyman limit and are therefore LLS with $\tau_{\mathrm{eff}}^{\mathrm{LL}} \gg 1$. Readers interested in DLAs are referred to the review article by [96] (and my DLA lecture). I include only a few key observational results here.

The frequency distribution $f(N_{\mathrm{HI}})$ of the DLAs has been well estimated from the large datasets of SDSS and BOSS. Using the SDSS DR5, [69] established that $f(N_{\mathrm{HI}})$ in the DLA regime follows a single power-law with $\alpha \approx -2$ until an observed break at $N_{\mathrm{HI}} \approx 10^{21.5}$ cm^{-2} (Fig. 2.47). This break may be naturally understood in a simple model of DLAs as a population of randomly inclined exponential disks [19, 95]. Assuming a face-on column density distribution with radius r

$$N_{\perp}(r) = N_{\perp,0} \exp(-r/r_d) \tag{2.163}$$

with $N_{\perp,0}$ the central column density and r_d a scale-length. For random, face-on disks (no inclination) the probability of intersection at r is proportional to r giving

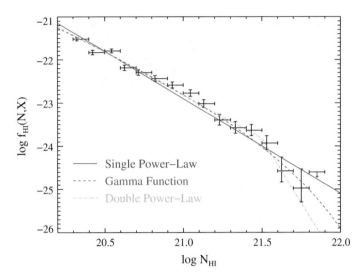

Fig. 2.47 Frequency distribution of N_{HI} for the damped Lyα systems at $z \approx 2.5$. These gas layers show a roughly power-law distribution with $\alpha \approx -2$ for lower N_{HI} values which breaks to a steeper function at $N_{HI} > 10^{21.5}$ cm^{-2}. Taken from [69]. ©AAS. Reproduced with permission

$$f(N_\perp, X) = \frac{2\pi \, cr_d^2 n_c}{H_0} \left[\frac{\ln(N_{\perp,0}/N_\perp)}{N_\perp} \right] \tag{2.164}$$

where n_c is the assumed comoving number density of DLAs. Lastly, allow for random inclination

$$f(N, X) = \int\limits_0^{min(N,N_{\perp,0})} dN_\perp \left[\frac{N_\perp^2 f(N_\perp)}{N^3} \right] \tag{2.165}$$

Evaluating

$$f(N, X) \propto \begin{cases} \frac{1}{N}\left[1 - 2\ln\left(\frac{N}{N_{\perp,0}}\right) \right] & N \leq N_{\perp,0} \\ \frac{N_{\perp,0}^2}{N^3} & N \geq N_{\perp,0} \end{cases} \tag{2.166}$$

The first term scales as N^{-2} for $N \ll N_{\perp,0}$ i.e., we expect a broken power-law!

A larger sample of DLAs has been analyzed by [57] from BOSS. They report an approximately N_{HI}^{-3} decline at the highest column densities and the discovery of DLAs towards quasars with $N_{HI} > 10^{22}$ cm^{-2}.

Surprisingly, the $f(N_{HI})$ distribution for DLAs shows very weak evolution in its shape across cosmic time. This includes $z = 0$, i.e. 21 cm observations [100], but see [11]. This implies a nearly non-evolving population of HI gas from $z = 2$ to 0.

Measurements of $\ell_{DLA}(X)$ constrain the evolution of HI gas within (and around) galaxies across cosmic time (Fig. 2.48). One observes a sharp evolution from

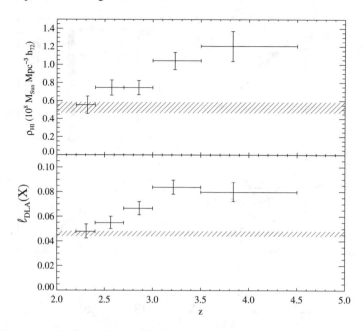

Fig. 2.48 Estimates for the incidence of DLA absorption (lower) and the total mass density in atomic gas contained within these systems ρ_{HI} (upper). The hatched bands express the same quantities measured at $z \sim 0$ from 21 cm observations [100]. Taken from [68]. ©AAS. Reproduced with permission

$z = 2-4$ which requires a $2\times$ decline in $n_c A_p$ in this ≈ 2 Gyr window. It remains an open question whether this evolution is associated with photoionization of the gas, its rapid accretion from the halo onto galaxies, and/or processes related to galaxy–galaxy mergers.

2.5.10 $f(N_{HI})$ Revisited

We may integrate the results described thus far in the chapter to model $f(N_{HI})$ for the IGM (approximately $N_{HI} < 10^{15}$ cm^{-2}), the optically thick gas traced by Lyman limit systems, the damped Lyα systems, the effective opacity of Lyα, and the mean free path λ_{mfp}^{912}. Prochaska et al. [74] have adopted a cubic Hermite spline model (constrained to decrease monotonically) to describe $f(N_{HI})$. This model is not physical, i.e. it is purely mathematical, but it well describes the data and is smoother than broken power-laws.

With MCMC techniques they derived the curve shown in Fig. 2.49. Perhaps the most obvious feature in $f(N_{HI})$ is the inflection at $N_{HI} \approx 10^{18}$ cm^{-2}. Such an inflection was predicted from theory [1, 99], due to the transition from optically thin to thick HI gas.

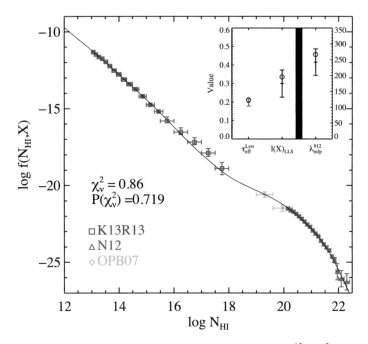

Fig. 2.49 Complete estimation of $f(N_{HI})$ for gas in the IGM ($N_{HI} < 10^{15}$ cm^{-2}), optically thick gas, and the damped Lyα systems. The curve is a Hermite spline fit to the data using MCMC techniques. The analysis includes integral constraints from the effective Lyα opacity, the incidence of LLS, and estimations of the mean free path. Taken from [74]. Reproduced by permission of MNRAS

Acknowledgements The author acknowledges the efforts of the IGM community which he has attempted to summarize in this Chapter. Without a doubt, however, many important additional works were not recognized.

References

1. Altay, G., Theuns, T., Schaye, J., Crighton, N.H.M., Dalla Vecchia, C.: Through thick and thin: H I absorption in cosmological simulations. ApJ **737**, L37 (2011). https://doi.org/10.1088/2041-8205/737/2/L37, arXiv:1012.4014
2. Bahcall, J.N.: Phenomenological limits on the absorbing regions of quasi-stellar sources. ApJ **149**, L7 (1967). https://doi.org/10.1086/180041
3. Bahcall, J.N., Peebles, P.J.E.: Statistical tests for the origin of absorption lines observed in quasi-stellar sources. ApJ **156**, L7+ (1969)
4. Bahcall, J.N., Salpeter, E.E.: Absorption lines in the spectra of distant sources. ApJ **144**, 847 (1966). https://doi.org/10.1086/148675
5. Bahcall, J.N., Spitzer, L.J.: Absorption lines produced by Galactic Halos. ApJ **156**, L63 (1969)
6. Bahcall, J.N., Wolf, R.A.: Fine-structure transitions. ApJ **152**, 701–+ (1968)

7. Bajtlik, S., Duncan, R.C., Ostriker, J.P.: Quasar ionization of Lyman-alpha clouds - the proximity effect, a probe of the ultraviolet background at high redshift. ApJ **327**, 570–583 (1988).
   https://doi.org/10.1086/166217
8. Becker, G.D., Hewett, P.C., Worseck, G., Prochaska, J.X.: A refined measurement of the mean
   transmitted flux in the Lyα forest over $2<z<5$ using composite quasar spectra. MNRAS **430**,
   2067–2081 (2013). https://doi.org/10.1093/mnras/stt031, arXiv:1208.2584
9. Bergeron, J., Boisse, P.: Properties of the galaxies giving rise to MgII quasar absorption systems.
   Adv. Space Res. **11**, 241–244 (1991). https://doi.org/10.1016/0273-1177(91)90496-7
10. Boksenberg, A., Carswell, R.F., Smith, M.G., Whelan, J.A.J.: The absorption-line spectrum of
    Q 1246–057. MNRAS **184**, 773–782 (1978). https://doi.org/10.1093/mnras/184.4.773
11. Braun, R.: Cosmological evolution of atomic gas and implications for 21 cm H I absorption.
    ApJ **749**, 87 (2012). https://doi.org/10.1088/0004-637X/749/1/87, arXiv:1202.1840
12. Brown, R.L., Roberts, M.S.: 21-centimeter absorption at z = 0.692 in the Quasar 3c 286.
    ApJ **184**, L7 (1973). https://doi.org/10.1086/181276
13. Burbidge, E.M., Lynds, C.R., Burbidge, G.R.: On the measurement and interpretation of ab-
    sorption features in the spectrum of the quasi-stellar object 3c 191. ApJ **144**, 447 (1966). https://
    doi.org/10.1086/148629
14. Busca, N.G., Delubac, T., Rich, J., Bailey, S., Font-Ribera, A., Kirkby, D., Le Goff, J.M.,
    Pieri, M.M., Slosar, A., Aubourg, É., Bautista, J.E., Bizyaev, D., Blomqvist, M., Bolton, A.S.,
    Bovy, J., Brewington, H., Borde, A., Brinkmann, J., Carithers, B., Croft, R.A.C., Dawson, K.S.,
    Ebelke, G., Eisenstein, D.J., Hamilton, J.C., Ho, S., Hogg, D.W., Honscheid, K., Lee, K.G.,
    Lundgren, B., Malanushenko, E., Malanushenko, V., Margala, D., Maraston, C., Mehta, K.,
    Miralda-Escudé, J., Myers, A.D., Nichol, R.C., Noterdaeme, P., Olmstead, M.D., Oravetz, D.,
    Palanque-Delabrouille, N., Pan, K., Pâris, I., Percival, W.J., Petitjean, P., Roe, N.A., Rollinde,
    E., Ross, N.P., Rossi, G., Schlegel, D.J., Schneider, D.P., Shelden, A., Sheldon, E.S., Simmons,
    A., Snedden, S., Tinker, J.L., Viel, M., Weaver, B.A., Weinberg, D.H., White, M., Yèche, C.,
    York, D.G.: Baryon acoustic oscillations in the Lyα forest of BOSS quasars. A&A **552**, A96
    (2013). https://doi.org/10.1051/0004-6361/201220724, arXiv:1211.2616
15. Cooksey, K.L., Thom, C., Prochaska, J.X., Chen, H.: The last eight-billion years of intergalac-
    tic C IV evolution. ApJ **708**, 868–908 (2010). https://doi.org/10.1088/0004-637X/708/1/868,
    arXiv:0906.3347
16. Davé, R., Tripp, T.M.: The statistical and physical properties of the low-redshift LYα forest
    observed with the hubble space telescope/STIS. ApJ **553**, 528–537 (2001). https://doi.org/10.
    1086/320977, arXiv:astro-ph/0101419
17. Dekel, A., Birnboim, Y.: Galaxy bimodality due to cold flows and shock heating. MNRAS **368**,
    2–20 (2006). https://doi.org/10.1111/j.1365-2966.2006.10145.x. arXiv:astro-ph/0412300
18. Duncan, R.C., Ostriker, J.P., Bajtlik, S.: Voids in the Ly-alpha forest. ApJ **345**, 39–51 (1989).
    https://doi.org/10.1086/167879
19. Fall, S.M., Pei, Y.C.: Obscuration of quasars by dust in damped Lyman-alpha systems. ApJ
    402, 479–492 (1993). https://doi.org/10.1086/172151
20. Faucher-Giguère, C.A., Kereš, D.: The small covering factor of cold accretion streams. MNRAS
    412, L118–L122 (2011). https://doi.org/10.1111/j.1745-3933.2011.01018.x, 1011.1693
21. Faucher-Giguère, C.A., Lidz, A., Hernquist, L., Zaldarriaga, M.: A flat photoionization rate
    at $2 <= z <= 4.2$: evidence for a stellar-dominated UV background and against a decline
    of cosmic star formation beyond $z \sim 3$. ApJ **682**, L9 (2008). https://doi.org/10.1086/590409,
    arXiv:0806.0372
22. Faucher-Giguère, C.A., Prochaska, J.X., Lidz, A., Hernquist, L., Zaldarriaga, M.: A direct
    precision measurement of the intergalactic Lyα opacity at $2 \leq z \leq 4.2$. ApJ **681**, 831–855
    (2008). https://doi.org/10.1086/588648. arXiv:0709.2382
23. Faucher-Giguère, C.A., Lidz, A., Zaldarriaga, M., Hernquist, L.: A new calculation of the
    ionizing background spectrum and the effects of He II reionization. ApJ **703**, 1416–1443
    (2009). https://doi.org/10.1088/0004-637X/703/2/1416, arXiv:0901.4554
24. Field, G.B.: An attempt to observe neutral hydrogen between the galaxies. ApJ **129**, 525 (1959).
    https://doi.org/10.1086/146652

25. Font-Ribera, A., Miralda-Escudé, J., Arnau, E., Carithers, B., Lee, K.G., Noterdaeme, P., Pâris, I., Petitjean, P., Rich, J., Rollinde, E., Ross, N.P., Schneider, D.P., White, M., York, D.G.: The large-scale cross-correlation of Damped Lyman alpha systems with the Lyman alpha forest: first measurements from BOSS. JCAP **11**, 059 (2012). https://doi.org/10.1088/1475-7516/2012/11/059, arXiv:1209.4596

26. Fumagalli, M., Prochaska, J.X., Kasen, D., Dekel, A., Ceverino, D., Primack, J.R.: Absorption-line systems in simulated galaxies fed by cold streams. MNRAS **418**, 1796–1821 (2011). https://doi.org/10.1111/j.1365-2966.2011.19599.x, arXiv:1103.2130

27. Fumagalli, M., O'Meara, J.M., Prochaska, J.X., Worseck, G.: Dissecting the properties of optically thick hydrogen at the peak of cosmic star formation history. ApJ **775**, 78 (2013). https://doi.org/10.1088/0004-637X/775/1/78, arXiv:1308.1101

28. Fynbo, J.P.U., Jakobsson, P., Prochaska, J.X., Malesani, D., Ledoux, C., de Ugarte, Postigo A., Nardini, M., Vreeswijk, P.M., Wiersema, K., Hjorth, J., Sollerman, J., Chen, H., Thöne, C.C., Björnsson, G., Bloom, J.S., Castro-Tirado, A.J., Christensen, L., De Cia, A., Fruchter, A.S., Gorosabel, J., Graham, J.F., Jaunsen, A.O., Jensen, B.L., Kann, D.A., Kouveliotou, C., Levan, A.J., Maund, J., Masetti, N., Milvang-Jensen, B., Palazzi, E., Perley, D.A., Pian, E., Rol, E., Schady, P., Starling, R.L.C., Tanvir, N.R., Watson, D.J., Xu, D., Augusteijn, T., Grundahl, F., Telting, J., Quirion, P.: Low-resolution spectroscopy of gamma-ray burst optical afterglows: biases in the swift sample and characterization of the absorbers. ApJS **185**, 526–573 (2009). https://doi.org/10.1088/0067-0049/185/2/526, arXiv:0907.3449

29. Gardner, J.P., Katz, N., Hernquist, L., Weinberg, D.H.: Simulations of Damped Lyα and lyman limit absorbers in different cosmologies: implications for structure formation at high redshift. ApJ **559**, 131–146 (2001). https://doi.org/10.1086/322403

30. Giallongo, E., Grazian, A., Fiore, F., Fontana, A., Pentericci, L., Vanzella, E., Dickinson, M., Kocevski, D., Castellano, M., Cristiani, S., Ferguson, H., Finkelstein, S., Grogin, N., Hathi, N., Koekemoer, A.M., Newman, J.A., Salvato, M.: Faint AGNs at $z > 4$ in the CANDELS GOODS-S field: looking for contributors to the reionization of the Universe. A&A **578**, A83 (2015). https://doi.org/10.1051/0004-6361/201425334, arXiv:1502.02562

31. Gunn, J.E., Peterson, B.A.: On the density of neutral hydrogen in intergalactic space. ApJ **142**, 1633–1641 (1965). https://doi.org/10.1086/148444

32. Haardt, F., Madau, P.: Radiative transfer in a clumpy universe. II. The ultraviolet extragalactic background. ApJ **461**, 20–+ (1996). https://doi.org/10.1086/177035, arXiv:astro-ph/9509093

33. Hui, L., Gnedin, N.Y.: Equation of state of the photoionized intergalactic medium. MNRAS **292**, 27–+ (1997). arXiv:astro-ph/9612232

34. Hui, L., Rutledge, R.E.: The B distribution and the velocity structure of absorption peaks in the LYalpha forest. ApJ **517**, 541–548 (1999). https://doi.org/10.1086/307202, arXiv:astro-ph/9709100

35. Inoue, A.K., Shimizu, I., Iwata, I., Tanaka, M.: An updated analytic model for attenuation by the intergalactic medium. MNRAS **442**, 1805–1820 (2014). https://doi.org/10.1093/mnras/stu936, arXiv:1402.0677

36. Kereš, D., Katz, N., Fardal, M., Davé, R., Weinberg, D.H.: Galaxies in a simulated ΛCDM Universe - I. Cold mode and hot cores. MNRAS **395**, 160–179 (2009). https://doi.org/10.1111/j.1365-2966.2009.14541.x, arXiv:0809.1430

37. Kim, T.S., Partl, A.M., Carswell, R.F., Müller, V.: The evolution of H I and C IV quasar absorption line systems at $1.9 < z < 3.2$. A&A **552**, A77 (2013). https://doi.org/10.1051/0004-6361/201220042, arXiv:1302.6622

38. Kirkman, D., Tytler, D.: Intrinsic properties of the $z = 2.7$ LY alpha forest from keck spectra of quasar HS 1946+7658. ApJ **484**, 672–+ (1997). https://doi.org/10.1086/304371, arXiv:astro-ph/9701209

39. Kirkman, D., Tytler, D., Suzuki, N., Melis, C., Hollywood, S., James, K., So, G., Lubin, D., Jena, T., Norman, M.L., Paschos, P.: The HI opacity of the intergalactic medium at redshifts $1.6 < z < 3.2$. MNRAS **360**, 1373–1380 (2005). https://doi.org/10.1111/j.1365-2966.2005.09126.x, arXiv:astro-ph/0504391

40. Lee, H.W.: Asymmetric deviation of the scattering cross section around Lyα by atomic hydrogen. ApJ **594**, 637–641 (2003). https://doi.org/10.1086/376867, arXiv:astro-ph/0308083
41. Lee, K.G., Suzuki, N., Spergel, D.N.: Mean-flux-regulated principal component analysis continuum fitting of sloan digital sky survey Lyα forest spectra. AJ **143**, 51 (2012). https://doi.org/10.1088/0004-6256/143/2/51, arXiv:1108.6080
42. Lee, K.G., Hennawi, J.F., Stark, C., Prochaska, J.X., White, M., Schlegel, D.J., Eilers, A.C., Arinyo-i-Prats, A., Suzuki, N., Croft, R.A.C., Caputi, K.I., Cassata, P., Ilbert, O., Garilli, B., Koekemoer, A.M., Le Brun, V., Le Fèvre, O., Maccagni, D., Nugent, P., Taniguchi, Y., Tasca, L.A.M., Tresse, L., Zamorani, G., Zucca, E.: Lyα Forest tomography from background galaxies: the first megaparsec-resolution large-scale structure map at z > 2. ApJ **795**, L12 (2014). https://doi.org/10.1088/2041-8205/795/1/L12, arXiv:1409.5632
43. Lee, K.G., Hennawi, J.F., Spergel, D.N., Weinberg, D.H., Hogg, D.W., Viel, M., Bolton, J.S., Bailey, S., Pieri, M.M., Carithers, W., Schlegel, D.J., Lundgren, B., Palanque-Delabrouille, N., Suzuki, N., Schneider, D.P., Yèche, C.: IGM constraints from the SDSS-III/BOSS DR9 Lyα forest transmission probability distribution function. ApJ **799**, 196 (2015). https://doi.org/10.1088/0004-637X/799/2/196, arXiv:1405.1072
44. López, S., D'Odorico, V., Ellison, S.L., Becker, G.D., Christensen, L., Cupani, G., Denney, K.D., Pâris, I., Worseck, G., Berg, T.A.M., Cristiani, S., Dessauges-Zavadsky, M., Haehnelt, M., Hamann, F., Hennawi, J., Iršič, V., Kim, T.S., López, P., Lund Saust, R., Ménard, B., Perrotta, S., Prochaska, J.X., Sánchez-Ramírez, R., Vestergaard, M., Viel, M., Wisotzki, L.: XQ-100: a legacy survey of one hundred 3.5<z<4.5 quasars observed with VLT/X-shooter. A&A **594**, A91 (2016). https://doi.org/10.1051/0004-6361/201628161, arXiv:1607.08776
45. Lowenthal, J.D., Koo, D.C., Guzman, R., Gallego, J., Phillips, A.C., Faber, S.M., Vogt, N.P., Illingworth, G.D., Gronwall, C.: Keck spectroscopy of redshift Z approximately 3 galaxies in the hubble deep field. ApJ **481**, 673–+ (1997). https://doi.org/10.1086/304092, arXiv:astro-ph/9612239
46. Lusso, E., Comastri, A., Simmons, B.D., Mignoli, M., Zamorani, G., Vignali, C., Brusa, M., Shankar, F., Lutz, D., Trump, J.R., Maiolino, R., Gilli, R., Bolzonella, M., Puccetti, S., Salvato, M., Impey, C.D., Civano, F., Elvis, M., Mainieri, V., Silverman, J.D., Koekemoer, A.M., Bongiorno, A., Merloni, A., Berta, S., Le Floc'h, E., Magnelli, B., Pozzi, F., Riguccini, L.: Bolometric luminosities and Eddington ratios of X-ray selected active galactic nuclei in the XMM-COSMOS survey. MNRAS **425**, 623–640 (2012). https://doi.org/10.1111/j.1365-2966.2012.21513.x, arXiv:1206.2642
47. Lusso, E., Worseck, G., Hennawi, J.F., Prochaska, J.X., Vignali, C., Stern, J., O'Meara, J.M.: The first ultraviolet quasar-stacked spectrum at z∼2.4 from WFC3. MNRAS **449**, 4204–4220 (2015). https://doi.org/10.1093/mnras/stv516, arXiv:1503.02075
48. Madau, P.: Radiative transfer in a clumpy universe: the colors of high-redshift galaxies. ApJ **441**, 18–27 (1995)
49. Madau, P., Haardt, F.: Cosmic reionization after planck: could quasars do it all? ApJ **813**, L8 (2015). https://doi.org/10.1088/2041-8205/813/1/L8, arXiv:1507.07678
50. McQuinn, M.: The evolution of the intergalactic medium. ARA&A **54**, 313–362 (2016). https://doi.org/10.1146/annurev-astro-082214-122355, arXiv:1512.00086
51. Meiksin, A., Madau, P.: On the photoionization of the intergalactic medium by quasars at high redshift. ApJ **412**, 34–55 (1993). https://doi.org/10.1086/172898
52. Meiksin, A.A.: The physics of the intergalactic medium. Rev. Modern Phys. **81**, 1405–1469 (2009). https://doi.org/10.1103/RevModPhys.81.1405, arXiv:0711.3358
53. Miralda-Escudé, J., Cen, R., Ostriker, J.P., Rauch, M.: The Ly alpha forest from gravitational collapse in the cold dark matter + lambda model. ApJ 471:582–+ (1996). https://doi.org/10.1086/177992, arXiv:astro-ph/9511013
54. Miralda-Escudé, J., Haehnelt, M., Rees, M.J.: Reionization of the inhomogeneous universe. ApJ **530**, 1–16 (2000). https://doi.org/10.1086/308330, arXiv:astro-ph/9812306
55. Møller, P., Jakobsen, P.: The Lyman continuum opacity at high redshifts - through the Lyman forest and beyond the Lyman valley. A&A **228**, 299–309 (1990)

56. Nestor, D.B., Shapley, A.E., Kornei, K.A., Steidel, C.C., Siana, B.: A refined estimate of the ionizing emissivity from galaxies at z $\sim$= 3: spectroscopic follow-up in the SSA22a field. ApJ **765**, 47 (2013). https://doi.org/10.1088/0004-637X/765/1/47, arXIv:1210.2393

57. Noterdaeme, P., Petitjean, P., Carithers, W.C., Pâris, I., Font-Ribera, A., Bailey, S., Aubourg, E., Bizyaev, D., Ebelke, G., Finley, H., Ge, J., Malanushenko, E., Malanushenko, V., Miralda-Escudé, J., Myers, A.D., Oravetz, D., Pan, K., Pieri, M.M., Ross, N.P., Schneider, D.P., Simmons, A., York, D.G.: Column density distribution and cosmological mass density of neutral gas: sloan digital sky survey-III data release 9. A&A **547**, L1 (2012). https://doi.org/10.1051/0004-6361/201220259, arXiv:1210.1213

58. O'Meara, J.M., Prochaska, J.X., Worseck, G., Chen, H.W., Madau, P.: The HST/ACS+WFC3 survey for lyman limit systems. II. Science. ApJ **765**, 137 (2013). https://doi.org/10.1088/0004-637X/765/2/137, arXiv:1204.3093

59. O'Meara, J.M., Lehner, N., Howk, J.C., Prochaska, J.X., Fox, A.J., Swain, M.A., Gelino, C.R., Berriman, G.B., Tran, H.: The first data release of the KODIAQ survey. AJ **150**, 111 (2015). https://doi.org/10.1088/0004-6256/150/4/111, arXiv:1505.03529

60. Ostriker, J.P., Cowie, L.L.: Galaxy formation in an intergalactic medium dominated by explosions. ApJ **243**, L127–L131 (1981). https://doi.org/10.1086/183458

61. Ostriker, J.P., Heisler, J.: Are cosmologically distant objects obscured by dust? - a test using quasars. ApJ **278**, 1–10 (1984). https://doi.org/10.1086/161762

62. Ostriker, J.P., Ikeuchi, S.: Physical properties of the intergalactic medium and the Lyman-alpha absorbing clouds. ApJ **268**, L63–L68 (1983). https://doi.org/10.1086/184030

63. Pâris, I., Petitjean, P., Rollinde, E., Aubourg, E., Busca, N., Charlassier, R., Delubac, T., Hamilton, J.C., Le Goff, J.M., Palanque-Delabrouille, N., Peirani, S., Pichon, C., Rich, J., Vargas-Magaña, M., Yèche, C.: A principal component analysis of quasar UV spectra at z $\sim$ 3. A&A **530**, A50 (2011). https://doi.org/10.1051/0004-6361/201016233, arXiv:1104.2024

64. Penton, S.V., Shull, J.M., Stocke, J.T.: The local Lyα forest. II. Distribution of H I absorbers, doppler widths, and Baryon content. ApJ **544**, 150–175 (2000). https://doi.org/10.1086/317179, arXiv:astro-ph/9911128

65. Petitjean, P., Webb, J.K., Rauch, M., Carswell, R.F., Lanzetta, K.: Evidence for structure in the H I column density distribution of QSO absorbers. MNRAS **262**, 499–505 (1993)

66. Planck (2015). https://doi.org/10.1051/0004-6361/201525830

67. Press, W.H., Rybicki, G.B., Schneider, D.P.: Properties of high-redshift Lyman-alpha clouds. I - statistical analysis of the Schneider-Schmidt-Gunn quasars. ApJ **414**, 64–81 (1993). https://doi.org/10.1086/173057, arXiv:astro-ph/9303016

68. Prochaska, J.X., Wolfe, A.M.: On the (Non)Evolution of H I gas in galaxies over cosmic time. ApJ **696**, 1543–1547 (2009). https://doi.org/10.1088/0004-637X/696/2/1543, arXiv:0811.2003

69. Prochaska, J.X., Herbert-Fort, S., Wolfe, A.M.: The SDSS damped Lyα survey: data release 3. ApJ **635**, 123–142 (2005). https://doi.org/10.1086/497287

70. Prochaska, J.X., Worseck, G., O'Meara, J.M.: A direct measurement of the intergalactic medium opacity to H I ionizing photons. ApJ **705**, L113–L117 (2009). https://doi.org/10.1088/0004-637X/705/2/L113, arXiv:0910.0009

71. Prochaska, J.X., O'Meara, J.M., Worseck, G.: A definitive survey for lyman limit systems at z $\sim$ 3.5 with the sloan digital sky survey. ApJ **718**, 392–416 (2010). https://doi.org/10.1088/0004-637X/718/1/392, arXiv:0912.0292

72. Prochaska, J.X., Hennawi, J.F., Lee, K.G., Cantalupo, S., Bovy, J., Djorgovski, S.G., Ellison, S.L., Wingyee Lau, M., Martin, C.L., Myers, A., Rubin, K.H.R., Simcoe, R.A.: Quasars probing quasars. VI. Excess H I absorption within one proper Mpc of z $\sim$ 2 quasars. ApJ **776**, 136 (2013). https://doi.org/10.1088/0004-637X/776/2/136, arXiv:1308.6222

73. Prochaska, J.X., Hennawi, J.F., Simcoe, R.A.: A substantial mass of cool, metal-enriched gas surrounding the progenitors of modern-day ellipticals. ApJ **762**, L19 (2013b). https://doi.org/10.1088/2041-8205/762/2/L19, arXiv:1211.6131

74. Prochaska, J.X., Madau, P., O'Meara, J.M., Fumagalli, M.: Towards a unified description of the intergalactic medium at redshift z $\sim$ 2.5. MNRAS **438**, 476–486 (2014). https://doi.org/10.1093/mnras/stt2218, arXiv:1310.0052

75. Prochaska, J.X., Werk, J.K., Worseck, G., Tripp, T.M., Tumlinson, J., Burchett, J.N., Fox, A.J., Fumagalli, M., Lehner, N., Peeples, M.S., Tejos, N.: The COS-halos survey: metallicities in the low-redshift circumgalactic medium. ApJ **837**, 169 (2017). https://doi.org/10.3847/1538-4357/aa6007, arXiv:1702.02618

76. Rahmati, A., Pawlik, A.H., Raicevic, M., Schaye, J.: On the evolution of the H I column density distribution in cosmological simulations. MNRAS **430**, 2427–2445 (2013). https://doi.org/10.1093/mnras/stt066, arXiv:1210.7808

77. Ribaudo, J., Lehner, N., Howk, J.C.: A hubble space telescope study of Lyman limit systems: census and evolution. ApJ **736**, 42–+ (2011). https://doi.org/10.1088/0004-637X/736/1/42, arXiv:1105.0659

78. Richards, G.T., Lacy, M., Storrie-Lombardi, L.J., Hall, P.B., Gallagher, S.C., Hines, D.C., Fan, X., Papovich, C., Vanden Berk, D.E., Trammell, G.B., Schneider, D.P., Vestergaard, M., York, D.G., Jester, S., Anderson, S.F., Budavári, T., Szalay, A.S.: Spectral energy distributions and multiwavelength selection of type 1 quasars. ApJS **166**, 470–497 (2006). https://doi.org/10.1086/506525, arXiv:astro-ph/0601558

79. Robertson, J.G.: Quantifying resolving power in astronomical spectra. PASA **30**, e048 (2013). https://doi.org/10.1017/pasa.2013.26, arXiv:1308.0871

80. Rudie, G.C., Steidel, C.C., Shapley, A.E., Pettini, M.: The column density distribution and continuum opacity of the intergalactic and circumgalactic medium at redshift langzrang = 2.4. ApJ **769**, 146 (2013). https://doi.org/10.1088/0004-637X/769/2/146, arXiv:1304.6719

81. Sargent, W.L.W., Young, P.J., Boksenberg, A., Tytler, D.: The distribution of Lyman-alpha absorption lines in the spectra of six QSOs - evidence for an intergalactic origin. ApJS **42**, 41–81 (1980). https://doi.org/10.1086/190644

82. Sargent, W.L.W., Steidel, C.C., Boksenberg, A.: A survey of Lyman-limit absorption in the spectra of 59 high-redshift QSOs. ApJS **69**, 703–761 (1989). https://doi.org/10.1086/191326

83. Schmidt, M.: 3C 273: a star-like object with large red-shift. Nature **197**, 1040 (1963)

84. Songaila, A., Cowie, L.L.: Approaching reionization: the evolution of the Ly α forest from z=4 to z=6. AJ **123**, 2183–2196 (2002). https://doi.org/10.1086/340079, arXiv:astro-ph/0202165

85. Steidel, C.C., Giavalisco, M., Pettini, M., Dickinson, M., Adelberger, K.L.: Spectroscopic confirmation of a population of normal star-forming galaxies at redshifts $z > 3$. ApJ **462**, L17+ (1996). https://doi.org/10.1086/310029, arXiv:astro-ph/9602024

86. Storrie-Lombardi, L.J., Irwin, M.J., McMahon, R.G.: APM $z > 4$ survey: distribution and evolution of high column density HI absorbers. MNRAS **282**, 1330–1342 (1996). arXiv:astro-ph/9608146

87. Suzuki, N.: Quasar spectrum classification with principal component analysis (PCA): emission lines in the Lyα forest. ApJS **163**, 110–121 (2006). https://doi.org/10.1086/499272

88. Telfer, R.C., Zheng, W., Kriss, G.A., Davidsen, A.F.: The rest-frame extreme-ultraviolet spectral properties of quasi-stellar objects. ApJ **565**, 773–785 (2002). https://doi.org/10.1086/324689, arXiv:astro-ph/0109531

89. Tytler, D.: QSO Lyman limit absorption. Nature **298**, 427–432 (1982). https://doi.org/10.1038/298427a0

90. Tytler, D.: The distribution of QSO absorption system column densities - evidence for a single population. ApJ **321**, 49–68 (1987). https://doi.org/10.1086/165615

91. van de Voort, F., Schaye, J.: Properties of gas in and around galaxy haloes. MNRAS **423**, 2991–3010 (2012). https://doi.org/10.1111/j.1365-2966.2012.20949.x, arXiv:1111.5039

92. Vanden Berk, D.E., Richards, GT, Bauer, A., Strauss, M.A., Schneider, D.P., Heckman, T.M., York, D.G., Hall, P.B., Fan, X., Knapp, G.R., Anderson, S.F., Annis, J., Bahcall, N.A., Bernardi, M., Briggs, J.W., Brinkmann, J., Brunner, R., Burles, S., Carey, L., Castander, F.J., Connolly, A.J., Crocker, J.H., Csabai, I., Doi, M., Finkbeiner, D., Friedman, S., Frieman, J.A., Fukugita, M., Gunn, J.E., Hennessy, G.S., Ivezić, Ž., Kent, S., Kunszt, P.Z., Lamb, D.Q., Leger, R.F., Long, D.C., Loveday, J., Lupton, R.H., Meiksin, A., Merelli, A., Munn, J.A., Newberg, H.J., Newcomb, M., Nichol, R.C., Owen, R., Pier, J.R., Pope, A., Rockosi, C.M., Schlegel, D.J., Siegmund, W.A., Smee, S., Snir, Y., Stoughton, C., Stubbs, C., SubbaRao, M., Szalay, A.S., Szokoly, G.P., Tremonti, C., Uomoto, A., Waddell, P., Yanny, B., Zheng, W.: Composite quasar

spectra from the sloan digital sky survey. AJ **122**, 549–564 (2011). https://doi.org/10.1086/321167

93. Vogt, S.S, Allen, S.L., Bigelow, B.C., Bresee, L., Brown, B., Cantrall, T., Conrad, A., Couture, M., Delaney, C., Epps, H.W., Hilyard, D., Hilyard, D.F., Horn, E., Jern, N., Kanto, D., Keane, M.J., Kibrick, R.I., Lewis, J.W., Osborne, J., Pardeilhan, G.H., Pfister, T., Ricketts, T., Robinson, L.B., Stover, R.J., Tucker, D., Ward, J., Wei, M.Z.: HIRES: the high-resolution Echelle spectrometer on the Keck 10-m Telescope. In: Proceedings of the SPIE Instrumentation in Astronomy VIII, Crawford, D.L., Craine, E.R., (eds.), vol. 2198, pp 362–+ (1994)

94. Wolfe, A.M., Turnshek, D.A., Smith, H.E., Cohen, R.D.: Damped Lyman-alpha absorption by disk galaxies with large redshifts. I - The Lick survey. ApJS **61**, 249–304 (1986). https://doi.org/10.1086/191114

95. Wolfe, A.M., Lanzetta, K.M., Foltz, C.B., Chaffee, F.H.: The large bright QSO survey for damped LY alpha absorption systems. ApJ **454**, 698–+ (1995). doi:https://doi.org/10.1086/176523

96. Wolfe, A.M., Gawiser, E., Prochaska, J.X.: Damped Lya systems. ARA&A **43**, 861–918 (2005)

97. Worseck, G., Prochaska, J.X., O'Meara, J.M., Becker, G.D., Ellison, S.L., Lopez, S., Meiksin, A., Ménard, B., Murphy, M.T., Fumagalli, M.: The Giant Gemini GMOS survey of $z_{em} >$ 4.4 quasars - I. Measuring the mean free path across cosmic time. MNRAS **445**, 1745–1760 (2014). https://doi.org/10.1093/mnras/stu1827, 1402.4154

98. Young, P.J., Sargent, W.L.W., Boksenberg, A., Carswell, R.F., Whelan, J.A.J.: A high-resolution study of the absorption spectrum of PKS 2126–158. ApJ **229**, 891–908 (1979). https://doi.org/10.1086/157024

99. Zheng, Z., Miralda-Escudé, J.: Self-shielding effects on the column density distribution of damped Lyα systems. ApJ **568**, L71–L74 (2002). https://doi.org/10.1086/340330, arXiv:astro-ph/0201275

100. Zwaan, M.A., van der Hulst, J.M., Briggs, F.H., Verheijen, M.A.W., Ryan-Weber, E.V.: Reconciling the local galaxy population with damped Lyman α cross-sections and metal abundances. MNRAS **364**, 1467–1487 (2005). https://doi.org/10.1111/j.1365-2966.2005.09698.x, arXiv:astro-ph/0510127

Chapter 3
Observations of Lyα Emitters at High Redshift

Masami Ouchi

3.1 Introduction

About two decades have passed since the observational discovery of Lyα emitters (LAEs) at high redshift. Before then, early theoretical studies focused on discussing young primordial galaxies with strong Lyα emission. However, after the discovery, observations have revealed a number of exciting characteristics of LAEs, some of which are beyond the theoretical predictions. In this section, I show the growing importance of LAE studies, overviewing the LAE observation history through the early theoretical predictions, the discovery, and new problems in this observational field. Throughout this lecture, magnitudes are in the AB system, if not otherwise specified. All physical values are calculated with the concordance cosmology of $H_0 = 100h \, \text{km} \, \text{s}^{-1} \, \text{Mpc}^{-1}$ with $h \simeq 0.7$, $\Omega_{\text{m}} \simeq 0.3$, $\Omega_{\Lambda} \simeq 0.7$, $\Omega_{\text{b}} h^2 \simeq 0.02$, $\sigma_8 \simeq 0.8$, and $n_s \simeq 1.0$ that are consistent with the latest Planck 2016 cosmology [268].

3.1.1 Predawn of the LAE Observation History

3.1.1.1 Theoretical Predictions

Partridge and Peebles [252] is the first well-known study that discusses galaxies emitting strong Lyα at high redshift, which are called LAEs today. Partridge and Peebles [252] predict that an early galaxy emits a strong hydrogen Lyα line through the recombination process in the inter-stellar medium (ISM) that is heated by young massive stars (Fig. 3.1). As much as 6–7% of the total galaxy luminosity can be converted to Lyα luminosity in a Milky Way mass halo. Assuming a high star-formation

M. Ouchi (✉)
ICRR, The University of Tokyo, 5-1-5 Kashiwanoha, Kashiwa 277-8882, Japan
e-mail: ouchims@icrr.u-tokyo.ac.jp

© Springer-Verlag GmbH Germany, part of Springer Nature 2019
M. Dijkstra et al., *Lyman-alpha as an Astrophysical and Cosmological Tool*,
Saas-Fee Advanced Course 46, https://doi.org/10.1007/978-3-662-59623-4_3

Fig. 3.1 Expected spectrum of a young galaxy [252]. Here, $\Delta v = 0.002v$ is assumed for the line width. ©AAS. Reproduced with permission

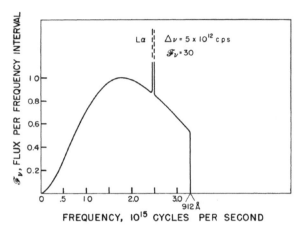

rate that converts 2% of hydrogen to metal within 3×10^7 yr in a Milky Way mass halo, Partridge and Peebles [252] suggest that such galaxies would have an extremely bright Lyα luminosity of $\sim 2 \times 10^{45}$ erg s^{-1} at $z \sim 10 - 30$. Interestingly, Partridge and Peebles [252] discuss the observability of those young Lyα emitting galaxies, taking the effects of cosmic reionization into account (see also a companion paper, Partridge and Peebles [253]). Here, Partridge and Peebles [252] introduce a possible strong free-electron scattering in the ionized IGM that smears the radiation from the young galaxies, which is an argument that differs from today's major discussion on the absorption of Lyα emission by the neutral IGM. Although their discussion of young galaxies' Lyα luminosities and reionization effects on the Lyα observability is very different from the present-day one, it is interesting to notice that the two major cosmological topics discussed today in relation with LAEs, namely galaxy formation and cosmic reionization, had already been studied theoretically in the 1960s.

3.1.1.2 Early Searches for LAEs

Since the theoretical predictions of Partridge and Peebles [252] were published, a number of observational projects searched for young LAEs at $z \sim 2 - 6$ (e.g. Koo and Kron [168], Pritchet and Hartwick [277, 278], Djorgovski and Thompson [72], Thompson et al. [340]). These observational searches were conducted in the 1980s and 1990s with 4 m-class optical telescopes including the Palomar 200-in Hale telescope that was the largest aperture telescope used for researches before the 10 m Keck I telescope became available. Although many candidates were pinpointed by narrowband imaging and slitless spectroscopy in these searches, no real young LAEs at $z \gtrsim 2$ were confirmed by spectroscopy. However, these null-detection results placed meaningful upper limits on the luminosity function of LAEs (Fig. 3.2).

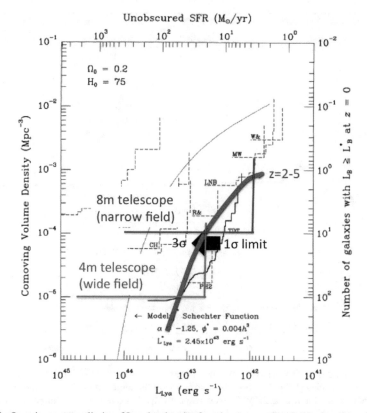

Fig. 3.2 One sigma upper limits of Lyα luminosity functions at $z = 2 - 9$ (black solid and dashed lines). The black arrow indicates the offset value between one and three sigma levels that can allow us to convert the one to three sigma upper limits. The red curve is the approximate Lyα luminosity function of LAEs at $z = 2 - 5$ obtained thus far (e.g. Gronwall et al. [102], Ouchi et al. [243], Cassata et al. [44]). The black curve is the model Lyα luminosity function. The blue (green) line represents a typical observational limit of a narrow-field (wide-field) imager mounted on an 8 m (4 m) class telescope. All numbers quoted in this figure hold for a cosmological model with $H_0 = 75\,\mathrm{km\,s^{-1}\,Mpc^{-1}}$, $\Omega_m = 0.2$, and $\Omega_\Lambda = 0.0$. This figure is adapted from Fig. 13 of Thompson et al. [340], ©AAS. Reproduced with permission

3.1.1.3 Discovery

Since 1996, LAE observations have entered a new era. Hu and McMahon [128] have identified two LAEs at $z = 4.55$ around the QSO BR2237-0607 using the Keck LRIS spectrograph (Fig. 3.3), after having detected them using deep UH88 imaging through a narrowband filter with the central wavelength tuned to the wavelength of Lyα emission at $z = 4.55$. At the same time, Pascarelle et al. [254] discovered 5 LAEs, including a weak AGN, around the radio galaxy 53W002 at $z = 2.39$ using the Hubble Space Telescope (HST) for broadband imaging, a ground based telescope for narrowband imaging, and the MMT for spectroscopy. Pascarelle et al. [255] claim

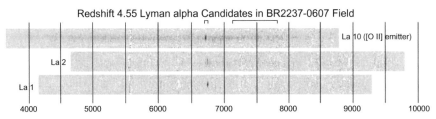

Fig. 3.3 Top: Images of three LAE candidates in the QSO field of BR2237-0607 at $z = 4.55$ marked with the small circles [128]. The QSO is indicated with the large circle. The left and right panels present narrowband and broadband images, respectively, that cover the redshifted Lyα emission and the rest-frame UV continua of the LAEs. Bottom: Two-dimensional spectra of the three LAE candidates [128]. Clear Lyα signals of the LAE candidates are found in the wavelength between 6500 and 7000 Å. The two objects in the two lower panels are LAEs, while the object in the top panel is classified as a low-z [OII] emitter. Reprinted by permission from Nature

that such LAEs form a galaxy group in the 53W002 region. These discovered LAEs have Lyα luminosities of a few times 10^{42} erg s^{-1} that is about $1/100$ of the young galaxies predicted by Partridge and Peebles [252].

These early studies found LAEs in the vicinity of AGNs that are thought to be signposts of high-z galaxy overdensities. LAEs in blank fields were first identified by Cowie and Hu [53] and Hu et al. [129] who carried out Keck LRIS narrowband imaging and spectroscopy in the SSA22 and Hubble Deep Field (HDF). Such deep field observations started to identify LAEs at $z \gtrsim 2$ in blank fields routinely with the high sensitivities of 8 m class telescopes and the large-area survey capabilities of 4–8 m class telescopes around the year 2000. These deep field observation programs include the Hawaii Survey [53], the Large Area Lyman Alpha Survey (LALA; Rhoads et al. [282]), the Subaru Surveys (e.g. Ouchi et al. [239]), and the Multiwavelength Survey by Yale-Chile (MUSYC; Gawiser et al. [94]), and recently an LAE search with a new technology has been demonstrated by the HETDEX Pilot Survey (HPS:

Adams et al. [2]).[1] Recent space based observations even find LAEs at $z \sim 0$ whose Lyα emission lines fall in the far UV wavelength range. There is an HST survey program, the so-called Lyman-Alpha Reference Sample (LARS; Östlin et al. [237]) that investigates the Lyα properties of star-forming galaxies originally selected as Hα emitters at $z \sim 0$. Moreover, far and near UV grism data from the Galaxy Evolution Explorer (GALEX) are used to detect LAEs at $z \sim 0 - 1$ and to build Lyα flux limited samples for studies of Lyα luminosity functions (Sect. 3.3.2; Deharveng et al. [64], Cowie et al. [54, 55]). For more details about $z \sim 0$ LAE observations, see M. Hayes' lectures in this course.

There is a question why no single LAE at high-z could be identified by the observations until 1996, about two decades after the predictions of Partridge and Peebles [252]. As shown in Fig. 3.2, the blank field surveys conducted until 1995 found no LAEs at $z > 2$ down to number densities of $\sim 10^{-4} \, \mathrm{Mpc}^{-3}$ at $L(\mathrm{Ly}\alpha) = 10^{43} \, \mathrm{erg \, s}^{-1}$ and $\sim 10^{-3} \, \mathrm{Mpc}^{-3}$ at $L(\mathrm{Ly}\alpha) = 10^{42} \, \mathrm{erg \, s}^{-1}$. These number-density limits just touch the Lyα luminosity functions at $z \sim 3 - 5$ that have been determined to date (see Sect. 3.3.2). In other words, one could have found LAEs in a blank field before 1996, if there was a one-more push of sensitivity or survey volume. However, such a one-more push was not made until 1996. In reality, there are two important approaches leading to these successful detections of LAEs. The first approach is to focus on AGN regions. The first LAE detections [128, 254] were accomplished in AGN regions, whose galaxy overdensities enhance the probability of bright LAEs existing in the survey area, resulting in successful selections and spectroscopic confirmations even with 2–4 m class telescopes. The second approach is to exploit the great sensitivity of 8 m-class telescopes newly available since the 1990s. In fact, the Keck deep narrowband observations by Cowie and Hu [53] successfully identified LAEs in blank fields. Interestingly, these two approaches provided successful detections almost at the same time in the late 1990s.

Around the year 2000, wide-field optical imagers started operation in 4 m-class telescopes (e.g. KPNO/MOSAIC), allowing the observers to detect LAEs in blank fields even with the moderately low sensitivity of 4 m-class telescopes (Rhoads et al. [282], Gawiser et al. [94]; Fig. 3.2). Moreover, after the first light of the wide-field optical imager Suprime-Cam on the 8 m-Subaru telescope in 1999, large LAE surveys cover wider sensitivity and volume ranges, in contrast with the previous narrow-field 8m-class observations (Fig. 3.2). The deep spectroscopic capabilities of the Keck, VLT, and Subaru telescopes are also key for confirmation of faint LAE candidates that are found in blank fields.

---

[1] In the early days of LAE observational studies, many names and abbreviations were used for LAEs, such as Lyα emitting galaxies, Lyα galaxies, LEGOs etc. The present-day established name, Lyα emitter, can be found in the early study of Hu and McMahon [128], and the abbreviation, LAE, was first used in Ouchi et al. [239].

3.1.1.4 Definition of LAEs

Here I introduce the definition of LAEs, although it should be noted that some detailed definitions depend on the study considered. Nowadays, the widely accepted definition of LAEs is: *LAEs are galaxies with a Lyα rest-frame equivalent width (EW_0) greater than $\simeq 20\,\text{Å}$.* The criterion of Lyα $EW_0 \gtrsim 20\,\text{Å}$ is historically determined by the realistic selection limit of narrowband imaging surveys for Lyα emitting galaxies at $z \sim 3$. By this definition, the main contribution to the LAE population consists of star-forming galaxies, some of which have AGN activity.

3.1.1.5 LAE Search Techniques

There are two popular techniques to search for LAEs. One is narrowband imaging. Figure 3.4 illustrates the idea. The redshifted Lyα emission of LAEs is identified by a flux excess in a narrowband image over other wavelength images (Fig. 3.4). The central wavelength of the narrowband filter, λ_c, determines the redshift of the target LAEs that is roughly given by $\lambda_c/1216 - 1$. The λ_c value of a narrowband filter is chosen by a scientific requirement (i.e., redshift of target LAEs) and/or

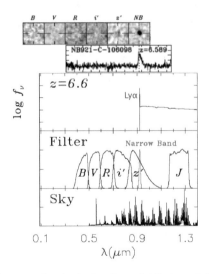

Fig. 3.4 Illustration of the narrowband selection for an LAE at $z = 6.6$ (NB921-C-106098). The top panel presents images of this LAE observed with broadbands (B, V, R, i', and z') and a narrowband whose central wavelength is $\sim$9200 Å. The second top panel is a spectrum of this LAE in the wavelength range of 9050–9275 Å. The third top panel shows the model spectrum of a LAE redshifted to $z \sim 6.6$. The second bottom panel exhibits the transmission curves of the broadbands and the narrowband. The bottom panel presents the atmospheric OH lines. Some images of this figure are taken from Ouchi et al. [246]. ©AAS. Reproduced with permission

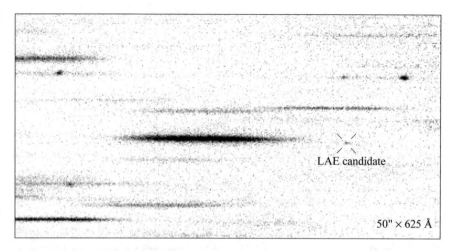

Fig. 3.5 An LAE candidate found in VLT/FORS grism data in a blank field [173], labeled on the image. An [OIII]5007 emitter is also found at the upper right of this image. Reproduced with permission ©ESO

observational constraints (e.g. avoiding weak night-sky OH emission lines). In most cases, λ_c is placed in an OH emission window (bottom panel of Fig. 3.4) to realize a high sensitivity.

The other technique is blind spectroscopy, including slitless spectroscopy. Figure 3.5 shows one example that uses a VLT/FORS grism targeting a blank sky field with no prior positional information of a LAE candidate [173]. The LAE candidate is found as a single-line emitter in the grism image. Although this technique provides positions and spectra of LAEs at the same time, the background sky level is high in the slitless data. Thus, HST grism observations are popular to perform slitless spectroscopic searches for LAEs, exploiting the low sky background in space [266]. The blind spectroscopy technique also includes slit spectroscopy such as long-slit spectroscopy conducted on positions of critical lines of lensing clusters searching for lensed LAEs [294]. Moreover, the recent advancement of integral field spectrographs (IFSs) allows blind spectroscopic searches for LAEs in reasonably large areas, keeping the background sky sufficiently low [13, 349].

Although these two techniques are major ones for identifying LAEs, recent deep spectroscopy has found continuum-selected galaxies (e.g. dropouts or Lyman break galaxies; LBGs) with a spectroscopic measurement of Lyα $EW_0 \gtrsim 20\,\text{Å}$ that are also classified as LAEs (e.g. Erb et al. [76]). In this series of lectures, LAEs include continuum-selected galaxies with Lyα $EW_0 \gtrsim 20\,\text{Å}$.

3.1.2 Progresses in LAE Observational Studies After the Discovery

Large survey programs have so far identified a total of more than 10^4 LAEs up to $z \sim 8$ photometrically (e.g. Yamada et al. [368], Konno et al. [166]), out of which about 10^3 have been spectroscopically confirmed (e.g. Hu et al. [132], Kashikawa et al. [152]). Due to the high abundance (i.e. number density) of LAEs, 10^{-3} Mpc^{-3} at $L_{Ly\alpha} \sim 10^{42} - 10^{43}$ erg s^{-1}, LAEs are thought to constitute one of the major populations of high-z galaxies. Below, I highlight progresses in LAE observations that are detailed in Sects. 3.3–3.6.

Deep photometric studies reveal the average spectral energy distribution (SED) of LAEs with deep optical and NIR photometric data. From comparisons with stellar population synthesis models, stellar population, one of the basic properties of galaxies, is studied (e.g. Gawiser et al. [94], Finkelstein et al. [86], Ono et al. [232, 233], Guaita et al. [104], Hagen et al. [107, 108]). Figure 3.6 compares LAEs' average stellar masses (M_s) and specific star-formation rates (sSFRs), defined as the star-formation rate (SFR) divided by stellar mass, with those of other galaxy populations: LBGs, distant-red galaxies (DRGs), and sub-millimeter galaxies (SMGs) at $z \sim 3$. The average M_s of LAEs is $10^8 - 10^9 M_\odot$, which falls in the lowest mass range among the high-z galaxy populations (Sect. 3.3.1). The low stellar masses of the LAEs suggest that LAEs are high-z analogs of local star-forming dwarf galaxies. The sSFR values of LAEs are comparable to or slightly higher than those of the other high-z galaxies, although the distribution of LAEs at the low-mass limit of Fig. 3.6 is biased by the observational selection limits.

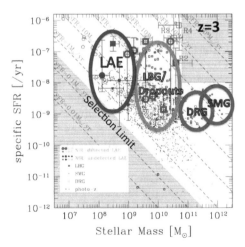

Fig. 3.6 sSFR as a function of stellar mass for typical LAEs (blue symbols), LBGs/dropouts (black open squares), DRGs (black open circles), and SMGs (star marks) at $z = 3$ [232]. The thick ovals indicate the approximated distributions of LAEs (blue), LBGs/dropouts (orange), DRGs (magenta), and SMGs (red). The red squares and circles with error bars denote peculiar LAEs with bright NIR-band fluxes. This figure is reproduced by permission of the Royal Astronomical Society

A narrowband imaging search for LAEs has serendipitously identified remarkable objects like Lyα blobs (LABs), many of which show no clear AGN signatures (Sect. 3.4.1), that were first found in the LBG overdensity region SSA22 (Fig. 3.7; Steidel et al. [328]). LABs consist of a large Lyα nebula with a spatial extent of ~ 10–$200\,\mathrm{kpc}$ and a bright total Lyα luminosity $L_{\mathrm{Ly}\alpha} \sim 10^{43} - 10^{44}\,\mathrm{erg\,s^{-1}}$ [201]. So far, a few tens of LABs are identified at $z \sim 2 - 7$ [244, 295, 371]. There are various models of LABs including HI scattering clouds and cooling radiation. However, the physical origins of the large Lyα nebulae are under debate.

Since the late 1990s when the early observations detected LAEs, LAEs have remained the most distant galaxies known to date (Fig. 3.8; Hu et al. [130, 131], Kodaira et al. [162], Iye et al. [144], Vanzella et al. [350], Ono et al. [234], Shibuya et al. [306], Finkelstein et al. [90], Oesch et al. [229], Zitrin et al. [374]), except for some examples of high-z dropouts whose redshifts are estimated with the Lyα continuum break with an accuracy $\Delta z = 0.1 - 0.2$ [230, 359]. Most of the highest redshift galaxies confirmed by spectroscopy are LAEs, because strong Lyα emission can be efficiently detected in a very faint source at high redshift. Some of the high-z galaxies show intrinsically large Lyα EW_0 values, suggestive of very young, population III (popIII)-like starbursts such as those predicted by Partridge and Peebles [252] (Sect. 3.1.1.1).

A number of LAEs have been spectroscopically identified at the epoch of reionization (EoR) at $z \gtrsim 6$ (Sect. 3.5). Because Lyα photons from LAEs are scattered by neutral hydrogen HI that exists in the IGM at EoR, the detectability of Lyα from LAEs depends on the fraction of HI in the IGM. In a statistical sense, weak Lyα emission of LAEs suggests more Lyα scattering in the IGM or lower Lyα production rates. Exploiting this dependence, LAEs are used as probes of cosmic reionization as well as galaxy formation (Fig. 3.9).

Cosmic reionization has been extensively investigated using LAEs, after isolating the effects of galaxy formation, namely the evolution of the Lyα luminosity, in conjunction with complementary observational constraints (Sects. 3.5–3.6; Malhotra and Rhoads [193], Kashikawa et al. [151, 152], Ouchi et al. [246], Pentericci et al. [262], Ono et al. [234], Schenker et al. [299], Treu et al. [347], Schenker et al. [301]).

3.1.3 Goals of This Lecture Series

The goal of this lecture series is to make the readers understand not only the established picture of LAEs, but also the cutting-edge results obtained from observations spanning the redshift range $z \sim 0 - 10$ covered to date. As shown in Sect. 3.1.2, today's major LAE studies address questions about the physical properties of high-z low mass galaxies, including popIII-like galaxies, sources of reionization, and the cosmic reionization history. In other words, most LAE observational studies discuss either galaxy formation or cosmic reionization. This lecture series thus covers

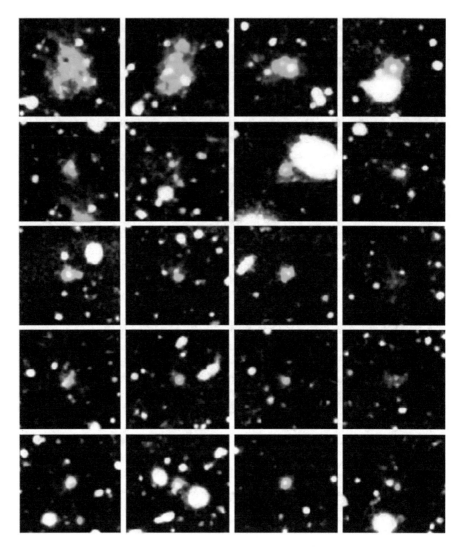

Fig. 3.7 Color composite images of LABs at $z = 3.1$ in the SSA22 field [201]. The green color indicates Lyα emission, while the blue and red colors are the continua bluer and redder than Lyα for $z \sim 3$ objects, respectively. The top left (second-left) panel shows LAB1 (LAB2), the first LAB in star-forming galaxies discovered by Steidel et al. [328]. Note that the Lyα emission nebula of LAB1 extends over $\sim$200 kpc. The size of each image is about 200 kpc $\times$ 200 kpc. This figure is adopted from http://www.naoj.org/Pressrelease/2006/07/26/index.html. Courtesy of the National Astronomical Observatory of Japan

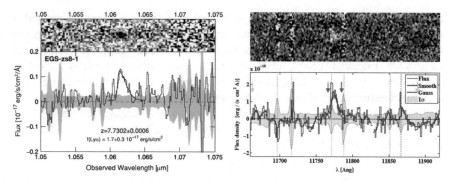

Fig. 3.8 Spectroscopically identified LAEs at $z = 7.73$ (left; Oesch et al. [229]) and $z = 8.68$ (right; Zitrin et al. [374]). The top and bottom panels present the two- and one-dimensional spectra, respectively. A clear asymmetric line typical for high-z LAEs is identified in the $z = 7.73$ LAE (left). ©AAS. Reproduced with permission

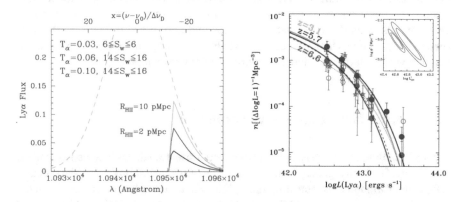

Fig. 3.9 Left: Theoretical predictions of Lyα line profiles of a Lyα emitter at $z = 8$ [66]. The black solid line represents the Lyα line from an LAE in the fully neutral IGM, while the gray dashed line denotes the intrinsic Lyα line. The red (gray) solid line shows the Lyα line of an LAE escaping from the center of an ionized bubble with a radius of 2 (10) physical Mpc in the neutral IGM. Right: Lyα luminosity functions at $z = 6.6$ (red), 5.7 (blue), and 3.1 (cyan) derived by observations [246]. The red and blue filled circles are the best-estimate luminosity functions that are the weighted average results of the luminosity functions estimated from LAEs in survey subfields, while all of the red open symbols are luminosity functions of the subfields. The inset panel presents the 68 and 90 percentile error contours for the Schechter function fitting of the best-estimate luminosity functions at $z = 6.6$ (red contours) and $z = 5.7$ (blue contours). ©Royal Astronomical Society and ©AAS. Reproduced with permission

(1) Galaxy formation (Sects. 3.2–3.4) and
(2) Cosmic reionization (Sects. 3.5–3.6).

Note that there are several promising studies of LAEs that are growing in this field. One is the Lyα emission distribution that traces the circum-galactic medium (CGM) extending along filaments of large-scale structures [40]. Because Lyα is a resonance line, it is used as a probe of the HI distribution of the underlying cosmological structures. The extended Lyα emission studies are detailed in Sect. 3.4, together with topics of Lyα blobs, diffuse Lyα halos, Lyα fluorescence, proto-clusters, and large-scale structures (LSSs), all of which are closely related to galaxy formation. Another important use of LAEs consists in probing properties of dark energy with accurate measurements of cosmic expansion history on the basis of baryon acoustic oscillations (BAO). Because no LAE studies have, so far, successfully detected BAO, an on-going LAE BAO cosmology study project is briefly touched in the section of future studies (Sect. 3.7).

3.2 Galaxy Formation I: Basic Theoretical Framework

One of the major scientific drivers of LAE studies is galaxy formation. In this section, I show the basic theoretical framework of galaxy formation and associated Lyα emission, and identify both established ideas and unresolved difficult issues. This section mainly targets first-year graduate students working on observations and those who know little about the modern picture of galaxy formation.

3.2.1 Basic Picture of Galaxy Formation

Figure 3.10 illustrates the basic picture of galaxy formation that is believed in modern astronomy. Generally, galaxy formation is made of two major processes, dark-matter (DM) halo formation and star formation [213]. First, DM halos are created from the initial density fluctuations, and then star formation takes place in the cold dense gas clouds made by radiative cooling in the DM halos. These two processes are detailed in the following subsections.

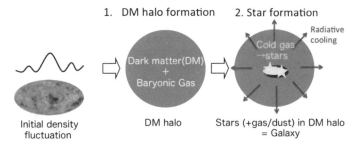

Fig. 3.10 Conceptual diagram of the galaxy-formation processes

3.2.1.1 DM Halo Formation

The standard cosmological model of Λ cold dark matter (ΛCDM) suggests that the initial density fluctuations in the early universe grow by gravity and produce cosmic structures [261]. DM halos, virialized systems of DM, with baryon gas are created by gravitational collapses. Low-mass DM halos are first made, and subsequently these low-mass DM halos increase their masses by merger and accretion processes. Because DM dominates the cosmic matter density, this sequence of the cosmological structure formation is governed by DM. DM physically interacts only by gravity, and the formation of cosmic structures including DM halos can be basically predicted with no serious systematics.

Exploiting the great performance of computers today, numerical simulations reproduce DM halos under the assumption that DM is composed of collisionless particles that follow Newton's law of gravitation. Figure 3.11 presents the DM-halo mass functions calculated by large cosmological simulations [321]. The state-of-the-art cosmological simulations (with a box size of a few-10 Mpc^3) have a good mass

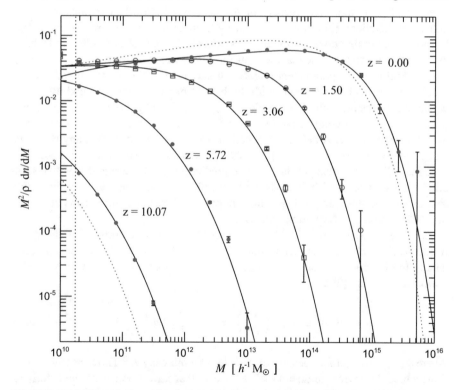

Fig. 3.11 DM-halo mass function as a function of mass and redshift obtained by numerical simulations [321]. The ordinate is the differential number density (dn/dM) multiplied by $M^2\rho^{-1}$ where ρ is the mean density of the universe. The solid lines are the best-fit functions to the mass functions with the analytical functional form of Jenkins et al. [149] (see also Sheth and Tormen [302]). The blue dashed lines are the Press–Schechter functions at $z = 10.07$ and 0 from left to right. Reprinted by permission from Nature

resolution, and already make DM halos with a mass of $\sim 10^7 M_\odot$ (Fig. 3.12; Ishiyama et al. [140, 141]) that is much smaller than those of most of the local dwarf galaxies and any high-z galaxies observed, to date. In other words, DM halos of galaxies are mostly recovered over cosmic time in numerical simulations. It is true that the physical origin of DM is poorly understood. However, under the collisionless DM particle assumption, the DM-halo formation is established today.

The DM halo formation is understood not only by numerical simulations, but also by analytic calculations. Starting from the Gaussian initial density fluctuations, one can derive an approximation of the DM-halo mass function based on linear structure growth and spherical collapse. The analytic form is referred to as the Press–Schechter function [274] that is,

$$n(M)dM = \frac{2}{\pi} \frac{a\rho_0}{M^{*2}} \left(\frac{M}{M^*} \right)^{a-2} \exp\left[-\left(\frac{M}{M^*} \right)^{2a} \right] dM \qquad (3.1)$$

where M^* is the characteristic mass, a is a power-law index of the mass fluctuations, and ρ_0 is the mean density of the universe. Figure 3.11 compares Press–Schechter functions (blue dashed lines) with numerical results (red points). The Press–Schechter functions reasonably approximate the numerical results, while there exist small departures. Note that theorists modify the analytic form of Eq. (3.1), with a few additional free parameters (e.g. Sheth and Tormen [302]), and obtain 'modified' Press–Schechter functions with the best-fit parameters determined by fitting the numerical results (solid lines in Fig. 3.11). A number of galaxy formation studies (including LAE modeling) exploit such modified Press–Schechter functions that are useful to reproduce DM-halo mass functions (mass vs. abundance) at any redshifts and cosmological parameter sets [195, 292]. It should be also noted that these analytic formalisms can also provide reliable predictions in clustering of DM halos. In other words, once a redshift, mass, and cosmological parameter set are given, the abundance and clustering of DM halos are predicted by these formalisms based on the ΛCDM structure formation scenario.

Because galaxies form in DM halos, galaxy luminosity functions and stellar-mass functions should have a functional shape similar to DM halo mass functions. Indeed, galaxy luminosity functions determined by observations can be fit well with the Schechter function [298],

$$\phi(L)dL = \phi^* \left(\frac{L}{L^*} \right)^\alpha \exp\left(-\frac{L}{L^*} \right) d\left(\frac{L}{L^*} \right) \qquad (3.2)$$

where $\phi(L)$ is the number density of galaxies at luminosity L.[2] The Schechter function includes three free parameters, ϕ^*, L^*, and α, that correspond to the characteristic number density, the characteristic luminosity, and the faint-end slope, respectively.

---

[2]This is the Schechter function on the luminosity basis. The magnitude-based Schechter function is shown in, e.g., Eq. 8 of Ouchi et al. [240].

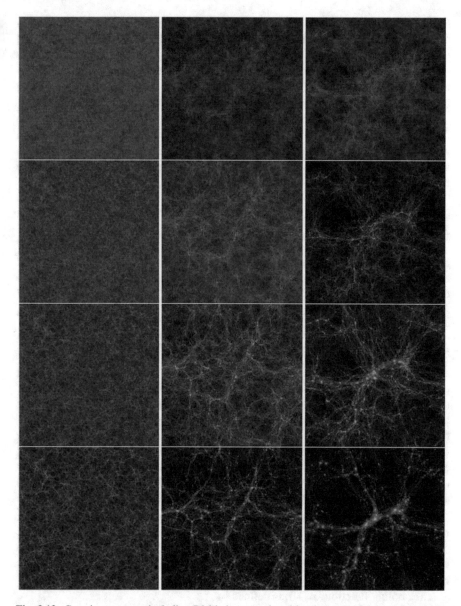

Fig. 3.12 Cosmic structures including DM halos reproduced by the state-of-the-art simulations [141]. From top to bottom, cosmic structures at $z = 7$, 3, 1, and 0 are shown. The left, center, and right panels represent the simulation results with three different DM-particle masses (box sizes), $2.2 \times 10^8 h^{-1} M_\odot$ ($1120 h^{-1}$ Mpc), $2.8 \times 10^7 h^{-1} M_\odot$ ($140 h^{-1}$ Mpc), and $3.4 \times 10^6 h^{-1} M_\odot$ ($70 h^{-1}$ Mpc), respectively. This figure is reproduced by permission of the ASJ

Similarly, stellar mass functions are expressed with stellar mass M_s and the characteristic stellar mass M_s^* that are in place of L and L^*, respectively, in Eq. (3.2) (Fig. 3.14).

It should be noted that Eq. (3.2) has a functional form of the product of an exponential cut-off and a power law for luminosity, which is the same as Eq. (3.1) where halo mass is replaced with luminosity.

The three free parameters of the Schechter function reflect differences from the DM-halo mass function, which depend on the baryonic processes of star-formation and feedback in galaxy formation (Sect. 3.2.1.2). In LAE studies, the Schechter function is used to approximate the Lyα luminosity function and the continuum luminosity function.

3.2.1.2 Star Formation

Star formation involves complicated physical processes of gas cooling and feedback as detailed below. Moreover, star-formation is induced by objects and matter outside of the galaxy by mergers and gas accretion. I choose gas cooling, feedback, and cold accretion that are key for understanding LAEs in the context of galaxy formation, and introduce these physical processes below.

Gas Cooling

Star formation requires a reservoir of cold dense gas in a DM halo. However, such cold dense gas cannot be easily produced in a DM halo. If an adiabatic gas contraction takes place in the DM halo, gas temperature increases. Then, the gas contraction stops by the thermal pressure, and dense gas cannot be produced. In this way, adiabatic contractions do not make dense gas that is necessary for star formation. Star-formation thus requires a gas contraction associated with radiative cooling that reduces thermal energy in the gas cloud (Fig. 3.10). By the radiative cooling, gas temperature should decrease from the virial temperature T of the DM halo ($\gtrsim 10^4$ K) to the temperature of molecular hydrogen H_2 clouds ($\lesssim 10^2$ K).

State-of-the-art simulations calculate the gas cooling processes numerically under realistic physical conditions, although these calculations cannot be described with simple analytical forms. Instead, I introduce a classic picture of gas cooling with the free-fall time of the spherical model [315] that helps the readers understand the idea of gas cooling.

The cooling function, $\Lambda(T)$, is defined by

$$|\dot{E}_{cool}| = n^2 \Lambda(T), \qquad (3.3)$$

where $\dot{E}_{cool}$ and n are the cooling rate (energy density divided by time) and the number density of particles, respectively. The cooling function is calculated based on quantum physics, and displayed in Fig. 3.13. It is clear that metal rich gas has a higher $\Lambda(T)$, because various atomic electron transitions are allowed for heavy elements that enhance the efficiency of radiative cooling. In zero-metal gas, there

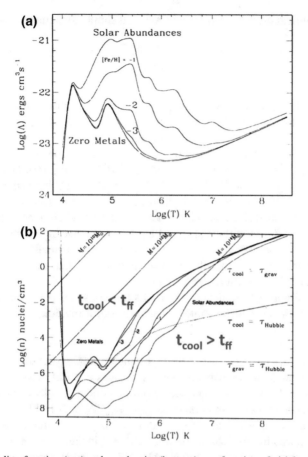

Fig. 3.13 Cooling function (top) and gas density (bottom) as a function of virial temperature in a spherical halo [315]. Top: The curves represent cooling functions with various metal abundances from zero to one solar metallicities that are indicated by the labels. Bottom: The solid curves denote $t_{cool} = t_{ff}$ for the metal abundances from zero to one solar metallicities (same labels as in the top panel). Beyond these curves, the cooling time is shorter than the free-fall time, and gas collapses with a negligible thermal pressure. The dashed curve (the horizontal line) indicates that the cooling time (free-fall time) is equal to the present Hubble time. Above the curve and the horizontal line, the gas contraction completes in the cosmic time. The diagonal lines show the density and virial temperature that correspond to the virial masses of the system, 10^{10}, 10^{12}, and 10^{14} $M_\odot$. This figure is reproduced by permission of the Physics Reports

are two peaks in $\Lambda(T)$ near 10^4 and 10^5 K that correspond to hydrogen and helium recombinations, respectively. The upturn of $\Lambda(T)$ from 10^6 to 10^8 K is explained by the cooling processes of Bremsstrahlung and Compton scattering.

In a virialized system, the kinetic energy E_K of gas is given by $E_K = 3nk_BT/2$, where k_B is the Boltzmann constant. Thus, the cooling time t_{cool} is given by

$$t_{cool} = \frac{E_K}{|\dot{E}_{cool}|} = \frac{3}{2} \frac{k_B T}{n \Lambda(T)}. \tag{3.4}$$

In the spherical model, the free-fall time t_{ff}, a simple dynamical time, is

$$t_{ff} = \sqrt{\frac{3\pi}{32G\rho_m}}, \qquad (3.5)$$

where ρ_m is the mass density that is proportional to n. If the radiative cooling is very efficient, the cooling time is shorter than the free-fall time, $t_{cool} < t_{ff}$. In this case, gas collapse takes place with a negligible thermal pressure, and cold dense gas necessary for star formation is produced. This condition of $t_{cool} < t_{ff}$ for gas collapse is presented in the T and n plot of Fig. 3.13 (bottom). Halo masses of $\sim 10^{12} M_\odot$ allow gas collapse down to the low gas densities, indicating an efficient gas cooling. It should be noted that the halo mass of $10^{12} M_\odot$ coincides with the mass of the Milky Way as well as the mass where the stellar-to-halo mass ratio is highest [22]. Metals ease the conditions of gas collapse in a massive halo with $> 10^{12} M_\odot$. In Fig. 3.13, gas halos with $t_{cool} > t_{ff}$ cannot collapse but cause a quasi-static contraction due to inefficient cooling, which can take time longer than the Hubble time. As shown in Sect. 3.3.8, typical LAEs have halo masses of 10^{10}–$10^{12} M_\odot$ and sub-solar metallicities. Figure 3.13 indicates that LAEs have physical parameters reasonably good for gas collapse, which enables subsequent star formation.

Feedback

Feedback is known as one of the most important physical processes involved in star formation. Figure 3.14 compares an observed galaxy stellar-mass function (filled squares) with a DM halo mass function from numerical simulations (dashed line). Because the cosmic baryon fraction f_b is $f_b \equiv \Omega_b/\Omega_m \simeq 0.16$ [267], the DM-halo mass function should be at least about an order of magnitude higher than the stellar-mass function, which can be clearly seen at $\sim 10^{11} M_\odot$ in Fig. 3.14 (see the mass values at a constant number density of $\log N/[\mathrm{Mpc}^3 M_\odot] \sim -14$). However, in Fig. 3.14, the stellar-mass function is flatter at the low-mass end and steeper at the massive end than the DM-halo mass function. These shape differences are thought to be made by feedback effects that suppress star-formation by gas heating and outflow associated with star-formation and AGN activities in a galaxy [32].

Theoretical studies assume two feedback mechanisms in the low and high mass regimes, energy- and momentum-driven feedback effects, respectively (Fig. 3.15; Muratov et al. [219]). The energy-driven feedback for low-mass galaxies is caused by thermal energy inputs from supernova (SN) explosions and stellar radiation. The momentum-driven feedback for high-mass galaxies is activated by kinetic energy inputs from stellar winds, radiative pressure, and AGN jets. Defining the mass-loading factor $\eta \equiv \dot{M}_{out}/SFR$ where $\dot{M}_{out}$ is the outflow rate, numerical simulations show

$$\eta \propto V_{circ}^{-2} \qquad (3.6)$$
$$\eta \propto V_{circ}^{-1} \qquad (3.7)$$

Fig. 3.14 An observed baryonic (stars + cold gas) mass function of local galaxies (squares and the black dotted line; Bell et al. [23]) and the DM-halo mass function (dashed line; Weller et al. [360]). The colored solid and dotted lines indicate baryonic mass functions of various Hubble type galaxies: ellipticals, spirals, irregulars, and dwarf ellipticals. This figure is adopted from Read and Trentham [283], and reproduced by permission of the Royal Society

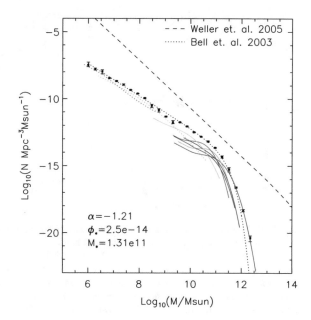

for the energy and momentum driven feedbacks, respectively. Here, V_{circ} is the circular velocity given by

$$V_{circ} = \sqrt{\frac{GM_h}{r_{vir}}},$$ (3.8)

where M_h and r_{vir} are the DM-halo mass and the virial radius, respectively. In Fig. 3.15, the energy and momentum driven feedbacks are seen at $M_h \lesssim 10^{10} M_\odot$ and $M_h \gtrsim 10^{10} M_\odot$, respectively. From observations of local galaxies, the relation of Eq. (3.7) is confirmed [121], while no observations reach $M_h \lesssim 10^{10} M_\odot$ to test the relation of the energy-driven feedback (Eq. 3.6). Because the average DM-halo mass of LAEs is estimated to be $M_h \sim 10^{11} M_\odot$ by clustering analysis (Sect. 3.3.8), feedbacks in typical LAEs are probably dominated by the momentum-driven feedback.

Note that some theoretical studies claim the existence of positive feedback effects that induce star-forming activities by, e.g., the shock cooling of AGN jets, radiation pressure, etc. [316, 357].

Cold Accretion

Another important mechanism for star-formation is cold accretion. In the standard picture of galaxy growth, gas infalling in a DM halo is heated to a virial temperature by shocks at around the DM-halo virial radius, and then reaches a quasi-hydrostatic equilibrium with $T \sim 10^6 (V_{circ}/167\,\mathrm{km\,s^{-1}})^2$ K. The hot virialized gas cools by cooling radiation, and forms a cold gas disk that produces stars at the DM-halo center [78, 281, 362]. Recent theoretical studies suggest that, in this galaxy growth process, the infalling gas can penetrate into the DM halo center through the diffuse shock-heated

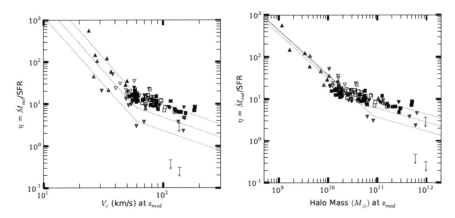

Fig. 3.15 Mass loading factor as a function of V_{circ} (left) and DM-halo mass (right) predicted by numerical simulations [219]. The red, blue, and black symbols represent galaxies at $z = 0 - 0.5$ ($z_{med} = 0.25$), $0.5 - 2$ (1.25), and $2 - 4$ (3), respectively, produced in the simulations. The red, blue, and black dotted lines indicate a broken-power law fit to the simulated galaxies at corresponding redshifts that roughly corresponds to the energy- and momentum-driven feedbacks (Eqs. 3.6 and 3.7), although the slope of the energy-driven feedback is slightly steeper than Eq. (3.6). This figure is reproduced by permission of the the Royal Astronomical Society

medium, if the infalling gas is dense ($\sim 1\,cm^{-3}$) and cold (a few 10^4 K; Fardal et al. [82], Kravtsov [171], Kereš et al. [159]), which is referred to as cold accretion, cold mode accretion, or cold stream. The theoretical studies predict that, beyond the virial radius of DM halos, there exist multiple cold dense gas streams to the DM-halo center through filaments of LSSs (Fig. 3.16). These cold dense gas streams collide at the DM-halo center, and cool very efficiently, which triggers intense star formation [65]. Such cold accretion is important, because about two-thirds of gas accretion in mass has a form of smooth gas flows, in contrast with the rest of the gas accretion taking place in a form of mergers with a $>1/10$-mass ratio [65, 155, 160]). This theoretical picture would explain high SFR galaxies with no merger signatures, such as bright LBGs and SMGs with an SFR of $\gtrsim 100 M_{\odot}\,yr^{-1}$, and could be an answer to the question why the number density of high SFR galaxies at $z \sim 2$ is significantly larger than those expected from merger events. Note that cold gas accretion is allowed in a massive halo only at $z \gtrsim 2$ (Fig. 3.16), when the accretion gas is sufficiently cold.

Because the cold accretion is a theoretical picture, in the past decade observers have searched for a signature of cold accretion in their observational data. In LBGs at $z \sim 2$, velocities of low ionization metal absorption lines are mostly blueshifted from the galaxy systemic velocities, indicating gas outflow associated with star-forming activities [330]. Although there are several reports of cold accretion object candidates in deep observational studies (e.g. Nilsson et al. [228], Rauch et al. [280]), no definitive observational evidence for cold accretion has been found so far. Because the cold accretion gas infalls along with filaments of LSSs, the covering fraction of cold accretion gas is very small, $\sim 1-2\%$ [83]. A large number of sightlines (i.e. a

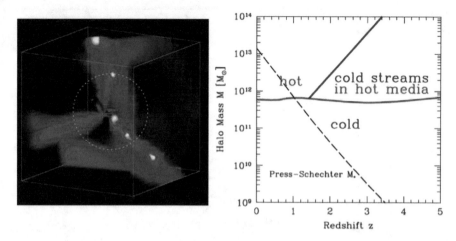

Fig. 3.16 Left: Radial flux of cold gas accretion into the center of a DM-halo predicted by numerical simulations [65]. The color scale indicates the inflow rate per solid angle. The box size is 320 kpc. The dotted line represents the virial radius of the DM halo. There are three cold accretion streams clearly found in this figure, two of which include gas clumps with a mass ten times lower than that of the central galaxy. Right: DM-halo mass as a function of redshift [65]. The red horizontal curve indicates the threshold mass above which infalling gas is shock-heated around the DM-halo virial radius. The blue line represents the limit of cold accretion of gas whose density (temperature) is high (low) enough to penetrate into the DM halo center through the diffuse shock-heated medium. The dashed line denotes the characteristic DM-halo mass of the Press–Schechter mass function (Eq. 3.1) at a given redshift. Reprinted by permission from Nature

large sample of galaxies) would be needed to prove or disprove the existence of cold accretion.

3.2.1.3   Role of Observations

As introduced in Sect. 3.2.1, the basic process of galaxy formation is DM-halo formation and star formation (Fig. 3.10). DM-halo formation is well understood with no large systematics by simple numerical simulations and analytic approximations (Sect. 3.2.1.1), while star-formation is poorly understood due to the complicated baryonic processes: gas cooling, feedback, and cold accretion as well as merger induced star-formation. The star-formation process involves a number of unknown parameters, such as gas metallicity, density, temperature, outflow, and inflow. Observations can obtain these key parameters tightly connected to star formation, and constrain free parameters of galaxy formation models. On the other hand, many cosmological simulations including those for LAEs assume a simple relation between halo-mass and galaxy luminosity (e.g. McQuinn et al. [211]) as well as an empirical relation between gas and star-formation density such as the Kennicutt–Schmidt (KS) law (Fig. 3.17). Such models with empirical relations can derive the star-formation sur-

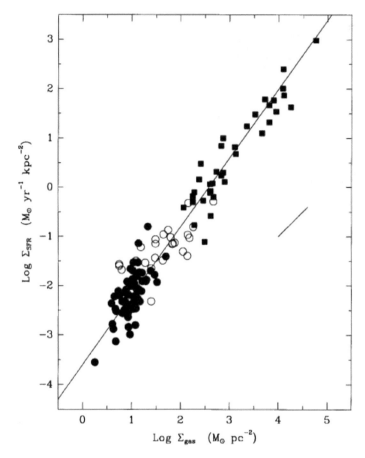

Fig. 3.17 Star-formation surface density as a function of gas surface density [158]. The gas surface density is defined by the average total surface density of atomic and molecular gas. The filled circles and squares represent normal-disk and starburst galaxies, respectively, in the local universe. The open circles denote the center of the normal-disk galaxies. The line is the least-squares fit to the data points, $\Sigma_{SFR} \propto \Sigma_{gas}^{1.4}$, which is known as the Kennicutt–Schmidt law. ©AAS. Reproduced with permission

face density Σ_{SFR} from the gas surface density Σ_{gas} that is predicted by numerical simulations and semi-analytic models, and aim to explain other various observational quantities of galaxies (e.g. Garel et al. [92]). In this way, key observational parameters and empirical relations are important to understand star-formation in galaxies. Thus, the goal of observations is to determine star-formation key parameters and empirical relations to develop a self-consistent physical picture of galaxy formation.

3.2.2   Origins of Lyα Emission from LAEs

Once a galaxy formation model is developed, Lyα emission of galaxies can be modeled. Theoretical studies suggest that in galaxies, Lyα emission can have five major origins, which probably explain the diversity of the spatial distribution of Lyα emission revealed by observations (Fig. 3.18).

Lyα emission following hydrogen recombination in the ISM near the center of a galaxy can result from two origins of ionizing sources: (i) star formation that makes HII regions and (ii) nuclear activities (i.e. AGN), if any, producing highly ionized broad and narrow-line regions in the galaxy center.

The remaining three origins are dominated by Lyα emission from the CGM to the outer halo: (iii) outflowing gas that collisionally excites hydrogen whose Lyα to Hα flux ratio is higher than the one of the optically thick case B recombination $f_{Ly\alpha}/f_{Ly\alpha} > 8.7$ [223]. (iv) cooling radiation in the hot halo gas (Sect. 3.2.1.2), and (v) fluorescence emission produced by the halo and IGM neutral hydrogen gas photoionized by UV background radiation supplied, e.g., by QSOs [164].

Although Lyα emission can be made by these five photo-ionization and collisional excitation processes (i)–(v), Lyα photons experience resonance scattering in the HI gas of the ISM, the CGM, and the IGM, due to the large Lyα cross section of HI. Lyα photons are re-distributed in space and wavelength by the resonance scattering. For this reason, scattered Lyα emission would dominate in the CGM, where the Lyα intensity of photo-ionization is relatively weak. It should be noted that observations identify Lyα photons last scattered by HI gas, and largely miss the original Lyα source position and gas dynamics information. However, this resonance nature of Lyα is also useful to probe the distribution and the kinematics of HI gas by observations via theoretical modeling (Sect. 3.3.5).

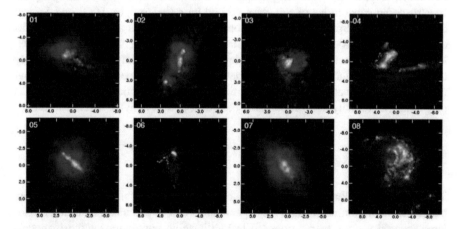

Fig. 3.18 False-color images of local Lyα emitters obtained by the LARS survey [120]. The blue, red, and green colors indicate Lyα emission, Hα emission, and far-UV continuum, respectively. The scales in units of kpc are shown in the vertical and horizontal axes. ©AAS. Reproduced with permission

3.2.3   Summary of Galaxy Formation I

This section overviews the basic theoretical framework of galaxy formation and Lyα production, targeting young observers with a limited theoretical background. This section explains that galaxy formation is made of two physical processes of DM-halo formation and star formation. DM-halo formation is well understood with a simple robust model of cosmic structure formation consistent with observations. However, star formation involves complicated baryonic processes of gas cooling, feedback, and cold accretion as well as mergers that include many physical parameters difficult to determine. It is concluded that observations should constrain important physical parameters and empirical relations that are key for filling in the missing piece of the picture of galaxy formation. To understand LAEs in the context of galaxy formation, one also needs physical models of Lyα emission. Five Lyα emission origins in theoretical models are introduced. Two origins are in ISM regions: (i) star formation (HII regions) and (ii) AGN (highly-ionized gas in a galaxy center). The other three dominate in the CGM and outer halo regions: (iii) outflowing gas, (iv) cooling radiation, and (v) fluorescence of UV background radiation. LAE observations should also reveal the origins of Lyα photons in parallel with the efforts to address the general galaxy formation issues.

3.3   Galaxy Formation II: LAEs Uncovered by Deep Observations

Since the discovery of LAEs in the late 1990s, various physical properties of LAEs have been revealed by exploiting deep optical to mid-infrared (MIR) imaging and spectroscopic capabilities of 8m-class ground based telescopes, HST, and Spitzer Space Telescope (Spitzer) in conjunction with observations at other wavelengths using the Chandra X-ray observatory (Chandra), GALEX, Herschel Space Observatory (Herschel), Atacama Large Millimeter/submillimeter Array (ALMA), and Very Large Array (VLA). In this section, I review key physical properties of LAEs uncovered by those observations: stellar population, luminosity function, morphology, ISM properties (metallicity, ionization parameter, dust), AGN activity, and clustering.

3.3.1   Stellar Population

The stellar population of galaxies is described by stellar mass, age, dust extinction, and some other parameters, and these parameters can be estimated by fitting broadband spectral energy distributions (SEDs) with stellar population synthesis models such as Bruzual and Charlot [35]. It is, however, difficult to investigate stellar populations of LAEs because most LAEs do not have detectable continuum emission even in deep images, although there do exist remarkably bright LAEs [175]. Making

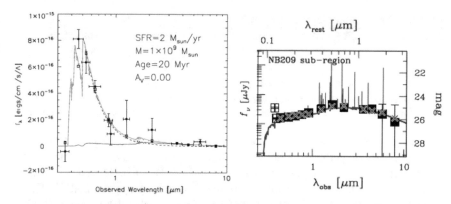

Fig. 3.19 Left: A composite SED of LAEs at $z = 3.1$ expressed in f_λ [94]. The thick solid line represents the best-fit SED that is made of an old (thin solid line) and young (dashed line) stellar components. Nebular emission is not included in modeling. Right: Same as the left panel, but for $z = 2.2$ in f_ν [222]. The blue line is the best-fit SED model that includes nebular emission. The blue crosses indicate the expected broadband photometry from the best-fit model. The two data points shown by open symbols are not used in the fitting in order to avoid a contamination from Lyα emission. ©AAS. Reproduced with permission

a composite (average) SED of a number of continuum-faint LAEs by image stacking, early studies have revealed that they have faint and blue SEDs on average. An example SED is shown in the left panel of Fig. 3.19 [94], which is explained by a model with a low stellar mass of $\sim 10^9 M_\odot$, a young stellar age of ~ 20 Myr, and a negligibly small dust extinction.

Because LAEs are young dust-poor star-forming galaxies, they often have strong nebular lines, such as Hα, Hβ, [OIII]5007, and [OII]3727, which contaminate continuum fluxes estimated from broadband photometry. Because the observed-frame equivalent width EW_{obs} of nebular lines increases with redshift as $EW_{obs} \propto (1 + z)$, nebular lines of high-z LAEs can cause serious systematic errors in broadband SEDs and hence in the calculation of stellar population parameters. Strong [OIII], Hβ, and [OII] lines near 4000 Å mimic a Balmer break that is an indicator of stellar age, and an over/underestimated age leads to an over/underestimated stellar mass. Schaerer and de Barros [297] introduce self-consistent population synthesis models with nebular lines where line ratios (as a function of metallicity) are fixed to the values of Galactic HII regions. Using a sample of $z \sim 6$ LBGs as an example, they claim that models without nebular lines overestimate stellar ages and masses by a factor of 3. Thus, considering nebular lines is critical to obtain stellar population parameters of high-z young star-forming galaxies including LAEs. It should also be noted that there is another important source of contamination, nebular *continuum*, that is the free-free/bound-free emission of hydrogen and helium and two photon continuum emission of hydrogen. Because nebular continuum emission significantly changes UV-continuum colors for very young stellar populations with a stellar age of $\lesssim 10$ Myr for instantaneous starbursts (see Figs. 3 and 4 of Bouwens et al. [30]), it is usually included in nebular emission modeling.

Table 3.1 General properties of typical LAEs[†]

Stellar mass ($M_\odot$)	$E(B - V)_s$[a]	SFR ($M_\odot\,\mathrm{yr}^{-1}$)	Stellar age (Myr)	Metallicity ($Z_\odot$)
10^7–10^{10}	0–0.2	1–100	1–100	0.1–0.5

[†]LAEs at $z \sim 2 - 3$ with a Lyα luminosity near $L^*_{\mathrm{Ly}\alpha} \simeq 10^{42} - 10^{43}\,\mathrm{erg\,s}^{-1}$
[a]Color excess due to stellar extinction. The color excess due to nebular extinction, $E(B - V)_{\mathrm{neb}}$, falls in the same range as $E(B - V)_s$ (Sect. 3.3.4.3). Calzetti's extinction law [37] is assumed

The right panel of Fig. 3.19 presents an average SED of $z = 2$ LAEs and its best-fit stellar population synthesis model with nebular emission. Table 3.1 summarizes the typical ranges of stellar population parameters of $z = 2 - 7$ LAEs that are obtained under the assumptions of constant star-formation history, a Salpeter IMF [291], and Calzetti extinction law (Calzetti et al. [37]; see Gawiser et al. [94], Ono et al. [232, 233], Guaita et al. [104], Hangen et al. [107, 108]). Although different samples give different parameter values, Table 3.1 shows that LAEs are low-stellar mass galaxies with a low dust extinction, a medium-low SFR, and a young stellar age.

Figure 3.20 compares LAEs (blue circles) with other galaxies in the stellar mass versus SFR plane. At $M_s \gtrsim 10^{10} M_\odot$ in Fig. 3.20, there is a star-formation (SF) main sequence, a tight positive correlation between M_s and SFR [59, 74]. LAEs fall in the low mass regime of $M_s \sim 10^7 - 10^{10} M_\odot$ slightly above an extrapolation of the SF main sequence found at $M \gtrsim 10^{10} M_\odot$ [107, 108], suggesting that typical LAEs are high-z dwarf galaxies in a weak burst mode. LAEs are located in a similar area in the stellar mass versus SFR plane to other emission line galaxies, i.e., [OII], Hβ, and [OIII] emitters, at $z \sim 2$ (green dots).

3.3.2 Luminosity Function

The luminosity function and its evolution over time is one of the most fundamental properties for any galaxy population. The Lyα luminosity function of LAEs has been derived at $z \sim 0 - 8$ by large survey programs (Sect. 3.1.1.3) since the discovery of LAEs in the late 1990s. The bottom and top panels of Fig. 3.21 present Lyα luminosity functions and their best-fit Schechter function parameters, respectively, from $z \sim 0$ to 6, where the Schechter function parameters are the characteristic Lyα luminosity $L^*_{\mathrm{Ly}\alpha}$ and the normalization $\phi^*_{\mathrm{Ly}\alpha}$ that determines the abundance.[3] Two evolutionary trends are seen in Fig. 3.21: a monotonic increase in the normalization from $z \sim 0$ to 3 and no evolution in either the normalization or the shape over $z \sim 3 - 6$. I explain details of these two trends in the following paragraphs.

The first evolutionary trend is an increase found at $z \sim 0 - 3$ [54, 64]. It is notable that the abundance of $z = 0.3$ LAEs is very low with $\phi^*_{\mathrm{Ly}\alpha} = 1 \times 10^{-4}\,\mathrm{Mpc}^{-3}$, about 50 times lower than that of $z \sim 0$ SDSS optical-continuum selected galaxies, $\phi^* = 5 \times 10^{-3}\,\mathrm{Mpc}^{-3}$ [25], meaning that LAEs are very rare in the local universe [64]. The top panel of Fig. 3.21 suggests that over $z = 0.3$ and $z = 2.2$ the

---

[3]Lyα luminosity functions above $z \sim 6$ are discussed in the cosmic reionization section (Sect. 3.5).

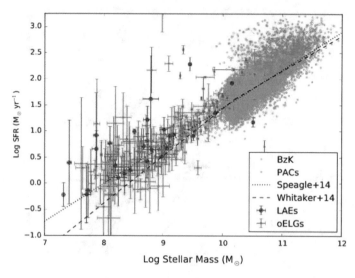

Fig. 3.20 LAEs compared with the star-formation main sequence on the SFR versus stellar mass plot [108]. The blue circles represent LAEs, while the green circles denote optical emission galaxies (oELGs). The gray and magenta circles indicate BzK and Herschel/PACS-detected galaxies, respectively [289]. The dotted and dashed lines show the star-formation main sequences at $z = 2$ obtained by Speagle et al. [320] and Whitaker et al. [361], respectively. ©AAS. Reproduced with permission

increase is statistically more significant in $L^*_{Ly\alpha}$ than in $\phi^*_{Ly\alpha}$.[4] The evolution of the Lyα luminosity function is also quantified with the Lyα luminosity density,

$$\rho_{Ly\alpha} = \int_{L^{lim}_{Ly\alpha}}^{\infty} L_{Ly\alpha}\phi_{Ly\alpha}(L_{Ly\alpha})dL_{Ly\alpha}, \tag{3.9}$$

where $L_{Ly\alpha}$ and $L^{lim}_{Ly\alpha}$ are the Lyα luminosity and the limiting Lyα luminosity, respectively. For reference, the UV continuum[5] luminosity density ρ_{UV} is defined by

$$\rho_{UV} = \int_{L^{lim}_{UV}}^{\infty} L_{UV}\phi_{UV}(L_{UV})dL_{UV}, \tag{3.10}$$

---

[4]In this panel, the data for $z = 0.9$ has a very low $\phi^*_{Ly\alpha}$ value that does not fall in the interpolation of the best-fit $\phi^*_{Ly\alpha}$ values between $z = 0.3$ and $z = 2 - 3$. Although the Lyα luminosity function may have a truly very low $\phi^*_{Ly\alpha}$ value at $z \sim 1$, there remains a possibility that the $z = 0.9$ Lyα luminosity function could be biased toward a high $L^*_{Ly\alpha}$, which gives a low $\phi^*_{Ly\alpha}$ value. In fact, this $z = 0.9$ Lyα luminosity function is derived only with bright LAEs [18]. The result of the Lyα luminosity function at $z \sim 1$ is still under debate.

[5]The wavelength of the UV continuum is often chosen at the far UV wavelength of ~ 1500 Å in the rest frame that is longer than the Lyα-line wavelength.

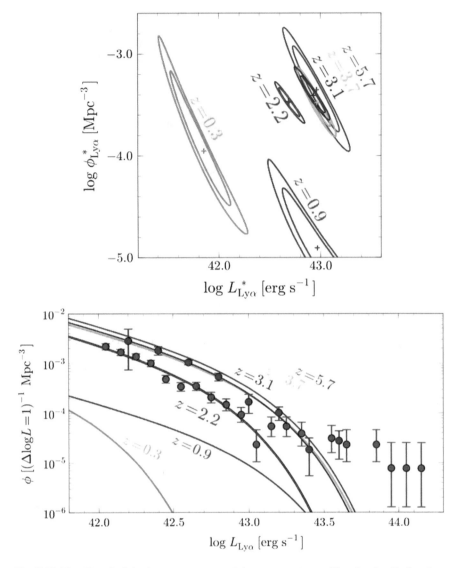

Fig. 3.21 Top: Best-fit Schechter parameters and the error contours of Lyα luminosity functions at $z \sim 0-6$ [166]. The pluses represent the best-fit values while the contours indicate their 68 and 90% confidence levels. Redshift is coded by color: orange, $z = 0.3$; magenta, 0.9; red, 2.2; blue, 3.1; cyan, 3.7; green, 5.7. Bottom: Lyα luminosity functions at $z \sim 0-6$ [166]. The curves indicate the best-fit Lyα luminosity functions, with the same color code as the top panel. The red and blue circles show the data points of the Lyα luminosity functions at $z = 2.2$ and 3.1, respectively. ©AAS. Reproduced with permission

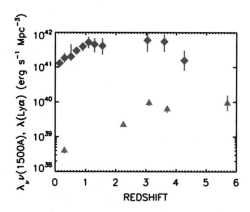

Fig. 3.22 Lyα-luminosity density (red triangles) and UV-continuum luminosity density (blue diamonds) as a function of redshift [54]. ©AAS. Reproduced with permission

where $\phi_{\mathrm{UV}}(L_{\mathrm{UV}})$, L_{UV}, and $L_{\mathrm{UV}}^{\mathrm{lim}}$ are the UV-continuum luminosity function, the UV-continuum luminosity, and the limiting UV-continuum luminosity, respectively. Figure 3.22 compares the evolutions of Lyα and UV-continuum luminosity densities, where the latter is derived with UV-continuum selected galaxies [346]. Figure 3.22 clearly shows that, from $z \sim 0$ to 3, the Lyα luminosity density increases by a factor of $\sim$20–30, which is significantly faster than the UV-continuum luminosity density evolution (a factor of $\sim$5–7; Deharveng et al. [64], Cowie et al. [54]). Similarly, in the same redshift range, the Lyα luminosity density increases even faster than the cosmic SFR density (a factor of $\sim$10) on the Madau–Lilly plot [190], indicating that the Lyα luminosity density evolution cannot be explained by the cosmic SFR density evolution alone.

The second evolutionary trend is that the Lyα luminosity function is nearly constant over $z \sim 3 - 6$ [243]. In contrast, the UV-continuum luminosity function of UV continuum-selected LBGs decreases from $z \sim 3$ to 6 and beyond, indicating that Lyα emitting galaxies dominate in number more at $z \sim 6$ than at $z \sim 3$ [243]. Indeed, deep spectroscopic surveys for UV-continuum selected LBGs suggest that the Lyα emitting ($\mathrm{EW}_0 > 25\,\text{Å}$) galaxy fraction $x_{\mathrm{Ly}\alpha}$ increases from $x_{\mathrm{Ly}\alpha} \sim 0.3$ to 0.5 over $z = 4 - 6$ for galaxies with an absolute UV magnitude M_{UV} range of $-20.25 < M_{\mathrm{UV}} < -18.75$ corresponding to $\lesssim L_{\mathrm{UV}}^{*}$ (the left panel of Fig. 3.23). In other words, about a half of $\lesssim L_{\mathrm{UV}}^{*}$-LBGs at $z \sim 6$ are LAEs with $\mathrm{EW}_0 > 25\,\text{Å}$. This $x_{\mathrm{Ly}\alpha}$ evolution result indicates an increase with redshift in either the fraction of Lyα emitting galaxies or the Lyα luminosity of galaxies or both. Allowing both $\phi_{\mathrm{Ly}\alpha}^{*}$ and $L_{\mathrm{Ly}\alpha}^{*}$ to evolve, one can derive the number- and luminosity-weighted average Lyα escape fraction $\left\langle f_{\mathrm{esc}}^{\mathrm{Ly}\alpha} \right\rangle$ from the Lyα luminosity density (Eq. 3.9) as:

$$\left\langle f_{\mathrm{esc}}^{\mathrm{Ly}\alpha} \right\rangle = \rho_{\mathrm{Ly}\alpha} / \rho_{\mathrm{Ly}\alpha}^{\mathrm{int}}, \tag{3.11}$$

where $\rho_{\mathrm{Ly}\alpha}^{\mathrm{int}}$ is the intrinsic Lyα luminosity density expected from the cosmic SFR density Ψ_{SFR} (e.g. Madau and Dickinson [190]). The intrinsic Lyα luminosity density

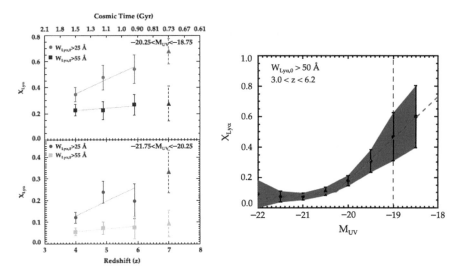

Fig. 3.23 Left: Lyα emitting galaxy fraction $x_{Ly\alpha}$ as a function of redshift [324]. UV-continuum faint $(-20.25 < M_{UV} < -18.75)$ and bright $(-21.75 < M_{UV} < -20.25)$ galaxy results are shown in the top and the bottom panels. The red and blue circles (lines) represent the data points (the best-fit power law functions) for $EW_0 > 25$ Å galaxies with faint and bright UV-continuum magnitudes, respectively, while the purple and green squares (lines) are for $EW_0 > 55$ Å galaxies. The triangles with dashed error bars at $z \sim 7$ are predictions from the best-fit functions. Right: $x_{Ly\alpha}$ as a function of UV-continuum magnitude for $EW_0 > 50$ Å galaxies at $z = 3 - 6.2$ [323]. The black circles represent measurements, while the red dashed line corresponds to the best-fit first-order polynomial to the data points. ©AAS. Reproduced with permission

can be estimated by $\rho_{Ly\alpha}^{int}$ [erg s^{-1} Mpc^{-3}] $= 1.1 \times 10^{42} \Psi_{SFR}$ [$M_\odot$ yr^{-1} Mpc^{-3}] under the assumption of the case B recombination ($L_{Ly\alpha}/L_{H\alpha} = 8.7$; Brocklehurst [34]) and the Hα luminosity $L_{H\alpha}$-SFR relation of Kennicutt [157]. Figure 3.24 presents $\left\langle f_{esc}^{Ly\alpha} \right\rangle$ as a function of redshift [118], and indicates a monotonic increase in $\left\langle f_{esc}^{Ly\alpha} \right\rangle$ from $z \sim 0$ to 6.

Here I address the issue whether all of these observational results at $z \sim 3 - 6$ are self-consistent. Observations of UV-continuum selected galaxies show that faint UV-continuum galaxies have a higher chance of emitting strong Lyα in this redshift range (right panel of Fig. 3.23; Ando et al. [7], Stark et al. [324]). In other words, a majority of LAEs are faint UV-continuum galaxies. Although the abundance of bright ($>L^*$) UV-continuum galaxies drops significantly, the abundance of faint UV-continuum galaxies does not largely decrease towards high-z, due to a steepening of the luminosity function slope α [31]. Because the abundance of LAEs is linked to the one of faint UV-continuum galaxies, the Lyα luminosity function of LAEs does not evolve largely over $z \sim 3 - 6$. In this way, all observational results suggest a self-consistent physical picture.

Because LAEs become a more dominant population at $z \sim 6$ than at $z \sim 3$, LAEs contribute much to the cosmic SFR density at $z \sim 6$. Figure 3.25 presents the evolution of the cosmic SFR density [50]. The contribution of LAEs is only 1/10 of

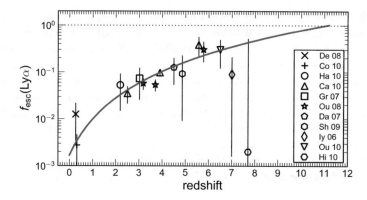

Fig. 3.24 Average Lyα escape fraction as a function of redshift [118]. The black symbols indicate average Lyα escape fraction estimates based on various observational results while the red line is the best-fit power law function to them. ©AAS. Reproduced with permission

Fig. 3.25 Cosmic SFR density as a function of redshift [50]. The data points with error bars represent the LAEs' contribution to the cosmic SFR density, while the gray region indicates the total cosmic SFR density estimated with LBGs corrected for dust extinction. ©AAS. Reproduced with permission

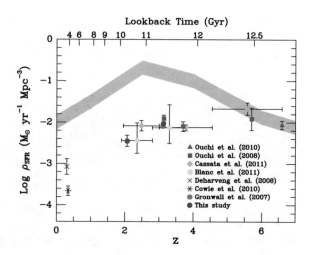

the total cosmic SFR density at $z \sim 3$ while it becomes the whole of it at $z \sim 6$. If this trend continues at $z \gtrsim 6$ (i.e. the EoR), LAEs may be a major population that emit ionizing photons for cosmic reionization. Thus, it is probably important to study LAEs to understand the physical properties of ionizing sources for cosmic reionization, although a large fraction of Lyα photons from LAEs (galaxies with *intrinsically* strong Lyα emission) may not reach observers due to absorption by neutral hydrogen in the IGM at the EoR (see Sect. 3.6 for details of reionization sources).

An interesting approach to estimate the cosmic SFR density has been proposed by Croft et al. [57]. It is based on the so-called intensity mapping technique, and consists in determining the power spectrum of diffuse Lyα emission from star-forming galaxies that are too faint to be detected individually, but numerous enough to yield a significant signal in a statistical sense. Croft et al. [57] have found that the cosmic SFR density at $z = 2 - 3.5$ estimated from diffuse Lyα emission is about 30 times higher than those by Ciardullo et al. [50] and comparable to (or higher than) the dust-extinction corrected total cosmic SFR density. Because some amount of Lyα emission should be absorbed by dust, this result may be overestimating the true cosmic SFR density. Although being a powerful important technique, intensity mapping requires a very careful evaluation of systematics. Recently, Croft et al. [58] have updated the analysis with the systematics removals, reducing the intensity measurement of the diffuse Lyα emission by a factor of 2. Croft et al. [58] find that there is no correlation between the diffuse Lyα emission and the Lyα forest, and show that the diffuse Lyα emission is not explained by faint star-forming galaxies, but fluorescence Lyα emission around QSOs in a scale up to $15h^{-1}$ Mpc.

3.3.3 Morphology

It has been known that LAEs are generally very compact since first revealed by HST in the mid 90s (Pascarelle et al. [254, 255]; Fig. 3.26). Deep HST images reveal small effective radii, r_e, in rest-frame UV and optical continua, ~ 1 kpc on average [194,

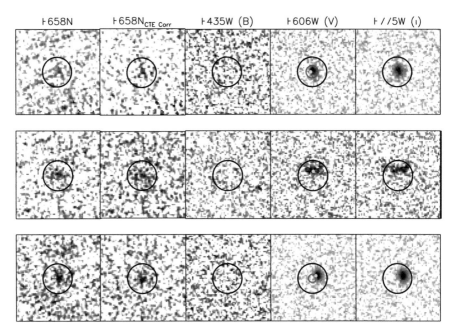

Fig. 3.26 HST images of three LAEs dubbed CHa-2, CH8-1, and CH8-2 at $z = 4.4$ [89]: from left to right, narrowband $F658N$, narrowband $F658N$ corrected for the charge-transfer effect, B_{435}, V_{606}, and i_{775}. The $F658N$ band includes the redshifted Lyα emission of the LAEs. The centers of the black circles indicate the centroids of the LAEs in the $F658N$ band. See Finkelstein et al. [89] for more details. ©AAS. Reproduced with permission

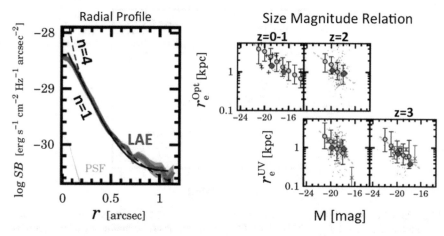

Fig. 3.27 Left: Radial profile of UV-continuum surface brightness obtained with the HST images [311]. The red shade represents the radial profile of the composite LAE with the 1σ uncertainty. The composite LAE consists of LAEs with $0.12 - 1L^*_{z=3}$ at $z = 0 - 7$, where $L^*_{z=3}$ is the characteristic luminosity of the Schechter function for the UV luminosity function at $z = 3$. The black solid and dashed lines indicate the Sérsic profiles with $n = 1$ and 4, respectively. The gray line denotes the radial profile of the point-spread function (PSF). Right: Size-magnitude (r_e-magnitude) relation [311]. The top two panels show the size-magnitude relations in the rest-frame optical wavelength, while the bottom two panels present those in the rest-frame UV wavelength. The red and cyan circles represent LAEs and continuum-selected star-forming galaxies, respectively, where the error bars indicate the 16th- and 84th-percentiles of the r_e distribution. The red dots indicate the size and magnitude measurements of individual LAEs. The cyan dotted lines are the power law functions best fit to the data points of the cyan circles. The gray symbols represent the measurements for LAEs obtained in the other studies. ©AAS. Reproduced with permission

256, 311]. The radial profiles of the rest-frame UV and optical continua typically show a disk morphology with a Sérsic index n of $n \simeq 1$, and follow the r_e-magnitude relation similar to the one of LBGs, indicating that faint continuum LAEs have a small size in r_e (Fig. 3.27; Paulino-Afonso et al. [256], Shibuya et al. [311]). Because a majority of LAEs have a faint continuum, the r_e-magnitude relation can explain the compact morphologies of LAEs.

Although previous HST studies claim no redshift evolution of r_e on average, a recent HST study [311] finds that the no-redshift evolution results may be produced by the sample selection bias. If a sample selection is not controlled, one can identify more LAEs with a faint continuum at low redshift. Because LAEs with a faint continuum have a r_e value smaller than LAEs with a bright continuum due to the r_e-magnitude relation, r_e measurements of low-redshift LAEs are typically small, diminishing the trend of the redshift evolution. Figure 3.28 shows the median r_e value as a function of redshift that is obtained with the controlled samples whose LAEs fall in the same continuum luminosity range [311]. The median r_e value monotonically decreases as $\sim(1 + z)^{-1}$ for a given continuum luminosity [311]. This evolutionary trend of LAEs is similar to the one of LBGs (e.g. Shibuya et al. [308]).

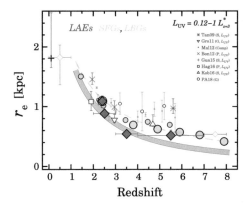

Fig. 3.28 Effective radius as a function of redshift for LAEs and LBGs with the same UV-continuum luminosity of $0.12 - 1\ L^*_{z=3}$ [311]. Here, $L^*_{z=3}$ is the characteristic luminosity of the Schechter function for the UV luminosity function at $z = 3$. The red filled diamonds (with the black circle) indicate LAEs in the rest-frame UV (optical) wavelength. The red open diamond at $z \sim 0.5$ ($z \sim 7.5$) represents the effective radius in the rest-frame optical (UV) wavelength estimated from the extrapolation of the size-magnitude relation (the small sample consisting of 3 LAEs). The small and large cyan circles denote the effective radii of galaxies selected by photo-z and Lyman break techniques, respectively. The red-broad and cyan-dashed curves show the power-law functions of $(1 + z)^{\beta}$ for $\beta \simeq -1$ that are best-fit to the data points of the red filled diamonds and the cyan filled circles, respectively. The gray and black symbols present the effective radii obtained by various studies whose magnitude limits are different. ©AAS. Reproduced with permission

The compact morphology of LAEs is not only found in continua, but also in Lyα emission by HST narrowband imaging studies [27, 89]. It should be noted that deeper narrowband-imaging and spectroscopic observations identify very diffuse extended ($\gtrsim 10$ kpc) Lyα halos around LAEs that are detailed in Sect. 3.4 [117, 258, 331, 364]. A combination of these observational studies indicates that the spatial structure of Lyα emission of LAEs is composed of a peaky Lyα core and a diffuse Lyα halo (see also Leclercq et al. [181]).

3.3.4 ISM Properties

Hydrogen in the ISM has three gas phases: H^+ ions, H atoms, and H_2 molecules. Corresponding to these three phases, the ISM is classified into three regions: HII regions (H^+), photodissociation regions (PDR; H), and molecular regions (H_2) whose gas temperatures are $\sim 10^4$, $10^2 - 10^3$, and $10^1 - 10^2$ K, respectively (Fig. 3.29).

In local galaxies, most of the ISM is in PDRs, where UV radiation from stars photodissociates molecules. However, PDRs have large spatial variations of temperature and density (Fig. 3.29), making it difficult to understand PDR properties by simple modeling. Moreover, there are many atomic and molecular transitions in PDRs as well as in molecular regions. In contrast, HII regions are moderately

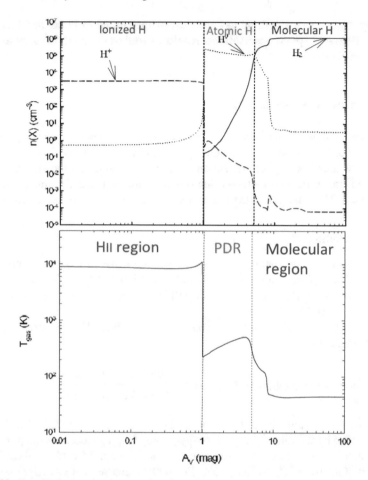

Fig. 3.29 Top: Hydrogen density as a function of visual extinction A_V that corresponds to the illuminated face (small A_V) to the shielded face (large A_V; Abel et al. [1]). The dashed, dotted, and solid curves represent the model predictions of the ionized, atomic, and molecular hydrogen densities, respectively. The two vertical lines indicate the hydrogen ionization front and the half-density molecular hydrogen point. Bottom: Predicted gas temperature as a function of A_V. ©AAS. Reproduced with permission

homogeneous media with a small number of ionization transitions, and thus can be modeled more simply than PDRs and molecular regions. It should also be noted that HII regions radiate emission lines falling in optical wavelengths where ground-based deep spectroscopy is possible. These emission lines enable us to constrain physical parameters of HII regions such as gas-phase metallicity, electron temperature T_e, ionization parameter q_{ion}, and electron density n_e.

Although it is difficult to characterize ISM properties of LAEs that are generally faint, recent LAE observations have constrained the gas-phase metallicity and ionization parameter in HII regions. Moreover, there are some useful observations to constrain parameters of atomic gas and dust mainly found in PDRs and molecular regions. Below I explain observational results as well as the methods used to probe the ISM properties.

3.3.4.1 Gas-Phase Metallicity

One of the most important ISM quantities that characterize galaxies is the gas-phase metallicity of HII regions. The metallicity of a galaxy is estimated from the ratio of appropriate lines using photoionization models. Depending on the strength of the lines used, there are two methods: the direct T_e method and the strong emission line method. Note that all of the line ratios discussed below are corrected for dust extinction.

Direct T_e Method

The direct T_e method mainly uses weak lines sensitive to electron temperature, OIII]1661,1666, [OIII]4363, [NII]5755, and [OII]7320,7330 etc. Because the most popular line among these is the auroral [OIII]4363 line, below I explain the direct T_e method with this line.

One can estimate T_e from [OIII]4363 and [OIII]4959,5007 line fluxes using the following equation:

$$(f_{\text{[OIII]}4959} + f_{\text{[OIII]}5007})/f_{\text{[OIII]}4363} = \frac{7.90 \exp(3.29 \times 10^4/T_e)}{1 + 4.5 \times 10^{-4}n_e/T_e^{1/2}}, \qquad (3.12)$$

with a small uncertainty depending on n_e (left panel of Fig. 3.30; [235]). The T_e value is determined by the ratio $(f_{\text{[OIII]}4959} + f_{\text{[OIII]}5007})/f_{\text{[OIII]}4363}$, because the [OIII]4363 ([OIII]4959,5007) flux increases (decreases) when the rate of collisional excitation (de-excitation) from $^1D_2$ to $^1S_0$ (from $^1D_2$ to ^3P) increases with T_e (right panel of Fig. 3.30).[6] Because metals are major coolants of the gas in HII regions, T_e is primarily determined by metallicity. Therefore, if T_e is estimated, one can reliably derive the metallicity based on photoionization models [145],

$$12 + \log \frac{O^+}{H^+} = \log \frac{f_{\text{[OII]}3727}}{f_{H\beta}} + 5.961 + \frac{1.676}{t_2}$$
$$-0.40 \log t_2 - 0.034 t_2 + \log(1 + 1.35x), \qquad (3.13)$$
$$12 + \log \frac{O^{2+}}{H^+} = \log \frac{f_{\text{[OIII]}4959} + f_{\text{[OIII]}5007}}{f_{H\beta}} + 6.200 + \frac{1.251}{t_3}$$
$$-0.55 \log t_3 - 0.014 t_3, \qquad (3.14)$$

---

[6]Note that electrons stay at $^1D_2$ significantly longer than ^3P.

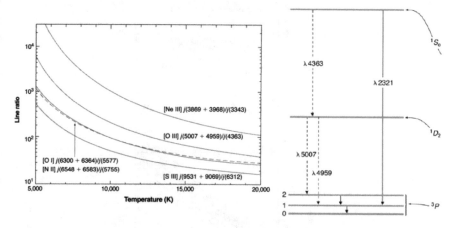

Fig. 3.30 Left: Line ratios as a function of electron temperature [235]. The second top curve indicates the line ratio $(f_{[OIII]4959} + f_{[OIII]5007})/f_{[OIII]4363}$, while the other curves show line ratios that are not discussed in the text. Right: Schematic diagram presenting the quantum states $^1S_0$, $^1D_2$, and ^3P of an O^{2+} ion and emission lines produced by transitions between two states [?]. The solid and dashed-line arrows represent permitted and forbidden lines, respectively. This figure is reproduced by permission of the University Science Books

where

$$t_2 = 10^{-4}T_e[\text{OII}], \tag{3.15}$$

$$t_3 = 10^{-4}T_e[\text{OIII}], \tag{3.16}$$

$$x = 10^{-4}n_e t_2^{-0.5}, \tag{3.17}$$

and O^+, O^{2+}, and H^+ are the abundances of singly-ionized oxygen, doubly-ionized oxygen, and ionized hydrogen, respectively; $T_e[\text{OII}]$ and $T_e[\text{OIII}]$ are the electron temperatures in O^+ and O^{2+} ion gas. For simplicity, one can assume the relation [39, 93]

$$t_2 = 0.7t_3 + 0.3 \tag{3.18}$$

that generally does not change the metallicity estimate. Because the last term of Eq. (3.13), $\log(1 + 1.35x)$, is negligibly small, this term can be practically omitted.

The oxygen abundance is calculated with

$$\frac{O}{H} = \frac{O^+}{H^+} + \frac{O^{2+}}{H^+}. \tag{3.19}$$

It should be noted that the contribution of O^{3+} and higher-order ionized oxygen is negligibly small, only $<1\%$, in HII regions heated by stars.

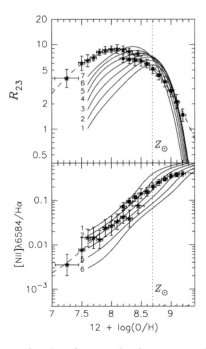

Fig. 3.31 $R23$ ($N2$) index as a function of oxygen abundance presented in the top (bottom) panel [221]. The star marks denote local galaxy averages obtained by the direct T_e method, and the dashed line represents an empirical fit to them. The solid lines show photoionization model calculation results with the normalized ionization parameter, $U \equiv q_{ion}/c$, of $\log U = -3.8, -3.5, -3.2, -2.9,$ $-2.6, -2.3,$ and -2.0 from bottom to top (top to bottom) in the top (bottom) panel. The dotted lines indicate solar metallicity. Reproduced with permission ©ESO

Strong Emission Line Method

The strong emission line method uses flux ratios of major emission lines of star-forming galaxies that include [OII]3727, Hβ4861, [OIII]5007, Hα6563, and [NII]6584. One of the most frequently used line ratios is the R_{23} index defined by:

$$R_{23} = \frac{f_{[OII]3727} + f_{[OIII]4959} + f_{[OIII]5007}}{f_{H\beta}}. \tag{3.20}$$

The top panel of Fig. 3.31 presents R_{23} as a function of oxygen abundance. The solid lines in this panel show R_{23}-oxygen abundance relations calculated by photoionization models with different ionization parameter (q_{ion}) values. Here, q_{ion} is defined by

$$q_{ion} = \frac{Q_{H^0}}{4\pi R_s^2 n_H}, \tag{3.21}$$

where Q_{H^0} is the number of hydrogen ionizing photons produced per unit time. R_s is the Strömgren radius, and n_H is the hydrogen density. Because R_{23} strongly depends on $q_{\rm ion}$ (see the top panel of Fig. 3.31), one needs to empirically calibrate the R_{23}-oxygen abundance relation with local galaxies that have R_{23} and oxygen abundance measurements from the direct T_e method. In the top panel of Fig. 3.31, the star marks with error bars denote the average values of the local galaxies, and the dashed line is an empirical relation that fits the star marks. In this way, a locally calibrated empirical relation is used to derive oxygen abundances from R_{23} measurements. However, it should be noted that the oxygen abundances of high-z galaxies estimated in this manner have systematic errors because ISM properties of high-z galaxies including $q_{\rm ion}$ are different from those of local galaxies [224].

The top panel of Fig. 3.31 shows a degeneracy in all R_{23}-oxygen abundance relations because there are basically two possible oxygen abundances for a given R_{23} measurement. This is because R_{23} increases with increasing metallicity up to $\sim 1 Z_\odot$, while at higher metallicities fine-structure cooling emission in far-infrared (FIR) wavelengths dominates and reduces the collisionally excited line fluxes of [OII] and [OIII] that are the numerators of R_{23}. To resolve this degeneracy, one can use the $N2$ index defined by:

$$N2 = \frac{f_{[\rm NII]6584}}{f_{\rm H\alpha}}. \tag{3.22}$$

The bottom panel of Fig. 3.31 displays photoionization model calculations (solid lines), local galaxy averages (star marks), and an empirical $N2$-oxygen abundance relation that fits the local galaxy averages (dashed line). Since the $N2$ index does not include oxygen line measurements i.e. only $f_{[\rm NII]6584}$ and $f_{\rm H\alpha}$, oxygen abundances from the $N2$ index have a systematic uncertainty due to a possible variation of the nitrogen-to-oxygen abundance ratio, implying that the $N2$ index alone is not a good estimator of oxygen abundance. However, one can obtain a coarse oxygen abundance estimate from an $N2$ index (Eq. 3.22) measurement that is useful to resolve the degeneracy of the R_{23}-oxygen abundance relation discussed above. Once the degeneracy is resolved by the $N2$ index, a single solution of oxygen abundance can be obtained from the R_{23} index.

Because the strong emission line method does not use weak lines such as the [OIII]4363 auroral line whose flux intensity is only $\sim 1/70$ of [OIII]5007 [145], it can efficiently estimate the metallicities of faint galaxies including LAEs. However, as described above, one should keep in mind that abundances based on the strong emission line method would be biased if the ISM properties of galaxies in question are different from those of local calibrators.

Metallicity Estimates of LAEs

Although it is difficult to determine the metallicity of LAEs due to their faintness, some constraints have been obtained for a small number of LAEs.[7] Finkelstein et al. [88] have obtained upper limits of $Z < 0.17 Z_\odot$ and $<0.28 Z_\odot$ for two LAEs with the spectroscopic $N2$ index (see also Guaita et al. [105]), while Nakajima et al. [222] have placed a lower limit of $Z > 0.09 Z_\odot$ for a stack of 105 LAE narrowband

---

[7]Here, a solar metallicity of $\log(Z/Z_\odot) = 12 + \log(\rm O/H) - 8.69$ is assumed [8].

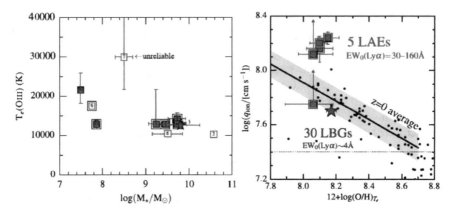

Fig. 3.32 Left: Electron temperature as a function of stellar mass [163]. The red filled/open squares marked with a black open square are one un-lensed LAE (ID 1) and five lensed LAEs (IDs 3, 4, 5, 6, and 11). Note that ID 6 does not have reliable line flux estimates. The other squares and the star mark indicate individually-identified LBGs and a stacked LBG, respectively. Right: Oxygen abundance and ionization parameter of LAEs and LBGs that are estimated by the reliable direct T_e method [163]. The red filled squares marked with a black square are LAEs (IDs 1, 3, 4, 5, and 11), where ID 6 with an unreliable result ($Z = 0.1Z_\odot$) is removed. The blue star mark indicates the stacked LBG. The black circles denote the results of local galaxy stacks, while the black line (the gray region) represents the local best-fit power law relation between oxygen abundance and ionization parameter (the typical scatter around the best-fit power law). This figure is reproduced by permission of the PASJ

images that cover [OII]3727, [NII]6584, and Hα lines based on a combination of two strong line methods, the $N2$ index and the [OII]/Hβ index [221]. Nakajima and Ouchi [224] have examined the metallicities of 6 LAEs at $z = 2 - 3$ with the $R23$ index and the ratio of [OIII]5007 to [OII]3727 fluxes, and found that they fall in the range of $12 + \log(O/H) = 7.98 - 8.81$, with an average of $Z \sim 0.5Z_\odot$. These metallicity constraints generally agree with the mass metallicity relation [88] and an extrapolation of the SFR-mass metallicity relation to low mass of star-forming galaxies at similar redshifts [222].

More recently, Kojima et al. [163] have measured the oxygen abundances of LAEs by the direct T_e method (left panel of Fig. 3.32), to show that one LAE (and five lensed LAEs) has (have) a metallicity (metallicity range) of $Z = 0.26^{+0.08}_{-0.07}Z_\odot$ ($Z = 0.1 - 0.3Z_\odot$). The right panel of Fig. 3.32 presents oxygen abundance and ionization parameter measurements for the one LAE and four lensed LAEs, after carefully removing one lensed LAE (ID 6) whose value, $Z = 0.1Z_\odot$, is based on unreliable flux estimates. These results by the direct T_e method and the strong line method suggest that the gas-phase metallicity of LAEs at $z \sim 2 - 3$ is typically $Z \simeq 0.1 - 0.5Z_\odot$ (Table 3.1). So far, no extremely metal poor LAEs with $Z \lesssim 0.01Z_\odot$ have been identified. However, because most of the LAEs with metallicity estimates are moderately bright, there remains a possibility that very metal poor LAEs are found in future studies targeting faint objects.

3.3.4.2 Ionization State

The ionization state of typical LAEs' HII regions is clearly different from those of other types of high-z galaxies. Recent optical-NIR spectroscopy has revealed that the O_{32} ratios of $z \sim 2-3$ LBGs and LAEs are significantly higher than those of local SDSS galaxies (left panel of Fig. 3.33; Nakajima et al. [223], Nakajima and Ouchi [224]), where the O_{32} ratio is defined by

$$O_{32} = \frac{f_{[\text{OIII}]5007}}{f_{[\text{OII}]3727}}. \tag{3.23}$$

Specifically, LAEs have extremely large values of $O_{32} \sim 10$, being about ~ 10–100 times higher than those of the local SDSS galaxies and even higher than those of the LBGs on average. In the left panel of Fig. 3.33, photoionization models with various metallicity and q_{ion} values are compared with LAEs. Although the models predict that O_{32} increases with decreasing metallicity, the high O_{32} values of LAEs cannot be explained by models that reproduce the local SDSS galaxies with $q_{\text{ion}} = 0.1 - 1 \times 10^8$ cm s^{-1}. The $z \sim 2-3$ LAEs are found to have $q_{\text{ion}} = 1 - 9 \times 10^8$ cm s^{-1}, about an order of magnitude larger than those of the local SDSS galaxies [224]. The left panel of Fig. 3.33 also shows that there exist local counterparts to LAEs, green pea galaxies (GPs; Cardamone et al. [36], Jaskot and Oey [148]), whose O_{32} and R_{23} values are comparable with those of LAEs.

The physical origin of the high q_{ion} values of LAEs is not well understood. The ionization parameter defined by Eq. (3.21) is rewritten as

$$q_{\text{ion}}^3 \propto Q_{\text{H}^0} n_{\text{H}} \varepsilon^2 \tag{3.24}$$

by the substitution of the Strömgren radius. Here, the Strömgren radius R_{s} is defined as

$$Q_{\text{H}^0} = \frac{4}{3} \pi R_{\text{s}}^3 n_{\text{H}}^2 \alpha_{\text{B}} \varepsilon \tag{3.25}$$

with the coefficient of the total hydrogen recombination to the $n > 1$ levels α_{B}, where ε is the volume filling factor of the Strömgren sphere. Equation (3.24) indicates that either Q_{H^0}, n_{H}, or ε needs to increase by a factor of 10^3, 10^3, or 30, respectively, to explain the high q_{ion} values of LAEs. With a moderately high SFR and metal-poor young stellar population, LAEs produce ionizing photons more efficiently than the local SDSS galaxies. However, it may not be possible that the Q_{H^0} of LAEs are $\sim 10^3$ times higher than those of the local SDSS galaxies. Some studies have reported an increase in the electron density from ~ 25 cm^{-3} ($z \sim 0$) to ~ 250 cm^{-3} ($z \sim 2$; Steidel et al. [332], Shimakawa et al. [312], Sanders et al. [293]), but these increase rates are not as high as 10^3 times. It is also unlikely that the average ε increases by a factor of 30 from $z \sim 0$ to 2. I discuss the issue of high q_{ion} at the end of this subsection.

Some other observations also suggest that LAEs have high q_{ion} values. LAEs with a large Lyα EW_0 tend to have high-ionization metal lines in rest-frame UV spectra.

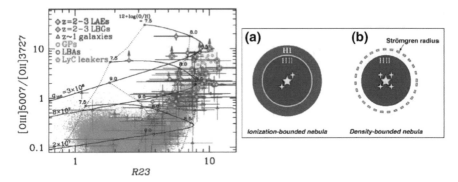

Fig. 3.33 Left: O_{32} ($= f_{[OIII]5007}/f_{[OII]3727}$) as a function of R_{23} [224]. The Red and blue diamonds indicate LAEs and LBGs, respectively, at $z = 2 - 3$. The green circles, magenta circles, and orange pentagons denote, respectively, green-pea galaxies, Lyman-break analogs, and Lyman continuum leaking galaxies in the local universe. The gray triangles and dots represent $z \sim 1$ galaxies and $z \sim 0$ SDSS galaxies, respectively. The three solid lines show photoionization model tracks with the ionization parameter of $q_{ion} = 3 \times 10^8$, 8×10^7, and 2×10^7 cm s^{-1} from top to bottom. The model tracks cover the oxygen abundance, from $12 + \log(O/H) = 7.5$ to 9.0 and beyond, with dotted lines connecting the same oxygen abundances. Right: Schematic illustrations of two types of HII regions: **a** an ionization bounded nebula and **b** a density-bounded nebula [224]. The blue and gray regions denote HII and HI gas regions, respectively. The star marks represent massive stars emitting ionizing photons. The cyan solid (dotted) lines indicate the (expected) positions of the Strömgren radius. Reproduced by permission of MNRAS

Stark et al. [325] have identified moderately strong CIII]1901,1909 lines in lensed LAEs at $z \sim 2$ by deep spectroscopy, and revealed a positive correlation between Lyα and CIII]1901,1909 EW_0. The left panel of Fig. 3.34 suggests that high-ionization lines CIII]1901,1909 are strong for large-Lyα EW_0 galaxies such as LAEs. Highly ionized gas containing C^{2+} is probably more abundant in LAEs compared to other types of galaxies. Subsequently, Stark et al. [326, 327] have reported the detections of moderately strong lines of CIII]1907,1909 lines in two LAEs at $z = 6 - 7$ and CIV1548 line in an LAE at $z = 7$, respectively. Although there still remains the possibility that the CIV1548 line is produced by a hidden AGN, not by young, massive stars (e.g. detection of NV1239 for the definitive AGN identification; Laporte et al. [178]), these spectroscopic results suggest that ionization state of LAEs at $z = 2 - 7$ is very high.

Most of the ALMA studies of LAEs have targeted the [CII]158 μm fine structure line that originates from low-ionization C^+ gas. Because C has a lower ionization potential than H, it is thought that the majority of [CII]158 μm photons are produced in PDRs that extend beyond HII regions. These ALMA studies have found that LAEs have significantly fainter [CII]158 μm luminosities $L_{[CII]}$ than local galaxies with similar SFRs (right panel of Fig. 3.34; Ouchi et al. [247], Ota et al. [238], Knudsen et al. [161], Pentericci et al. [264]; cf. Maiolino et al. [191]). In fact, Harikane et al. [113] report a significant anti-correlation between a [CII]-luminosity to SFR ratio ($L_{[CII]}/SFR$) and Lyα EW_0. The local galaxy relation in the right panel of

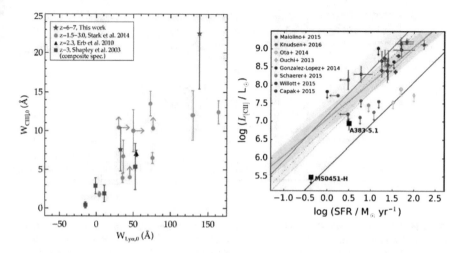

Fig. 3.34 Left: CⅢ]1909 rest-frame equivalent width as a function of Lyα rest-frame equivalent width [326]. The red stars, orange circles, black triangles, and blue squares represent galaxies including LAEs at $z = 6 - 7$, $1.5 - 3.0$, 2.3, and 3, respectively. Right: [CⅡ]158 μm luminosity as a function of SFR [161]. The cyan diamond/circles, purple circles, orange circles, green circles, and black squares indicate LAEs. The blue and red circles are continuum-bright LBGs, some of which show Lyα emission. This figure is reproduced by permission of MNRAS

Fig. 3.34 indicates that high SFR galaxies have bright [CⅡ]158 μm emission, because of high production rates of carbon ionizing photons from massive stars. In this sense, LAEs' faint [CⅡ]158 μm luminosities are puzzling, because they also have high SFRs. Although [CⅡ]158 μm emission is relatively weak for galaxies with an AGN, there is no hint of AGN in the LAEs observed by ALMA. Since [CⅡ]158 μm is a forbidden line, it can be weakened by collisional de-excitation in high density gas, but no hint of a high gas density has been obtained for LAEs. On the contrary, $z \sim 7$ LAEs show a hint of strong emission of the [OⅢ]4959,5007 forbidden line (left panel of Fig. 3.35; Roberts-Borsani et al. [288]). Moreover, recent ALMA studies have identified [OⅢ]88 μm fine-structure line emission in an LAE at $z = 7$ (right panel of Fig. 3.35; Inoue et al. [137]).

Regarding the ISM state of LAEs, the physical origins of two properties, high O_{32} ratios (i.e. high q_{ion}) and weak [CⅡ]158 μm emission remain open questions as detailed above. While there are no clear answers to them, it is suggested that these two properties can be consistently explained if the HⅡ regions of LAEs are density-bounded [224]. Figure 3.33 shows a conceptual diagram of a density-bounded nebula, compared with an ionization-bounded nebula that is the standard picture of HⅡ regions. In the standard picture, the size of an ionized nebula is determined by the number of ionizing photons, which corresponds to the radius of the Strömgren sphere (Eq. 3.25). On the other hand, the size of a density-bounded nebula is determined by the amount of atomic gas around the ionizing source. In contrast with a ionization-bounded nebula, a density-bounded nebula does not have an outer shell of ionized

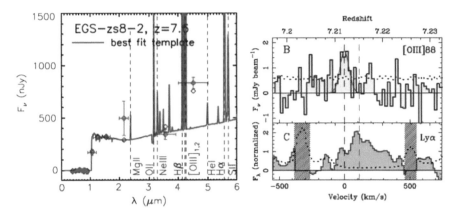

Fig. 3.35 Left: Observed SED of EGS-zs8-2 at $z = 7.48$ (red circles; Roberts-Borsani et al. [288]). The blue line denotes the best-fit SED model with nebular emission lines. The open diamonds are the expected photometry in the photometric bands for the best-fit SED. The vertical dashed lines mark the wavelengths of major nebular emission lines. The observed flux density at 4.5 μm shows an excess that is probably made by strong [OIII]4959,5007 and Hβ line emission. Right: [OIII]88 μm (top) and Lyα lines of SXDF-NB1006-2 [137]. The dotted line denotes the 1σ noise level. The blue and purple dashed lines represent the line-peak velocities of [OIII]88 μm and Lyα, respectively. The red curve in the top panel indicates the best-fit Gaussian function to the [OIII]88 μm line. The gray regions in the bottom panel show the wavelength range with large sky subtraction systematics. ©AAS and ©AAAS. Reproduced with permission

hydrogen gas emitting low-ionization lines such as [OII]3727, but an inner shell of ionized hydrogen gas producing high-ionization lines such as CIII]1907,1909, [OIII]5007, and [OIII]88 μm. Moreover, PDRs, major sources of [CII]158 μm emission, are not well developed. The density-bounded nebula scenario explains the high O_{32} ratio (i.e. high q_{ion}) and the weak [CII]158 μm emission. If this scenario applies to LAEs, ionizing photons escape easily from the ISM of LAEs. Such ionizing photons can be major sources of cosmic reionization (Nakajima and Ouchi [224], Jaskot and Oey [148]; Sect. 3.6.3.2). Although this scenario should be tested by theoretical models and more observations, it is interesting that the ISM state of LAEs may be important for the understanding of cosmic reionization.

3.3.4.3 Dust and Extinction

Stellar population analyses of LAEs suggest that the dust extinction of stellar continuum emission is as low as $E(B-V)_s \simeq 0 - 0.2$ on average under the assumption of Calzetti's extinction law (Calzetti et al. [37]; Sect. 3.3.1). One can also estimate the color excess of nebular lines, $E(B-V)_{neb}$, with the Balmer decrement. Note that $E(B-V)_{neb}$ is not necessarily the same as $E(B-V)_s$, because nebular lines originate from star-forming regions that are generally dustier than other regions in the galaxy. Calzetti et al. [37] claim that local starbursts have

Fig. 3.36 Top: β as a function of UV magnitude [323]. The red triangles denote LAEs with $EW_0 > 50\,\text{Å}$, while the purple circles indicate dropout galaxies at $z \sim 4$ with $EW < 50\,\text{Å}$. The green squares and blue squares represent dropout galaxies at $z \sim 4$. Bottom: Typical uncertainties in β measurements. ©AAS. Reproduced with permission

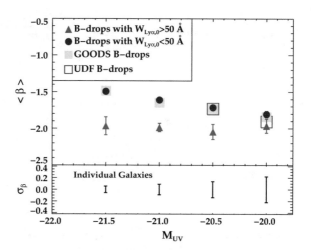

$E(B - V)_{\text{neb}} = E(B - V)_{\text{s}}/0.44$, although the relation between $E(B - V)_{\text{neb}}$ and $E(B - V)_{\text{s}}$ for high-z galaxies is poorly understood.

With a Balmer decrement measurement, Hα/Hβ, the dust extinction of nebular lines is estimated with

$$E(B - V)_{\text{neb}} = \frac{2.5}{k_{\text{H}\beta} - k_{\text{H}\alpha}} \log\left(\frac{\text{H}\alpha/\text{H}\beta}{2.86}\right), \tag{3.26}$$

where $k_{\text{H}\alpha}$ and $k_{\text{H}\beta}$ are coefficients depending on the dust extinction law. The Calzetti extinction law gives $k_{\text{H}\alpha} = 3.325$ and $k_{\text{H}\beta} = 4.598$ [154, 214]. Here, Hα/Hβ = 2.86 is an intrinsic (i.e., dust-free) line ratio at $T_{\text{e}} = 10^4$ K and $n_{\text{e}} = 10^2\,\text{cm}^{-3}$ for the case B recombination [236]. Since the Hβ fluxes of LAEs are generally too faint to detect (e.g. Guaita et al. [105]), only a small number of LAEs have $E(B - V)_{\text{neb}}$ measurements; they fall in the range of $E(B - V)_{\text{neb}} = 0 - 0.2$ [163]. So far, there are no studies of $E(B - V)_{\text{neb}}$ statistics nor of the relation between $E(B - V)_{\text{neb}}$ and $E(B - V)_{\text{s}}$ for LAEs (cf. Erb et al. [77]). A statistical study addressing these issues of $E(B - V)_{\text{neb}}$ should be conducted for LAEs in the near future.

The dust extinction of high-z galaxies is characterized with the UV-continuum slope, β, defined by

$$f_\lambda = \lambda^\beta, \tag{3.27}$$

where f_λ is the UV-continuum spectrum of the galaxy in the wavelength range $\simeq 1300 - 3000\,\text{Å}$ [38]. The β value is used as an indicator of the amount of dust extinction for moderately young star-forming galaxies such as LBGs and LAEs whose intrinsic UV-continuum slope is $\beta \simeq -2.2$.

Figure 3.36 presents the average β values of LAEs with high Lyα equivalent widths $EW_0 > 50\,\text{Å}$, and compares them with those of LBGs. The LAEs have $\beta \simeq -2$ that is significantly smaller than those of the LBGs at the same UV luminosity, suggesting

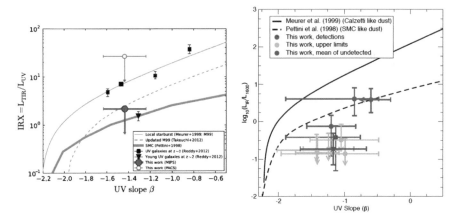

Fig. 3.37 IRX as a function of the UV slope β for LAEs at $z \sim 2$ (left; Kusakabe et al. [174]) and at $z \sim 5 - 6$ (right; Capak et al. [42]). Left: The red filled and open circles indicate, respectively, the upper limits of IRX estimated by the stack of Spitzer/MIPS and PACS images centered at the sky positions of 213 LAEs at $z = 2.2$. The gray thin-solid, dashed, and thick-solid lines denote, respectively, the dust extinction relations of Calzetti, Takeuchi [336], and the SMC. The squares (inverse-triangle) represent UV-continuum bright (and young) galaxies at $z \sim 2$. Right: The red, orange, and blue circles show the detections, the upper limits, and the average of the no-detection data for galaxies (including 3 LAEs) at $z \sim 5 - 6$. The solid and dashed lines are the IRX-β relations of Calzetti and the SMC. ©AAS and ©Nature. Reproduced with permission

that LAEs are generally dust poor. Figure 3.36 also indicates that UV-continuum faint LAEs and LBGs with $M_{UV} \sim -20$ have similar β values, and probably similar extinction properties, supporting the idea that a high fraction of faint LBGs are LAEs (Sect. 3.3.2).

To evaluate the dust extinction law of a galaxy, one can use the IRX ratio,

$$IRX = \frac{L_{IR}}{L_{UV}}, \tag{3.28}$$

where L_{IR} and L_{UV} are the total infrared (IR; $3 - 1000\,\mu m$) and UV ($\sim 1500\,\text{Å}$) luminosities, respectively. The values of L_{IR} can be estimated from, e.g., Spitzer/MIPS, Herschel/SPIRE, APEX/LABOCA, and ALMA photometry [42, 174, 358]. Figure 3.37 presents the IRX-β relation for LAEs and LBGs at $z \sim 2$ and $5 - 6$, together with the model curves of Calzetti and SMC dust extinction. It is clear that LAEs have low IRX values at a given β on average. The left panel of Fig. 3.37 indicates that LAEs at $z \sim 2$ have an extinction curve similar to that of the SMC and different from those of Calzetti's local starbursts. There are three LAEs at $z = 5 - 6$ with IRX-β measurements shown in the right panel of Fig. 3.37. This panel suggests that these three LAEs have IRX values which fall close to or even below the SMC curve, although these extremely low IRX estimates are still under debate. However, there is a consensus based on deep ALMA observations that LAEs at $z > 5$ have faint ~ 1 mm flux densities [42, 161, 191, 238, 247].

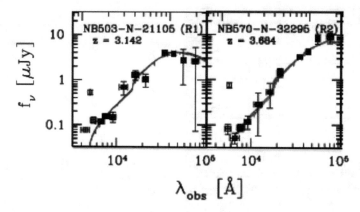

Fig. 3.38 SEDs of two red LAEs, R1 and R2, at $z = 3.142$ and 3.684, respectively [232]. The squares denote photometric data obtained by deep optical and near-infrared observations. The red and blue curves represent the best-fit SED models of exponentially-decaying and constant star-formation histories, respectively; data shown by open squares are not used for the fitting because they are contaminated by either strong Lyα emission or the IGM Lyα forest absorption. This figure is reproduced by permission of MNRAS

On average, LAEs have low extinction and low dust masses. However, there exists a rare population of dusty LAEs with red stellar SEDs and bright submm luminosities. Figure 3.38 shows the SEDs of two spectroscopically-confirmed LAEs (dubbed R1 and R2) at $z = 3 - 4$ which have red SEDs and strong Lyα emission [232]. It should be noted that some SMGs have strong Lyα emission that can be used for redshift determination [41, 46]. How Lyα photons can escape from those dusty starbursts without significant extinction is an open question. Dusty LAEs might have dust-poor star-forming regions that are spatially separated from usual dust-rich star-forming regions.

3.3.5   Outflow and Lyα Profile

Using[8] deep optical and near-infrared spectra, many researchers have investigated the velocities of the Lyα line, the low-ionization UV metal absorption lines, and the nebular emission lines in LAEs. The average outflow velocity V_{out} of LAEs at $z \sim 2$ is estimated to be $V_{out} \simeq 200 \, \mathrm{km \, s^{-1}}$ with low-ionization UV metal absorption lines blueshifted from the systemic velocity (left panel of Fig. 3.39; Hashimoto et al. [114], Shibuya et al. [307]). Here, the systemic velocity is determined by strong nebular lines such as Hα that originate from HII regions. Blueshifted absorption lines are thought to form in the outflowing gas. Another important feature in line velocities

---

[8]In the Saas Fee lectures, the topics of this subsection were originally included in Sect. 3.4.2. For the readers' convenience, I have moved these topics here.

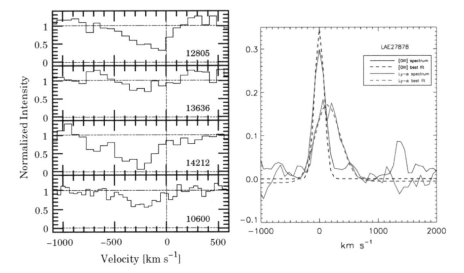

Fig. 3.39 Left: Average UV low ionization absorption lines for four LAEs [307] obtained by stacking the Si II1260, CII1334, and Si II1526 lines. The systemic velocity is determined with nebular emission lines. Right: Line profiles of Lyα (red solid line) and [OIII]5007 (black solid line) observed by McLinden et al. [209]. The red and black dotted lines represent the best-fit model profiles for the Lyα and [OIII] lines, respectively. ©AAS. Reproduced with permission

Fig. 3.40 Lyα offset velocity $\Delta V_{\mathrm{Ly}\alpha}$ for LAEs (red histograms) and LBGs (blue histograms) [307]. ©AAS. Reproduced with permission

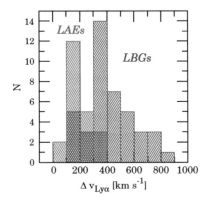

is that the Lyα line peak is generally redshifted from the systemic velocity (right panel of Fig. 3.39). The Lyα line offset $\Delta V_{\mathrm{Ly}\alpha}$ is defined as the offset velocity of the Lyα line peak with respect to the systemic velocity. The average Lyα line offset of $z \sim 2 - 3$ LAEs is $\Delta V_{\mathrm{Ly}\alpha} \simeq 200\,\mathrm{km\,s}^{-1}$ [76, 114, 209, 307]. This average Lyα offset velocity is comparable with the average outflow velocity, $\Delta V_{\mathrm{Ly}\alpha} \simeq V_{\mathrm{out}}$.

Interestingly, typical LBGs ($L_{\mathrm{UV}} \gtrsim L*$) at $z \sim 2 - 3$ have $V_{\mathrm{out}} \simeq 200\,\mathrm{km\,s}^{-1}$ on average [265, 330], being comparable with that of LAEs. However, the average Lyα offset velocity of LBGs is $\Delta V_{\mathrm{Ly}\alpha} \simeq 400\,\mathrm{km\,s}^{-1}$, about twice as large as LAEs' $\Delta V_{\mathrm{Ly}\alpha}$ (Fig. 3.40). Note that there is a negative correlation between $\Delta V_{\mathrm{Ly}\alpha}$ and Lyα EW_0

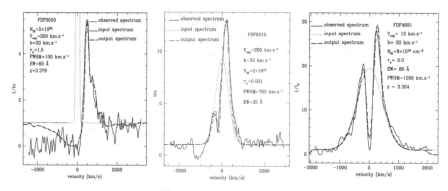

Fig. 3.41 Lyα profiles obtained by observations (black lines) and ES modeling (blue and red lines) for three LAEs [353]. The blue and red lines represent the best-fit spectra and the intrinsic spectra predicted by the ES models, respectively. Reproduced with permission ©ESO

(e.g. Fig. 7 of Hashimoto et al. [114]). In contrast with LAEs, LBGs show $\Delta V_{\mathrm{Ly}\alpha} \simeq 2V_{\mathrm{out}}$. The $\Delta V_{\mathrm{Ly}\alpha} - V_{\mathrm{out}}$ relation is key to understanding the physical differences in LAEs and LBGs via theoretical modeling as discussed below.

Detailed Lyα profiles of LAEs at $z \sim 2 - 3$ are investigated by medium-high resolution spectroscopy. Such spectroscopic efforts have revealed that LAEs have a variety of Lyα profiles [115, 339, 369]. Among those, three typical profiles are a single asymmetric/symmetric line, an asymmetric line with a weak blue peak, and a double-peak line, as presented in Fig. 3.41.

Lyα profiles depend on physical parameters relating to HI resonance scattering of Lyα, such as the HI density, gas dynamics (including outflows), and dust extinction, and are quantitatively investigated by modeling. One of the most popular models for Lyα profiles is the expanding shell (ES) model [4, 352]. This model assumes a galaxy-scale spherical shell of outflowing gas around the Lyα source that is described with four parameters: the HI column density N_{HI}, the expansion (outflow) velocity corresponding to V_{out}, the doppler (thermal) velocity of gas in the shell b, and the optical depth of dust extinction τ_a. The assumption that LAEs have an ES is supported by the fact that nearby starbursts have a galaxy-scale supershell made by multiple SNe in star-forming regions [169, 196, 197].

Figure 3.42 illustrates the ES model and predicted profiles of Lyα emission escaping to the observer. The physical origins of the individual Lyα profiles are explained below. The light path "3" produces the profile "3", where Lyα photons travel straight to the observer. Note, however, that the profile is slightly redshifted because the blue side of the Lyα emission is efficiently scattered off by the HI gas of the ES. The light path "1b" is back-scattered once by the ES, providing a strong, redshifted peak in the predicted profile. The velocity of the peak, $\sim 2V_{\mathrm{out}}$, is accomplished by two effects: (i) Lyα photons are scattered by the gas receding with V_{out} and hence their wavelengths are redshifted by V_{out} as seen from the gas, and (ii) the gas is receding from the observer by V_{out}. The light path "1c" indicates multiple scattering of Lyα

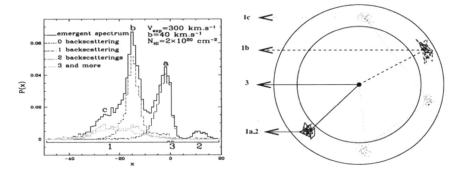

Fig. 3.42 Left: Lyα profile predicted by the ES model [352]. Here, the abscissa axis is in units of the normalized velocity of the shell defined by $x = -2V_{out}/b$ (see text). A negative x value indicates a wavelength longer than the systemic velocity ($x = 0$). The black line shows the total Lyα line profile that can be observed. The blue, red, green, and cyan lines indicate the spectral components corresponding to the number of scattering events of none, once, twice, and 3 times. The labels 1, 3, and 2 below the spectra represent the regimes of long, systemic, and short wavelengths. Right: Illustration of the ES model [352]. The central dot indicates the Lyα source position, while the annulus made of the two black circles represents the HI gas shell. The solid, dashed, and dotted line arrows show examples of Lyα photon light paths (towards the observer) that produce the spectral components whose labels correspond to those in the left panel. Reproduced with permission ©ESO

photons that gives the highly redshifted Lyα profile, but its contribution to the total flux is small in the reasonable range of HI column density.

Back-scattered light dominates the total Lyα flux when the HI column density is higher than $N_{HII} \sim 10^{20}$ cm^{-2}. Therefore, the velocity offset of the total flux, $\Delta V_{Ly\alpha}$, changes with N_{HII} from $\Delta V_{Ly\alpha} \sim 0$ to $\sim 2V_{out}$. The value of $\Delta V_{Ly\alpha} \sim 0$ is found in low N_{HI} where the majority of Lyα photons take the path "3", while $\Delta V_{Ly\alpha}$ has $\sim 2V_{out}$ for $N_{HI} \gtrsim 10^{20}$ cm^{-2}.

As demonstrated in Fig. 3.41, the best-fit ES models reproduce the variety of Lyα profiles with the only four physical parameters.

The ES models also explain the $\Delta V_{Ly\alpha}$-V_{out} relations of LAEs and LBGs. Because LAEs have the relation of $\Delta V_{Ly\alpha} \simeq V_{out}$, the ES models suggest that their HI column density is low, $N_{HII} \lesssim 10^{20}$ cm^{-2}, which produces weak back-scattered Lyα emission [114, 115, 307]. In contrast, LBGs have the relation of $\Delta V_{Ly\alpha} \simeq 2V_{out}$. The ES models indicate that back-scattered Lyα emission dominates in LBGs, and that their HI column density is $N_{HII} \gtrsim 10^{20}$ cm^{-2} on average that is higher than those of LAEs.

The low HI column densities of LAEs may explain the large increase in the average Lyα escape fraction from $z \sim 0$ to 6 shown in Fig. 3.24, because the fraction of LAEs in the entire galaxy population increases with redshift (Fig. 3.23). There are six possible mechanisms that control the Lyα escape fraction: (1) IGM absorption, (2) stellar population, (3) outflow velocity, (4) gas-cloud clumpiness (Neufeld effect), (5) simple dust extinction, and (6) HI gas resonance scattering in the ISM with dust. Because the IGM absorption is stronger at higher z, mechanism (1) cannot explain the increase in Lyα escape fraction towards high z. The stellar population and outflow velocity of the mechanisms (2) and (3) evolve little for LAEs in the range $z \sim 3 - 6$

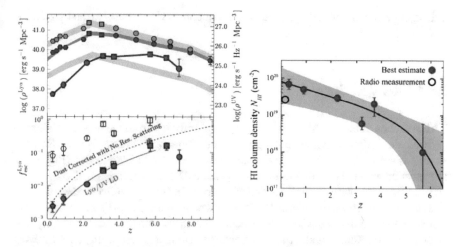

Fig. 3.43 Top left: Evolution of the Lyα and UV luminosity densities whose absolute values are presented in the labels in the left and right ordinate axes, respectively. The red symbols and line indicate the Lyα luminosity density. The blue and orange symbols/shades represent the observed and dust-corrected UV luminosity densities, respectively. For comparison, the gray shade denotes the dust-corrected UV luminosity density scaled to the position of the Lyα luminosity density at $z \sim 3$. Bottom left: Evolution of the Lyα escape fraction derived from the observed Lyα and dust-corrected luminosity densities (red filled symbols). The best-fit function to the Lyα escape fraction evolution is shown with the magenta and the black-dashed lines obtained by Konno et al. [166] and Hayes et al. [118], respectively. The blue symbols represent the Lyα escape fraction corrected for dust extinction under the simple assumption of no Lyα resonance scattering. Right: Evolution of the average HI column density of galaxies inferred from the Lyα escape fraction and the ES models (red circles). The black solid line and gray shade are the best-fit function with the outflow velocity of $150 \, \mathrm{km \, s^{-1}}$ and the uncertainty raised by the different assumptions of the outflow velocity ranging from 50 to $200 \, \mathrm{km \, s^{-1}}$. The black open circle shows the average HI column density obtained by radio observations. All of these plots are taken from Konno et al. [166]. ©AAS. Reproduced with permission

(Sects. 3.3.1 and 3.3.5), which are not large enough to explain the evolution of two orders of magnitude of the Lyα escape fraction in the range $z \sim 0 - 6$ (Fig. 3.24). The mechanism (4) has been ruled out by recent theoretical studies (Sect. 3.4.2; Laursen et al. [179], Duval et al. [73]).

The top left panel of Fig. 3.43 presents the evolution of the Lyα escape fraction corrected for dust extinction with a simple screen model (Eq. 3.34). This simple dust extinction evolution of the mechanism (5) is found to predict only one order of magnitude evolution of the Lyα escape fraction. Instead, the large evolution of the Lyα escape fraction can be probably explained by the mechanism (6). The mechanism (6) involves HI gas evolution with the effect of dust extinction in the ISM via Lyα resonance scattering. To reproduce the large Lyα escape fraction evolution based on the ES models, it is suggested that the HI column density should decrease by 1–2 orders of magnitude (Fig. 3.43) from $z \sim 0$ to 6 that reduces the effect of selective Lyα absorption for LAEs towards $z \sim 6$. In this way, the HI column density evolution is a major parameter of LAEs that determines not only $\Delta V_{\mathrm{Ly}\alpha}$, but also the Ly$\alpha$ escape fraction.

3.3.6 AGN Activity

LAE searches often find AGNs with strong Lyα emission from the presence of broad ($\gtrsim 500\,\mathrm{km\,s^{-1}}$) emission lines and high ionization lines such as C IV 1548 and N V 1240 as well as strong X-ray, far-UV, radio, and short-wavelength IR emission (Fig. 3.44). Diagnostics with nebular line ratios such as the BPT [20] diagram are also used [88, 105, 223]. AGNs with strong Lyα emission are referred to as AGN-LAEs.

Early studies claim that about 1% of narrowband-selected LAEs at $z \sim 2 - 3$ are AGN-LAEs, and that the AGN-LAE fraction increases with the Lyα luminosity [94, 243]. Recently, Konno et al. [166] have obtained statistical results on AGN-LAEs at $z = 2$ based on a narrowband survey (see also Matthee et al. [199]). The top panel of Fig. 3.45 presents the Lyα luminosity function of LAEs that clearly shows an excess over the best-fit Schechter function at $\log L_{\mathrm{Ly}\alpha} \gtrsim 43.4\,\mathrm{erg\,s^{-1}}$. Almost all objects in this luminosity range are bright in either X-ray, far-UV, or radio, thus being classified as AGN-LAEs. The bottom panel of Fig. 3.45 is the AGN UV luminosity function derived with these AGN-LAEs, where the moderately large error bars include the systematic uncertainty raised by the incompleteness of AGN missed from the LAE selection because of weak or no Lyα emission. The AGN UV luminosity function estimated from the AGN-LAE sample agrees well with the faint-end UV luminosity function given by the SDSS study. In summary, the bright end of the Lyα luminosity

Fig. 3.44 Example spectra of AGN-LAEs [243]. ©AAS. Reproduced with permission

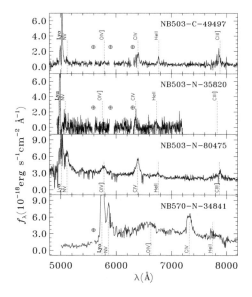

function ($\log L_{Ly\alpha} \gtrsim 43.4 \, \text{erg} \, \text{s}^{-1}$) is dominated by AGN-LAEs, and AGN-LAEs may be used to estimate the AGN UV luminosity function at the faint end.

3.3.7 Overdensity and Large-Scale Structure

As explained in Sect. 3.1.1.3, early LAE searches targeted fields centered on an AGN, especially a massive radio galaxy, that is thought to be a signpost of a high-z galaxy overdensity. I present one of the LAE overdensity examples in Fig. 3.46 that shows narrowband selected LAEs around a luminous radio galaxy, TN J1338 1942, at $z \sim 4.1$ [351]. The number density of LAEs in this overdense region is about 15 times higher than the one in the average blank field. The number-density peak is clearly found at the redshift of TN J1338 1942, while the narrowband is capable to detect LAEs in a moderately broad redshift range (bottom panel of Fig. 3.46). Such an overdensity of high-z galaxies is often refereed to as a 'proto-cluster' [328]. There are about 30 overdensities of high-z galaxies in the range $z \sim 2 - 8$ reported to date (see Table 5 of Chiang et al. [47]). About a half of them are LAE overdensities like the one around TN J1338 1942.

Some systematic narrowband surveys of LAEs have covered a contiguous field with a size of > 100 comoving Mpc, and discovered filamentary LSSs in a flanking field of a 'proto-cluster' at $z \simeq 3$ (Fig. 3.47; Yamada et al. [368]) and in a blank field at $z \simeq 6$ [241]. It is interesting that narrowband surveys can efficiently map out a high-z galaxy distribution in a large scale. Specifically, a mapping observation in an unbiased blank field allows to obtain average features of LAE clustering that can be used to constrain properties of the dark-matter halos hosting LAEs, with structure formation models (Sect. 3.3.8).

3.3.8 Clustering

Properties of galaxy-hosting DM halos are key to understanding galaxy formation (Fig. 3.10; Sect. 3.2.1). Invisible DM halos are not directly observed, but there are various indirect methods to estimate their masses, M_h. One promising approach is to use gravitational lensing that provides reliable estimates of M_h. Other approaches include abundance matching and clustering.[9] The abundance matching and clustering methods statistically estimate an average M_h for a given galaxy population, by comparing the abundance and the clustering amplitude, respectively, between galaxies and DM halos. The abundance and the clustering amplitude are chosen as an indicator of M_h because they are reliably calculated as a function of M_h using the standard

[9]There are a number of methods to characterize hosting DM halos: e.g., satellite kinematics, X-ray luminosities, and Sunyaev–Zel'dovich effects. Here, I highlight only lensing, abundance matching, and clustering that require optical and NIR imaging data alone.

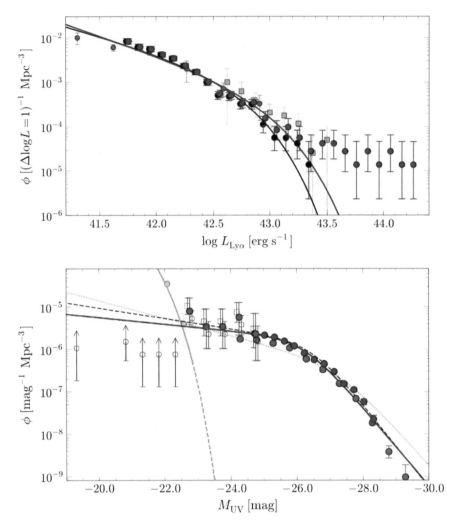

Fig. 3.45 Top: Lyα luminosity function at $z = 2$. The blue and magenta (orange) circles (squares) represent Lyα luminosity functions derived by three different studies. The black circles are the same as the blue circles, but based on LAEs with no AGN signature, i.e., with neither strong X-ray, far-UV, nor radio emission. The blue and black curves are the best-fit Schechter functions with the blue and black circle data points, respectively. Bottom: UV luminosity function of AGN at $z = 2$. The red circles indicate the AGN UV luminosity function estimated with the LAE-AGN whose errors include the systematic uncertainty raised by the incompleteness of AGN missed from the LAE selection because of weak or no Lyα emission. The blue and green circles represent the AGN UV luminosity functions derived with the SDSS and 2dF-SDSS LRG data, respectively. The red curve shows the best-fit Schechter function. The cyan circles and curve denote the UV luminosity function of LBGs. These two plots are taken from Konno et al. [166]. ©AAS. Reproduced with permission

ΛCDM structure formation model (abundance: Fig. 3.12; clustering: Fig. 3.48) at any redshifts (Sect. 3.2.1.1). The basic idea of the abundance matching (clustering) method is to identify the ensemble of model dark-matter halos whose number density (clustering amplitude) is equal to that of observed galaxies. Modern abundance matching methods take account of the contribution of sub-halos to explain satellite galaxies as well as star-formation and merger histories, and increase the reliability of M_h estimates (e.g. Behroozi et al. [22]). Similarly, for the clustering method, it is popular to use clustering predictions of the halo occupation distribution (HOD) of central and satellite galaxies (e.g. Zheng et al. [373]). Figure 3.49 compares M_h of local galaxies estimated by various techniques, and indicates that the results of the three techniques, lensing, clustering, and abundance matching, agree very well in the galaxy DM halo mass scale up to $M_h \sim 10^{13} M_\odot$. Table 3.2 summarizes the three techniques.

Fig. 3.46 Overdensity of LAEs around the radio galaxy TN J1338 1942. Top: VLT narrowband image centered at TN J1338 1942. The green box and the blue circles indicate the positions of TN J1338 1942 and LAEs, respectively. The image is taken from https://www.eso.org/public/news/eso0212/. Bottom: Histogram of velocity differences Δ_v between LAEs and TN J1338 1942 [351]. Reproduced with permission ©ESO

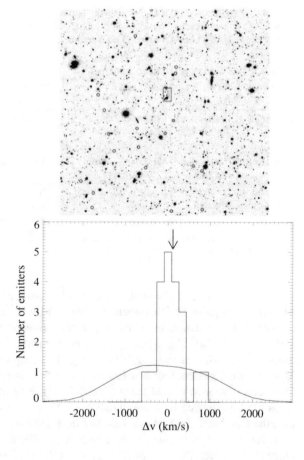

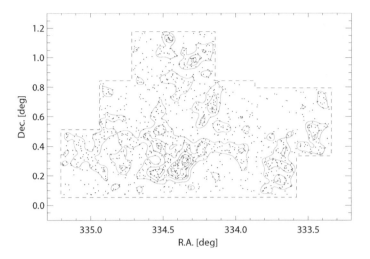

Fig. 3.47 Large-scale (>100 Mpc) sky distribution of LAEs in and around the SSA22 overdensity of galaxies at $z = 3.1$ [368]. The dots indicate LAEs, while the green contours represent the surface density of LAEs. ©AAS. Reproduced with permission

Table 3.2 DM-halo mass estimate techniques

Technique	Key quantity	Advantage	Disadvantage (Requirement)	Redshift range[a]
Lensing	Background object shear	Moderately simple gravity model	Large galaxy sample of high spatial resolution imaging data	0–1
Clustering	Correlation function	Virtually free from duty-cycle systematics	Large galaxy sample	0–7
Abundance matching	Luminosity function	Small galaxy sample	Many parameters constrained by star-formation/merger histories	0–8

[a]Redshift range that is covered by observations, to date

Figure 3.50 presents the stellar to DM-halo mass ratio (SHMR) as a function of M_h for $z \sim 0$ galaxies. This result is obtained by the combination of these three techniques that enhances the reliability of M_h estimates. In Fig. 3.50, the SHMR has a peak at $M_h \sim 10^{12} M_{\odot}$, meaning that stars are most efficiently formed in DM halos whose present-day mass is $M_h \sim 10^{12} M_{\odot}$. The shape of the SHMR plot reflects M_h-dependent gas cooling and feedback and thus is essential to understand star-formation processes in DM halos (Sect. 3.2.1.2). Properties of galaxy-hosting DM halos are investigated by a combination of these three techniques up to $z \sim 1$ in, e.g., the COSMOS field where wide and deep HST data are available [180]. However, for more distant galaxies at $z \gtrsim 2$, it is difficult to apply the lensing technique that requires shear measurements of a large number of background objects in

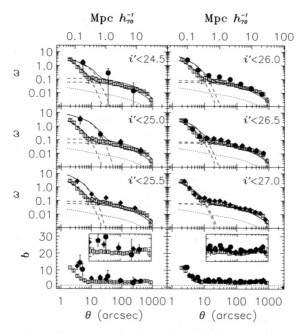

Fig. 3.48 Angular correlation function (top three panels) and clustering bias (bottom panel) as a function of angular distance for $z \sim 4$ LBGs [242]. The open squares denote the angular correlation function of objects with $i' < 27.5$, while the filled circles are those for different magnitude limits indicated by labels. The solid lines are the best-fit HOD models, while the dashed lines indicate their breakdowns into one-halo and two-halo terms that are the correlations of galaxies in single halos and two different halos, respectively. The dotted lines represent the DM angular correlation function (bias is unity, i.e. $b_g = 1$). ©AAS. Reproduced with permission

high-sensitivity and high-spatial resolution images. In contrast, the clustering and abundance matching techniques can still be used at $z \gtrsim 2$. Figure 3.51 presents the evolution of the SHMR obtained with HST and Subaru data by these two techniques. Although the DM-halo mass range is limited, a clear evolution of the SHMR is identified at $M_h \sim 10^{11} M_\odot$. Beyond $z \sim 7$, there are no clustering measurements obtained to date, because clustering analyses require a large number of galaxies. Only requiring the number density of galaxies, the abundance matching technique has been applied up to $z \sim 8$ to date [22]. Figure 3.51 compares DM-halo masses estimated by the abundance matching alone to those from the clustering+abundance matching, and suggests that the abundance matching provides good estimates of M_h at $z \gtrsim 4$ within an uncertainty of a factor of ~ 3 [111].

DM-halo masses of LAEs have not been estimated by either lensing or abundance matching. Lensing analyses cannot be performed for high-z LAEs at $z \gtrsim 2$ due to the limited quality and amount of imaging data. Moreover, abundance matching does not work because LAEs have a very small duty cycle of strong Lyα emission and

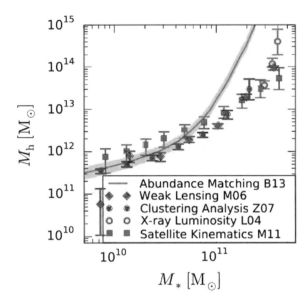

Fig. 3.49 DM-halo masses estimated by five methods plotted as a function of stellar mass [110]. The red diamonds represent estimates by the reliable lensing technique. The blue circles, green open circles, and gray squares denote DM-halo masses estimated from clustering, X-ray luminosity, and satellite kinematics. The green line with a green shade indicates the DM-halo mass estimated by the abundance matching technique. This figure is reproduced by permission of the University of Tokyo

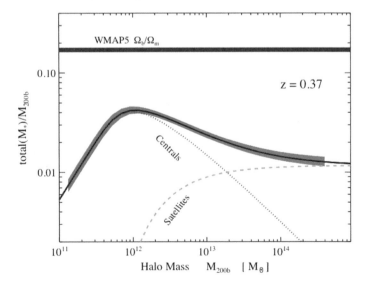

Fig. 3.50 SHMR as a function of DM-halo mass at $z \sim 0$ [180]. The black line and the gray shade represent the total SHMR and its uncertainty, while the red-dotted and orange-dashed lines indicate the SHMRs contributed by central and satellite galaxies, respectively. The green horizontal line denotes the cosmic baryon mass to total mass ratio. ©AAS. Reproduced with permission

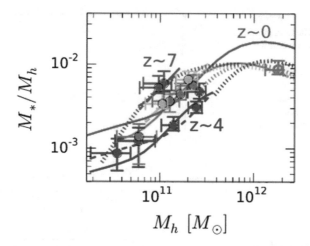

Fig. 3.51 Redshift evolution of the SHMR over $z = 0 - 7$ (Harikane et al. [111]; see also Harikane et al. [112]). The red, orange, green, and blue circles and solid lines indicate SHMRs at $z \sim 7$, 6, 5, and 4, respectively, derived from a combination of the clustering and abundance matching techniques. The dotted lines are the same as the solid lines, but for SHMRs obtained by the abundance matching technique alone. The gray solid line denotes the SHMR at $z \sim 0$. ©AAS. Reproduced with permission

hence have a significantly smaller abundance than DM halos (Sect. 3.3.9). Thus, only the clustering method has been used for LAEs.

The clustering amplitude of a given galaxy sample can be evaluated with the angular correlation function (ACF), $\omega_{obs}(\theta)$, defined as the excess probability of finding galaxies in two solid angles $d\Omega_1$ and $d\Omega_2$ separated by the angular distance θ,

$$dP = N^2[1 + \omega_{obs}(\theta)]d\Omega_1 d\Omega_2, \tag{3.29}$$

where dP is the probability finding galaxies and N the mean galaxy density per steradian [103]. Large-area surveys have derived ACFs of LAEs with the Landy and Szalay [177] estimator [94, 170, 239]. As an example, Fig. 3.52 presents ACF measurements for $z = 3.1$ LAEs. Although the statistics is not very good due to moderately small sample sizes, it is well known that the clustering of LAEs is weak, $b_g \sim 2$, at $z \sim 2 - 3$. Here, b_g is the large scale galaxy bias defined by

$$b_g^2 = \omega_{obs}/\omega_{DM}, \tag{3.30}$$

where ω_{DM} is the dark-matter ACF predicted by the structure formation model (e.g. Peacock and Dodds [260]). ACFs and b_g have been derived for LAEs up to $z = 6.6$. Figure 3.53 summarizes the bias of $z \simeq 2 - 7$ LAEs with $L_{Ly\alpha} \gtrsim$ a few $\times 10^{42}$ erg s^{-1}. Including the moderately large errors, the estimated halo masses of LAEs are typically $M_h = 10^{11\pm1} M_\odot$ over $z \simeq 2 - 7$ [246]. In Fig. 3.53, b_g increases with redshift, suggesting that LAEs in earlier universes ($z \sim 5 - 7$) form from higher density fluc-

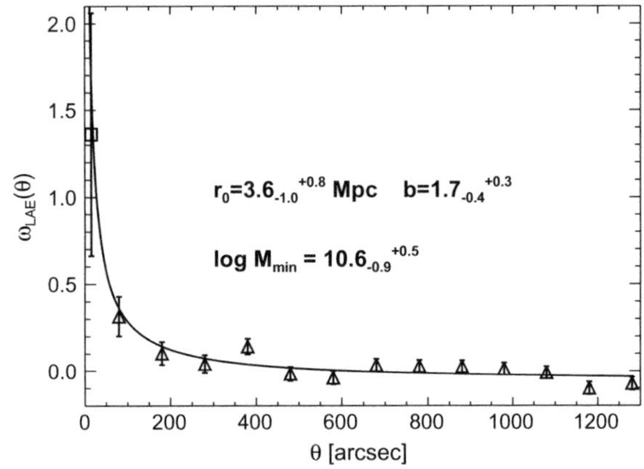

Fig. 3.52 ACF of LAEs at $z = 3.1$ [94]. The data points and the best-fit power law function are shown with triangles and a solid line, respectively. The square at <30" is the data point that is not used for the fitting. ©AAS. Reproduced with permission

Fig. 3.53 Bias of clustering b_g as a function of redshift [246]. The star marks represent b_g measurements of LAEs with $\gtrsim L^*_{Ly\alpha}$ in the range $z = 2 - 7$. The solid lines represent b_g for DM halos with a mass of 10^8, 10^9, 10^{10}, 10^{11}, and $10^{12} M_\odot$ predicted by the model of Sheth and Tormen [302], while the gray area indicates the DM halo mass range of 10^{10}–$10^{12} M_\odot$ corresponding to the typical DM halos of LAEs. The dotted lines denote the evolutionary tracks of b_g for the galaxy-number conserving model. ©AAS. Reproduced with permission

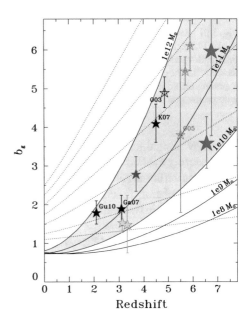

tuation peaks and are progenitors of present-day massive elliptical galaxies [246]. On the other hand, the small b_g of LAEs at $z \sim 2 - 3$ indicate that LAEs at $z \sim 2 - 3$ may be progenitors of today's Milky-Way like galaxies based on the average evolution of b_g (Fig. 3.53; Gawiser et al. [94], Ouchi et al. [246]). Note that, due to the relatively small samples, the ACF measurements of LAEs still include large statistical errors as found from a comparison of Fig. 3.52 and the bottom right panel of Fig. 3.48. So far, no studies of LAE clustering have identified the one-halo term made by LAEs residing in single halos (see Fig. 3.48 and the caption). There remains an open question whether LAEs have a moderately strong one-halo term similar to continuum-selected galaxies (Fig. 3.48).

3.3.9 Lyα Duty Cycle

Although stars can be formed in all DM halos (except in the least massive ones), not all halos with stars can be observed as LAEs for the following two reasons. First, if galaxies tend to have an intermittent star-formation history as suggested by theoretical models, they can produce strong Lyα emission only over a limited fraction of cosmic time. Second, it is not easy for Lyα photons produced in a galaxy to escape from it because of their resonant nature. To quantify the first effect, let us introduce the duty cycle of strong Lyα emission, $DC_{\mathrm{Ly}\alpha}$:

$$DC_{\mathrm{Ly}\alpha}(M_h) = \frac{n^{\mathrm{model}}_{\mathrm{Ly}\alpha}}{n^{\mathrm{model}}_{\mathrm{All}}}, \tag{3.31}$$

where $n^{\mathrm{model}}_{\mathrm{Ly}\alpha}$ is the number density of DM halos with M_h which are producing strong enough Lyα emission to be observed as LAEs if all Lyα photons escape, and $n^{\mathrm{model}}_{\mathrm{All}}$ is the number density of all DM halos with the same mass calculated from the DM halo mass function.

Nagamine et al. [220] have used cosmological numerical simulations to study the effects of $DC_{\mathrm{Ly}\alpha}$ on the Lyα luminosity function (Fig. 3.54) and the ACF. In the simulated luminosity functions, one can find two trends due to changing $DC_{\mathrm{Ly}\alpha}$

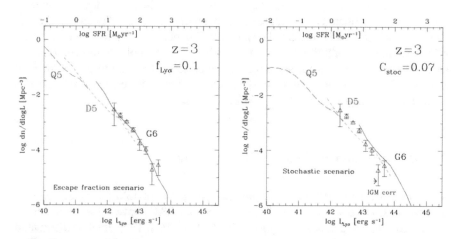

Fig. 3.54 Lyα luminosity functions at $z = 3$ obtained by observations (triangles) and cosmological simulations (curves, color-coded by mass resolution and simulation box size: Nagamine et al. [220]). Left: The model Lyα luminosity functions are calculated by converting SFRs into Lyα luminosities with the Lyα escape fraction $\left\langle f^{\mathrm{Ly}\alpha}_{\mathrm{esc}} \right\rangle = 0.1$ and the duty cycle of strong Lyα emission $DC_{\mathrm{Ly}\alpha} = 1$. Right: Same as the left panel, but with $\left\langle f^{\mathrm{Ly}\alpha}_{\mathrm{esc}} \right\rangle = 1$ and $DC_{\mathrm{Ly}\alpha} = 0.07$. It should be noted that the simulation results shown in the two panels are indistinguishable despite very different $\left\langle f^{\mathrm{Ly}\alpha}_{\mathrm{esc}} \right\rangle$ and $DC_{\mathrm{Ly}\alpha}$ values. In other words, the observed Lyα luminosity function (triangles) can be explained by adjusting only one of these two parameters. This figure is reproduced by permission of the PASJ

and $\left\langle f_{\text{esc}}^{\text{Ly}\alpha} \right\rangle$ (Sect. 3.3.2). Lowering $DC_{\text{Ly}\alpha}$ decreases the number density of LAEs irrespective of their luminosity, thus uniformly lowering the luminosity function. On the other hand, the number density of LAEs also decreases by reducing $\left\langle f_{\text{esc}}^{\text{Ly}\alpha} \right\rangle$ because of a uniform reduction of Lyα luminosities. These results mean that the luminosity function alone cannot distinguish a change in $DC_{\text{Ly}\alpha}$ from a change in $\left\langle f_{\text{esc}}^{\text{Ly}\alpha} \right\rangle$. However, this degeneracy can be resolved with clustering measurements. In a galaxy-formation model, LAEs are populated from the most-massive DM halos to low-mass DM halos until the LAE number density becomes as large as the one given by observations. If a $DC_{\text{Ly}\alpha}$ value is high (low), LAEs are hosted by high-mass (low-mass) DM halos on average for a given LAE number density, which show a strong (weak) LAE clustering signal. Although various combinations of $\left\langle f_{\text{esc}}^{\text{Ly}\alpha} \right\rangle$ and $DC_{\text{Ly}\alpha}$ can explain the LAE number density (or Lyα luminosity function), the choice of $DC_{\text{Ly}\alpha}$ changes the LAE clustering signal that can be tested with observational results. Nagamine et al. [220] have made two competing LAE models: a high $DC_{\text{Ly}\alpha}$ ($= 1$) and a low $\left\langle f_{\text{esc}}^{\text{Ly}\alpha} \right\rangle = 0.1$ (left panel of Fig. 3.54) and a low $DC_{\text{Ly}\alpha}$ ($= 0.07$) and a high $\left\langle f_{\text{esc}}^{\text{Ly}\alpha} \right\rangle = 1$ (right panel of Fig. 3.54), both of which well reproduce the observed Lyα luminosity function.[10] They find that the LAEs in the former model are clustered much more strongly than observed ones (Sect. 3.3.8), while those in the latter model reproduce the observed weak clustering. These models suggest that $DC_{\text{Ly}\alpha}$ is an important parameter less than unity. With the same idea, the value of $DC_{\text{Ly}\alpha}$ can be estimated by a simple comparison of number density and b_g (i.e. the clustering strength). Figure 3.55 shows that the number density of observed LAEs $n_{\text{Ly}\alpha}^{\text{obs}}$ is smaller than $n_{\text{All}}^{\text{model}}$ with the same b_g by two orders of magnitude. This is a sharp contrast to LBGs at the same redshift (Fig. 3.55). Based on results of the kind shown in Fig. 3.55, Gawiser et al. [94] and Ouchi et al. [246] estimate $DC_{\text{Ly}\alpha}$ to be about 1%, replacing $n_{\text{Ly}\alpha}^{\text{model}}$ with $n_{\text{Ly}\alpha}^{\text{obs}}$ in Eq. (3.31).

3.3.10 Summary of Galaxy Formation II

Section 3.3 has reviewed the basic physical properties of LAEs characterized by observations. Observed SEDs indicate that typical LAEs have a low stellar mass ($\sim 10^7$–$10^{10} M_\odot$) and a moderately low SFR (~ 1–$10 M_\odot$). They are thus distributed in the lowest mass regime of the star-formation main sequence (Sect. 3.3.1). The Lyα luminosity function of LAEs rapidly increases from $z \sim 0$ to 3, but shows no significant evolution over $z \sim 3 - 6$. On the other hand, the UV luminosity function of UV-continuum selected galaxies (i.e. dropouts) shows a moderate increase from $z \sim 0$ to 3, followed by a decrease to $z \sim 6$ and beyond. These Lyα and UV luminosity function evolution results suggest a monotonic increase in the Lyα escape fraction $f_{\text{esc}}^{\text{Ly}\alpha}$

---

[10]Nagamine et al. [220] refer to the Lyα duty cycle as the stochasticity of LAEs.

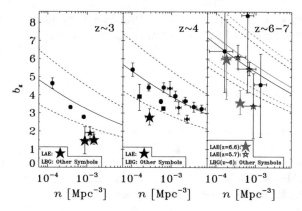

Fig. 3.55 b_g as a function of abundance [246] at $z \sim 3$ (left), $z \sim 4$ (center), and $z \sim 6 - 7$ (right). The star marks represent LAEs, while the other symbols (e.g. circles) denote dropout galaxies. The solid lines are for DM halos. The dashed lines are the same as the solid lines, but the abundance is multiplied by 1/10 and 10. In the right panel, black and red colors correspond to $z \sim 6$ and 7, respectively. Note that the black star marks for $z \sim 6$ correspond to a highly clustered region of LAEs whose b_g values are probably higher than the average. ©AAS. Reproduced with permission

from $z \sim 0$ to 6 (Sect. 3.3.2). The morphology of LAEs is very compact on average, with $r_e \sim 1$ (Sect. 3.3.3). Showing strong high ionization lines such as [OIII]5007 and CIII]1907,1909, typical LAEs are metal-poor ($\sim 0.3\ Z_\odot$) and highly ionized ($\log q$ [s cm^{-1}] $\simeq 8 - 9$) star-forming galaxies with negligibly small dust extinction ($A_V \sim 0$; Sect. 3.3.4). Deep spectra of LAEs show a signature of an outflow that is as strong as that of LBGs. Through theoretical modeling, Lyα profiles and luminosities are useful to constrain the outflow velocity, hydrogen column density, dust extinction, and the spatial distribution of gas clouds (Sect. 3.3.5). Multi-wavelength data suggest that an AGN is found in about 1% of LAEs in a given unbiased sample, while a majority of bright LAEs with $\log L_{\mathrm{Ly\alpha}} \gtrsim 43.5$ erg s^{-1} host an AGN at $z \sim 2 - 3$ (Sect. 3.3.6). Various studies use LAEs as low-mass galaxies associated with proto-clusters and LSSs to probe the high-z galaxy distribution. Clustering analyses of LAEs suggest that LAEs are more weakly clustered than typical LBGs. The masses of LAE-hosting DM halos are estimated to be $M_h \sim 10^{11\pm1}\ M_\odot$, about an order of magnitude smaller than for typical LBGs. Because LAEs are $\sim 10^2$ times less abundant than DM halos with the same bias value (or the same halo mass), the duty cycle of the LAE phase (i.e. the phase when a galaxy is observed as a dust-poor star-forming galaxy with strong Lyα) is only $\sim 1\%$ (Sects. 3.3.7–3.3.9).

3.4 Galaxy Formation III: Challenges of LAE Observations

There exist many open questions about the observational properties and the physical origins of LAEs. In this section, I highlight three important questions about LAEs that are being actively discussed: extended Lyα halos, Lyα escape mechanisms, and the connection between LAEs and pop III star formation. I explain major observational and theoretical progresses achieved to date about these issues.

3.4.1 Extended Lyα Halos

Deep observations in the 2000s (1980s–1990s) discovered $\sim$10–100-kpc large Lyα nebulae associated with star-forming galaxies (and AGNs) at $z \gtrsim 2$ that spatially extend beyond the stellar components. These nebulae are categorized into two classes, Lyα blobs (LABs) and diffuse Lyα halos (LAHs), according to their size and luminosity, with the former being larger and brighter.

3.4.1.1 Lyα Blobs

The first LABs discovered are Blob1 and Blob2 (dubbed LAB1 and LAB2) in a LBG overdensity field at $z = 3$ in the SSA22 field (Steidel et al. [328]; Sect. 3.1.2). LAB1 and LAB2 are each a huge ($>$100 kpc in physical length), bright (10^{-15} erg s^{-1}) Lyα nebula belonging to the largest class of LABs. Although similar extended Lyα emission has been found around high-z radio-loud galaxies since the 1980s (e.g. McCarthy et al. [205], van Ojik et al. [354]), these two LABs are accompanied only by star-forming galaxies with no clear AGN signature (see below for more details about the connection between LABs and AGNs). The sizes and luminosities of these two LABs are, respectively, two and one order(s) of magnitude larger than those of L^* LAEs at $z \sim 3$, $\sim$1 kpc and $\sim$10^{43} erg s^{-1} (Figs. 3.28 and 3.21, respectively). Now a few tens of LABs are known (e.g. Matsuda et al. [201, 202]; Fig. 3.7). The definition of LABs has not been quantitatively determined yet, but galaxies with a spatially extended Lyα halo with a size of $>$10 kpc are usually referred to as LABs in high-z galaxy studies. The LABs found so far have large diversities in Lyα size and luminosity. Moreover, it should be noted that the Lyα sizes and luminosities follow a continuous distribution extending from regular LAEs to the largest LABs such as LAB1 and LAB2 (Fig. 3.7). Although LABs are so far identified at $z \sim 2 - 7$ (Fig. 3.56), their physical origins are under debate. Three possible physical origins are suggested; AGN photoionization (Sect. 3.3.6), cooling radiation (Sect. 3.2.1.2), and Lyα scattering HI clouds (Sect. 3.3.5).

It is known that some AGNs are surrounded by an extended Lyα nebula, but the question is whether all LABs owe their luminosity to an AGN. Deep X-ray follow-up observations find that a number of LABs host an AGN, and that about $\sim$20% of LABs show AGN activities [21, 95]. Conversely, a large fraction of LABs including LAB1 have no AGN signature. Although bright AGNs would contribute to making large extended Lyα nebulae in some cases, there should exist other physical mechanisms to create LABs without an AGN.

Theoretical studies claim that cooling radiation can be the origin of LABs [68, 82, 99]. Although the Lyα luminosity of cooling radiation around a galaxy is usually

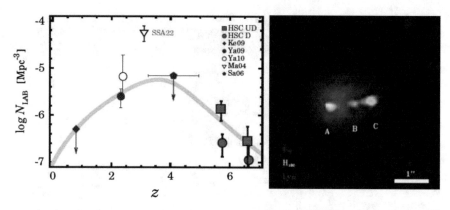

Fig. 3.56 Left: Abundance of LABs identified at $z \sim 2 - 7$ [309]. The inverse triangle is for LABs in the overdensity region SSA22, and the other symbols for those in field regions. The grey curve illustrates a possible abundance evolution of LABs. Because the selection criteria of LABs are heterogeneous, this evolutionary trend is not conclusive. Right: Color composite image of an LAB at $z = 6.6$ dubbed Himiko [247]. The red color shows a Lyα nebula extending over 17 kpc, while the blue and green colors represent rest-frame UV continua observed in the HST J and H bands, respectively. The labels A, B, and C, indicate the three continuum clumps in Himiko. ©AAS. Reproduced with permission

fainter than the one of young stars in it, these two luminosities are comparable in massive galaxies. Moreover, cooling radiation would dominate in the outer region of a galaxy, because it is produced primarily in the outer DM halo where Lyα photons are not absorbed by dust [82]. Goerdt et al. [99] suggest that the luminosity and morphology of LABs are reproduced by models that produce Lyα by collisional excitation in cold accretion gas. Although theoretical studies reproduce the characteristics of LABs with cooling radiation, so far there is no observational evidence that clearly supports the cooling radiation scenario. Yang et al. [370] claim that the HeII1640 line is useful to test if LABs at $z \sim 2 - 3$ originate from cooling radiation, because narrow-line (<400 km s^{-1}) HeII1640 emission can be produced neither by strong galactic outflow nor by population-II star photoionization, but only by cooling radiation. However, HeII1640 emission alone is not sufficient to distinguish cooling radiation from a narrow-line (type II) AGN and population-III star formation (Sect. 3.4.3).

Recent observations have advanced the understanding of LABs. Hayes et al. [119] have detected a tangential polarization signal of 0–20% in the Lyα emission of LAB1 (Fig. 3.57). Because it is predicted that resonance scattering of Lyα in HI clouds makes a tangential polarization [67], the polarization signal in LAB1 suggests that extended Lyα nebulae are produced by Lyα resonance scattering in HI clouds around galaxies. However, there remains a question about the source of Lyα photons. The HI cloud scattering scenario usually assumes that Lyα photons are produced in the central galaxy of the LABs [119], but HI cloud scattering also takes place for Lyα photons produced in situ by gas cooling. Trebitsch et al. [345] have performed radiative hydrodynamics simulations for Lyα photons from the central galaxy and gas cooling,

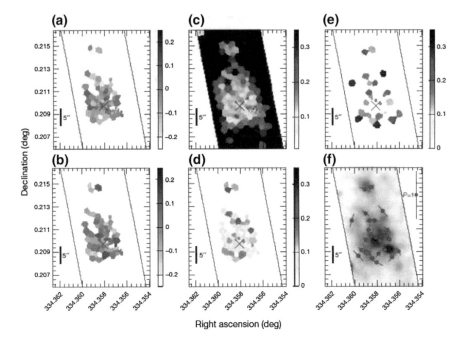

Fig. 3.57 Polarization observation results of LAB1 [119]. Panels **a** and **b** present the Stokes Q and U parameters, respectively. Panel **c** shows a map of polarization fraction measurements beyond the 2σ confidence level, with errors color-coded by values. Panels **d** and **e** are maps of the polarization fraction P for signals with an absolute (2σ) error smaller than 16%, and with a relative error smaller than 50% respectively. Panel **f** indicates an intensity image (gray scale) with polarization vectors (red bars) whose lengths correspond to the amount of polarization. The blue cross denotes the central position of LAB1. Reprinted by permission from Nature

and calculated the polarization and surface brightness (SB) of Lyα emission that are shown in Fig. 3.58. If LAB1 is made by the HI scattering of Lyα photons from the central galaxy, the polarization signal is larger than 20% at >40 kpc, which is significantly larger than the observational results. Moreover, Trebitsch et al. [345] find that Lyα photons from the cooling radiation are also scattered and polarized to a level of 10–15%. To explain the moderately small polarization and the large SB values, Trebitsch et al. [345] suggest that a significant contribution from cooling radiation is necessary. In a way like this, the origins of LABs are still being actively discussed.

Moreover, there is an interesting problem not only about the origin of LABs without AGN signatures, but also of LABs harboring an AGN. Cantalupo et al. [40] report the discovery of a gigantic LAB around a radio-quiet QSO, UM287, at $z = 2.3$ (Fig. 3.59). This Lyα halo is 460 kpc in size that is larger than the virial diameter, ∼280 kpc, of the DM halo hosting this QSO, and may even extend to a filament of the LSS. If this Lyα nebula is produced by Lyα photons from the recombination of a large cloud that was initially ionized by the QSO, and subsequent scattering in

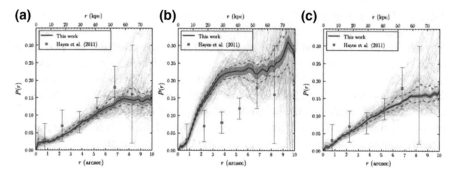

Fig. 3.58 Lyα photon polarization as a function of the distance from the central galaxy [345]. The orange lines represent polarization values along various line of sights in the simulation box. In each panel, the red line denotes the median of the orange lines, while the red region and the red-dashed lines indicate the 1σ dispersion and the first/third quartiles, respectively. The green data points are the observational measurements of Hayes et al. [119]. The left and central panels present polarization profiles for Lyα photons produced in the extragalactic gas and the galaxy, respectively, while the right panel shows the sum of these two components. Reproduced with permission ©ESO

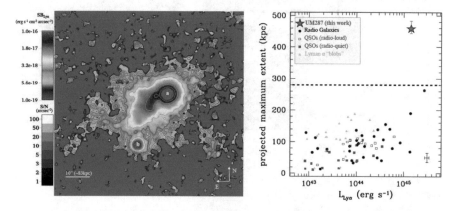

Fig. 3.59 Left: Large Lyα nebula around the radio-quiet quasar UM287 that is dubbed Slug nebula [40]. This is the continuum-subtracted Lyα image, presenting the Lyα surface brightness. The bottom-left scale bar represents 10" (∼80 kpc). Right: Projected maximum sizes of the Lyα nebulae as a function of Lyα luminosity [40]. The size and luminosity of the Slug nebula is indicated with the red star mark. The black circles and the blue squares are radio galaxies and QSOs, respectively. The green triangles denote LABs with or without AGN mainly identified by narrowband imaging. Reprinted by permission from Nature

the now neutral cloud, then a very high gas mass or clumping factor is required to explain its high Lyα SB. It is, however, not clear whether this object truly has such a very high gas mass or clumping factor. Thus, the physical origin of this gigantic LAB is also under debate. There is also a report of the identification of large LABs with intermediate sizes of 100–300 kpc around radio quiet QSOs [28]. These large LABs fill the gap between small and gigantic LABs and may facilitate our understanding of the whole LAB zoo.

3.4.1.2   Diffuse Lyα Halos

Hayashino et al. [117] and Steidel et al. [331] have identified diffuse Lyα halos
(LAHs) around star-forming galaxies at $z \sim 2 - 3$ by deep spectroscopy and
narrowband-image stacking analyses (Fig. 3.60). To date, LAHs are found in star-
forming galaxies including LAEs in a wide-redshift range, $z \sim 2 - 7$ [84, 203, 215].
LAHs extend to a scale of a few $\times 10$ kpc with a Lyα SB of $\lesssim 1 \times 10^{-18}$ erg s^{-1} cm^{-2}
arcsec^{-2} that is about $10 - 100$ times fainter than that of LABs. Their Lyα SB profiles
roughly follow a power law (Fig. 3.60) with the LAH scale length r_n,

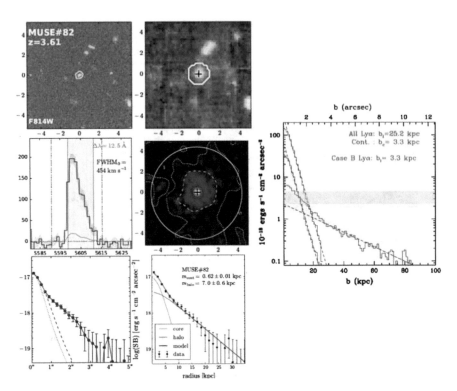

Fig. 3.60 Left six panels: Images, Lyα spectrum, and radial profiles of a galaxy, MUSE#82, at
$z = 3.61$ [181]. The top left and right panels present continuum images, an HST $F814W$ image and
a MUSE white-light image, respectively. The middle left and right panels show a Lyα spectrum of
the central part of this galaxy, and a Lyα image, respectively. The bottom left panel displays a Lyα
SB radial profile (blue circles) and a UV continuum profile (green line) together with the PSF (red
line). The bottom right panel presents the best-fit model (red line) and its decomposition to core
(green line) and LAH (blue line) components, to compare with the observed profile (black circles).
Here the core component is modeled with an exponential profile. Right panel: Stacked SB radial
profiles of galaxies at $z \sim 2.7$ [331]. The red and blue solid lines denote Lyα and UV-continuum
SB radial profiles, respectively. The green solid line indicates an estimated Lyα SB radial profile for
no LAH case that is calculated with the UV-continuum SB radial profile under the assumption of
the Case B recombination. The dashed lines are the best-fit exponential functions of these profiles.
Reproduced with permission ©ESO and ©AAS

$$S(r) = C_n \exp(-r/r_n), \tag{3.32}$$

where $S(r)$, r, and C_n are the Lyα SB, radius, and normalization factor, respectively.

The parameter r_n characterizes the size of an LAH. Figure 3.61 presents the relations between r_n and several physical properties of LAEs. Matsuda et al. [203] claim that r_n positively correlates with the local LAE surface density δ_{LAE}. The result of Matsuda et al. [203] would imply that galaxies in a dense environment have a large Lyα halo (cf. Xue et al. [367]). Momose et al. [216] find that r_n negatively correlates with the Lyα luminosity of the main body of galaxies, $L_{Ly\alpha}^{cent}$, at $r < 8$ kpc ($r < 1"$). Because $L_{Ly\alpha}^{cent}$ depends on H I column density through resonant scattering, galaxies with the ISM rich in H I would have a faint $L_{Ly\alpha}^{cent}$ and a large extended Lyα halo. It is suggestive that LAEs (galaxies with a bright $L_{Ly\alpha}^{cent}$) have the ISM whose H I column density is lower than that of LBGs (galaxies with a faint $L_{Ly\alpha}^{cent}$).

Because r_n depends on some galaxy properties, the evolution of r_n should be investigated carefully with uniformly selected samples at different redshifts. The red star marks in Fig. 3.62 indicate the r_n values of field ($\delta_{LAE} \lesssim 1$) LAEs with $L_{Ly\alpha}^{cent} \gtrsim 2 \times 10^{42}$ erg s^{-1} at $z = 2.2 - 6.6$. It is found that r_n is nearly constant over $z = 2.2 - 5.7$, falling in the range $r_n = 5 - 10$ kpc [215]. There is a hint of an increase

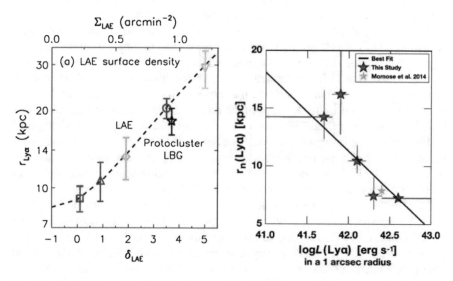

Fig. 3.61 Left: LAH scale length as a function of the LAE sky overdensity at $z = 3$ in the SSA22 region [203]. The star mark denotes the median LAH scale length of LBGs in the galaxy overdense region, while the other symbols represent those of LAEs. The dashed line is the best-fit quadratic function to all data points. (Note the non-regular scale on the y-axis.) Right: LAH scale length as a function of Lyα luminosity for LAEs at $z = 2$ [216]. The star marks are measurements by Momose et al. [215, 216], and the solid line is the best-fit linear function to them. This figure is reproduced by permission of MNRAS

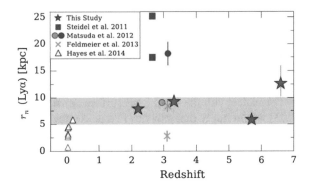

Fig. 3.62 Redshift evolution of the LAH scale length [215]. The red star marks represent the median LAH scale lengths of LAEs with $L_{Ly\alpha} \gtrsim 2 \times 10^{42} \, \mathrm{erg \, s^{-1}}$ at $z = 2.2 - 6.6$. The gray region indicates the range of $5 - 10 \, \mathrm{kpc}$ where the LAH scale lengths of LAEs at the post reionization epoch ($z = 2.2 - 5.7$) fall. The blue squares and the orange and blue circles denote the LAH scale lengths of LAEs in the overdense region of SSA 22. The orange crosses are the measurements of bright LAEs. The open triangles show Lyα Petrosian radii of local LAEs. This figure is reproduced by permission of MNRAS

in r_n from $z = 5.7$ to 6.6 that could be relevant to cosmic reionization, but the error bars are too large to conclude whether this is a real signature.

The physical origin of LAHs is not well understood yet. There are four possible scenarios, (i) CGM's HI gas scattering Lyα photons that originate from star-forming regions, (ii) cooling radiation, (iii) unresolved dwarf satellite galaxies, and (iv) fluorescence. Lake et al. [176] perform radiative transfer calculations in hydrodynamical simulations (Fig. 3.63), and find that the scenario (i) cannot explain high Lyα SB at large radii found by observations (red line in the left panel of Fig. 3.63). The high Lyα SB at large radii requires either the mechanism (ii) or (iii) (right panel of Fig. 3.63). To distinguish the contributions of (ii) and (iii), UV-continuum SB profiles in stacked broadband images are useful. This is because the mechanism (iii) produces stellar UV-continuum emission, while the mechanism (ii) creates only negligible UV-continuum emission. It is, however, difficult to investigate UV-continuum SB profiles in the stacked images due to systematic errors in sky subtraction [216]. Recent deep spectroscopic observations with VLT/MUSE find a diffuse LAH on an individual basis with no use of stacking data (Fig. 3.60; Wisotzki et al. [364], Leclercq et al. [181]). This discovery rules out the possibility that moderately large unresolved dwarf satellite galaxies mimic a diffuse LAH in the scenario (iii). Studies of LAHs are proceeding rapidly, and much progress can be expected in the coming few years.

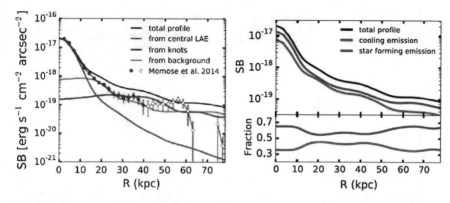

Fig. 3.63 Average Lyα SB profiles of nine model LAEs at $z = 3$ obtained by numerical simulations [176]. Left: The black curve represents the total Lyα SB profile, while the red, blue, and green curves denote a decomposition into Lyα photons originating from the central star-forming regions, surrounding knotty star-forming regions, and background regions, respectively. The purple filled circles show the observational data of Momose et al. [215]. The open circles are the same as the filled circles, but for data points potentially with large systematic errors. The purple line simply connects all data points. Right: Same as the left panel, but for a decomposition into cooling radiation (red line) and star-formation radiation (blue line) in the top panel. The fractional contributions of these two radiation components (to the total Lyα SB profile) are shown in the bottom panel. ©AAS. Reproduced with permission

3.4.2 Lyα Escape Fraction

One[11] of the most important questions about LAEs is how they emit strong Lyα light. To discuss this question, let us introduce the Lyα escape fraction, $f_{\rm esc}^{\rm Ly\alpha}$, defined by

$$f_{\rm esc}^{\rm Ly\alpha} = \frac{L_{\rm Ly\alpha}^{\rm obs}}{L_{\rm Ly\alpha}^{\rm int}}, \tag{3.33}$$

where $L_{\rm Ly\alpha}^{\rm obs}$ and $L_{\rm Ly\alpha}^{\rm int}$ are observed and intrinsic Lyα luminosities, respectively. Intrinsic Lyα luminosities can be estimated from a UV continuum or an Hα line luminosity. Note that $f_{\rm esc}^{\rm Ly\alpha}$ is similar to the number/luminosity average Lyα escape fraction, $\left\langle f_{\rm esc}^{\rm Ly\alpha} \right\rangle$ (Eq. 3.11), but that $f_{\rm esc}^{\rm Ly\alpha}$ is defined for one individual galaxy. The left panel of Fig. 3.64 presents the Lyα escape fraction as a function of color excess for LAEs at $z \sim 0$ and $2 - 4$. There is a clear anti-correlation between these two quantities, suggesting that a certain fraction of Lyα photons are absorbed by dust in the ISM. In the left panel of Fig. 3.64, the $f_{\rm esc}^{\rm Ly\alpha}$ values of LAEs are compared with the amount of dust extinction predicted by a simple screen model that includes no Lyα resonance scattering effects,

---

[11]In the Saas Fee lecture, this subsection dealt with outflows and Lyα profiles as well. For the readers' convenience, I have moved these topics to Sect. 3.3.5.

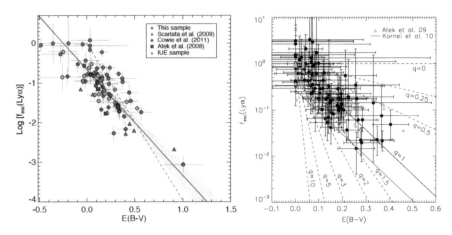

Fig. 3.64 Lyα escape fraction $f_{\text{esc}}^{\text{Lyα}}$ as a function of color excess E(B-V). Left: The data points are all measured at $z = 0 - 0.3$ [9]. The black line with a yellow shade indicates the best-fit linear function with the 1σ fitting error. The red dashed line represents a simple attenuation with the Cardelli et al. [43] extinction law. Right: The black circles and the green triangles denote LAEs at $z \sim 2 - 4$ and 0.3, respectively [24]. The black lines present correlations suggested from Eq. 3.35 with various q values, with a solid one corresponding to $q = 1$. The red line is the best-fit linear function to LBGs at $z \sim 2$. Reproduced with permission ©ESO and ©AAS

$$f_{\text{esc}}^{\text{Lyα,screen}} = 10^{-0.4k_{1216}E(B-V)_{\text{neb}}}, \tag{3.34}$$

where k_{1216} is the extinction coefficient at 1216 Å. Calzetti's law provides $k_{1216} = 12.0$ [166]. There is a hint of an excess of $f_{\text{esc}}^{\text{Lyα}}$ beyond the dust screen model for some LAEs, albeit with large measurement errors. The right panel of Fig. 3.64 shows various model lines with different q values [87] defined as

$$q = \frac{\tau(\text{Lyα})}{\tau_{1216}}, \tag{3.35}$$

where $\tau(\text{Lyα})$ and τ_{1216} are the optical depths for the Lyα line and 1216 Å UV-continuum emission. In the right panel of Fig. 3.64, the $q = 1$ model line corresponds to Eq. (3.34).

LAEs with a $f_{\text{esc}}^{\text{Lyα}}$ excess have $q < 1$. A selectively large Lyα extinction ($q > 1$) is simply explained, if Lyα photons cross a long effective distance in a dusty ISM due to a number of Lyα resonance scatterings, which enhances the probability of absorption by dust.

However, a selectively small Lyα extinction ($q < 1$) is difficult to understand.

Clumpy gas clouds in the ISM may explain $q < 1$, as has been originally suggested by Neufeld [226]. Figure 3.65 illustrates this idea. If the ISM is made of clumpy gas clouds, Lyα (resonance) photons that encounter clumpy clouds are scattered on their surface with a negligible dust absorption. On the other hand, UV-continuum (non-resonance) photons can go into the clouds and are eventually absorbed by dust inside

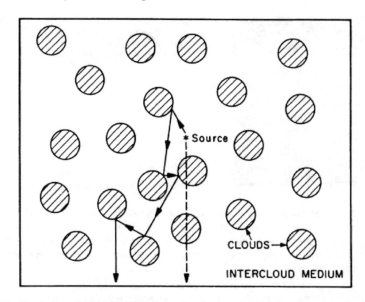

Fig. 3.65 Illustration of Neufeld's [226] clumpy cloud model [226]. In this model, the ISM is made of gas clumps whose central regions contain dust (in cold molecular gas) that can absorb UV-continuum photons. The solid-line arrows indicate the light path of a Lyα (resonance) photon from the source. The Lyα photon is scattered on the surfaces of gas clumps with negligible dust absorptions. The dashed-line arrow represents the light path of an UV-continuum (non-resonance) photon that penetrates clumps. ©AAS. Reproduced with permission

them. Through these scattering and absorption processes, Lyα photons are absorbed less than UV-continuum photons, resulting in a very high Lyα EW_0. The clumpy cloud model is sometime referred to as Neufeld's [226] effect. This model predicts narrow Lyα line widths because Lyα photons experience only a small number of resonant scattering before escaping from galaxies [109, 226].

Although Neufeld [226] investigated this model only in a simple case of static and very clumpy/dusty media, recent studies have used radiative transfer simulations to test this model in realistic ISM conditions (Laursen et al. [179], Duval et al. [73]; cf. Hansen and Oh [109]). These simulations have found that Neufeld's [226] effect is seen (Fig. 3.66) only under special conditions: a low outflow velocity ($<200\,\mathrm{km\,s^{-1}}$), very high extinction ($E(B-V) > 0.3$), and an extremely clumpy gas distribution with a density contrast larger than 10^7 (i.e. most gas is locked up in clumps), many of which do not meet the observed properties of LAEs (Table 3.1). Moreover, under these special conditions, observed Lyα lines can have neither a velocity shift nor an asymmetric profile. Laursen et al. [179] and Duval et al. [73] have concluded that while it is true that Neufeld's [226] effect is working to some degree, this effect cannot explain the fact that a large fraction of LAEs have high $f_{\mathrm{esc}}^{\mathrm{Ly\alpha}}$ values (i.e. $q < 1$). In summary, the physical origin of the high $f_{\mathrm{esc}}^{\mathrm{Ly\alpha}}$ values found for LAEs is still under debate (see Sect. 3.4.3 for more discussion).

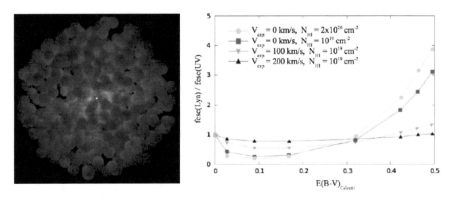

Fig. 3.66 Left: ISM gas geometry of 500 clumpy clouds with a radius of 350 pc produced in the simulations by Laursen et al. [179]. Right: Ratio of Lyα to UV-continuum photon escape fractions, $f_{\text{esc}}^{\text{Lyα}}/f_{\text{esc}}^{\text{UV}}$, as a function of color excess E(B-V) for models of the clumpy cloud ISM [73]. The cyan circles, blue squares, gray inverse-triangles, and black triangles represent models with outflow velocities of $V_{\text{exp}} = 0 - 200 \, \text{km s}^{-1}$ and HI column densities of $N_{\text{HI}} = 10^{19} - 2 \times 10^{20} \, \text{cm}^{-2}$ indicated in the legend. An enhancement of Lyα photon escape, $f_{\text{esc}}^{\text{Lyα}}/f_{\text{esc}}^{\text{UV}} > 1$, is achieved only with $V_{\text{exp}} = 0-100 \, \text{km s}^{-1}$ and $E(B - V) \gtrsim 0.3$. Reproduced with permission ©ESO

3.4.3 Large Lyα and HeII Equivalent Widths: Pop III in LAEs?

LAEs with large Lyα EW_0 values are potentially important objects that have excessive Lyα emission at a given stellar continuum. Malhotra and Rhoads [192] claim the existence of LAEs with Lyα $EW_0 \gtrsim 240 \, \text{Å}$ that cannot be explained by young star formation with the solar metallicity and a Salpeter IMF [291]. Figure 3.67 presents a Lyα EW_0 histogram of $z = 4.5$ LAEs. Although observational EW estimates include large uncertainties and systematics due to weak or undetected continua (see, e.g., Fig. 14 of Shimasaku et al. [313]), LAE studies have shown that ∼10–30% of LAEs in a narrowband-selected sample have large ($\gtrsim 200-300 \, \text{Å}$) Lyα EW_0 at $z \sim 2 - 7$ [61, 153, 243, 313]. The physical origins of the large Lyα EW_0 objects are not well understood. These LAE studies discuss the possibilities of the Neufeld effect (Sect. 3.4.2), cooling radiation (Sect. 3.4.1.1), and pop III star formation. The relation between large Lyα EW_0 and pop III star formation is presented in the left panel of Fig. 3.68. This panel shows theoretically calculated Lyα EW_0 as a function of stellar age for various stellar populations. For a star-formation history of an instantaneous burst with solar metallicity, Salpeter IMF, and a mass range of 1–100 $M_{\odot}$, the Lyα EW_0 does not exceed ∼200–300 Å even at the birth time. A stellar population with a larger number of massive young stars that produce more ionizing photons has a higher Lyα EW_0. Lyα EW_0 is thus sensitive to the shape of the IMF and metallicity as well as stellar age. It is predicted that a top heavy IMF is realized in metal poor gas clouds because they contain only a small amount of coolants that are needed for low-mass gas clumps to collapse. Moreover, metal poor stars efficiently produce ionizing

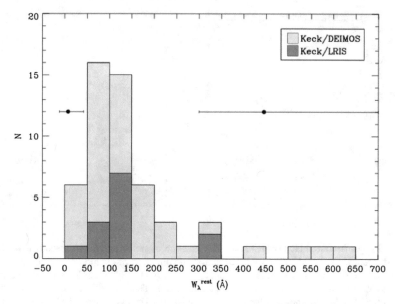

Fig. 3.67 Lyα EW_0 distribution for LAEs at $z = 4.5$ [61]. The light and dark gray histograms are obtained by Keck/DEIMOS and LRIS spectroscopy, respectively. The two filled circles with horizontal error bars show the typical errors at two Lyα EW_0 values. ©AAS. Reproduced with permission

photons, because ionizing photons are not absorbed by metals in the stellar atmosphere. The left panel of Fig. 3.68 indicates that Lyα EW_0 can reach $\sim$500–1500 Å for galaxies having metal poor instantaneous star-formation with top heavy IMFs. As demonstrated in this panel, LAEs with large Lyα EW_0 values can be candidates of pop III galaxies, although not definitive ones (e.g. Yang et al. [370]). Moreover, the production rate of ionizing photons is sensitive not only to IMF, metallicity, and stellar age, but also to the binary fraction of massive stars and many other physical conditions such found in the BPASS model [75].

Thus, another test is necessary to isolate pop III star formation from the candidates. HeII1640 is an ideal emission line for such a test.[12] Because He^+ has a high ionization potential of 54.4 eV, He^+ can be ionized by hard spectra of very massive young stars that can be found in HII regions of pop III star formation. The right panel of Fig. 3.68 presents HeII1640 EW_0 as a function of stellar age for instantaneous star-formation, and suggests that a large HeII1640 EW_0 ($\gtrsim$10 Å) is indicative of pop III. Although the hot outflowing gas from a WR star and the broad-line region of an AGN can also produce HeII1640 emission with $EW_0 \gtrsim 10$ Å, both lines are predicted to be much broader, with a line-width velocity of $\sim$1000 km s^{-1}, than those from HII regions of pop III star formation (a few hundred km s^{-1}). However, the HeII1640 emission of narrow-line (type II) AGNs has similarly small line widths. To isolate pop III stars from such AGNs, one needs to investigate high ionization lines such as Nv, Ovi,

---

[12] The HeII1640 line corresponds to HeII Hα. Note that the HeII304 line corresponding to HeII Lyα cannot be easily observed.

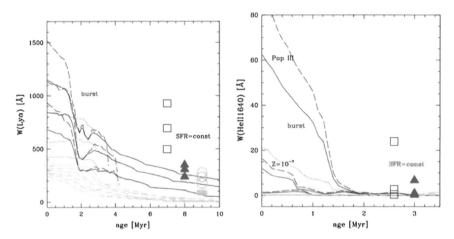

Fig. 3.68 Lyα (left) and HeII (right) EW_0 as a function of stellar age for an instantaneous burst of star-formation predicted by the stellar evolution and photoionization models of Schaerer [296]. The three blue dashed lines represent metallicity $Z = 0$ models with a Saltpeter IMF and mass ranges of 50–500, 1–500, and 1–100$M_\odot$ from top to bottom. The blue solid and dotted lines are the same as the blue dashed lines, but for $Z = 10^{-7}$ and 10^{-5}. The cyan dashed lines denote models with $Z = 0.0004, 0.001, 0.004, 0.008, 0.020$, and 0.040 from top to bottom. The three squares show constant SFR models with a 50–500$M_\odot$ Salpeter IMF for $Z = 0, 10^{-7}$, and 10^{-5} from top to bottom. The red triangles and green circles are the same as the squares, but for the mass ranges 1–500 and 1–100$M_\odot$, respectively. The green short lines are the same as the green circles but with metallicities of $Z = 0.0004, 0.001, 0.004, 0.008, 0.020$, and 0.040 from top to bottom. Note that, in the right panel, HeII $EW_0 \gtrsim 5$ Å is reached only for models with very low metallicities of Z $\lesssim 10^{-7}$, except for the mass range 1–100$M_\odot$. Reproduced with permission ©ESO

and strong X-ray emission that cannot be produced by the photoionization by very massive stars (cf. fast radiative shocks; Thuan and Izotov [341]).

A strong narrow HeII line is found in a $z = 2$ LAE with an extended Lyα halo, named PRG1 (Fig. 3.69; Prescott et al. [271]). The HeII line of PRG1 is strong, HeII $EW_0 = 37 \pm 10$ Å, and the ratio of HeII to Lyα fluxes is HeII/Lyα = 0.12. Metal lines are not detected with upper limits of CIV1548/Lyα and CIII]1909/Lyα $\lesssim 0.03$. These properties of a strong narrow HeII and very weak (or no) metal lines are suggestive of pop III star formation. However, the subsequent deep spectroscopy of Prescott et al. [273] has clearly detected CIV and CIII] lines with a line ratio of CIV/HeII, CIII]/HeII~ 0.5, indicating that PRG1 is photoionized by an AGN, not by pop III stars.

More recently, Sobral et al. [318] have reported a detection of a strong narrow HeII emission in an LAE at $z = 6.6$ that is dubbed CR7. The left and right panels of Fig. 3.70 present the SED and the HeII spectrum of CR7. This object is made of three stellar components whose total SED exhibits a mature stellar population with a Balmer break. Although the SEDs of the three stellar components are not clearly distinguished, there is a possibility that one component, A, (Fig. 3.70) would have a very young population with a blue SED. Sobral et al. [318] report that CR7 has a very large

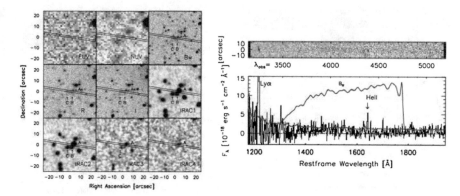

Fig. 3.69 Left: Thumbnail images of PRG1 taken from GALEX (FUV and NUV), Kitt-Peak 4 m (B_w, R, and I), and Spitzer (3.6, 4.5, 5.8, and 8.0 μm; Prescott et al. [271]). Spectroscopy slit positions are shown with lines. The labels A, B, and C denote the positions of three IRAC counterparts. Right: Two dimensional (top) and one-dimensional (bottom) spectra of PRG1 [271]. The black line presents the spectrum of PRG1, while the red line denotes 1σ errors. The Lyα and HeII lines are detected. The blue line indicates the B_w band transmission curve. ©AAS. Reproduced with permission

HeII equivalent width of $EW_0 = 80 \pm 20$ Å as well as a large Lyα equivalent width of $EW_0 = 211 \pm 20$ Å. The reported line ratio of HeII/Lyα$= 0.22$ is about twice as large as the one of PRG1. No metal lines are detected in VLT/X-Shooter spectra covering the entire NIR wavelength range accessible from the ground. Some theoretical studies suggest that CR7 is a candidate of a direct collapse black hole because of a strong HeII line without detection of metal lines from moderately-massive stellar components [69, 250]. Recently, Shibuya et al. [310] present reanalysis results of the CR7 X-Shooter spectra, and find no HeII line, placing only an upper limit. A similar upper limit is also reported by Sobral et al. [319]. Moreover, ALMA observations reveal the metal [CII] line [200]. To summarize, although CR7 was a promising candidate of pop III star formation or a direct collapse black hole, subsequent studies find no such evidence.

Spectral hardness measurements are useful to diagnose the presence of pop III star formation in a galaxy. Figure 3.71 shows theoretical predictions of the spectral hardness Q_{He^+}/Q_H as a function of metallicity [296], where Q_{He^+} and Q_H are the fluxes of ionizing photons for He$^+$ (>54.4 eV) and H (>13.6 eV), respectively. This spectral hardness can be estimated from observed HeII and Lyα fluxes, $f_{HeII1640}$ and $f_{Lyα}$:

$$\frac{f_{HeII1640}}{f_{Lyα}} \sim 0.55 \frac{Q_{He^+}}{Q_H}. \tag{3.36}$$

Reported $f_{HeII1640}/f_{Lyα}$ measurements give $Q_{He^+}/Q_H \sim 0.22$ for PRG1 [271] and ~ 0.42 for CR7 [318]. In Fig. 3.71, the estimated Q_{He^+}/Q_H values are larger than the predictions for pop III star formation by an order of magnitude even for a top heavy IMF with a mass range of 50–500$M_\odot$. The large Q_{He^+}/Q_H value of PRG1 is

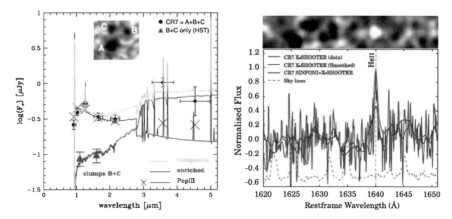

Fig. 3.70 Left: The SED (main panel) and a thumbnail image (inset panel) of CR7 [318]. The thumbnail image is an HST NIR image that resolves the three continuum components of CR7 named clumps A, B, and C, where clump A is the candidate of pop III star formation reported by Sobral et al. [318]. In the main panel, the black circles represent total flux densities, while the magenta triangles denote HST NIR photometry of clumps B+C. The black and magenta lines are the pop III and $Z = 0.2Z_\odot$ SED models best-fit to the 1–2 μm data points of the black circles and the magenta triangles, respectively, and the green line indicates the sum of these two model SEDs. The red crosses are the photometry predicted by the best-fit pop III model. Right: Two-dimensional (top) and one-dimensional (bottom) spectra of CR7 reported by Sobral et al. [318] (cf. Shibuya et al. [310], Sobral et al. [319]). The black and blue lines represent X-Shooter spectra with and without a spectrum smoothing process, respectively. The green line denotes a stack of the X-Shooter and VLT/SINFONI spectra. The red line indicates the sky background spectrum. ©AAS. Reproduced with permission

explained by the existence of an AGN, while that of CR7 is probably explained by the recent reanalysis results of no HeII line detection [310, 319]. In addition to analyses on an individual basis, one can also measure an average $f_{\mathrm{HeII1640}}/f_{\mathrm{Ly}\alpha}$ from stacked LAE spectra. Composite spectra using large LAE samples show no clear detection of HeII emission, placing upper limits of $f_{\mathrm{HeII1640}}/f_{\mathrm{Ly}\alpha} < 2\%$ and 20% at $z = 3$ and 5, respectively [60, 243]. Although the upper limit for $z = 5$ LAEs is not strong enough to give a meaningful constraint ($Q_{\mathrm{He}^+}/Q_{\mathrm{H}} \lesssim 0.1$), the one for $z = 3$ LAEs ($Q_{\mathrm{He}^+}/Q_{\mathrm{H}} \lesssim 0.01$) indicates that on average $z = 3$ LAEs do not have star-formation dominated by pop III with a top heavy IMF with a mass cut of 50–500 or 1–500 $M_\odot$.

3.4.4 Summary of Galaxy Formation III

Section 3.4 has presented three important questions about LAEs: extended Lyα halos, Lyα escape mechanisms, and the connection between LAEs and pop III star formation. Deep observations have revealed largely extended Lyα nebulae, dubbed LABs, Lyα filaments, and LAHs, with a size of about a few 10 kpc to 500 kpc around high-z

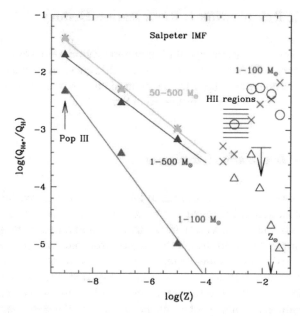

Fig. 3.71 Spectral hardness Q_{He^+}/Q_H as a function of metallicity [296]. The red, blue, and green filled triangles are the spectral hardness values for three very metal poor cases with mass cuts of 1–100, 1–500, and 50–500$M_\odot$, respectively, predicted with the stellar evolution and photoionization models of Schaerer [296]. The red, blue, and green lines indicate the linear functions in the log-log plot that are best fit to the red, blue, and green filled triangles, respectively. The red crosses are the same as the red filled triangles, but for metal rich cases. The blue open circles and open triangles are other model predictions (see Schaerer [296]). The shaded region and the upper limit denote the spectral hardness of HII regions estimated with observational data. Reproduced with permission ©ESO

star-forming galaxies and AGNs. Because tangential polarization signals are detected in the Lyα blob LAB1, HI gas scattering of Lyα photons should exist in LABs. However, it is not clear what is (are) the major source(s) of the Lyα photons. Proposed candidate sources are HII regions in the ISM, cooling radiation, and unresolved dwarf satellite galaxies. There exist LAEs with a Lyα EW_0 as large as a few 100 Å. Recent theoretical calculations suggest that a clumpy ISM made of discrete clouds would boost Lyα EW_0, but that the boosting is only found in physical conditions (high extinction and low outflow velocities) that are clearly different from those seen in typical LAEs. The mechanism producing large EW_0 values is still unknown. Several observational studies have reported LAEs with narrow and strong HeII emission lines. These LAEs may be candidates of galaxies with pop III star formation whose young massive stars emit moderately high energy photons ionizing He$^+$. However, these pop III star-formation candidates can also be narrow-line AGNs or may include erroneous HeII emission measurements. The spectral hardness Q_{He^+}/Q_H is useful to diagnose pop III star formation and AGNs, although, to date, Q_H is often estimated from a Lyα flux that includes a large uncertainty in the Lyα escape fraction.

3.5 Cosmic Reionization I: Reionization History

There are two major questions about cosmic reionization: reionization history and reionization sources. This section addresses the first question with an emphasis on LAE studies, starting with a brief introduction to cosmic reionization. The second question, reionization sources, is discussed in Sect. 3.6.

3.5.1 What Is Cosmic Reionization?

Cosmic reionization is a cosmic event that took place at a high redshift (Fig. 3.72). By the recombination of hydrogen at $z \sim 10^3$, the early universe with hot plasma gas evolved into one filled with neutral gas (i.e. atomic hydrogen gas); the last photon scattering surface made by this transition is observed as the cosmic microwave background (CMB). On the other hand, today's universe does not contain abundant neutral gas, but harbors fully ionized gas in the inter-galactic space that makes no Lyα absorption lines at $z \sim 0$ in UV spectra of QSOs [16]. These two pieces of evidence suggest that hydrogen atoms that became neutral at $z \sim 10^3$ were ionized again by today. This event is known as cosmic reionization (see the review of Fan et al. [81]). In this lecture, I focus on hydrogen reionization that is deeply related to LAEs and galaxy formation. See, e.g., Worseck et al. [365] for observational progresses in helium reionization studies. Hereafter, 'reionization' indicates hydrogen reionization, if not otherwise specified.

Cosmic reionization is driven by ionizing photon radiation, γ, the origin of which being stellar or non-stellar or both:

$$H + \gamma \rightarrow H^+ + e^-, \tag{3.37}$$

where H^+ is an ionized hydrogen (proton) and e^- is an electron.

Although the cosmic reionization process and reionization sources are not well understood, many theoretical studies suggest a picture where massive stars in galaxies provide the majority of ionizing photons and first ionize the IGM in the vicinity of galaxies (Fig. 3.73). Ionized regions made around galaxies are called ionized bubbles

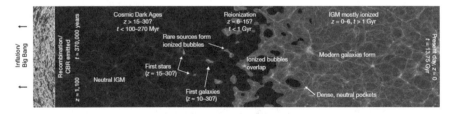

Fig. 3.72 Picture of cosmic reionization in the cosmic history [285]. Reprinted by permission from Nature

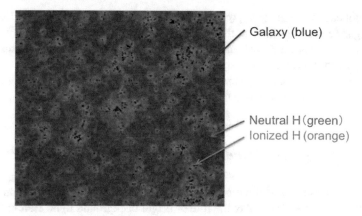

Fig. 3.73 A snapshot of the simulations of cosmic reionization by Iliev et al. [135]. The blue dots indicate galaxies emitting ionizing photons. The green and orange regions represent, respectively, regions with a neutral and ionized IGM. The orange regions are called ionized bubbles. This figure is reproduced by permission of the MNRAS

or cosmic HII regions. Ionized bubbles grow by time and merge, eventually making the universe fully ionized. In this process, star-formation in low-mass galaxies is suppressed by heating of their cold gas by background ionizing photons (UV background radiation) if they are located in ionized bubbles (e.g. Susa and Umemura [334], Wyithe and Loeb [366]). This physical picture indicates that galaxies drive cosmic reionization by supplying ionizing photons, while star-formation in galaxies is strongly influenced by the UV background radiation. Thus, cosmic reionization and galaxy formation have a tight physical relation. Because the UV background radiation in ionized bubbles is originally produced by galaxies, one can also find that cosmic reionization is a cosmological-scale feedback process for galaxies. In reionization studies, one of the observational goals is to test this physical picture.

The key quantity for describing cosmic reionization is the neutral hydrogen fraction,

$$x_{\rm HI} = \frac{n_{\rm HI}}{n_{\rm H}},\tag{3.38}$$

where $n_{\rm HI}$ is the neutral hydrogen density and $n_{\rm H}$ the neutral+ionized hydrogen density.[13] The evolution of the IGM ionization, i.e., the history of reionization, is described by $x_{\rm HI}$ as a function of redshift. Note that the ionized hydrogen fraction

$$Q_{\rm HII} = 1 - x_{\rm HI}\tag{3.39}$$

is often used in place of $x_{\rm HI}$.

---

[13]This is the volume-averaged fraction. There is another definition of the neutral hydrogen fraction, the mass-averaged neutral hydrogen fraction, that is sometime used. Because the mass-averaged neutral hydrogen fraction is difficult to evaluate in observations, the volume-averaged fraction is referred to as the neutral hydrogen fraction in most observational studies.

It is not easy to estimate x_{HI} (or Q_{HII}). Emission from ionized gas (e.g. Lyα) and neutral gas (e.g. 21 cm line) in the IGM is too diffuse to be directly detected even with today's technology. Instead, one needs to detect an absorption or scattering signal by neutral gas imprint in spectra of bright background sources such as QSOs, CMB, LAEs, and gamma ray bursts (GRBs). The following subsections summarize x_{HI} estimates obtained with this method.

3.5.2 Probing Reionization History I: Gunn Peterson Effect

The classic method for estimating x_{HI} is to use HI Lyα absorption lines in high-z QSO spectra, where QSOs play the role of bright background light. Before the end of reionization, neutral hydrogen of the IGM makes a complete absorption trough in QSO spectra at wavelengths shorter than Lyα, which is called the Gunn Peterson effect (Gunn and Peterson [106]; see also Field [85], Shklovskii [303], Bahcall and Salpeter [15]). The strength of this effect for a given QSO spectrum is evaluated with the Gunn–Peterson optical depth τ_{GP}

$$I/I_0 = e^{-\tau_{GP}}, \tag{3.40}$$

where I and I_0 are the observed and intrinsic QSO continuum flux densities at the wavelength of the Lyα absorption of the redshifted IGM neutral hydrogen. Here, I_0 is estimated by a power-law extrapolation of the observed QSO continuum at > 1216 Å. The relation between τ_{GP} and x_{HI} is written as

$$\tau_{GP}(z) = 4.9 \times 10^5 \left(\frac{\Omega_m h^2}{0.13} \right)^{-1/2} \left(\frac{\Omega_b h^2}{0.02} \right) \left(\frac{1+z}{7} \right)^{3/2} x_{HI}(z) \tag{3.41}$$

Fan et al. [81]. One can use this equation to estimate x_{HI} from τ_{GP}. Figure 3.74 presents optical spectra of QSOs at $z = 5.7 - 6.4$. In this figure, QSO residual fluxes escaping from the IGM absorption are found at $\lesssim 8000$ Å, while no significant continuum fluxes remain at $\gtrsim 8000$ Å up to rest-frame 1216 Å (i.e. Lyα).[14] These QSO spectra indicate large τ_{GP} values at $\gtrsim 8000$ Å, or at $z \gtrsim 6$. Figure 3.75 shows τ_{GP}-based x_{HI} measurements over $z \sim 5$–6.5, indicating that x_{HI} rapidly increases at $z \gtrsim 5.7$. This rapid increase in x_{HI} suggests that cosmic reionization is completing at $z \sim 6$.

In this figure, only lower limits of x_{HI} are obtained at $z \gtrsim 5.7$. This is because this method can estimate x_{HI} only when the IGM is highly ionized with $x_{HI} \lesssim 10^{-4}$. Since Lyα is a resonance line, τ_{GP} cannot be accurately measured for the IGM even with moderately small neutral fractions of $x_{HI} \sim 10^{-4} - 1$, owing to the saturation of

---

[14]Except at wavelengths very close to the QSO Lyα. These wavelengths correspond to the proximity region where hydrogen is completely ionized by strong UV radiation from the QSO.

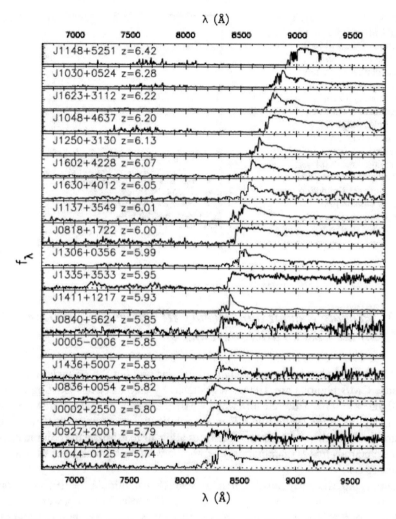

Fig. 3.74 Optical spectra of QSOs at $z = 5.7 - 6.4$ [80]. ©AAS. Reproduced with permission

Lyα absorption. In other words, this method is useful only at the final stage of cosmic reionization, i.e., $z \lesssim 6$. To probe x_{HI} higher than $\sim 10^{-4}$, one can use Lyβ and Lyγ lines whose absorption is weaker than that of Lyα by factors of 3 and 5, respectively. With Lyα absorption lines in QSO spectra, there is another technique to evaluate x_{HI}, which measures the wavelength range over which the spectrum is completely absorbed. This measurement is called the dark gap [81]. However, this additional technique extends the redshift range of x_{HI} measurements only up to $z \sim 6.5$.

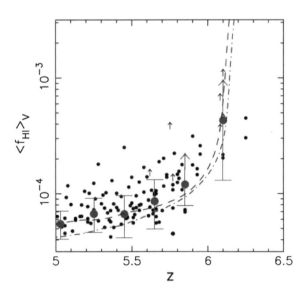

Fig. 3.75 Volume-averaged neutral hydrogen fraction as a function of redshift [80]. The small circles (arrows) indicate neutral hydrogen fraction estimates (lower limits) from QSO τ_{GP} measurements, while the large circles (large circles with an arrow) represent average values (lower limits) for individual redshift bins. The dashed and dot-dashed lines denote the simulation results of Gnedin [98]. ©AAS. Reproduced with permission

3.5.3 Probing Reionization History II: Thomson Scattering of the Cosmic Microwave Background

The CMB is another background light useful for probing the reionization history. Because CMB photons are Thomson-scattered by free electrons existing between $z = 0$ and $z = 1100$ (the redshift when CMB photons are created), one can identify signatures of the scattering in E-mode polarization and temperature fluctuation smearing seen in the CMB. These signatures allow us to estimate the column density of free electrons that is quantitatively expressed with the optical depth of Thomson scattering τ_e.

Here, τ_e is particularly sensitive to large-scale (low multipole $\ell < 10$) anisotropies of CMB polarization. The top panel of Fig. 3.76 presents auto-power spectra of CMB E-mode polarization anisotropies with various τ_e values, demonstrating that measurements of CMB polarization can constrain τ_e. The auto-power spectra of CMB polarization depend not only on τ_e but also on the cosmic reionization history, i.e., redshift evolution of x_{HI}. However, the dependence on the latter is much smaller than the uncertainties in τ_e measurements to date [268]. Thus, when deriving τ_e, one can safely assume that the universe is instantaneously ionized at a redshift that is referred to as z_{re}. The bottom panel of Fig. 3.76 shows posterior probability distributions of τ_e given by Planck 2016 observations. The best-estimate τ_e from the Planck 2016 study is $\tau_e = 0.058 \pm 0.012$.

Fig. 3.76 Top: E-mode polarization auto-power spectrum [268]. The colored curves represent power spectra for τ_e values indicated with the color code on the right hand side, while the gray region denotes the cosmic variance for full sky observations in the case of $\tau_e = 0.06$. Bottom: Posterior distributions of τ_e for various combinations of Planck data [268]. The gray region indicates the range of τ_e that is ruled out by observational constraints from the QSO Gunn–Peterson effect. Reproduced with permission ©ESO

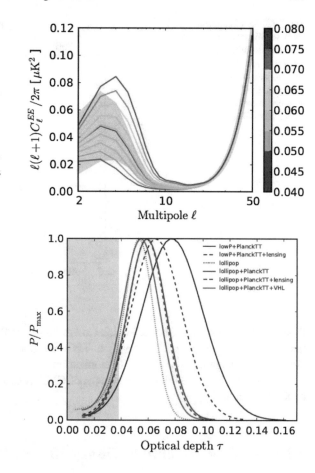

The Thomson scattering optical depth up to a given redshift is expressed as:

$$\tau_e(z) = \sigma_T \int_0^z n_e(z') \frac{dl(z')}{dz'} dz', \qquad (3.42)$$

where σ_T is the cross section of Thomson scattering and n the number density of free electrons. Setting z to 1100 gives the total optical depth between today and the time when CMB photons are created. In the standard picture, there is a negligible contribution to τ_e before the formation of the first stars ($z \gtrsim 20$).

In the case of instantaneous reionization, Eq. (3.42) is simplified to

$$\tau_e(z_{re}) \simeq 0.07 \left(\frac{h}{0.7}\right) \left(\frac{\Omega_b}{0.04}\right) \left(\frac{\Omega_m}{0.3}\right)^{-1/2} \left(\frac{1+z_{re}}{10}\right)^{3/2}. \qquad (3.43)$$

Fig. 3.77 Lyα scattering cross section as a function of wavelength (or velocity). The cross section profile consists of two components: an exponential profile made by thermal motions and a power-law tail by natural broadening

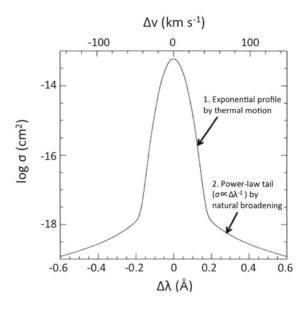

Based on the τ_e estimate above, Planck Collaboration et al. [268] obtain the instantaneous reionization redshift to be $z_{re} \simeq 7.8 - 8.8$ that is a moderately late epoch. However, note again that the cosmic reionization history cannot be constrained well by this method because the power spectra of CMB polarization are not sensitively dependent on it.[15]

3.5.4 Probing Reionization History III: Lyα Damping Wing

Sections 3.5.2 and 3.5.3 have reviewed two methods to probe the cosmic reionization history. The method using the Gunn Peterson effect can pinpoint the completion epoch of cosmic reionization at $z \sim 6$, but it cannot probe $z \gtrsim 6$ due to the saturation of Lyα absorptions. The method using CMB Thomson scattering, on the other hand, can probe the entire cosmic history with free electrons between $z = 0$ and the CMB epoch, but it has not been able to clearly distinguish different cosmic reionization histories.

To probe x_{HI} at the missing epoch of $z \gtrsim 6$, one can use HI Lyα damping wing (DW) absorptions seen in LAE, GRB, and QSO spectra. Briefly, Lyα DW is the tail of Lyα absorption. Figure 3.77 presents the profile of Lyα cross section. The Lyα absorption profile consists of two components. One is the main component with

---

[15]There is another probe for the cosmic reionization history that uses Kinetic Sunyaev–Zeldovich effects of the CMB temperature anisotropies made by the bulk motion of free electrons at the EoR. However, the constraints on the cosmic history, so far obtained, are not strong (see the summary of Planck Collaboration et al. [268]).

an exponential profile produced by thermal motions, and the other a weak, power-law component due to natural broadening (i.e., quantum mechanics's uncertainty principle in energy and time). Lyα DW corresponds to the latter. The Lyα DW absorption is more than 5 orders of magnitude weaker than the peak of the main absorption component. Moreover, since the DW absorption has a power-law shape ($\sigma \propto \Delta\lambda^{-2}$), it can extend to much redder wavelengths beyond 1216 Å than the main component. Because the Lyα DW absorption is significantly weaker than that by the Gunn–Peterson effect, the DW absorption allows us to investigate the IGM with a moderately high neutral hydrogen fraction, $x_{HI} \sim 0.1 - 1.0$. Moreover, the extended profile of the DW absorption is useful to study continua at >1216 Å free from the Gunn–Peterson effect. Below, I detail x_{HI} constraints from the Lyα DW absorption so far obtained with GRB, QSO, and LAE spectra.

3.5.4.1 GRBs

To date, four GRBs at $z \sim 6 - 7$ have been used to constrain x_{HI}: GRB 050904 at $z = 6.3$ [342], GRB 080913 at $z = 6.7$ [257], GRB 130606A at $z = 5.9$ [48, 343, 344], and GRB 140515A at $z = 6.3$ [49]. Figure 3.78 presents an observed spectrum of GRB 050904 and the best-fit continuum model with the Lyα DW absorption. Here, the intrinsic spectrum shortward of Lyα is an extrapolation of a power-law function fitted to the observed continuum longward of Lyα. Note that Lyα DW absorption

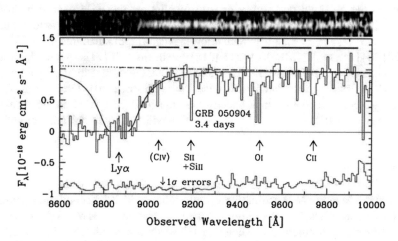

Fig. 3.78 GRB 050904 afterglow spectrum [342]. Top: Two dimensional spectrum. Middle: One dimensional spectrum. The solid curve represents the best-fit model of the absorption by the neutral hydrogen of the host galaxy in the case of no IGM absorption. The dotted line is an intrinsic power-law spectrum determined from the observed continuum in the wavelength ranges indicated by the horizontal lines at $F_\lambda \simeq 1.35 \times 10^{-18}\,\mathrm{erg\,cm^{-2}\,s^{-1}\,\mathring{A}^{-1}}$. The dashed line denotes a model spectrum when only the IGM absorption shortward of Lyα is taken into account; a very low neutral fraction of $x_{HI} = 10^{-3}$ is assumed here. Bottom: One sigma errors in the observed spectrum. This figure is reproduced by permission of the PASJ

modeling should consider not only IGM neutral hydrogen but also that in the GRB host galaxy. The Lyβ line is also used to resolve the degeneracy between the IGM and host-galaxy components [342]. Among the four GRBs, one gives an estimate of $x_{\rm HI}$ of a few percent, while the others only place an upper or lower limit of $x_{\rm HI} \lesssim 0.1 - 0.7$ at $z \sim 6 - 7$. Although there are many GRBs found at the EoR, $z \gtrsim 6$, that include GRB 090423 at $z = 8.2$, the most distant GRB confirmed to date [338], all of them are too faint to identify the Lyα DW absorption.

3.5.4.2   QSOs

DW absorptions in QSO spectra also provide constraints on $x_{\rm HI}$. As in the GRB DW absorption analyses, one needs to assume an intrinsic spectrum for the QSO in question. While the major uncertainty in GRB analyses is the DW absorption by HI gas in the host galaxy, QSO analyses include potential systematic uncertainties in the Lyα emission profile and the QSO near-zone size that impact on the shape of the intrinsic QSO spectrum before IGM absorption. To mitigate these systematic uncertainties, one can use a low-z QSO spectrum template to estimate the intrinsic spectrum. Figure 3.79 presents the spectrum of QSO ULAS J1342+0928 at $z = 7.5$. The bottom left panel of Fig. 3.79 is a close-up around the Lyα wavelength of this object, overplotted with the best-estimate intrinsic spectrum that is a composite of SDSS QSO spectra (thick red line). The spectrum with the DW absorption by neutral hydrogen is presented with a thick blue line. The bottom right panel of Fig. 3.79 indicates the obtained probability density function of the IGM neutral hydrogen fraction, from which the best-estimate IGM neutral hydrogen fraction is found to be $x_{\rm HI} = 0.56^{+0.21}_{-0.18}$ at $z = 7.5$ [17]. With a lower-redshift QSO than this object, ULAS J112010641 at $z = 7.1$, Mortlock et al. [218] obtain $x_{\rm HI} > 0.1$. This neutral hydrogen constraint at $z = 7.1$ is a lower limit because of an uncertain contribution by a damped Lyα (DLA) system associated with this object. Subsequently, Greig et al. [101] carefully reconstruct the intrinsic spectrum of this QSO from SDSS BOSS data, and obtain $x_{\rm HI} = 0.40^{+0.21}_{-0.19}$ at $z = 7.1$ in conjunction with patchy reionization modeling.

3.5.4.3   LAEs

LAEs play a unique role in estimating $x_{\rm HI}$. In contrast with bright continuum sources, i.e., GRBs and QSOs, DW absorptions in the Lyα emission lines of LAEs are used to quantify $x_{\rm HI}$ (e.g. Malhotra and Rhoads [193], Kashikawa et al. [151, 152], Ouchi et al. [246]). Although spectra of LAEs are too faint to be modeled with a comparable accuracy as those of GRBs and QSOs, the abundance of LAEs is orders of magnitude higher than those of GRBs and QSOs. Thus, LAEs can probe the HI of the IGM with a large number of sightlines, which reduces the field variance systematics. Moreover, one can evaluate the IGM absorption amount of Lyα DW with simple comparisons of Lyα statistics, exploiting the large statistics given by abundant LAEs.

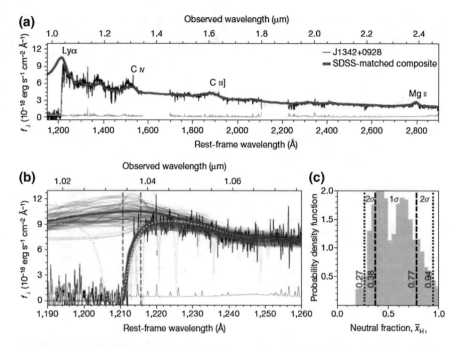

Fig. 3.79 Spectra of QSO ULAS J1342+0928 at $z = 7.5$ and x_{HI} estimation [17]. Top: Observed spectrum of ULAS J1342+0928 (black line) and the best-matched SDSS QSO composite spectrum (red line). The gray line represents the 1σ error. Bottom left: Same as the top panel, but for a close-up view around the Lyα wavelength. The thick blue line represents the best-matched composite spectrum with the DW absorption by the IGM whose neutral fraction is set to the average value over the redshift range between $z = 7$ and the end of the QSO proximity zone (blue dashed lines). The green dotted line denotes a spectrum with a single absorber, such as a foreground galaxy, that appears different from the observed spectrum. Bottom right: Probability density distribution (blue histogram) of x_{HI} values that are obtained by a fitting of the composite spectrum with DW absorptions. The 1 and 2σ ranges of x_{HI} are indicated with vertical dashed and dotted lines, respectively. Reprinted by permission from Nature

Figure 3.80 shows the evolution of the Lyα luminosity function and $\rho_{\text{Ly}\alpha}$ over $z = 5.7 - 7.3$ obtained by deep narrowband imaging and spectroscopic surveys. Both the Lyα luminosity function and $\rho_{\text{Ly}\alpha}$ decrease from $z = 5.7$ towards higher z. Moreover, it is also suggested that $\rho_{\text{Ly}\alpha}$ evolution is accelerated at $z \gtrsim 7$, and that the decrease in $\rho_{\text{Ly}\alpha}$ at $z > 7$ is clearly faster than that in ρ_{UV} that represents the star-formation rate density evolution. This accelerated evolution of $\rho_{\text{Ly}\alpha}$, which cannot be explained by any observed evolutionary trends in the star-formation properties of LAEs, suggests high neutral hydrogen fractions of $x_{HI} = 0.2 \pm 0.2$ ($z = 6.6$), $x_{HI} = 0.25 \pm 0.25$ ($z = 7.0$), and $x_{HI} = 0.55 \pm 0.25$ ($z = 7.3$) from comparisons of various reionization models, where the errors include model variances [142, 165, 246]. To estimate x_{HI}, the Lyα emitting galaxy fraction $X_{\text{Ly}\alpha}$ is also measured by deep follow-up spectroscopy of dropout galaxies. Here, $X_{\text{Ly}\alpha}$ is defined by the ratio of

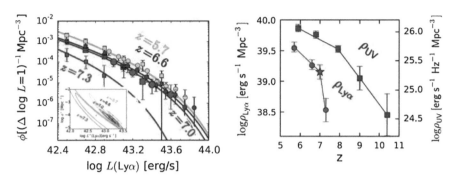

Fig. 3.80 Left: Evolution of the Lyα luminosity function over $z = 5.7 - 7.3$ [142]. The cyan, blue, red, and magenta data points (curves) present Lyα luminosity functions (the best-fit Schechter functions) at $z = 5.7, 6.6, 7.0$, and 7.3, respectively. The inset panel presents the error contours of the Schechter parameters ϕ^* and $L_{Ly\alpha}^*$ at the 68 and 90% confidence levels. Right: Redshift evolution of the Lyα luminosity density (red symbols and lines) and the UV luminosity density (blue symbols and lines) [142]. The Lyα luminosity density drops at $z \gtrsim 7$ faster than the UV continuum density, suggestive of strong Lyα absorption by the IGM with a moderately high x_{HI} at $z \gtrsim 7$. The left and right ordinate axes indicate the Lyα and UV-continuum luminosity densities, respectively. ©AAS. Reproduced with permission

galaxies with Lyα emission to all galaxies down to a given UV-continuum magnitude limit. In contrast to the LAE selection, the UV-continuum selection does not depend on cosmic reionization (i.e., the value of x_{HI}). Figure 3.81 indicates that $X_{Ly\alpha}$ peaks at $z \sim 6$ and decreases towards higher z [234, 262, 263, 299, 301, 324, 347]. The x_{HI} value is estimated to be $x_{HI} = 0.39^{+0.08}_{-0.09}$ ($z \sim 7$) and $x_{HI} > 0.64$ ($z \sim 8$; Schenker et al. [301]). All of these Lyα emission observations suggest that the Lyα emissivity of galaxies decreases from $z \sim 6$ towards higher z, and that the neutral hydrogen fraction is moderately high at $z \sim 7 - 8$.

3.5.5 Reionization History

Figure 3.82 summarizes x_{HI} estimates given by the Gunn Peterson effect, Thomson scattering of the CMB, and Lyα DW absorptions of GRBs, QSOs, and LAEs. This figure clarifies that the measurements of the Gunn Peterson effect reveal the completion epoch of cosmic reionization at $z \sim 6$ (Sect. 3.5.2). The Lyα DW absorption measurements suggest a moderately high x_{HI} at $z \sim 6 - 8$, indicative of late reionization, albeit with large uncertainties (Sect. 3.5.4). Although the CMB Thomson scattering results have no time resolution, they also imply a moderately late reionization epoch of $z_{re} \simeq 7.8 - 8.8$ for the case of instantaneous reionization (Sect. 3.5.3). However, because all the $x_{HI}(z)$ data plotted here have large uncertainties, the duration of cosmic reionization, by which the reionization process is characterized, e.g., as being sharp or extended, remains to be determined [139].

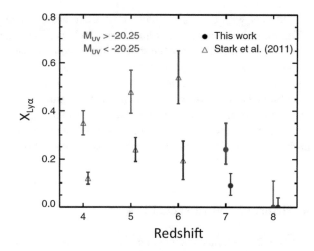

Fig. 3.81 Lyα emitting galaxy fraction $X_{Ly\alpha}$ as a function of redshift [301]. Lyα emitting galaxies are defined as those with Lyα $EW \geq 25$ Å. The red and blue data points represent $X_{Ly\alpha}$ for UV-continuum faint ($M_{UV} > -20.25$) and bright ($M_{UV} < -20.25$) dropout galaxies, respectively. ©AAS. Reproduced with permission

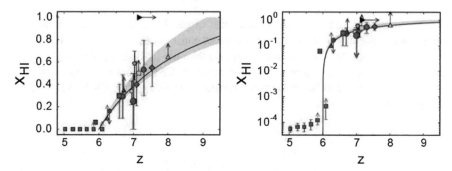

Fig. 3.82 Redshift evolution of x_{HI} [142]. The left and right panels are the same, but with linear and log-scale ordinate axes, respectively. The green squares represent estimates from QSO Gunn Peterson optical depths, while the other green symbols indicate results of QSO DW measurements. The magenta and red data points denote x_{HI} estimated from LAE DW absorptions. The cyan symbols are given by Lyα emitting galaxy fractions. The blue symbols show the results of GRB DW measurements. The orange pentagon presents an estimate obtained with the Lyα EW distribution of dropout galaxies. The black triangle with an arrow indicates the 1σ lower limit of z_{re} obtained by Planck Collaboration et al. [268]. The black curve and gray shade show x_{HI} and its uncertainty suggested from the evolution of the UV luminosity function [139]. ©AAS. Reproduced with permission

3.5.6 H I 21 cm Observations: Direct Emission from the IGM

Sections 3.5.2–3.5.4 introduce studies of cosmic reionization probed with bright background radiation. Although it is difficult to directly detect emission from the diffuse IGM at the EoR with the technology today, there are many efforts to find such a signal. The most important direct signal is the H I hyperfine structure line of 21 cm wavelength that is produced when the spins of the proton and electron in a neutral hydrogen atom flip from antiparallel to parallel. Catching this emission produced at the EoR ($z \sim 10$) needs a low frequency radio observation because it is redshifted to $\sim$100 MHz. Future 21 cm emission data will allow us to study the cosmic reionization history $x_{HI}(z)$ and the topology of H I distribution that depends on ionizing sources (i.e., galaxies vs. AGNs; ionizing photons from these two populations have different mean-free paths against neutral hydrogen gas because of different spectral shapes).

3.5.6.1 Basic Picture of EoR H I 21 cm Emission

The strength of the 21 cm emission depends on the spin temperature T_s (from Maxwell–Boltzmann equation) that is defined by

$$\frac{n_{\uparrow\uparrow}}{n_{\uparrow\downarrow}} = 3\exp\left(-\frac{h\nu_{21\,cm}}{kT_s}\right), \tag{3.44}$$

where $n_{\uparrow\uparrow}$ and $n_{\uparrow\downarrow}$ are the number densities of parallel and antiparallel hydrogen atoms, h the Planck constant, k the Boltzmann constant, and $\nu_{21\,cm}$ the frequency of 21 cm wavelength. When T_s is lower (higher) than the CMB temperature T_{CMB}, the 21 cm line is observed as absorption (emission) in the CMB spectrum. The observable is thus an increment of brightness temperature relative to the CMB, δT_B, that is described as

$$\delta T_B \simeq \frac{T_s - T_{CMB}}{1+z}\tau_{21\,cm} \tag{3.45}$$

$$\simeq 7(1+\delta)x_{HI}\left(1 - \frac{T_{CMB}}{T_s}\right)(1+z)^{1/2}\quad \text{mK}, \tag{3.46}$$

where δ is the baryon overdensity and $\tau_{21\,cm}$ the H I optical depth at 21 cm [81, 275]. Figure 3.83 presents a theoretical prediction of the brightness temperature increment evolution, together with an H I map illustration. At $z \sim 150$, baryons and CMB decouple because collisions and cooling dominate in baryon gas. After $z \sim 80$, the cosmic baryon density is sufficiently low that the collisional cooling is inefficient. After the first stars and QSOs, i.e. galaxies, form at $z \sim 20 - 30$, Lyα photons from galaxies are scattered by H I gas in the IGM. This process redistributes the two spin states of H I, and enlarges the difference between the spin and CMB temperatures (Lyα cooling aka Wouthuysen-Field effect). Then, the H I of the IGM is heated by X-ray emission from objects. At $z \sim 15$, the reionization begins, and the brightness temperature increment becomes small, due to an increase in ionized regions in the IGM. In this way, the evolution of δT_B is predicted. The spatial fluctuations of δT_B also vary with evolutionary phase due to differences in heating and cooling sources.

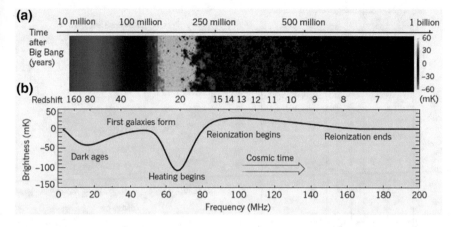

Fig. 3.83 Evolution of brightness temperature from $z = 200$ to 6 suggested by a theoretical model [276]. Top: Brightness temperature with spatial fluctuations as a function of redshift (cosmic age). The brightness temperature is color-coded following the color bar on the right hand side. Bottom: Sky-averaged brightness temperature as a function of redshift (observed frequency). Also indicated are the epochs of several major cosmic events that change the brightness temperature of the 21 cm line. Reprinted by permission from Nature

3.5.6.2 Early Hı 21 cm Observation Results and Expectations

Measuring the brightness temperature of the IGM at the EOR requires radio observations in low frequencies ($\sim$100 MHz) (see Fig. 3.83), and several programs have conducted such observations: Giant Metrewave Radio Telescope (GMRT; Paciga et al. [249]), Precision Array for Probing the Epoch of Reionization (PAPER; Parsons et al. [251], Ali et al. [5]), LOw Frequency ARray (LOFAR; Yatawatta et al. [372], Jelić et al. [150]), and Murchison Widefield Array (MWA; Dillon et al. [70]). The left panel of Fig. 3.84 summarizes 21-cm power spectrum ($z \sim 8 - 9$) results from these programs. So far, no programs have identified a signal of the 21-cm emission from the EoR, and only upper limits have been obtained. The left panel of Fig. 3.84 indicates that the 2σ upper limits are about 2 orders of magnitude higher than predicted signals. Although some programs have expected sensitivities high enough to detect the EoR 21 cm emission, in practice the expected sensitivities cannot be reached due to difficulties in subtraction of bright foreground emission. There are many Galactic and telluric foreground sources. One of the most challenging foregrounds is ionospheric radio emission that varies with sky position and time. Thus, the most important challenge in detecting 21 cm signals from the EOR is to properly model the foreground emission that is a few orders of magnitude brighter. Because such foreground emission dominates in specific Fourier spaces and wavelengths, it would be possible to isolate the EoR 21 cm emission in the parameter space that is referred to as the "EoR window" [63], free from the foreground emission.

Although EoR 21-cm emission signals have not been detected yet, there is a report of EoR 21-cm absorption detection. Bowman et al. [33] have conducted low-

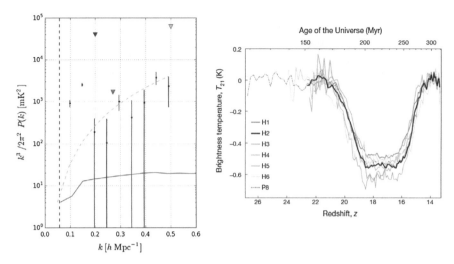

Fig. 3.84 Left: Observed and predicted power spectra of 21 cm emission [5]. All observational data points are upper limits. The yellow, magenta, and green triangles show 2σ upper limits at $z = 8.6$, 9.5, and 7.7 that are given by GMRT, MWA, and PAPER experiments, respectively. The magenta curve represents the model 21 cm power spectrum for the 50% reionization case predicted by Lidz et al. [185]. The black dots, the cyan dashed line, and the black dashed vertical line indicate the results of the PAPER experiment shown in Ali et al. [5], while these results are negated by Ali et al. [6] due to the underestimations of the signal uncertainties. ©AAS. Reproduced with permission. Right: 21 cm absorption profiles at $z = 17$ best-fitted to the EDGES data [33]. The eight lines with different colors indicate absorption profiles for different observational set-up data, among which the thick black line corresponding to the highest signal-to-noise ratio case. Reprinted by permission from Nature

frequency radio observations with the Experiment to Detect the Global Epoch of Reionization Signature (EDGES) low-band instruments, and found an absorption at 78 MHz corresponding to $z = 17$ (right panel of Fig. 3.84). This absorption may be a signature of first stars and QSOs whose Lyα photons lower the brightness temperature by the Wouthuysen-Field effect. However, the observed absorption is significantly stronger than model predictions. In other words, the hydrogen gas at $z \sim 17$ is suggested to be colder than the gas kinetic temperature as well as the CMB temperature. Barkana [19] claims that cold hydrogen gas can be produced by the interaction of hydrogen gas with dark matter whose temperature is low enough to explain the low brightness temperature suggested by the EDGES observations. Because the detection of an absorption by the EDGES has a great impact on our understanding of the thermal history of the universe, it should be confirmed by independent projects.

Theoretical models predict that the cross-correlation function between H I 21 cm emission and LAEs is key for understanding the reionization process. If cosmic reionization proceeds from high to low density regions, so called in the inside-out manner, star-forming galaxies including LAEs exist preferentially in cosmic ionized bubbles. Moreover, cosmic ionized bubbles allow Lyα photons escaping from LAEs

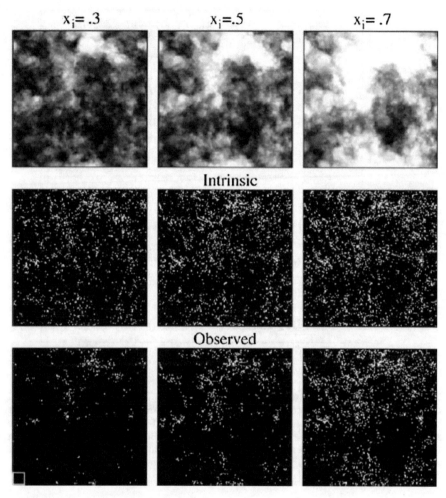

Fig. 3.85 Spatial distributions of ionized gas (top panels), all LAEs (middle panels), and observed LAEs (bottom panels) over a cosmological volume ($94 \times 94 \times 35\,\mathrm{Mpc}^3$) with average ionized fractions of 0.3 (or $x_{\mathrm{HI}} = 0.7$; left column), 0.5 (middle column), and 0.7 ($x_{\mathrm{HI}} = 0.3$; right column) predicted by numerical simulations [211]. In the top panels, ionized regions are indicated in white. In the middle and bottom panels, LAEs are shown with white dots. This figure is reproduced by permission of MNRAS

to survive, thus enhancing the observed overdensity of LAEs in ionized bubbles (Fig. 3.85; McQuinn et al. [211]). It is thus expected that Hɪ 21 cm emission and LAEs anti-correlate strongly. Figure 3.86 presents theoretical predictions of the cross-power spectrum between Hɪ 21 cm emission and LAEs [186]. The distance scale where the sign of the correlation changes can be used to constrain the typical size of ionized bubbles. The distance scale becomes large towards the end of reionization (i.e. small x_{HI}; Fig. 3.86).

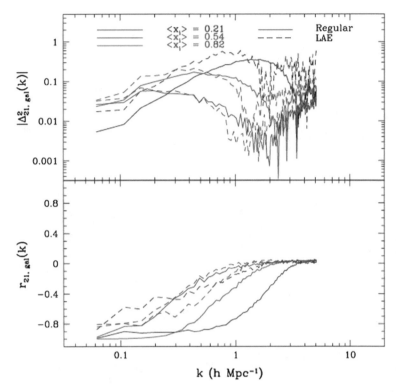

Fig. 3.86 Top: Galaxy-21 cm cross-power spectra predicted by Lidz et al. [186]. The solid and dashed lines denote the 21 cm cross-power spectra for all galaxies and LAEs, respectively. The black, red, and blue colors mean the ionized fractions of 0.21, 0.54, and 0.82 (corresponding to $x_{HI} = 0.79, 0.46,$ and 0.18) at $z = 8.3, 7.3,$ and 6.9, respectively. Bottom: Same as the top panel, but for the cross-correlation coefficients between 21 cm radiation and galaxies. ©AAS. Reproduced with permission

3.5.7 Summary of Cosmic Reionization I

This section has introduced the basic physical picture of cosmic reionization, and showcased various techniques to probe the cosmic reionization history, discussing constraints obtained by those techniques on the neutral hydrogen fraction x_{HI} (or ionized hydrogen fraction Q_{HII}) as a function of redshift. There are three major techniques to estimate x_{HI} that require bright background light. One uses the Gunn–Peterson effect of background QSO spectra, while the other two measure, respectively, the Thomson optical depth of the CMB and Lyα DW absorptions seen in background GRB, QSO, and LAE spectra. The measurements of the Gunn Peterson effect suggest that cosmic reionization ended at $z \sim 6$ (Sect. 3.5.2). The Thomson scattering optical depth of the CMB obtained from the recent Planck 2016 data is small ($\tau_e \sim 0.06$) (Sect. 3.5.3), and the Lyα DW absorption strengths indicate a moderately high x_{HI} at $z \sim 6 - 8$ (Sect. 3.5.4). These two pieces of information support late reionization. Since the CMB Thomson scattering results available to date have no time resolution and the Lyα DW constraints on x_{HI} are not very strong, the duration of cosmic reionization (i.e. sharp or extended reionization) has been constrained only weakly.

3.6 Cosmic Reionization II: Sources of Reionization

This section presents progresses in observations for understanding sources of reionization that is one of the two major questions of cosmic reionization (see the first paragraph of Sect. 3.5).

3.6.1 What Are the Major Sources Responsible for Reionization?

There are several candidates for sources of cosmic reionization that supply ionizing photons at the EoR. These candidates include galaxies, AGNs, high-mass X-ray binaries (HMXBs), primordial blackholes (PBHs), and dark-matter annihilation.

Although it is obvious that galaxies and AGNs should contribute to cosmic reionization because they are bright in UV, the question is the relative contributions of individual candidate populations. AGNs produce not only UV ionizing photons, but also X-ray photons whose mean-free paths in the HI IGM are as large as the sizes of LSSs. If the X-ray emission of AGNs dominates in reionizing the universe, the structures of ionized regions should be smooth. HMXBs can also contribute via X-ray radiation, but it is not yet clear whether they play a major role because the observed X-ray background is mostly explained by known AGNs at redshifts up to $z \sim 6$ [123]. PBHs would emit Hawking radiation that would heat the IGM, but observational studies place moderately tight upper limits on the fraction of the total mass of PBHs to dark matter that is less than $\sim 10\%$ over the PBH masses of $\sim 10^{-20}$ to $1 M_\odot$ [227]. It is predicted that dark-matter particles annihilate into high energy particles including neutrinos and gamma rays that produce X-ray radiation. However, this happens only if dark matter is made of supersymmetric particles such like axions.

In this lecture, I only consider reionization by UV ionizing photons and discuss two promising reionization sources, galaxies, and AGNs.

3.6.2 Ionization Equation for Cosmic Reionization

The key quantity for sources of reionization is the production rate of ionizing photons $\dot{n}_{ion}$ that is defined by the number of ionizing photons per volume and time. The $\dot{n}_{ion}$ values should be estimated by observations for galaxies and AGNs. The $\dot{n}_{ion}$ value is related with the ionized hydrogen fraction of the IGM Q_{HII} (Eq. 3.39) via the simple one zone model of the ionization equation [138, 189, 286, 287],

$$\dot{Q}_{\mathrm{HII}} = \frac{\dot{n}_{\mathrm{ion}}}{\langle n_{\mathrm{H}} \rangle} - \frac{Q_{\mathrm{HII}}}{t_{\mathrm{rec}}}, \tag{3.47}$$

where $\langle n_{\mathrm{H}} \rangle$ and t_{rec} are the average hydrogen number density and the recombination time, respectively, given by

$$\langle n_{\mathrm{H}} \rangle = \frac{X_{\mathrm{p}} \Omega_{\mathrm{b}} \rho_{\mathrm{c}}}{m_{\mathrm{H}}} \tag{3.48}$$

$$t_{\mathrm{rec}} = \frac{1}{C_{\mathrm{HII}} \alpha_{\mathrm{B}}(T)(1 + Y_{\mathrm{p}}/4X_{\mathrm{p}}) \langle n_{\mathrm{H}} \rangle (1 + z)^3}. \tag{3.49}$$

Here, X_{p} (Y_{p}), ρ_{c}, and m_{H} are the primordial mass fraction of hydrogen (helium), the critical density, and the mass of the hydrogen atom, respectively. In Eq. (3.49), $\alpha_{\mathrm{B}}(T)$ is the case B hydrogen recombination coefficient for the IGM temperature T at a mean density. The value of C_{HII} is the clumping factor,

$$C_{\mathrm{HII}} = \frac{\langle n_{\mathrm{HII}}^2 \rangle}{\langle n_{\mathrm{HII}} \rangle^2}, \tag{3.50}$$

where n_{HII} is the density of ionized hydrogen gas in the IGM. With the brackets, $\langle n_{\mathrm{HII}}^2 \rangle$ and $\langle n_{\mathrm{HII}} \rangle^2$ are the spatially averaged values. One can derive the evolution of Q_{HII} with the observational estimates of $\dot{n}_{\mathrm{ion}}$ via Eq. (3.47), where most of the parameters are determined by physics and cosmology. By this technique, the budget of ionizing photons is evaluated (see Sect. 3.6.3).

For the specific case of ionization equilibrium, one can substitute $\dot{Q}_{\mathrm{HII}} = 0$ and $Q_{\mathrm{HII}} = 1$ in Eq. (3.47), and obtain

$$\dot{n}_{\mathrm{ion}} = \frac{\langle n_{\mathrm{H}} \rangle}{t_{\mathrm{rec}}} \tag{3.51}$$

$$= 10^{50.0} C_{\mathrm{HII}} \left(\frac{1 + z}{7} \right)^3 \quad \mathrm{s}^{-1} \, \mathrm{Mpc}^{-3}. \tag{3.52}$$

This condition of $\dot{n}_{\mathrm{ion}}$ gives the lower limit of the ionizing photon production rate that can keep the ionized universe (e.g. Bolton and Haehnelt [26], Ouchi et al. [245]).

One of the free parameters in the ionization equation (Eq. 3.47) is the clumping factor C_{HII} (Eq. 3.53) that determines the recombination rate of ionized hydrogen in the IGM. Based on the ionizing photon emissivity measurements from QSO absorption line data, the clumping factor is estimated to be as low as $C_{\mathrm{HII}} \sim 3$ at $z \sim 6$ [26]. Because the universe becomes homogeneous $C_{\mathrm{HII}} = 1$ with negligibly small fluctuations at the Big Bang epoch, the clumping factor is low, $C_{\mathrm{HII}} \sim 1 - 3$, over the EoR ($z > 6$). In fact, cosmological numerical simulations with the QSO UV background radiation predict monotonically decreasing values of C_{HII} towards high-z with $C_{\mathrm{HII}} \sim 1 - 3$ (Fig. 3.87; Shull et al. [314]; see also Pawlik et al. [259]). The numerical simulation results of Fig. 3.87 are approximated by the power law,

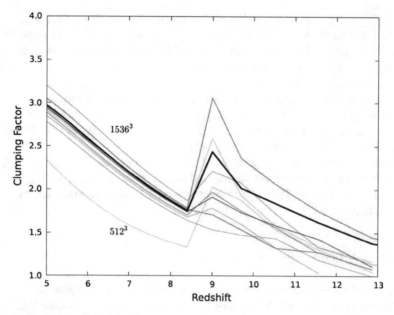

Fig. 3.87 Evolution of clumping factor predicted by the numerical simulations [314]. The black solid line indicates the results of the simulations with 1536^3 cells in the $50h^{-1}$ Mpc box, which are approximated with the function of Eq. (3.53) at $z = 5 - 9$. The rest of the solid lines are the same as the black solid lines, but for 768^3-cell sub-volumes. The dotted line represents the 512^3-cell simulations. ©AAS. Reproduced with permission

$$C_{\mathrm{HII}}(z) = 2.9 \left(\frac{1+z}{6} \right)^{-1.1} \tag{3.53}$$

at $z = 5 - 9$ [314]. Although there remain systematic uncertainties related with the mass resolution and the radiative transfer implementation, the majority of theoretical models agree with these small clumping factors ($C_{\mathrm{HII}} \sim 1 - 3$) at the EoR. If it is true, the uncertainties of clumping factors are not as large as those of the other free parameters (see Sect. 3.6.3).

3.6.3 Galaxy Contribution

To evaluate the ionizing photon contribution of galaxies to the cosmic reionization with Eq. (3.47), one needs to estimate $\dot{n}_{\mathrm{ion}}$ of galaxies. The value of $\dot{n}_{\mathrm{ion}}$ is calculated by

$$\dot{n}_{\mathrm{ion}} = \int_{-\infty}^{M_{\mathrm{trunc}}} f_{\mathrm{esc}}^{\mathrm{ion}}(M_{\mathrm{UV}})\xi_{\mathrm{ion}}(M_{\mathrm{UV}})\phi(M_{\mathrm{UV}})L(M_{\mathrm{UV}})dM_{\mathrm{UV}} \tag{3.54}$$

$$= f_{\mathrm{esc}}^{\mathrm{ion}}\,\xi_{\mathrm{ion}}\,\rho_{\mathrm{UV}}(M_{\mathrm{trunc}}) \qquad \text{[for no } M_{\mathrm{UV}} \text{ dependences]}, \tag{3.55}$$

where $f_{\mathrm{esc}}^{\mathrm{ion}}$ and ξ_{ion} are the ionizing photon escape fraction (Eq. 3.56; Sect. 3.6.3.2) and the ionizing photon production efficiency (Eq. 3.60; Sect. 3.6.3.3), respectively. Here, for simplicity, it is assumed that $f_{\mathrm{esc}}^{\mathrm{ion}}$ and ξ_{ion} do not depend on M_{UV}. It should be noted that $f_{\mathrm{esc}}^{\mathrm{ion}}$ is the escape fraction of ionizing photons that is different from the escape fraction of Lyα photons (Eq. 3.33). The value of $\rho_{\mathrm{UV}}(M_{\mathrm{trunc}})$ is the UV luminosity density defined with Eq. (3.10), where M_{trunc} is the limiting magnitude for the integration, a.k.a. the truncation magnitude [138]. The truncation magnitude indicates how faint galaxies can exist, which depends on the gas cooling and feedback efficiencies in a faint (i.e. low-mass) galaxy.

There are three major parameters for $\dot{n}_{\mathrm{ion}}$, i.e. $\rho_{\mathrm{UV}}(M_{\mathrm{trunc}})$, $f_{\mathrm{esc}}^{\mathrm{ion}}$, and ξ_{ion}.[16] These three parameters are constrained by observations. In the following sections (Sects. 3.6.3.1–3.6.3.3), I introduce constraints on the parameters obtained by observations, to date.

3.6.3.1 UV Luminosity Density

A number of deep optical and NIR imaging surveys have derived luminosity functions of UV continuum at $\sim$1500 Å at the EoR, and estimated ρ_{UV} (e.g. McLure et al. [210], Schenker et al. [300], Oesch et al. [229], Bouwens et al. [31]). These surveys provide good measurements of UV luminosity functions at $z = 6 - 10$, and reveal that the faint-end slopes of the UV luminosity functions are as steep as $\alpha \simeq 2$ [31]. The steep faint-end slopes imply that the ρ_{UV} ($\propto \dot{n}_{\mathrm{ion}}$) value is significantly contributed by faint galaxies ($M_{\mathrm{UV}} \gtrsim -15$) that are not luminous but abundant (see, e.g., Robertson et al. [285]). Because these conventional deep surveys only reach the moderately bright magnitude limit of $M_{\mathrm{UV}} \sim -17$ at $z \sim 7$ even in the Hubble Ultra Deep Field (HUDF) program, a ρ_{UV} value estimate requires an extrapolation of the UV luminosity function from $M_{\mathrm{UV}} \sim -17$ to $M_{\mathrm{UV}} > -15$ to obtain ρ_{UV} via Eq. (3.10). Moreover, the limiting magnitude of M_{trunc} is unknown, requiring an assumption such as $M_{\mathrm{trunc}} = -13$ [286]. The major uncertainty in the ρ_{UV} determination is the extrapolation of the UV luminosity function at the faint end below the detection limit.

To determine ρ_{UV} at the EoR with the measurements of the faint-end UV luminosity function, the Hubble Frontier Fields (HFF) project is conducted [188]. The HFF project has performed ultra-deep optical and NIR imaging with HST/ACS and WFC3-IR, respectively, in six massive galaxy clusters at $z \sim 0.3 - 0.5$, and targeted intrinsically very faint background galaxies lensed by the clusters. Exploiting the lensing magnifications, one can probe the UV luminosity functions down to the detection limit deeper than the one of the HUDF program by a few magnitudes (e.g. Atek et al. [10, 11], Ishigaki et al. [138, 139], Coe et al. [51], Oesch et al. [229, 231], McLeod et al. [207, 208], Livermore et al. [187]). The left panel of Fig. 3.88 presents the UV luminosity function at $z \sim 7$ thus obtained. Although it reaches

---

[16]To understand the sources of reionization, one needs to solve the Eq. (3.47). In this case, there are four major parameters, the three parameters (ρ_{UV}, $f_{\mathrm{esc}}^{\mathrm{ion}}$, ξ_{ion}) for $\dot{n}_{\mathrm{ion}}$ and one parameter (C_{HII}) for t_{rec}.

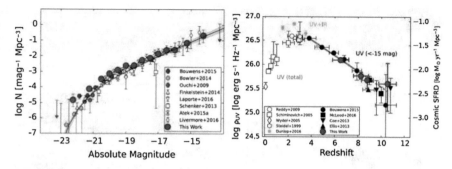

Fig. 3.88 Left: UV luminosity function up to $M_{UV} \sim -14$ [139]. The red circles (and the black open diamonds and crosses) are the luminosity functions derived from the HFF data. The other symbols including blue circles are the luminosity functions obtained without HFF data, which reach up to $M_{UV} \sim -17$. The black line and the gray shade denote the best-fit Schechter function and the fitting error, respectively. Right: Evolution of UV luminosity density that is derived under the assumption of $M_{trunc} = -15$ [139]. The ordinate axis on the right-hand side indicates the cosmic SFR density. The red and black symbols represent the observational data points calculated with UV luminosity functions. The orange circles show the cosmic SFR density estimated from the sum of the UV and FIR luminosity densities. ©AAS. Reproduced with permission

$M_{UV} \sim -14$ mag, no signature of the truncation of the luminosity function is found. The truncation magnitudes would exist at even fainter magnitudes. Nevertheless, the HFF project has revealed the UV luminosity function up to $M_{UV} \sim -14$ mag with no extrapolations, thereby imposing the constraint on the truncation magnitude, that it must be fainter than $M_{UV} \sim -14$ mag at $z \sim 7$. The results of the HFF project significantly reduce the uncertainty on ρ_{UV} estimates that is given by the extrapolation and the assumed M_{trunc} value. The right panel of Fig. 3.88 shows the redshift evolution of ρ_{UV} calculated from these UV luminosity functions under the assumption of $M_{trunc} = -15$. In this panel, ρ_{UV} monotonically decreases from $z \sim 2$ towards high-z. The values of ρ_{UV} at $z \sim 9 - 10$ still include moderately large statistical uncertainties due to the small number of galaxies identified at these redshifts.

3.6.3.2 Escape Fraction of Ionizing Photon

The ionizing photon escape fraction is measured at $\simeq 900$ Å with the ratio of the observed flux f_{900}^{obs} to intrinsic Lyman-continuum (LyC) flux (f_{900}^{int});

$$f_{esc}^{ion} = \left(f_{900}^{obs} / f_{900}^{int} \right) e^{\tau_{900}}, \tag{3.56}$$

where $e^{\tau_{900}}$ is the line-of-sight average IGM opacity to the LyC photons that is determined by QSO absorption line observations [329]. Because the estimate of f_{900}^{int} includes large uncertainties with assumptions, observers introduce the relative escape fraction

$$f_{esc,rel}^{ion} = \left(f_{900}^{obs} / f_{1500}^{obs} \right) e^{\tau_{900}}, \tag{3.57}$$

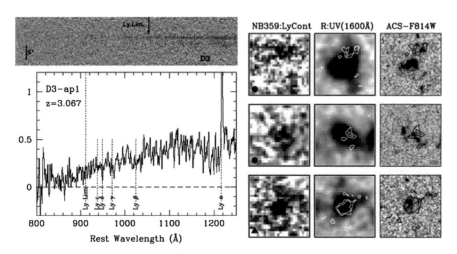

Fig. 3.89 Left: LyC spectra of a galaxy at $z = 3$ [305]. The two- and one-dimensional spectra are shown in the top and bottom panels, respectively. LyC emission is found at wavelengths shorter than the Lyman limit corresponding to the label "Ly Lim" on the left-hand side. Right: Ground-based imaging data of three galaxies at $z = 3$ taken with a narrowband covering the LyC wavelengths (left panels; Iwata et al. [143]). The central panels are the same as the left panels, but with the broadband (R band) images. The right panels display the HST broadband I_{814} images. The contours indicate the LyC emission detected in the narrowband data. ©AAS and ©PASJ. Reproduced with permission

where f_{1500}^{obs} is the observed 1500 Å UV continuum flux [136, 305, 329]. Because the observation study papers discuss f_{esc}^{ion} (i.e. the absolute escape fraction) and $f_{esc,rel}^{ion}$ (i.e. the relative escape fraction), one needs to carefully check the definition of the ionizing photon escape fraction. Hereafter, the absolute escape fraction f_{esc}^{ion} is discussed, unless otherwise specified.

A determination of f_{esc}^{ion} requires very deep observations of galaxies for LyC detections. There are two observational approaches to detect LyC of galaxies, extremely deep spectroscopy and narrowband imaging for star-forming galaxies. The left and right panels of Fig. 3.89 present LyC emission of $z \sim 3$ galaxies found in the spectrum and the narrowband image, respectively. With the spectra and narrowband images, the average f_{esc}^{ion} value is estimated to be $\sim 5\%$ for LBGs and $\sim 20\%$ for LAEs at $z \sim 3$ (Shapley et al. [305], Iwata et al. [143], Nestor et al. [225]; see the other cases in Vanzella et al. [355], de Barros et al. [62]). Interestingly, the average f_{esc}^{ion} of LAEs is higher than the one of LBGs, suggesting a positive correlation between f_{esc}^{ion} and Lyα EW in the moderately low Lyα EW regime (left panel of Fig. 3.90). Moreover, the average f_{esc}^{ion} of $z \sim 3$ galaxies is significantly higher than that of most well known star-forming galaxies ($f_{esc}^{ion} < 3\%$) such as Haro11 and Tol 1247-232 [183, 184]. In addition to the high estimated value of f_{esc}^{ion} at $z \sim 3$, the LyC emission in narrowband images show a spatial offset from the intensity peak of the UV (~ 1500 Å) continuum (right panel of Fig. 3.89). Although the spatial offset of the LyC emission may indicate that the major LyC emitting region is different from the UV-continuum emitting

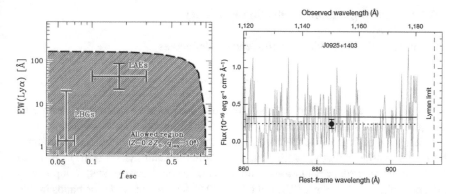

Fig. 3.90 Left: Rest-frame Lyα EW and f_{esc}^{ion} for LAEs (red) and LBGs (blue) at $z = 3$ [224]. The dashed curve is the results of the CLOUDY model calculations with a metallicity 0.2 times solar and an ionization parameter $q_{ion} = 10^8$ cm s^{-1}, indicating the upper limits of Lyα EW. The shaded region represents the parameter space allowed for the model. Right: LyC spectrum of a local star-forming galaxy (gray line; Izotov et al. [146]). Both the filled circle and the dotted line represent the LyC flux averaged over the range 860–913 Å. The error bar associated with the filled circle shows the 3σ uncertainty for the average LyC flux measurement. The horizontal solid line is the same as the dotted one, but corrected for the Milky Way extinction. This figure is reproduced by permission of the MNRAS and Nature

region, there is a possibility that the spatial offset would be a signature of the chance alignment of a foreground (low-z) object whose rest-frame UV continuum can mimic the LyC. However, the probability for such chance alignments is estimated to be only 2–3% [143]. With this small probability, one cannot explain all of the LyC emitting galaxy candidates by chance alignments of foreground objects.

Although foreground contamination objects may not be the reason for the high observational value of f_{esc}^{ion} at $z \sim 3$, there remains the question why observations do not find star-forming galaxies at $z \sim 0$ with $f_{esc}^{ion} \simeq 5 - 20\%$ that is as high as that of $z \sim 3$ LBGs and LAEs. Recent HST/COS observations identify a total of 5 star-forming galaxies at $z \sim 0$ with a high escape fraction, $f_{esc}^{ion} \simeq 6 - 13\%$ [146, 147], which is the definitive evidence that there exist local galaxies with a high f_{esc}^{ion} value comparable with those of high-z galaxies (right panel of Fig. 3.90).[17] These 5 star-forming galaxies are selected with the criterion of $O_{32} \gtrsim 5$ [147]. Because there is a possibility that such a high O_{32} value indicates a high ionization parameter and perhaps a density-bounded nebula, there would exist a positive correlation between f_{esc}^{ion} and O_{32} [224]. In this case, one can easily understand that high-z galaxies, especially LAEs, have the high escape fraction of $\simeq 5 - 20\%$, because high-z galaxies have a large value of $O_{32} \sim 10$ that is significantly larger than the average O_{32} value of local galaxies (Fig. 3.33; Sect. 3.3.4.2). The question of the high f_{esc}^{ion} value for high-z galaxies is being answered by recent studies.

---

[17]It should be noted that Borthakur et al. [29] have identified a local star-forming galaxy with an absolute escape fraction of $f_{esc}^{ion} \simeq 1\%$. This absolute escape fraction corresponds to 21%, if one does not include dust extinction effects.

3.6.3.3 Ionizing Photon Production Efficiency

The ionizing photon production efficiency is defined as

$$\xi_{ion} = \dot{n}^0_{ion}/L_{UV}, \tag{3.58}$$

where $\dot{n}^0_{ion}$ and L_{UV} are the intrinsic ionizing photon production rate (before the escape from the ISM) and the UV ($\sim 1500\,\text{Å}$) continuum luminosity, respectively. The left pane of Fig. 3.91 presents ξ_{ion} as a function of UV spectral slope predicted by stellar synthesis models with various metallicities and IMFs [286]. In this way, ξ_{ion} depends on stellar populations. Although many parameters of stellar populations are constrained with galaxy SEDs in UV ($>1216\,\text{Å}$), optical, and NIR bands including the UV spectral slope, there remain large differences of ξ_{ion} for a given SED shape (see Fig. 3.91 for a given UV slope). This is because ξ_{ion} is very sensitive to the metallicity, IMF, and star-formation history that are key parameters for ionizing photon production, while the observable galaxy SED at $>1216\,\text{Å}$ does not change. Moreover, there is another large uncertainty in the choice of stellar synthesis models for ionizing production rates that depend on the physical properties of massive binary stars [75]. Nevertheless, the galaxy SED approach can suggest a value of

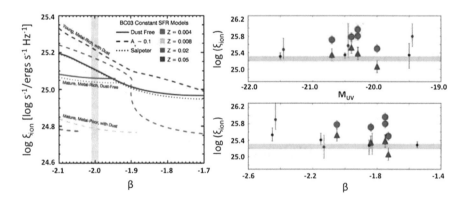

Fig. 3.91 Left: Model predictions for ξ_{ion} as a function of β [286]. The solid and dashed curves represent dust-free and dusty ($A_V \sim 0.1$) galaxies, respectively, that are calculated with the stellar population synthesis model of Bruzual and Charlot [35] under the assumptions of the Chabrier [45] IMF and constant star-formation history. The red, orange, blue, and purple colors of the curves indicate the metallicities of $Z = 0.004, 0.008, 0.02$, and 0.05, respectively. The dotted curves are the same as the solid curves, but for the Salpeter [291] IMF. The calculations stop at 7.8×10^8 yr corresponding to the cosmic age at $z \sim 7$. The gray shade denotes the β range for the average $z \sim 7$ galaxy that is obtained by observations. Right: ξ_{ion} as a function of M_{UV} (top) and β (bottom; Schaerer et al. [304]). The red circles (blue triangles) show ξ_{ion} with (without) a dust attenuation correction of the UV continuum for five star-forming galaxies at $z \sim 0.3$. Note that two blue triangles are indistinguishable in this plot. The black and magenta circles represent LBGs at $z = 3.8 - 5$ and $5.1 - 5.4$, respectively. The cyan region indicates the canonical value for the dust-extinction corrected ξ_{ion}. ©AAS and ©ESO. Reproduced with permission

$\log \xi_{\rm ion}[{\rm erg}^{-1}\,{\rm Hz}] \sim 25.2$ that should include systematic uncertainties by a factor of a few (Robertson et al. [286]; Fig. 3.91). To determine $\xi_{\rm ion}$ with no such systematics, one needs other approaches. A promising method to estimate $\xi_{\rm ion}$ is to use hydrogen Balmer lines such as Hα and Hβ. Because Balmer lines are produced in HII regions via photoionization, one can estimate intrinsic ionizing photon production rates with Balmer line fluxes with the simple analytical relation,

$$\dot{n}_{\rm ion}^0 = 2.1 \times 10^{12}(1 - f_{\rm esc}^{\rm ion})^{-1}L_{{\rm H}\beta}, \qquad (3.59)$$

$$\log(\xi_{\rm ion}) = \log(\dot{n}_{\rm ion}^0) + 0.4\,M_{\rm UV} - 20.64, \qquad (3.60)$$

where $L_{{\rm H}\beta}$ and $M_{\rm UV}$ are the extinction-corrected Hβ luminosity (erg s^{-1}) and the extinction-corrected UV continuum magnitude, respectively [304, 333]. In Eq. (3.59), the term of $(1 - f_{\rm esc}^{\rm ion})$ subtracts the ionizing photons escaping from the galaxy to the IGM. The right panel of Fig. 3.91 shows $\xi_{\rm ion}$ as a function of UV magnitude or UV slope that is obtained by the Balmer line method. The right panel of Fig. 3.91 indicates $\log \xi_{\rm ion}[{\rm erg}^{-1}\,{\rm Hz}] \simeq 25.2 - 25.4$ for $z = 4 - 5$ LBGs [31] and $z \sim 0$ Lyman-continuum leaking galaxies of Izotov et al. [146, 147] that are similar to (or slightly higher than) the value determined by the galaxy SED study [304]. There would be a trend of increasing $\xi_{\rm ion}$ towards small UV slopes (i.e. blue UV continuum; Bouwens et al. [31]). Hereafter, the value of $\log \xi_{\rm ion}[{\rm erg}^{-1}\,{\rm Hz}] \simeq 25.2 - 25.4$ is referred to as the fiducial value.

3.6.3.4 Galaxy Contribution to Reionization: Comparisons of $Q_{\rm HII}(z)$ and τ_e

Because the observational studies have constrained the three major parameters of galaxies, i.e. $\rho_{\rm UV}(M_{\rm trunc})$, $f_{\rm esc}^{\rm ion}$, and $\xi_{\rm ion}$ (Sects. 3.6.3.1–3.6.3.3), one can determine $\dot{n}_{\rm ion}$ via Eq. (3.55) that is the amount of the ionizing photon contribution of galaxies.

The $\dot{n}_{\rm ion}$ value is used in the ionization equation, Eq. (3.47). In Eq. (3.47), $\dot{n}_{\rm ion}$ should include not only the ionizing photons from galaxies but also those from the other ionizing sources such as AGNs. However, $\dot{n}_{\rm ion}$ values of AGNs are poorly determined as discussed in Sect. 3.6.4. Here, one can first assume that the contribution to $\dot{n}_{\rm ion}$ from the other ionizing sources is negligible (i.e. galaxies' $\dot{n}_{\rm ion}$ is dominant), and obtain the evolution of the ionized fraction $Q_{\rm HII}(z)$ and the inferred τ_e with the Eqs. (3.47) and (3.42) via (3.39). Comparing these $Q_{\rm HII}(z)$ and τ_e values with those from the direct measurements shown in Sect. 3.5, one can test whether galaxies can reionize the universe or the other sources of reionization are necessary. Below, I discuss the galaxy contribution to reionization based on these arguments.

The $\dot{n}_{\rm ion}$ value of galaxies is estimated with the three parameters, $\rho_{\rm UV}$ shown in Fig. 3.88 (Sect. 3.6.3.1), $f_{\rm esc}^{\rm ion} \simeq 20\%$ (Sect. 3.6.3.2), and $\log \xi_{\rm ion}[{\rm erg}^{-1}\,{\rm Hz}] \sim 25.2$ (Sect. 3.6.3.3), where no redshift evolutions of $f_{\rm esc}^{\rm ion}$ and $\xi_{\rm ion}$ are included. Assuming the value of $C_{\rm HII} = 3$ (cf. Eq. 3.53), one can derive $Q_{\rm HII}(z)$ with Eq. (3.47). The left panel of Fig. 3.92 presents $Q_{\rm HII}(z)$ calculated with the galaxies' $\dot{n}_{\rm ion}$ thus

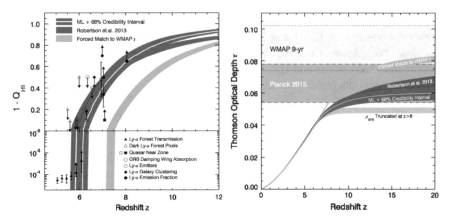

Fig. 3.92 Neutral hydrogen fraction $(1 - Q_{HII})$ as a function of redshift (left) and Thomson scattering optical depth integrated over redshift (right; Robertson et al. [287]). The red region indicates the best estimates obtained from the cosmic star-formation history determined by observations in the case of $f_{esc}^{ion} = 0.2$ and $\log \xi_{ion}/[\mathrm{erg}^{-1}\mathrm{Hz}] = 53.14$. The blue, orange, and cyan regions are the results of the previous study [286], the forced-match to the WMAP data [126], and a different cosmic star-formation history with a rapid decrease at $z \sim 8$, respectively. In the left panel, the data points and the arrows represent the neutral hydrogen fraction estimates obtained from the Gunn–Peterson effect and Lyα damping wing measurements. In the right panel, the dark and light-gray shades denote the Thomson scattering optical depth measurements given by Planck 2015 [267] and the nine-year WMAP [126]. ©AAS. Reproduced with permission

obtained. Although the direct measurements of $Q_{HII}(z)$ have large uncertainties, the inferred $Q_{HII}(z)$ is consistent with the existing direct measurements. The right panel of Fig. 3.92 shows τ_e as a function of redshift. Again, the inferred τ_e value agrees with the direct CMB measurement of τ_e given by Planck Collaboration et al. [267].[18] Although the direct measurements have large statistical errors in $Q_{HII}(z)$ and potentially significant systematic errors in τ_e, the $\dot{n}_{ion}$ value of galaxies alone explain both the direct measurements of $Q_{HII}(z)$ and τ_e. These results may suggest that galaxies are major sources of cosmic reionization.

However, it should be noted that there are a factor of $\gtrsim 2$ uncertainties in the relatively poor determinations of the three parameters as well as the direct measurements. There remain possibilities that these large errors would not allow us to identify an inconsistency between the inferred value and the direct measurement of $Q_{HII}(z)$ or τ_e. One can conclude whether galaxies are major sources of cosmic reionization or not, after the errors on these parameters and the direct measurements become considerably small.

---

[18]The up-to-date best measurement of τ_e is systematically smaller than the value of Planck Collaboration et al. [267] beyond the 1 sigma error level, $\tau_e = 0.058 \pm 0.012$ [268]. The value is even smaller in the latest Planck result, $\tau_e = 0.0561 \pm 0.0071$ [269].

3.6.4 AGN Contribution

Recent galaxy studies suggest that a majority of ionizing photons for reionization would be supplied by galaxies (Sect. 3.6.3). However, these study results still include large systematic uncertainties. One needs to test whether the other sources can contribute to cosmic reionization. AGNs are prominent sources supplying ionizing photons, because a large amount of ionizing photons are efficiently produced in AGNs and escape from them.

The top panel of Fig. 3.93 presents the number density evolution of QSOs, where QSOs are defined as AGNs brighter than -27.6 magnitude. The number density of QSOs peaks at $z \sim 2 - 3$, and decreases towards high-z. At $z \sim 6$, the QSO number density is only $10^{-9}\,\mathrm{Mpc}^{-3}$ [79, 284]. Similarly, UV luminosity functions of QSOs decrease very rapidly towards $z \sim 6$. Based on the UV luminosity functions, the production rate of ionizing photons of QSOs are estimated in the same manner as those of galaxies (Sect. 3.6.3). Although it is assumed that QSO spectra have an ionizing photon escape fraction of unity, and that QSOs have a power law spectrum, these assumptions are plausible for QSOs whose LyC escape and production are well understood in contrast with galaxies (Sect. 3.6.3). The QSO contribution is estimated to be only $\sim$1–10% of ionizing photons at $z \sim 6$ [322]. The ionizing photon production rate of QSOs is negligibly small, due to the very small number density of QSOs at $z \sim 6$. A further decrease of QSO number density is suggested from $z \sim 6$ to $z \sim 7$ [356]. These results indicate that the ionizing photon production rate of QSOs becomes smaller towards high-z.

Although QSOs (i.e. bright AGNs) do not significantly contribute to cosmic reionization, there remains the possibility that faint AGNs could be major contributors of cosmic reionization. As shown in the bottom four panels of Fig. 3.93, the number density of faint AGNs is larger than QSOs by orders of magnitudes. Given the efficient ionizing photon production with the power-law continuum of AGNs, faint AGNs would be important in cosmic reionization. Moreover, the number density of faint AGNs does not drop as steeply as those of bright AGNs towards the epoch of reionization.

Properties of faint AGNs at the EoR are not well understood, due to the difficulties in identifying faint AGNs at such a high redshift. Treister et al. [348] report the $>5\sigma$ detections in the stacked X-ray spectra of dropout galaxies at $z \sim 6$, suggesting the existence of faint AGNs in dropout galaxies.[19] However, subsequent studies identify no X-ray emission in a similar stacked X-ray data of $z \sim 6$ dropout galaxies [91, 363]. Cowie et al. [56] argue that the background subtraction of the X-ray data would produce wrong detections, and that there are no signatures of faint AGNs in dropout galaxies at $z \sim 6$ on average. Nevertheless, the contribution of faint AGNs is still under debate.

Giallongo et al. [96] show a number of dropouts at $z \sim 4 - 6$ with X-ray detections on the individual basis, and claim a steep faint-end slope of AGN UV luminosity

---

[19] Because these faint AGNs are not identified in UV but X-ray, Treister et al. [348] claim that these faint AGNs do not contribute to cosmic reionization due to the obscuration of UV photons.

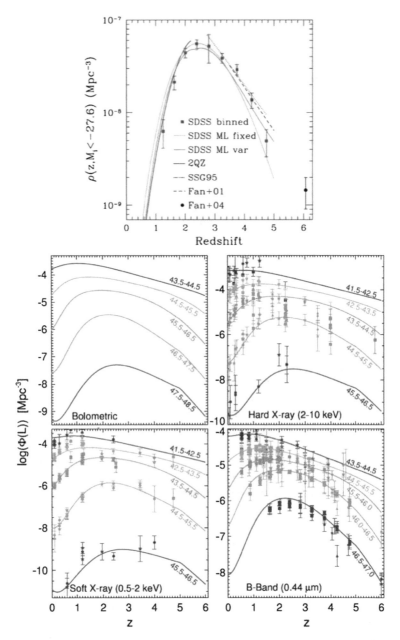

Fig. 3.93 Top panel: Evolution of the QSO number density [284]. The red squares and the black circle are the number densities integrated up to $i = -27.6$ mag in QSO luminosity functions obtained from the SDSS data. The lines indicate the fitting results of these number densities and previous studies. Bottom four panels: Evolution of number densities for AGNs with different luminosities [127]. The top left, top right, bottom left, and bottom right panels present the redshift evolution of AGNs for a given luminosity interval of bolometric luminosity, hard X-ray (2–10 keV), soft X-ray (0.5–2 keV), and B band (0.44 μm), respectively, where the luminosity intervals are indicated with the labels. The lines represent the results of the best-fit evolving double power-law models. ©AAS. Reproduced with permission

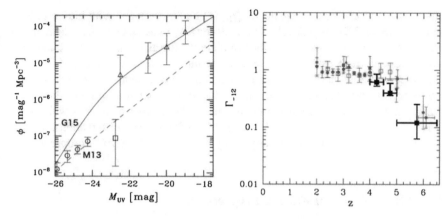

Fig. 3.94 Left: UV luminosity function of AGN at $z \sim 5$. The triangles, circles, and square are the AGN luminosity functions obtained by Giallongo et al. [96], McGreer et al. [206], Ikeda et al. [134], respectively. The solid and dashed lines are the double-power laws best-fit to the data of Giallongo et al. [96] and McGreer et al. [206], respectively. Right: Redshift evolution of the cosmic photoionization rate Γ_{-12} in units of $10^{-12}\,\mathrm{s}^{-1}$ [96]. The black squares are the AGN contribution to Γ_{-12} (connected to $\dot{n}_{\mathrm{ion}}$ of AGNs), which are obtained by Giallongo et al. [96], while the red data points denote the total Γ_{-12} estimated from QSO Lyα absorption line systems. Reproduced with permission ©ESO

functions (left panel of Fig. 3.94). If this steep faint-end slope is true, ionizing photons are mostly originated from faint AGNs at $z \sim 6$ (Giallongo et al. [96]; right panel of Fig. 3.94). However, the faint-end slope value of the AGN UV luminosity function at $z \sim 4 - 6$ remains an open question. Although the UV magnitude range of the Giallongo et al. [96] luminosity function has only a small overlap with the one of the previous study at $M_{\mathrm{UV}} \sim -22$ [206], the AGN number density of Giallongo et al. [96] is about an order of magnitude higher than that of McGreer et al. [206] (left panel of Fig. 3.94). Moreover, the ionizing photon escape fraction of faint AGNs is not well understood [100], while QSOs have an escape fraction as high as unity that is indicated by QSO proximity effects. In this way, the contribution of faint AGNs is not clearly understood yet.

3.6.5   Summary of Cosmic Reionization II

This section has discussed what are the major sources that reionize the universe with the latest observational progresses, explaining the procedures to quantify the budget of ionizing photons. Starting with the galaxy contribution to cosmic reionization, I clarify three major parameters, $\rho_{\mathrm{UV}}(M_{\mathrm{trunc}})$, $f_{\mathrm{esc}}^{\mathrm{ion}}$, and ξ_{ion}, to evaluate the production rate of ionizing photons, i.e. $\dot{n}_{\mathrm{ion}}$. There is another parameter, the clumping factor of the ionized hydrogen IGM (C_{HII}). With the latest observational results for these parameters, the ionizing photon production rate of galaxies alone agrees with the direct observational measurements of $Q_{HII}(z)$ and τ_e obtained with high-z objects and CMB (Sect. 3.5), respectively. These results may indicate that galaxies are major sources of cosmic reionization, although the agreements are not strong, due to the large errors on these parameters. Because the number density of bright AGNs, i.e. QSOs, is very small at the EoR, the ionizing photon contribution of QSOs is negligibly small. However, there is a possibility that ionizing photons of faint AGNs may significantly contribute to cosmic reionization. The number density as well as the escape fraction of the faint AGNs is not clearly understood by observations, to date, and various observational studies are trying to determine these AGN properties.

3.7 On-Going and Future Projects

This section summarizes the open questions about LAEs discussed in the previous sections, and introduces on-going and future projects potentially answering them.

3.7.1 Open Questions

Major open questions about LAEs discussed in this lecture are listed below.

Galaxy Formation

– What are the physical reasons for the LAE's star-formation and ISM characteristics that distinguish LAEs from the other galaxy populations? For example, are the high-ionization parameters of LAEs due to high ionization production rate, to high electron density, or to density-bounded ISM?

– What are the Lyα blobs and diffuse halos? Are they related to cold accretion? Where is the cold accretion that produces Lyα emission?

– What makes the large Lyα EWs ($\gtrsim 200$ Å) of LAEs? What is the major physical reason for the $f_{esc}^{Ly\alpha}$ increase from $z \sim 0$ to 6?

– Have we already identified real popIII star-formation in $z \lesssim 7$ LAEs with strong HeII emission and no detectable metal lines?

Cosmic Reionization

– How did cosmic reionization proceed? Is it true that the reionization occurred late, as suggested by the LAEs and recent CMB studies? Is the evolution of Q_{HII} extended or sharp? How can some Lyα photons escape from $z > 8$ where the IGM is highly neutral (see Sect. 3.7.2.2)?

– Are star-forming galaxies major sources of reionization, especially for low-mass galaxies most of which have intrinsically strong Lyα emission? Do faint AGNs play an important role in supplying ionizing photons? Can we conclude what are the major sources of cosmic reionization, given the large uncertainties on the parameters, $\rho_{UV}(M_{trunc})$, f_{esc}^{ion}, and ξ_{ion} and C_{HII}?

3.7.2 New Projects Addressing the Open Questions

3.7.2.1 On-Going Projects

There are three major on-going projects that can study LAEs up to $z \lesssim 7$, Subaru/Hyper Suprime-Cam (HSC; Miyazaki et al. [212]) + Prime-Focus Spectrograph (PFS; Tamura et al. [337]), Hobby–Eberly Telescope Dark Energy Experiment (HETDEX; Hill et al. [124]), and VLT/Multi-Unit Spectroscopic Explorer (MUSE; Bacon et al. [12]). These three projects can target LAEs, covering the complementary parameter space of LAEs in depth and redshift (Fig. 3.95). Moreover, these three projects are complementary in the covering areas. Subaru/HSC+PFS and HETDEX cover large areas of a few 10–100 deg², while VLT/MUSE will observe small fields with an area up to a few 10 arcmin².

Subaru/HSC+PFS

Subaru/HSC is a wide-field optical imager with a field-of-view (FoV) of 1.5 deg-diameter circle (Fig. 3.96). The FoV of HSC is seven times larger than that of its wide-field imager predecessor at Subaru, Suprime-Cam. The HSC large program, which is the Subaru Strategic Program (SSP) survey, has started in March 2014; it is a collaborative project involving Japanese institutes, Princeton, and Taiwanese

Fig. 3.95 Parameter space of Lyα luminosity versus redshift that is covered by HSC+PFS (red), HETDEX (blue), and MUSE (green) surveys. The HSC+PFS observations will cover $z \sim 2$ and $z \sim 5 - 7$

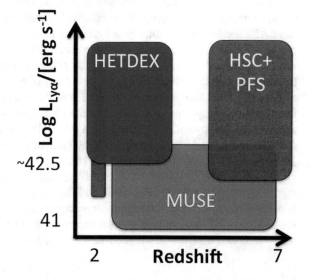

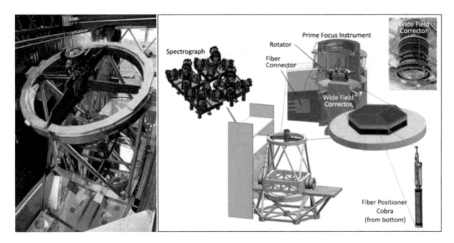

Fig. 3.96 Left: Subaru/HSC mounted on the Subaru telescope (Courtesy: Y. Utsumi). Right: Subaru/PFS conceptual design [335]. The PFS instrument with its 2400 fibers will be installed at the Subaru prime focus, just like the HSC instrument. The massive spectrographs fed by the fibers are placed on the floor of the Subaru dome. This figure is reproduced by permission of the PASJ

institutes. A total of 300 nights are allocated to the HSC SSP survey that will be completed around 2019. The HSC SSP survey has the wedding cake type survey design: it is planned to obtain g, r, i, z, and y band imaging data down to the 5 sigma limiting magnitudes $i \sim 26$ in the wide layer ($1400 \deg^2$), $i \sim 27$ in the deep layer ($30 \deg^2$), and $i \sim 28$ in the ultra-deep layer ($3 \deg^2$). The deep and ultra-deep layers are covered by narrowband images with 5 sigma limiting magnitudes of 25–26. With these imaging data, the HSC SSP survey has a wide variety of scientific goals from cosmology to solar-system objects. For LAE studies, the HSC SSP survey will uncover about 20,000 LAEs and 1,000 Lyα blobs down to $L_{\mathrm{Ly}\alpha} \sim 3 \times 10^{42}\,\mathrm{erg\,s^{-1}}$ at $z = 5.7$ and 6.6. Additional samples of $z = 2.2$ and 7.3 LAEs will be gathered in the deep and ultra-deep layers, respectively. It should be noted that the $z = 5.7$ and 6.6 LAEs are found in a total area nearing 1 comoving Gpc^2, allowing studies of $z \sim 6 - 7$ LAEs in a cosmological large scale. These LAE samples will allow the researchers to address a number of issues of galaxy formation and cosmic reionization by analyses with Lyα luminosity functions and LAE clustering measurements. Some of the early HSC SSP survey results on LAEs have been published. For examples, the evolution of Lyα blobs is determined on the basis of an unprecedentedly large sample (e.g. Shibuya et al. [309]), and the constraints on x_{HI} are being obtained with the goal of an x_{HI} determination accuracy comparable to the model uncertainty of 10% for the reionization history [167, 248].

The wide-field spectrograph Subaru/PFS complements the wide-field imaging capability of Subaru/HSC (Fig. 3.96). PFS is a multi-object fiber spectrograph that is being developed by a consortium including Japan, Princeton, JHU, Caltech/JPL, LAM, Brazil, ASIAA, and many other contributors. PFS accommodates 2400 fibers with a diameter of $1''0 - 1''1$ that cover a FoV of $1.3 \deg^2$, sharing the Subaru wide-

field optical corrector with HSC. There are three arms for blue, red, and NIR bands
that take spectra at $0.38 - 0.65, 0.63 - 0.97$, and $0.94 - 1.26 \,\mu$m, respectively, with
spectral resolutions of $R = 2400 - 4200$. The first light of PFS is planned around
2020. There is a plan of large galaxy survey with PFS, and the strategy of the galaxy
survey is being built. Follow-up spectroscopy for the LAEs detected by the Sub-
aru/HSC SSP survey is envisaged; it will provide unique data sets that will significantly enhance our understanding of galaxy formation and cosmic reionization.

HETDEX

HETDEX is a survey with the Hobby Eberly Telescope (HET) and its visible integral-
field replicable unit spectrograph (VIRUS, Hill et al. [125]). The schematic view of
HET and VIRUS is shown in Fig. 3.97. The VIRUS instrument on HET accommodates 78 integral-field units (IFUs) each of which holds 448 fibers. The total number
of the VIRUS fibers is over 30 thousands. Because HET/VIRUS is optimized to target
redshifted Lyα emission at $z = 1.9 - 3.5$ very efficiently to detect BAO and to constrain the equation of state of dark energy, HET/VIRUS has the narrow wavelength
coverage of 3500–5500 Å and the low spectral resolution of $R \sim 700$. The VIRUS
instrument in part saw the first light in the middle of 2016. The VIRUS instrument
is being built, while the HETDEX survey is conducted. The HETDEX survey will
identify 0.8 million LAEs at $z = 1.9 - 3.5$ in a total area of 434 deg^2 ($\sim$20% filling area) down to $\sim 4 \times 10^{-17}$ erg s^{-1} cm^{-2}. Although the major scientific goal of
HETDEX is understanding the state of equation of dark energy by BAO analysis
with the LAEs, a number of statistical studies are being planned with the HETDEX

Fig. 3.97 Schematic view of HETDEX [156]. Left: Hobby Eberly Telescope seen though the
dome. Right: Schematics of the prime-focus instrument and spectrographs of VIRUS mounted on
the prime focus and the dome, respectively. The IFU fibers are connected from the prime-focus
instrument to the spectrographs. This figure is reproduced by permission of the SPIE

LAEs; for examples, the LAE luminosity function evolution and environment effect in conjunction with clustering analysis. Because the HETDEX LAE data are large and unique, the HETDEX survey will impact many areas of LAE studies.

It should be noted that galaxy-LSS connections can be investigated by such large-volume spectroscopic surveys with HETDEX as well as Subaru/PFS (e.g. Adelberger et al. [3], Rudie et al. [290], Rakic et al. [279], Lee et al. [182], Mawatari et al. [204], Mukae et al. [217]). Deep spectroscopy with Subaru/PFS would reveal the IGM H I and metal gas distribution of LSS with Lyα and metal absorption lines found in spectra of background bright AGNs and galaxies. The dense IFU spectroscopy of HETDEX will give the spatial distribution of LAEs. The combination of these surveys may provide the three-dimensional maps of the IGM gas and LAEs addressing the question where LAEs form in the LSSs.

VLT/MUSE

VLT/MUSE is an IFS with a contiguous field coverage of 1 arcmin^2 that is orders of magnitude larger than those of the other existing IFSs (top panel of Fig. 3.98). MUSE has a reasonably wide range of wavelength coverage (4650–9300 Å) and a medium high spectral resolution ($R \sim 3000$). The image slicers of MUSE keep the total throughputs high, allowing high sensitivity observations. A number of exciting MUSE observation results have been reported since 2015. One of the early MUSE observations consists in a 27-h integration in the central 1 arcmin^2 area of the Hubble Deep Field South (HDF-S). The MUSE HDF-S observations reach the 5σ detection limit of $\simeq 5 \times 10^{-19}$ erg s^{-1} cm^{-2}, and increase the number of the spectroscopically identified galaxies by an order of magnitude (bottom panel of Fig. 3.98; Bacon et al. [13]). Moreover, there are 26 MUSE-identified LAEs whose continuum is not detected in the HST HDF-S images with the detection limit of $I \simeq 29.5$. Further deep and wide surveys are being conducted in HUDF [14] and Chandra Deep Field South [122]. As stated in Sects. 3.4.1.1 and 3.4.1.2, MUSE observations are playing important roles in various Lyα studies, e.g. diffuse Lyα emitters in blank fields and Lyα emission around QSOs.

It should be noted that there is a counterpart IFS that is developed for Keck telescopes, Keck Cosmic Web Imager (KCWI; Martin et al. [198]) that is moved from the Palomar 5 m telescope [198]. The uniqueness of KCWI is the blue-band coverage below 4650 Å that is not covered by VLT/MUSE. The blue-band coverage will allow the studies of the LAEs-IGM relation (see above) at $z \sim 2$ that is an optimal redshift for IGM H I detections with a high S/N from the ground.

3.7.2.2 Future Projects

Although the three major LAE surveys of Subaru/HSC+PFS, HETDEX, and VLT/ MUSE cover a wide range of parameter space, these LAE surveys only target redshifts up to $z \sim 7$ (Fig. 3.95), due to the limited wavelength coverages of optical bands (Sect. 3.7.2.1). There remains the unexplored redshift of $z \gtrsim 8$ for LAE studies.

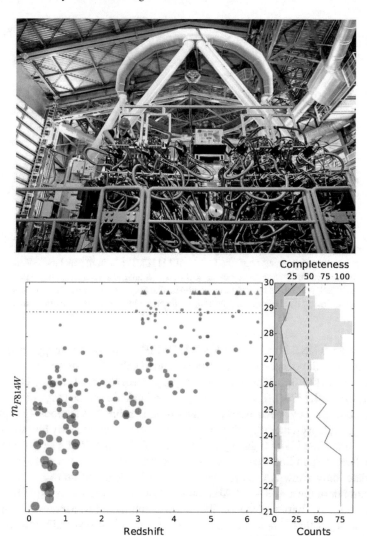

Fig. 3.98 Top: Picture of the MUSE instrument that is installed at the Nasmyth platform (Courtesy: ESO). Bottom: HST I_{814}-band continuum magnitude as a function of redshift (left) and source number/completeness (right) for LAEs identified by the deep MUSE observations in the HDF-S [13]. Left: The brown circles indicate the spectroscopically confirmed objects newly identified by the MUSE observations. The green circles are the same as the brown circles, but for objects confirmed by previous spectroscopy. The sizes of the circles represent the continuum size obtained with the HST I_{814}-band image. The red triangles show the MUSE spectroscopically-confirmed objects with no HST counterparts. The dashed horizontal line presents the 3σ detection limit in the HST I_{814}-band image. Right: The gray histogram denotes the magnitude distribution for all of the HST I_{814}-band detected objects. The light-blue histogram (with the shade at $I_{814} = 29.5 - 30.0$ mag) indicates the magnitude distribution of the MUSE spectroscopically-confirmed objects with (no) HST counterparts. The blue curve represents the completeness of the MUSE spectroscopic confirmation. The dashed vertical line marks 50% of completeness. Reproduced with permission ©ESO

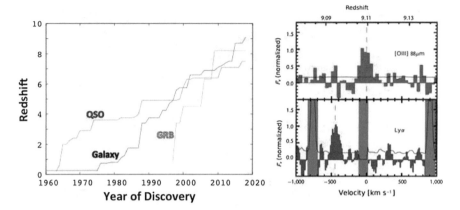

Fig. 3.99 Left: Observational redshift limit versus year of discovery for the redshift limit object. The purple, green, and cyan lines represent a galaxy, QSO, and GRB, respectively. Right: [OIII]88 μm emission of a galaxy (top panel) confirmed at $z = 9.1$ [116]. The bottom panel presents the reported Lyα emission significantly blueshifted by $\simeq 450\,\mathrm{km\,s^{-1}}$ from the systemic ([OIII]88 μm emission) velocity. The black solid lines indicate the 1σ noise. The grey regions show the wavelength ranges strongly contaminated by the night sky emission. Reprinted by permission from Nature

Recent deep HST imaging and Keck/MOSFIRE NIR spectroscopy have pushed the redshift frontier of spectroscopically-identified galaxies from $z \sim 7$ to $8 - 9$ with Lyα emission (Fig. 3.8; Finkelstein et al. [90], Schenker et al. [301], Oesch et al. [229], Zitrin et al. [374]). These observations find galaxies in the range of the highest redshift confirmed by spectroscopy with a strong Lyα line as a signpost (left panel of Fig. 3.99). Moreover, there is an interesting report of the significantly blueshifted Lyα emission associated with a galaxy at a spectroscopic redshift of $z = 9.1$ confirmed with an [OIII]88 μm line (right panel of Fig. 3.99; Hashimoto et al. [116]).

To date, only a few LAEs are spectroscopically identified up to $z \sim 8 - 9$. The small number of the $z \sim 8 - 9$ LAE identifications would be partly explained by the limits of the sensitivities of the existing observation facilities. Moreover, there is an important effect that Lyα emission of LAEs at $z > 6$ is weakened by the damping wing absorption of the IGM HI at the EoR. In fact, the Lyα luminosity function evolution suggests strong dimming of Lyα luminosity from $z \sim 6$ to 7.3 (Fig. 3.80).

If one assumes that both the IGM and the LAEs are uniformly distributed and static, Lyα emission of LAEs cannot escape from the universe as early as the epoch of first galaxies. However, clustering and peculiar motions of LAEs would help Lyα photons escape from the highly neutral hydrogen IGM. Theoretical models suggest that up to 10% of Lyα fluxes can escape from an LAE in galaxy-clustered regions at the highly neutral epoch (Fig. 3.100; Gnedin and Prada [97]). To understand LAEs' physical properties as well as the evolution of x_{HI}, it is important to know how much fraction of high-redshift galaxies show Lyα emission escaping from the highly neutral hydrogen IGM. In other words, at what redshift LAEs disappear in the observations,

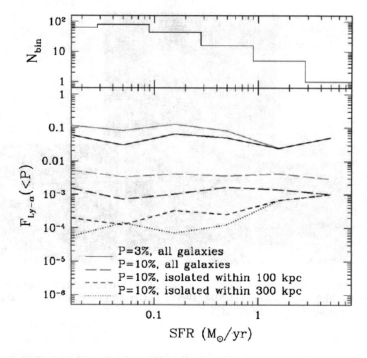

Fig. 3.100 Bottom: Fraction of observed to intrinsic Lyα fluxes as a function of central galaxy SFR that is predicted by numerical simulations [97]. The black (gray) solid and dashed lines indicate the top 3 and 10% Lyα flux survival, respectively, in all galaxies at $z = 9$, for the case of the Lyα offset velocity of 150 (200) km s^{-1}. The short-dashed and dotted lines are the same as the dashed lines, but for galaxies with no companion galaxy within a distance of 100 and 300 physical kpc, respectively. Top: Number of galaxies in each SFR bin that are used for the predictions shown in the bottom panel. ©AAS. Reproduced with permission

while continuum-selected star-forming galaxies are still seen at the same redshift in sufficiently deep observations.

Next generation large telescopes useful for observations of LAEs at $z \gtrsim 8 - 9$ are being built (Fig. 3.101). In space, James Webb Space Telescope (JWST) is planned to be launched in or after 2021. JWST is a successor of HST that covers optical to IR bands (0.6–28 μm) with a large 6.5 m segmented primary mirror. On the ground, there are three projects of extremely large telescopes for optical-IR observations; the European Extremely Large Telescope (E-ELT), the Giant Magellan Telescope (GMT), and the Thirty Meter Telescope (TMT) that have segmented primary mirrors with 39, 24, and 30 m effective diameters, respectively. The E-ELT and the GMT are being constructed in Cerro Armazones and Las Campanas Chile, respectively, in the southern hemisphere, while the TMT is planned to be placed in Mauna Kea, Hawaii in the northern hemisphere. These three projects will cover both the northern and southern hemispheres. There exists a possibility that the TMT may move out from Mauna Kea to La Palma on the Canary Islands, Spain, due to difficult issues of the

Fig. 3.101 Artist rendering images of JWST (top), E-ELT (bottom left), GMT (bottom center), and TMT (bottom right). Figure courtesy: NASA (JWST), ESO (E-ELT), GMTO Corporation (GMT), and TMT International Observatory (TMT)

construction site permission in the Hawaii Island. All three projects plan first light in the middle/late 2020s.

It is expected that the first-light of JWST comes earlier than these three ground-based next generation telescopes. Here I introduce two JWST science cases for LAEs.

The first JWST science case is to probe rest-frame optical nebular lines of LAEs at $z \gtrsim 5$ that are redshifted beyond the K band, which require a space telescope for deep observations. For the test of popIII star-formation (Fig. 3.71; Sect. 3.4.3), one can measure the spectral hardness of Q_{He^+}/Q_H for popIII LAE candidates with flux measurements of an HeII line and a Balmer line such as Hα and/or Hβ. The Balmer lines fall in the JWST's Near-Infrared Spectrograph (NIRSpec) wavelength window of 1–5 μm for high sensitivity spectroscopy. As discussed in Sect. 3.4.3, Q_H estimated from Lyα-line measurements have large systematic uncertainties raised by the Lyα escape fraction. If one can constrain the nebular extinction with a Balmer decrement of Hα/Hβ (or Hβ/Hγ for high-z sources), the Balmer-line method provides a reliable Q_{He^+}/Q_H value that is critical for popIII star-formation tests. It should be noted that the other JWST LAE studies will be also conducted with rest-frame optical nebular lines including [OII]3727 and [OIII]5007, indicators of metallicity and ionization parameter (Sects. 3.3.4.1–3.3.4.2), and that systemic velocities from the nebular lines are key for understanding the relation between Lyα photon escape and reionization [114].

The second JWST science case is to conduct an unbiased search for LAEs at a very high redshift. Such an LAE search can be performed with JWST/Near Infrared Imager and Slitless Spectrograph (NIRISS) that is packaged with the guide camera. NIRISS offers the wide-field slitless spectroscopy (WFSS) with grisms whose FoV and spectral resolution are $4\,\text{arcmin}^2$ and $R \sim 150$, respectively, at 0.8–$2.2\,\mu\text{m}$. The deep slitless spectroscopy can target Lyα emission redshifted to $z \sim 6 - 17$. Figure 3.102 presents simulations of deep (10-h integration) NIRISS/WFSS grism spectroscopy in the lensing galaxy cluster MACS J0647+7015 [71]. This simulation predicts the detections of 180 LBGs and LAEs with $F200W = 26 - 28$ mag at $z \sim 6 - 15$ with the help of gravitational lensing magnification, assuming a uniform random redshift distribution of LAEs. Investigating the number of detected LAEs as a function of redshift, one can address the problem of Lyα emission escaping from the highly neutral hydrogen IGM (Fig. 3.100).

In addition to these next generation projects for optical-IR observations, there are many multiwavelength projects useful for LAE studies. Programs of 21 cm observations are especially important for understanding a reionization-LAE connection. There are two major projects for next generation 21 cm observations, Hydrogen Epoch of Reionization Array (HERA; Pober et al. [270]) and Square Kilometer Array (SKA[20]).

HERA will be a 350-element interferometer that is composed of 14-m parabolic dishes covering 50–250 MHz. HERA is under construction. In 2016, a few percent of the dishes are already built on the site of South Africa [63]. HERA is not an interferometer that is significantly larger than the existing 21 cm instruments. However, HERA aims to accomplish detections of the weak EoR signals, improving calibration and foreground isolation accuracies with redundant baselines for the "EoR window" (Sect. 3.5.6.2) that are designed with the experiences of PAPER and MWA. HERA expects a 21 cm power spectrum detection with an S/N of ~ 20 at $z \sim 9$.

SKA will consist in radio interferometers whose total photon collecting area covers one square kilometer. Towards the completion of the interferometers, SKA has the two-phase approach: Phase 1 for the initial deployment starting in the early 2020s and Phase 2 for the full operation in later years. In the Phase 1, SKA integrates two precursor telescopes of MeerKAT[21] in Karoo, South Africa and Australian Square Kilometre Array Pathfinder (ASKAP[22]) in Murchison, Australia. The SKA Phase 1 instrument consists of three elements of SKA1-Mid (Karoo), SKA1-Survey, and SKA1-Low (Murchison; Fig. 3.103). Among the three elements, SKA1-Low is a low frequency array consisting of more than 100 thousand antennas covering 50–350 MHz corresponding to the wavelength of the redshifted EoR 21 cm emission. Before the full SKA construction, it is expected that SKA1-Low would provide 21 cm data useful for the cross-correlation analysis with LAEs [133, 172, 317]. Figure 3.103 presents the LAE-21 cm brightness temperature cross correlation functions predicted with reionization models [317]. The models assume LAEs from the Subaru/HSC survey and 100–1000 h SKA1-Low observations, and suggest that the two different cases with $x_{\text{HI}} = 0$ and 0.5 will be clearly distinguished over the uncertainties. Although the large systematic uncertainties (such as foreground subtraction

---

[20]https://www.skatelescope.org.

[21]http://www.ska.ac.za/gallery/meerkat/.

[22]http://www.csiro.au/en/Research/Facilities/ATNF/ASKAP.

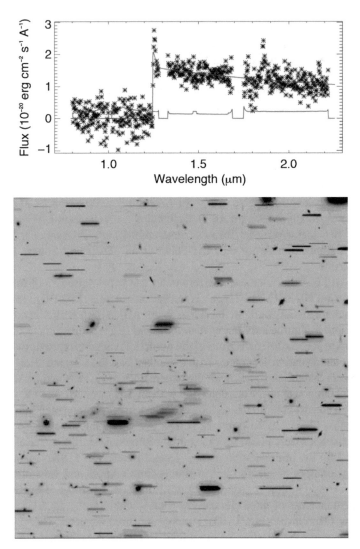

Fig. 3.102 Simulation results of JWST/NIRISS observations [71]. Bottom: Simulated grism GR150R image with the F200W filter for MACS J0647+7015. The assumed integration time is 10 h. Top: One-dimensional spectrum of a galaxy at $z = 9.3$ retrieved from the simulated grism image. The black symbols and the red line represent the simulated data points and the error, respectively. The blue curve denotes the best-fit model with a Lyα emission line

errors) are serious, as discussed in Sect. 3.5.6.2, there is a possibility that a new probe of LAE-21 cm cross correlation (Sect. 3.5.6.2) may be powerful, due to the capability for removing the systematics in the cross correlation analysis.

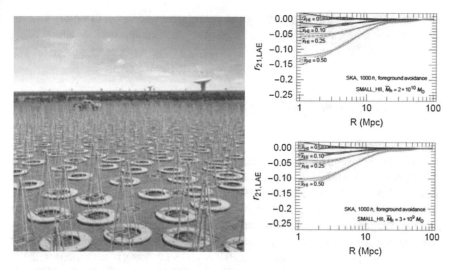

Fig. 3.103 Left: Artist impression image of SKA low frequency array (courtesy: SKA 2018). Right: LAE-21 cm cross correlation functions for the average neutral hydrogen fractions of $x_{HI} = 0$ (blue dot-dashed lines), 0.10 (red long-dashed lines), 0.25 (orange dashed lines), and 0.50 (green short-dashed lines) that are predicted by numerical simulations [317]. The shades indicate the total uncertainties of theoretical models and observations including foreground effects. These calculations assume SKA1-Low 1000 h observations for the 21 cm data and the HSC survey for the LAE data [248]. The top and bottom panels present the results of LAEs' host halo masses of 2×10^{10} and $3 \times 10^9 \, M_\odot$, respectively. This figure is reproduced by permission of the MNRAS

3.7.2.3 Trying Other Approaches

In Sects. 3.7.2.1 and 3.7.2.2, I have introduced the next-generation powerful instruments in the on-going and future programs. These next-generation instruments cover the unexplored parameter space, and are useful for resolving the issues of the open questions. However, even with no next-generation instruments, one can make a breakthrough with the present-generation instruments by a new approach. There are two observation examples in such breakthrough studies. One is the polarization observations for LABs conducted by Hayes et al. [119] and Prescott et al. [272]. Hayes et al. [119] have performed the polarization observations with VLT/FORS for a bright LAB, and detected the tangential polarization indicating the Lyα photon scattering in the LAB (Sect. 3.4.1.1). Another example is the identifications of luminous SNe (SNe IIn) hosted by high-z galaxies, $z \sim 2 - 4$ LBGs and LAEs [52]. Cooke et al. [52] have investigated variables in these high-z galaxies selected in the CFHT Legacy Survey data. Although it was well known that the popular bright SNe, SNe Ia, at $z \sim 2 - 4$ were too faint to be detected with the present data, Cooke et al. [52] successfully found luminous SNe in the $z \sim 2 - 4$ galaxies. These are two good examples of making a breakthrough discovery with the existing facilities and data, if one takes new approaches. I would suggest young astronomers not to simply wait for next-generation instruments, but to try new observational approaches with existing telescopes and data.

3.7.3 Summary of On-Going and Future Projects

I have showcased the five major open questions related to LAEs about galaxy formation and cosmic reionization. For galaxy formation, there are four questions about (i) the LAE's distinguishing characteristics of star-formation and ISM, (ii) physical origins of LABs and diffuse halos, (iii) physical reasons for the large Lyα EW + Lyα escape fraction, and (iv) the LAE-popIII connection. For cosmic reionization, two questions are presented: (a) reionization history (late or early/sharp or extended reionization?) and (b) major sources of reionization (star-forming galaxies and/or faint AGN?). These open questions are being addressed by on-going observations with the three new instruments, Subaru/HSC(+PFS), HETDEX, and VLT/MUSE. Although the observing programs of these three new instruments cover the complementary survey parameter space in redshift and depth, these programs can investigate LAEs only at $z \sim 2-7$. Beyond $z \sim 7-8$, one needs next generation large telescopes with great sensitivities, JWST, E-ELT, GMT, and TMT in the optical-IR wavelength range. Among these next generation projects, forthcoming JWST projects will probe rest-frame optical nebular lines of LAEs at $z \gtrsim 5$ for testing popIII, and conduct the unbiased search for LAEs at $z \sim 6-17$. Beyond the optical-IR wavelength range, 21 cm observations with HERA and SKA will provide important results. Theoretical studies suggest that the early-phase low-frequency array of SKA1-Low and Subaru/HSC will identify a signal of spatial anti-correlation between 21 cm emission and LAEs. Although these new facilities are key for exciting discoveries, there exist examples that new observational approaches with existing facilities can also make a breakthrough. With these examples in hand, I encourage young astronomers not to simply wait for next-generation instruments, but try new observational approaches for LAE studies with the existing facilities.

3.8   Grand Summary for This Series of Lectures

Observations of high-z LAEs and the related scientific subjects are detailed in this series of lecturers. After I explain the background of LAE discoveries, I clarify that LAEs are important objects for studying galaxy formation and cosmic reionization (Sect. 3.1). This is because LAEs are unique probes for high-z/low-mass (popIII-like) galaxies and the IGM H$_I$ gas via resonance scattering and damping wing absorptions. I then overview the basic theoretical ideas of galaxy formation including structure formation, gas cooling/feedback, and the five Lyα emission mechanisms (Sect. 3.2). In Sect. 3.3, I summarize the physical properties of LAEs, known to date, stellar populations, luminosity functions, morphologies, ISM state, LAE-AGN connection, and hosting halos. Section 3.4 discusses the observational challenges of LAE properties; extended Lyα halos, Lyα escape fraction, and LAE-popIII connection. This series of lectures moves to cosmic reionization science in Sect. 3.5. Evolution of neutral hydrogen fraction (i.e. cosmic reionization history) is constrained by the estimates of the IGM Lyα absorption including the Lyα damping wing absorption. The combination of the Lyα absorption, CMB, and HI 21 cm data provides the rough picture of reionization history. In Sect. 3.6, I explain simple analytic formulae to evaluate sources of reionization with three parameters for the ionizing photon production rate and one parameter for the IGM spatial distribution, and discuss whether star-forming galaxies can be major sources of reionization. The observational results suggest that star-forming galaxies alone can explain cosmic reionization, although the contribution of faint AGNs and the other sources are poorly understood, to date. Section 3.7 lists up the five open questions about LAEs, and explains three on-going projects for LAEs at $z \sim 2 - 7$ (Subaru/HSC+PFS, HETDEX, and VLT/MUSE) and various future optical-IR telescope projects for LAEs at $z \gtrsim 7 - 8$ (JWST and ground-based extremely large telescopes), together with low-frequency 21-cm observations (HERA and SKA), including some examples of science cases.

In the past two decades, I find that high-z LAE studies are one of the most important driving forces for the state-of-the-art observational facilities. The importance of Lyα emission is undoubted, due to the fact that Lyα is the strongest emission for the most abundant element in the universe, namely hydrogen. Moreover, the resonance nature of Lyα photons allows us to probe the distribution and kinematics of neutral hydrogen in any cosmic structures of galaxies and LSSs, from the ISM to the CGM and the IGM. In the coming decade, Lyα observations will continue leading the field of high-z observations with the on-going and future facilities. I hope that some of the readers will conduct new Lyα studies inspired by this series of lectures, and produce exciting results very soon.

Acknowledgements I am grateful to Anne Verhamme, Pierre North, Sebastiano Cantalupo, Hakim Atek, and Myriam Burgener Frick for inviting me to give these lectures in the historic and prestigious Saas Fee Advanced Course. I thank my fellow lectures, Mark Dijkstra, J. Xavier Prochaska, and Matthew Hayes for allowing to accommodate my lectures in the early days of the week for resolving my scheduling issue. I also thank the participants of the course who are enthusiastic about discussing scientific problems and presenting their ideas in the discussion time. I acknowledge Kazuhiro Shimasaku and Pierre North for editing these lecture notes and giving me a lot of valuable inputs. Finally, I thank my close colleagues, Yuichi Harikane, Yoshiaki Ono, and Takatoshi Shibuya for improving their figures and giving permissions to show them in this lecture note.

References

1. Abel, N.P., Ferland, G.J., Shaw, G., van Hoof, P.A.M.: ApJS **161**, 65 (2005)
2. Adams, J.J., Blanc, G.A., Hill, G.J., et al.: ApJS **192**, 5 (2011)
3. Adelberger, K.L., Steidel, C.C., Shapley, A.E., Pettini, M.: ApJ **584**, 45 (2003)
4. Ahn, S.-H.: ApJ **601**, L25 (2004)

5. Ali, Z.S., Parsons, A.R., Zheng, H., et al.: ApJ **809**, 61 (2015)
6. Ali, Z.S., Parsons, A.R., Zheng, H., et al.: ApJ **863**, 201 (2018)
7. Ando, M., Ohta, K., Iwata, I., et al.: ApJ **645**, L9 (2006)
8. Asplund, M., Grevesse, N., Sauval, A.J., Scott, P.: ARA&A **47**, 481 (2009)
9. Atek, H., Kunth, D., Schaerer, D., et al.: A&A **561**, A89 (2014)
10. Atek, H., Richard, J., Kneib, J.-P., et al.: ApJ **786**, 60 (2014)
11. Atek, H., Richard, J., Kneib, J.-P., et al.: ApJ **800**, 18 (2015)
12. Bacon, R., Accardo, M., Adjali, L., et al.: Proc. SPIE **7735**, 773508 (2010)
13. Bacon, R., Brinchmann, J., Richard, J., et al.: A&A **575**, A75 (2015)
14. Bacon, R., Conseil, S., Mary, D., et al.: A&A **608**, A1 (2017)
15. Bahcall, J.N., Salpeter, E.E.: ApJ **142**, 1677 (1965)
16. Bahcall, J.N., Jannuzi, B.T., Schneider, D.P., et al.: ApJ **377**, L5 (1991)
17. Bañados, E., Venemans, B.P., Mazzucchelli, C., et al.: Nature **553**, 473 (2018)
18. Barger, A.J., Cowie, L.L., Wold, I.G.B.: ApJ **749**, 106 (2012)
19. Barkana, R.: Nature **555**, 71 (2018)
20. Baldwin, J.A., Phillips, M.M., Terlevich, R.: PASP **93**, 5 (1981)
21. Basu-Zych, A., Scharf, C.: ApJ **615**, L85 (2004)
22. Behroozi, P.S., Wechsler, R.H., Conroy, C.: ApJ **770**, 57 (2013)
23. Bell, E.F., McIntosh, D.H., Katz, N., Weinberg, M.D.: ApJ **585**, L117 (2003)
24. Blanc, G.A., Adams, J.J., Gebhardt, K., et al.: ApJ **736**, 31 (2011)
25. Blanton, M.R., Dalcanton, J., Eisenstein, D., et al.: AJ **121**, 2358 (2001)
26. Bolton, J.S., Haehnelt, M.G.: MNRAS **382**, 325 (2007)
27. Bond, N.A., Feldmeier, J.J., Matković, A., et al.: ApJ **716**, L200 (2010)
28. Borisova, E., Cantalupo, S., Lilly, S.J., et al.: ApJ **831**, 39 (2016)
29. Borthakur, S., Heckman, T.M., Leitherer, C., Overzier, R.A.: Science **346**, 216 (2014)
30. Bouwens, R.J., Illingworth, G.D., Oesch, P.A., et al.: ApJ **708**, L69 (2010)
31. Bouwens, R.J., Illingworth, G.D., Oesch, P.A., et al.: ApJ **803**, 34 (2015)
32. Bower, R.G., Benson, A.J., Malbon, R., et al.: MNRAS **370**, 645 (2006)
33. Bowman, J.D., Rogers, A.E.E., Monsalve, R.A., Mozdzen, T.J., Mahesh, N.: Nature **555**, 67 (2018)
34. Brocklehurst, M.: MNRAS **153**, 471 (1971)
35. Bruzual, G., Charlot, S.: MNRAS **344**, 1000 (2003)
36. Cardamone, C., Schawinski, K., Sarzi, M., et al.: MNRAS **399**, 1191 (2009)
37. Calzetti, D., Armus, L., Bohlin, R.C., et al.: ApJ **533**, 682 (2000)
38. Calzetti, D.: PASP **113**, 1449 (2001)
39. Campbell, A., Terlevich, R., Melnick, J.: MNRAS **223**, 811 (1986)
40. Cantalupo, S., Arrigoni-Battaia, F., Prochaska, J.X., Hennawi, J.F., Madau, P.: Nature **506**, 63 (2014)
41. Capak, P.L., Riechers, D., Scoville, N.Z., et al.: Nature **470**, 233 (2011)
42. Capak, P.L., Carilli, C., Jones, G., et al.: Nature **522**, 455 (2015)
43. Cardelli, J.A., Clayton, G.C., Mathis, J.S.: ApJ **345**, 245 (1989)
44. Cassata, P., Le Fèvre, O., Garilli, B., et al.: A&A **525**, A143 (2011)
45. Chabrier, G.: PASP **115**, 763 (2003)
46. Chapman, S.C., Blain, A.W., Smail, I., Ivison, R.J.: ApJ **622**, 772 (2005)
47. Chiang, Y.-K., Overzier, R., Gebhardt, K.: ApJ **779**, 127 (2013)
48. Chornock, R., Berger, E., Fox, D.B., et al.: ApJ **774**, 26 (2013)
49. Chornock, R., Berger, E., Fox, D.B., et al.: arXiv:1405.7400 (2014)
50. Ciardullo, R., Gronwall, C., Wolf, C., et al.: ApJ **744**, 110 (2012)
51. Coe, D., Bradley, L., Zitrin, A.: ApJ **800**, 84 (2015)
52. Cooke, J., Sullivan, M., Gal-Yam, A., et al.: Nature **491**, 228 (2012)
53. Cowie, L.L., Hu, E.M.: AJ **115**, 1319 (1998)
54. Cowie, L.L., Barger, A.J., Hu, E.M.: ApJ **711**, 928 (2010)
55. Cowie, L.L., Barger, A.J., Hu, E.M.: ApJ **738**, 136 (2011)
56. Cowie, L.L., Barger, A.J., Hasinger, G.: ApJ **748**, 50 (2012)

57. Croft, R.A.C., Miralda-Escudé, J., Zheng, Z., et al.: MNRAS **457**, 3541 (2016)
58. Croft, R.A.C., Miralda-Escudé, J., Zheng, Z., Blomqvist, M., Pieri, M.: MNRAS **481**, 1320 (2018)
59. Daddi, E., Dickinson, M., Morrison, G., et al.: ApJ **670**, 156 (2007)
60. Dawson, S., Rhoads, J.E., Malhotra, S., et al.: ApJ **617**, 707 (2004)
61. Dawson, S., Rhoads, J.E., Malhotra, S., et al.: ApJ **671**, 1227 (2007)
62. de Barros, S., Vanzella, E., Amorín, R., et al.: A&A **585**, A51 (2016)
63. DeBoer, D.R., Parsons, A.R., Aguirre, J.E., et al.: PASP **129**, 045001 (2017)
64. Deharveng, J.-M., Small, T., Barlow, T.A., et al.: ApJ **680**, 1072 (2008)
65. Dekel, A., Birnboim, Y., Engel, G., et al.: Nature **457**, 451 (2009)
66. Dijkstra, M., Lidz, A., Wyithe, J.S.B.: MNRAS **377**, 1175 (2007)
67. Dijkstra, M., Loeb, A.: MNRAS **386**, 492 (2008)
68. Dijkstra, M., Loeb, A.: MNRAS **400**, 1109 (2009)
69. Dijkstra, M., Gronke, M., Sobral, D.: ApJ **823**, 74 (2016)
70. Dillon, J.S., Liu, A., Williams, C.L., et al.: Phys. Rev. D **89**, 023002 (2014)
71. Dixon, W.V., Ravindranath, S., Willott, C.: IAU General Assembly **22**, 2250490 (2015)
72. Djorgovski, S., Thompson, D.J.: IAUS **149**, 337 (1992)
73. Duval, F., Schaerer, D., Östlin, G., Laursen, P.: A&A **562**, A52 (2014)
74. Elbaz, D., Daddi, E., Le Borgne, D., et al.: A&A **468**, 33 (2007)
75. Eldridge, J.J., Stanway, E.R., Xiao, L., et al.: PASA **34**, e058 (2017)
76. Erb, D.K., Steidel, C.C., Trainor, R.F., et al.: ApJ **795**, 33 (2014)
77. Erb, D.K., Pettini, M., Steidel, C.C., et al.: ApJ **830**, 52 (2016)
78. Fall, S.M., Efstathiou, G.: MNRAS **193**, 189 (1980)
79. Fan, X., Hennawi, J.F., Richards, G.T., et al.: AJ **128**, 515 (2004)
80. Fan, X., Strauss, M.A., Becker, R.H., et al.: AJ **132**, 117 (2006)
81. Fan, X., Carilli, C.L., Keating, B.: ARA&A **44**, 415 (2006)
82. Fardal, M.A., Katz, N., Gardner, J.P., et al.: ApJ **562**, 605 (2001)
83. Faucher-Giguère, C.-A., Kereš, D.: MNRAS **412**, L118 (2011)
84. Feldmeier, J.J., Hagen, A., Ciardullo, R., et al.: ApJ **776**, 75 (2013)
85. Field, G.B.: ApJ **129**, 551 (1959)
86. Finkelstein, S.L., Rhoads, J.E., Malhotra, S., Pirzkal, N., Wang, J.: ApJ **660**, 1023 (2007)
87. Finkelstein, S.L., Rhoads, J.E., Malhotra, S., Grogin, N., Wang, J.: ApJ **678**, 655–668 (2008)
88. Finkelstein, S.L., Hill, G.J., Gebhardt, K., et al.: ApJ **729**, 140 (2011)
89. Finkelstein, S.L., Cohen, S.H., Windhorst, R.A., et al.: ApJ **735**, 5 (2011)
90. Finkelstein, S.L., Papovich, C., Dickinson, M., et al.: Nature **502**, 524 (2013)
91. Fiore, F., Puccetti, S., Grazian, A., et al.: A&A **537**, A16 (2012)
92. Garel, T., Blaizot, J., Guiderdoni, B., et al.: MNRAS **422**, 310 (2012)
93. Garnett, D.R.: AJ **103**, 1330 (1992)
94. Gawiser, E., Francke, H., Lai, K., et al.: ApJ **671**, 278 (2007)
95. Geach, J.E., Alexander, D.M., Lehmer, B.D., et al.: ApJ **700**, 1 (2009)
96. Giallongo, E., Grazian, A., Fiore, F., et al.: A&A **578**, A83 (2015)
97. Gnedin, N.Y., Prada, F.: ApJ **608**, L77 (2004)
98. Gnedin, N.Y.: ApJ **610**, 9 (2004)
99. Goerdt, T., Dekel, A., Sternberg, A., et al.: MNRAS **407**, 613 (2010)
100. Grazian, A., Giallongo, E., Boutsia, K., et al.: A&A **613**, A44 (2018)
101. Greig, B., Mesinger, A., Haiman, Z., Simcoe, R.A.: MNRAS **466**, 4239 (2017)
102. Gronwall, C., Ciardullo, R., Hickey, T., et al.: ApJ **667**, 79 (2007)
103. Groth, E.J., Peebles, P.J.E.: ApJ **217**, 385 (1977)
104. Guaita, L., Acquaviva, V., Padilla, N., et al.: ApJ **733**, 114 (2011)
105. Guaita, L., Francke, H., Gawiser, E., et al.: A&A **551**, A93 (2013)
106. Gunn, J.E., Peterson, B.A.: ApJ **142**, 1633 (1965)
107. Hagen, A., Ciardullo, R., Gronwall, C., et al.: ApJ **786**, 59 (2014)
108. Hagen, A., Zeimann, G.R., Behrens, C., et al.: ApJ **817**, 79 (2016)
109. Hansen, M., Oh, S.P.: MNRAS **367**, 979 (2006)

110. Harikane: Master thesis, The University of Tokyo (2016)
111. Harikane, Y., Ouchi, M., Ono, Y., et al.: ApJ **821**, 123 (2016)
112. Harikane, Y., Ouchi, M., Ono, Y., et al.: PASJ **70**, S11 (2018)
113. Harikane, Y., Ouchi, M., Shibuya, T., et al.: ApJ **859**, 84 (2018)
114. Hashimoto, T., Ouchi, M., Shimasaku, K., et al.: ApJ **765**, 70 (2013)
115. Hashimoto, T., Verhamme, A., Ouchi, M., et al.: ApJ **812**, 157 (2015)
116. Hashimoto, T., Laporte, N., Mawatari, K., et al.: Nature **557**, 392 (2018)
117. Hayashino, T., Matsuda, Y., Tamura, H., et al.: AJ **128**, 2073 (2004)
118. Hayes, M., Schaerer, D., Östlin, G., et al.: ApJ **730**, 8 (2011)
119. Hayes, M., Scarlata, C., Siana, B.: Nature **476**, 304 (2011)
120. Hayes, M., Östlin, G., Schaerer, D., et al.: ApJ **765**, L27 (2013)
121. Heckman, T.M., Alexandroff, R.M., Borthakur, S., Overzier, R., Leitherer, C.: ApJ **809**, 147 (2015)
122. Herenz, E.C., Urrutia, T., Wisotzki, L., et al.: A&A **606**, A12 (2017)
123. Hickox, R.C., Markevitch, M.: ApJ **661**, L117 (2007)
124. Hill, G.J., Gebhardt, K., Drory, N., et al.: American Astronomical Society Meeting 219, 424.01 (2012)
125. Hill, G.J., Tuttle, S.E., Lee, H., et al.: Proc. SPIE **8446**, 84460N (2012b)
126. Hinshaw, G., Larson, D., Komatsu, E., et al.: ApJS **208**, 19 (2013)
127. Hopkins, P.F., Richards, G.T., Hernquist, L.: ApJ **654**, 731 (2007)
128. Hu, E.M., McMahon, R.G.: Nature **382**, 231 (1996)
129. Hu, E.M., Cowie, L.L., McMahon, R.G.: ApJ **502**, L99 (1998)
130. Hu, E.M., McMahon, R.G., Cowie, L.L.: ApJ **522**, L9 (1999)
131. Hu, E.M., Cowie, L.L., McMahon, R.G., et al.: ApJ **568**, L75 (2002)
132. Hu, E.M., Cowie, L.L., Barger, A.J., et al.: ApJ **725**, 394 (2010)
133. Hutter, A., Dayal, P., Müller, V., Trott, C.: ApJ **836**, 176 (2017)
134. Ikeda, H., Nagao, T., Matsuoka, K., et al.: ApJ **756**, 160 (2012)
135. Iliev, I.T., Mellema, G., Pen, U.-L., et al.: MNRAS **369**, 1625 (2006)
136. Inoue, A.K., Iwata, I., Deharveng, J.-M.: MNRAS **371**, L1 (2006)
137. Inoue, A.K., Tamura, Y., Matsuo, H., et al.: Science **352**, 1559 (2016)
138. Ishigaki, M., Kawamata, R., Ouchi, M., et al.: ApJ **799**, 12 (2015)
139. Ishigaki, M., Kawamata, R., Ouchi, M., et al.: ApJ **854**, 73 (2018)
140. Ishiyama, T., Rieder, S., Makino, J., et al.: ApJ **767**, 146 (2013)
141. Ishiyama, T., Enoki, M., Kobayashi, M.A.R., et al.: PASJ **67**, 61 (2015)
142. Itoh, R., Ouchi, M., Zhang, H., et al.: ApJ **867**, 46 (2018)
143. Iwata, I., Inoue, A.K., Matsuda, Y., et al.: ApJ **692**, 1287 (2009)
144. Iye, M., Ota, K., Kashikawa, N., et al.: Nature **443**, 186 (2006)
145. Izotov, Y.I., Stasińska, G., Meynet, G., Guseva, N.G., Thuan, T.X.: A&A **448**, 955 (2006)
146. Izotov, Y.I., Orlitová, I., Schaerer, D., et al.: Nature **529**, 178 (2016)
147. Izotov, Y.I., Schaerer, D., Thuan, T.X., et al.: MNRAS **461**, 3683 (2016)
148. Jaskot, A.E., Oey, M.S.: ApJ **791**, L19 (2014)
149. Jenkins, A., Frenk, C.S., White, S.D.M., et al.: MNRAS **321**, 372 (2001)
150. Jelić, V., de Bruyn, A.G., Mevius, M., et al.: A&A **568**, A101 (2014)
151. Kashikawa, N., Shimasaku, K., Malkan, M.A., et al.: ApJ **648**, 7 (2006)
152. Kashikawa, N., Shimasaku, K., Matsuda, Y., et al.: ApJ **734**, 119 (2011)
153. Kashikawa, N., Nagao, T., Toshikawa, J., et al.: ApJ **761**, 85 (2012)
154. Kashino, D., Silverman, J.D., Rodighiero, G., et al.: ApJ **777**, L8 (2013)
155. Katz, N., Keres, D., Dave, R., Weinberg, D.H.: Ap&SSL Conf. Ser. **281**, 185 (2003)
156. Kelz, A., Jahn, T., Haynes, D., et al.: Proc. SPIE **9147**, 914775 (2014)
157. Kennicutt Jr., R.C.: ARA&A **36**, 189 (1998)
158. Kennicutt Jr., R.C.: ApJ **498**, 541 (1998)
159. Kereš, D., Katz, N., Weinberg, D.H., Davé, R.: MNRAS **363**, 2 (2005)
160. Kereš, D., Katz, N., Fardal, M., Davé, R., Weinberg, D.H.: MNRAS **395**, 160 (2009)
161. Knudsen, K.K., Richard, J., Kneib, J.-P., et al.: MNRAS **462**, L6 (2016)

162. Kodaira, K., Taniguchi, Y., Kashikawa, N., et al.: PASJ **55**, L17 (2003)
163. Kojima, T., Ouchi, M., Nakajima, K., et al.: PASJ **69**, 44 (2017)
164. Kollmeier, J.A., Zheng, Z., Davé, R., et al.: ApJ **708**, 1048 (2010)
165. Konno, A., Ouchi, M., Ono, Y., et al.: ApJ **797**, 16 (2014)
166. Konno, A., Ouchi, M., Nakajima, K., et al.: ApJ **823**, 20 (2016)
167. Konno, A., Ouchi, M., Shibuya, T., et al.: PASJ **70**, S16 (2018)
168. Koo, D.C., Kron, R.T.: PASP **92**, 537 (1980)
169. Kothes, R., Kerton, C.R.: A&A **390**, 337 (2002)
170. Kovač, K., Somerville, R.S., Rhoads, J.E., Malhotra, S., Wang, J.: ApJ **668**, 15 (2007)
171. Kravtsov, A.V.: ApJ **590**, L1 (2003)
172. Kubota, K., Yoshiura, S., Takahashi, K., et al.: MNRAS **479**, 2754 (2018)
173. Kurk, J.D., Cimatti, A., di Serego Alighieri, S., et al.: A&A **422**, L13 (2004)
174. Kusakabe, H., Shimasaku, K., Nakajima, K., Ouchi, M.: ApJ **800**, L29 (2015)
175. Lai, K., Huang, J.-S., Fazio, G., et al.: ApJ **674**, 70 (2008)
176. Lake, E., Zheng, Z., Cen, R., et al.: ApJ **806**, 46 (2015)
177. Landy, S.D., Szalay, A.S.: ApJ **412**, 64 (1993)
178. Laporte, N., Nakajima, K., Ellis, R.S., et al.: ApJ **851**, 40 (2017)
179. Laursen, P., Duval, F., Östlin, G.: ApJ **766**, 124 (2013)
180. Leauthaud, A., Tinker, J., Bundy, K., et al.: ApJ **744**, 159 (2012)
181. Leclercq, F., Bacon, R., Wisotzki, L., et al.: A&A **608**, A8 (2017)
182. Lee, K.-G., Hennawi, J.F., Stark, C., et al.: ApJ **795**, L12 (2014)
183. Leitet, E., Bergvall, N., Piskunov, N., Andersson, B.-G.: A&A **532**, A107 (2011)
184. Leitet, E., Bergvall, N., Hayes, M., Linné, S., Zackrisson, E.: A&A **553**, A106 (2013)
185. Lidz, A., Zahn, O., McQuinn, M., Zaldarriaga, M., Hernquist, L.: ApJ **680**, 962 (2008)
186. Lidz, A., Zahn, O., Furlanetto, S.R., et al.: ApJ **690**, 252 (2009)
187. Livermore, R.C., Finkelstein, S.L., Lotz, J.M.: ApJ **835**, 113 (2017)
188. Lotz, J.M., Koekemoer, A., Coe, D., et al.: ApJ **837**, 97 (2017)
189. Madau, P., Haardt, F., Rees, M.J.: ApJ **514**, 648 (1999)
190. Madau, P., Dickinson, M.: ARA&A **52**, 415 (2014)
191. Maiolino, R., Carniani, S., Fontana, A., et al.: MNRAS **452**, 54 (2015)
192. Malhotra, S., Rhoads, J.E.: ApJ **565**, L71 (2002)
193. Malhotra, S., Rhoads, J.E.: ApJ **617**, L5 (2004)
194. Malhotra, S., Rhoads, J.E., Finkelstein, S.L., et al.: ApJ **750**, L36 (2012)
195. Mao, J., Lapi, A., Granato, G.L., de Zotti, G., Danese, L.: ApJ **667**, 655 (2007)
196. Marlowe, A.T., Heckman, T.M., Wyse, R.F.G., Schommer, R.: ApJ **438**, 563 (1995)
197. Martin, C.L.: ApJ **506**, 222 (1998)
198. Martin, C., Moore, A., Morrissey, P., et al.: Proc. SPIE **7735**, 77350M (2010)
199. Matthee, J., Sobral, D., Best, P., et al.: MNRAS **465**, 3637 (2017)
200. Matthee, J., Sobral, D., Boone, F., et al.: ApJ **851**, 145 (2017)
201. Matsuda, Y., Yamada, T., Hayashino, T., et al.: AJ **128**, 569 (2004)
202. Matsuda, Y., Yamada, T., Hayashino, T., et al.: MNRAS **410**, L13 (2011)
203. Matsuda, Y., Yamada, T., Hayashino, T., et al.: MNRAS **425**, 878 (2012)
204. Mawatari, K., Inoue, A.K., Yamada, T., et al.: MNRAS **467**, 3951 (2017)
205. McCarthy, P.J., Spinrad, H., Djorgovski, S., et al.: ApJ **319**, L39 (1987)
206. McGreer, I.D., Jiang, L., Fan, X., et al.: ApJ **768**, 105 (2013)
207. McLeod, D.J., McLure, R.J., Dunlop, J.S., et al.: MNRAS **450**, 3032 (2015)
208. McLeod, D.J., McLure, R.J., Dunlop, J.S.: MNRAS **459**, 3812 (2016)
209. McLinden, E.M., Finkelstein, S.L., Rhoads, J.E., et al.: ApJ **730**, 136 (2011)
210. McLure, R.J., Dunlop, J.S., Bowler, R.A.A., et al.: MNRAS **432**, 2696 (2013)
211. McQuinn, M., Hernquist, L., Zaldarriaga, M., Dutta, S.: MNRAS **381**, 75 (2007)
212. Miyazaki, S., Komiyama, Y., Kawanomoto, S., et al.: PASJ **70**, S1 (2018)
213. Mo, H., van den Bosch, F.C., White, S.: Galaxy Formation and Evolution. Cambridge University Press, Cambridge (2010)
214. Momcheva, I.G., Lee, J.C., Ly, C., et al.: AJ **145**, 47 (2013)

215. Momose, R., Ouchi, M., Nakajima, K., et al.: MNRAS **442**, 110 (2014)
216. Momose, R., Ouchi, M., Nakajima, K., et al.: MNRAS **457**, 2318 (2016)
217. Mukae, S., Ouchi, M., Kakiichi, K., et al.: ApJ **835**, 281 (2017)
218. Mortlock, D.J., Warren, S.J., Venemans, B.P., et al.: Nature **474**, 616 (2011)
219. Muratov, A.L., Kereš, D., Faucher-Giguère, C.-A., et al.: MNRAS **454**, 2691 (2015)
220. Nagamine, K., Ouchi, M., Springel, V., Hernquist, L.: PASJ **62**, 1455 (2010)
221. Nagao, T., Maiolino, R., Marconi, A.: A&A **459**, 85 (2006)
222. Nakajima, K., Ouchi, M., Shimasaku, K., et al.: ApJ **745**, 12 (2012)
223. Nakajima, K., Ouchi, M., Shimasaku, K., et al.: ApJ **769**, 3 (2013)
224. Nakajima, K., Ouchi, M.: MNRAS **442**, 900 (2014)
225. Nestor, D.B., Shapley, A.E., Kornei, K.A., Steidel, C.C., Siana, B.: ApJ **765**, 47 (2013)
226. Neufeld, D.A.: ApJ **370**, L85 (1991)
227. Niikura, H., Takada, M., Yasuda, N., et al.: arXiv:1701.02151 (2017)
228. Nilsson, K.K., Fynbo, J.P.U., Møller, P., Sommer-Larsen, J., Ledoux, C.: A&A **452**, L23 (2006)
229. Oesch, P.A., van Dokkum, P.G., Illingworth, G.D., et al.: ApJ **804**, L30 (2015)
230. Oesch, P.A., Brammer, G., van Dokkum, P.G., et al.: ApJ **819**, 129 (2016)
231. Oesch, P.A., Bouwens, R.J., Illingworth, G.D., Labbé, I., Stefanon, M.: ApJ **855**, 105 (2018)
232. Ono, Y., Ouchi, M., Shimasaku, K., et al.: MNRAS **402**, 1580 (2010)
233. Ono, Y., Ouchi, M., Shimasaku, K., et al.: ApJ **724**, 1524 (2010)
234. Ono, Y., Ouchi, M., Mobasher, B., et al.: ApJ **744**, 83 (2012)
235. Osterbrock, D.E.: Astrophysics of Gaseous Nebulae and Active Galactic Nuclei, 2nd edn. University Science Books, ISBN 0-935702-22-9, 408pp (1989). See https://ui.adsabs.harvard.edu/abs/1989agna.book.....O/abstract
236. Osterbrock, D.E., Ferland, G.J.: Astrophysics of Gaseous Nebulae and Active Galactic Nuclei, 2nd edn. University Science Books, Sausalito (2006)
237. Östlin, G., Hayes, M., Duval, F., et al.: ApJ **797**, 11 (2014)
238. Ota, K., Walter, F., Ohta, K., et al.: ApJ **792**, 34 (2014)
239. Ouchi, M., Shimasaku, K., Furusawa, H., et al.: ApJ **582**, 60 (2003)
240. Ouchi, M., Shimasaku, K., Okamura, S., et al.: ApJ **611**, 660 (2004)
241. Ouchi, M., Shimasaku, K., Akiyama, M., et al.: ApJ **620**, L1 (2005)
242. Ouchi, M., Hamana, T., Shimasaku, K., et al.: ApJ **635**, L117 (2005)
243. Ouchi, M., Shimasaku, K., Akiyama, M., et al.: ApJS **176**, 301 (2008)
244. Ouchi, M., Ono, Y., Egami, E., et al.: ApJ **696**, 1164 (2009)
245. Ouchi, M., Mobasher, B., Shimasaku, K., et al.: ApJ **706**, 1136 (2009)
246. Ouchi, M., Shimasaku, K., Furusawa, H., et al.: ApJ **723**, 869 (2010)
247. Ouchi, M., Ellis, R., Ono, Y., et al.: ApJ **778**, 102 (2013)
248. Ouchi, M., Harikane, Y., Shibuya, T., et al.: PASJ **70**, S13 (2018)
249. Paciga, G., Chang, T.-C., Gupta, Y., et al.: MNRAS **413**, 1174 (2011)
250. Pallottini, A., Ferrara, A., Pacucci, F., et al.: MNRAS **453**, 2465 (2015)
251. Parsons, A.R., Liu, A., Aguirre, J.E., et al.: ApJ **788**, 106 (2014)
252. Partridge, R.B., Peebles, P.J.E.: ApJ **147**, 868 (1967)
253. Partridge, R.B., Peebles, P.J.E.: ApJ **148**, 377 (1967)
254. Pascarelle, S.M., Windhorst, R.A., Keel, W.C., Odewahn, S.C.: Nature **383**, 45 (1996)
255. Pascarelle, S.M., Windhorst, R.A., Driver, S.P., Ostrander, E.J., Keel, W.C.: ApJ **456**, L21 (1996)
256. Paulino-Afonso, A., Sobral, D., Ribeiro, B., et al.: MNRAS **476**, 5479 (2018)
257. Patel, M., Warren, S.J., Mortlock, D.J., Fynbo, J.P.U.: A&A **512**, L3 (2010)
258. Patrício, V., Richard, J., Verhamme, A., et al.: MNRAS **456**, 4191 (2016)
259. Pawlik, A.H., Schaye, J., van Scherpenzeel, E.: MNRAS **394**, 1812 (2009)
260. Peacock, J.A., Dodds, S.J.: MNRAS **280**, L19 (1996)
261. Peebles, P.J.E.: Principles of Physical Cosmology. Princeton University Press, Princeton (1993). ISBN: 978-0-691-01933-8
262. Pentericci, L., Fontana, A., Vanzella, E., et al.: ApJ **743**, 132 (2011)

263. Pentericci, L., Vanzella, E., Fontana, A., et al.: ApJ **793**, 113 (2014)
264. Pentericci, L., Carniani, S., Castellano, M., et al.: ApJ **829**, L11 (2016)
265. Pettini, M., Shapley, A.E., Steidel, C.C., et al.: ApJ **554**, 981 (2001)
266. Pirzkal, N., Xu, C., Malhotra, S., et al.: ApJS **154**, 501 (2004)
267. Planck Collaboration, Ade, P.A.R., Aghanim, N., et al.: A&A **594**, A13 (2016)
268. Planck Collaboration, Adam, R., Aghanim, N., et al.: A&A **596**, A108 (2016)
269. Planck Collaboration, Akrami, Y., Arroja, F., et al.: arXiv:1807.06205 (2018)
270. Pober, J.C., Liu, A., Dillon, J.S., et al.: ApJ **782**, 66 (2014)
271. Prescott, M.K.M., Dey, A., Jannuzi, B.T.: ApJ **702**, 554 (2009)
272. Prescott, M.K.M., Smith, P.S., Schmidt, G.D., Dey, A.: ApJ **730**, L25 (2011)
273. Prescott, M.K.M., Martin, C.L., Dey, A.: ApJ **799**, 62 (2015)
274. Press, W.H., Schechter, P.: ApJ **187**, 425 (1974)
275. Pritchard, J.R., Loeb, A.: Phys. Rev. D **82**, 023006 (2010)
276. Pritchard, J., Loeb, A.: Nature **468**, 772 (2010)
277. Pritchet, C.J., Hartwick, F.D.A.: ApJ **320**, 464 (1987)
278. Pritchet, C.J., Hartwick, F.D.A.: ApJ **355**, L11 (1990)
279. Rakic, O., Schaye, J., Steidel, C.C., Rudie, G.C.: ApJ **751**, 94 (2012)
280. Rauch, M., Becker, G.D., Haehnelt, M.G., et al.: MNRAS **418**, 1115 (2011)
281. Rees, M.J., Ostriker, J.P.: MNRAS **179**, 541 (1977)
282. Rhoads, J.E., Malhotra, S., Dey, A., et al.: ApJ **545**, L85 (2000)
283. Read, J.I., Trentham, N.: Philos. Trans. R. Soc. Lond. Ser. A **363**, 2693 (2005)
284. Richards, G.T., Strauss, M.A., Fan, X., et al.: AJ **131**, 2766 (2006)
285. Robertson, B.E., Ellis, R.S., Dunlop, J.S., McLure, R.J., Stark, D.P.: Nature **468**, 49 (2010)
286. Robertson, B.E., Furlanetto, S.R., Schneider, E., et al.: ApJ **768**, 71 (2013)
287. Robertson, B.E., Ellis, R.S., Furlanetto, S.R., Dunlop, J.S.: ApJ **802**, L19 (2015)
288. Roberts-Borsani, G.W., Bouwens, R.J., Oesch, P.A., et al.: ApJ **823**, 143 (2016)
289. Rodighiero, G., Daddi, E., Baronchelli, I., et al.: ApJ **739**, L40 (2011)
290. Rudie, G.C., Steidel, C.C., Trainor, R.F., et al.: ApJ **750**, 67 (2012)
291. Salpeter, E.E.: ApJ **121**, 161 (1955)
292. Samui, S., Srianand, R., Subramanian, K.: MNRAS **398**, 2061 (2009)
293. Sanders, R.L., Shapley, A.E., Kriek, M., et al.: ApJ **816**, 23 (2016)
294. Santos, M.R., Ellis, R.S., Kneib, J.-P., Richard, J., Kuijken, K.: ApJ **606**, 683 (2004)
295. Scarlata, C., Colbert, J., Teplitz, H.I., et al.: ApJ **706**, 1241 (2009)
296. Schaerer, D.: A&A **397**, 527 (2003)
297. Schaerer, D., de Barros, S.: A&A **502**, 423 (2009)
298. Schechter, P.: ApJ **203**, 297 (1976)
299. Schenker, M.A., Stark, D.P., Ellis, R.S., et al.: ApJ **744**, 179 (2012)
300. Schenker, M.A., Robertson, B.E., Ellis, R.S., et al.: ApJ **768**, 196 (2013)
301. Schenker, M.A., Ellis, R.S., Konidaris, N.P., Stark, D.P.: ApJ **795**, 20 (2014)
302. Sheth, R.K., Tormen, G.: MNRAS **308**, 119 (1999)
303. Shklovskii, I.S.: AZh **41**, 801 (1964)
304. Schaerer, D., Izotov, Y.I., Verhamme, A., et al.: A&A **591**, L8 (2016)
305. Shapley, A.E., Steidel, C.C., Pettini, M., Adelberger, K.L., Erb, D.K.: ApJ **651**, 688 (2006)
306. Shibuya, T., Kashikawa, N., Ota, K., et al.: ApJ **752**, 114 (2012)
307. Shibuya, T., Ouchi, M., Nakajima, K., et al.: ApJ **788**, 74 (2014)
308. Shibuya, T., Ouchi, M., Harikane, Y.: ApJS **219**, 15 (2015)
309. Shibuya, T., Ouchi, M., Konno, A., et al.: PASJ **70**, S14 (2018)
310. Shibuya, T., Ouchi, M., Harikane, Y., et al.: PASJ **70**, S15 (2018)
311. Shibuya, T., Ouchi, M., Harikane, Y., Nakajima, K.: arXiv:1809.00765 (2018)
312. Shimakawa, R., Kodama, T., Steidel, C.C., et al.: MNRAS **451**, 1284 (2015)
313. Shimasaku, K., Kashikawa, N., Doi, M., et al.: PASJ **58**, 313 (2006)
314. Shull, J.M., Harness, A., Trenti, M., Smith, B.D.: ApJ **747**, 100 (2012)
315. Silk, J., Wyse, R.F.G.: Phys. Rep. **231**, 293 (1993)
316. Silk, J.: ApJ **772**, 112 (2013)

317. Sobacchi, E., Mesinger, A., Greig, B.: MNRAS **459**, 2741 (2016)
318. Sobral, D., Matthee, J., Darvish, B., et al.: ApJ **808**, 139 (2015)
319. Sobral, D., Matthee, J., Brammer, G., et al.: MNRAS **482**, 2422 (2019)
320. Speagle, J.S., Steinhardt, C.L., Capak, P.L., Silverman, J.D.: ApJS **214**, 15 (2014)
321. Springel, V., White, S.D.M., Jenkins, A., et al.: Nature **435**, 629 (2005)
322. Srbinovsky, J.A., Wyithe, J.S.B.: MNRAS **374**, 627 (2007)
323. Stark, D.P., Ellis, R.S., Chiu, K., Ouchi, M., Bunker, A.: MNRAS **408**, 1628 (2010)
324. Stark, D.P., Ellis, R.S., Ouchi, M.: ApJ **728**, L2 (2011)
325. Stark, D.P., Richard, J., Siana, B., et al.: MNRAS **445**, 3200 (2014)
326. Stark, D.P., Richard, J., Charlot, S., et al.: MNRAS **450**, 1846 (2015)
327. Stark, D.P., Walth, G., Charlot, S., et al.: MNRAS **454**, 1393 (2015)
328. Steidel, C.C., Adelberger, K.L., Shapley, A.E., et al.: ApJ **532**, 170 (2000)
329. Steidel, C.C., Pettini, M., Adelberger, K.L.: ApJ **546**, 665 (2001)
330. Steidel, C.C., Erb, D.K., Shapley, A.E., et al.: ApJ **717**, 289 (2010)
331. Steidel, C.C., Bogosavljević, M., Shapley, A.E., et al.: ApJ **736**, 160 (2011)
332. Steidel, C.C., Rudie, G.C., Strom, A.L., et al.: ApJ **795**, 165 (2014)
333. Storey, P.J., Hummer, D.G.: MNRAS **272**, 41 (1995)
334. Susa, H., Umemura, M.: ApJ **600**, 1 (2004)
335. Takada, M., Ellis, R.S., Chiba, M., et al.: PASJ **66**, R1 (2014)
336. Takeuchi, T.T., Yuan, F.-T., Ikeyama, A., Murata, K.L., Inoue, A.K.: ApJ **755**, 144 (2012)
337. Tamura, N., Takato, N., Shimono, A., et al.: Proc. SPIE **9908**, 99081M (2016)
338. Tanvir, N.R., Fox, D.B., Levan, A.J., et al.: Nature **461**, 1254 (2009)
339. Tapken, C., Appenzeller, I., Noll, S., et al.: A&A **467**, 63 (2007)
340. Thompson, D., Djorgovski, S., Trauger, J.: AJ **110**, 963 (1995)
341. Thuan, T.X., Izotov, Y.I.: ApJS **161**, 240 (2005)
342. Totani, T., Kawai, N., Kosugi, G., et al.: PASJ **58**, 485 (2006)
343. Totani, T., Aoki, K., Hattori, T., et al.: PASJ **66**, 63 (2014)
344. Totani, T., Aoki, K., Hattori, T., Kawai, N.: PASJ **68**, 15 (2016)
345. Trebitsch, M., Verhamme, A., Blaizot, J., Rosdahl, J.: A&A **593**, A122 (2016)
346. Tresse, L., Ilbert, O., Zucca, E., et al.: A&A **472**, 403 (2007)
347. Treu, T., Schmidt, K.B., Trenti, M., Bradley, L.D., Stiavelli, M.: ApJ **775**, L29 (2013)
348. Treister, E., Schawinski, K., Volonteri, M., Natarajan, P., Gawiser, E.: Nature **474**, 356 (2011)
349. van Breukelen, C., Jarvis, M.J., Venemans, B.P.: MNRAS **359**, 895 (2005)
350. Vanzella, E., Pentericci, L., Fontana, A., et al.: ApJ **730**, L35 (2011)
351. Venemans, B.P., Kurk, J.D., Miley, G.K., et al.: ApJ **569**, L11 (2002)
352. Verhamme, A., Schaerer, D., Maselli, A.: A&A **460**, 397 (2006)
353. Verhamme, A., Schaerer, D., Atek, H., Tapken, C.: A&A **491**, 89 (2008)
354. van Ojik, R., Roettgering, H.J.A., Miley, G.K., Hunstead, R.W.: A&A **317**, 358 (1997)
355. Vanzella, E., de Barros, S., Vasei, K., et al.: ApJ **825**, 41 (2016)
356. Venemans, B.P., Findlay, J.R., Sutherland, W.J., et al.: ApJ **779**, 24 (2013)
357. Vitale, M., Fuhrmann, L., García-Marín, M., et al.: A&A **573**, A93 (2015)
358. Wardlow, J.L., Malhotra, S., Zheng, Z., et al.: ApJ **787**, 9 (2014)
359. Watson, D., Christensen, L., Knudsen, K.K., et al.: Nature **519**, 327 (2015)
360. Weller, J., Ostriker, J.P., Bode, P., Shaw, L.: MNRAS **364**, 823 (2005)
361. Whitaker, K.E., Franx, M., Leja, J., et al.: ApJ **795**, 104 (2014)
362. White, S.D.M., Rees, M.J.: MNRAS **183**, 341 (1978)
363. Willott, C.J.: ApJ **742**, L8 (2011)
364. Wisotzki, L., Bacon, R., Blaizot, J., et al.: A&A **587**, A98 (2016)
365. Worseck, G., Prochaska, J.X., Hennawi, J.F., McQuinn, M.: ApJ **825**, 144 (2016)
366. Wyithe, J.S.B., Loeb, A.: Nature **441**, 322 (2006)
367. Xue, R., Lee, K.-S., Dey, A., et al.: ApJ **837**, 172 (2017)
368. Yamada, T., Nakamura, Y., Matsuda, Y., et al.: AJ **143**, 79 (2012)
369. Yamada, T., Matsuda, Y., Kousai, K., et al.: ApJ **751**, 29 (2012)
370. Yang, Y., Zabludoff, A.I., Davé, R., et al.: ApJ **640**, 539 (2006)
371. Yang, Y., Zabludoff, A., Tremonti, C., Eisenstein, D., Davé, R.: ApJ **693**, 1579 (2009)
372. Yatawatta, S., de Bruyn, A.G., Brentjens, M.A., et al.: A&A **550**, A136 (2013)
373. Zheng, Z., Berlind, A.A., Weinberg, D.H., et al.: ApJ **633**, 791 (2005)
374. Zitrin, A., Labbé, I., Belli, S., et al.: ApJ **810**, L12 (2015)

Chapter 4
Lyman Alpha Emission and Absorption in Local Galaxies

Matthew Hayes

4.1 Introduction: Key Concepts, Observables, Definitions and Methods

4.1.1 Drivers for Low-z Lyman Alpha Observations

When introducing a topic such as observational studies of Lyman alpha (Lyα) in the local universe, one may immediately ask the question *why do such a thing?* After all, Lyα observations in the local restframe require expensive space-based telescopes to observe Lyα in the far ultraviolet (UV) and, as we will discuss at length, Lyα is a resonance line and is difficult to interpret. There may be better observables than Lyα to use in order to understand local galaxies.

4.1.1.1 The Potential Power of Lyα

Indeed the primary motivations for such observations—and therefore these lecture notes—come from the high redshift (z) universe. Astrophysics at high-z is extensively covered by lecture notes in this series by M. Ouchi. Lyα has a rest-frame wavelength of 1215.67 Å, and therefore redshifts into the centre of the U-band ($\lambda \approx 3600$ Å) at redshift $z \approx 2$, and remains in the optical domain until redshifting out of the z-band at $z \gtrsim 6.5$. Thus Lyα remains easily accessible to ground-based optical telescopes from the peak of cosmic star formation until the epoch of reionization, and is also available in the near infrared (NIR) until higher redshifts still. This potential power of the Lyα emission line is so well-recognized that it has provided one of the primary science drivers for the development of instrumentation on the largest telescopes: the

M. Hayes (✉)
Department of Astronomy, AlbaNova University Centre, Stockholm University,
SE-106 91 Stockholm, Sweden
e-mail: matthew@astro.su.se

© Springer-Verlag GmbH Germany, part of Springer Nature 2019
M. Dijkstra et al., *Lyman-alpha as an Astrophysical and Cosmological Tool*,
Saas-Fee Advanced Course 46, https://doi.org/10.1007/978-3-662-59623-4_4

Hobby Eberly Dark Energy Experiment (HETDEX) [57], the *Hyper Suprime-Cam* at the *Subaru* telescope [99], the *Multi Unit Spectroscopic Explorer* (MUSE) at the *Very Large Telescope* [5]. Many other major investments all cite mid-to-high redshift Lyα observations among their key science objectives.

Lyα is intrinsically the brightest recombination line in ionized nebulae, and like Hα and all other hydrogen recombination lines, its intrinsic intensity is directly proportional to the star formation rate (SFR) of a galaxy. Note that the observational aspects of Lyα emission in this series is heavily focussed on photoionized gas in star-forming galaxies; only minor attention is paid to non-thermal energy sources such as accretion onto supermassive black holes in Active Galactic Nuclei (AGN), or collisionally excited 'cooling' radiation.

The easiest way to illustrate the observational advantages that the Lyα line brings is with a simple calculation. Adopting the standard conversion factors between Hα luminosity and SFR for a Salpeter initial mass function (IMF) of solar metallicity stars [73], and an intrinsic Lyα/Hα ratio of 8.7 (Sect. 4.1.3), we determine that

$$L_{Ly\alpha} \text{ [erg s}^{-1}] = 1.10 \times 10^{42} \text{ SFR [M}_\odot \text{ yr}^{-1}] \tag{4.1}$$

where $L_{Ly\alpha}$ is the Lyα luminosity in erg s^{-1}. If we further place a hypothetical galaxy at $z = 6$ and assign it a modest SFR of 5 M$_\odot$ yr^{-1}, we would receive a flux of $\approx 10^{-17}$ erg s^{-1} cm^{-2} on earth. With a low-dispersion optical spectrograph on an 8 m class telescope (VLT/FORS2 is taken as an example), this Lyα feature can be detected at signal-to-noise ratio (SNR) of 10 in a single hour of observation.

We can compute the expected UV continuum luminosity by following a similar approach [73]. By dividing the Lyα luminosity by this continuum luminosity density, we compute the equivalent width (EW). The Lyα EW is 115 Å for these assumptions of IMF and metallicity, which implies we are looking at a particularly strong emission line. From these simple calculations it becomes clear that Lyα is both very luminous and has a high contrast against the continuum.

Indeed the use of Lyα to identify and study primeval galaxies forming in the early universe was suggested almost half a century ago [110] for these very reasons. Lyα should be a very efficient and observationally straightforward tool by which to identify galaxies in the distant universe, and out to the reionization epoch at $z \gtrsim 6.5$. Moreover, as a narrow spectral line, its spectroscopic detection also gives relatively precise (although not perfect) redshift determinations, and serves as a means of confirming/rejecting the true high-z nature of galaxies selected by other means. Currently the community has obtained Lyα observations for around 10^4 galaxies at $z \gtrsim 2$.

4.1.1.2 The Potential Problem of Lyα

The problem arises when we consider that Lyα is a resonance line. Being produced by spontaneous de-excitation between the $^2P \rightarrow {}^1S$ energy levels of atomic hydrogen, it will also be absorbed by the inverse of the this transition. See lecture notes in this series by M. Dijkstra and X. Prochaska. The optical depth in the centre of the transition ('*line centre*') is given by the expression

$$\tau = 1.041 \times 10^{-13} T_4^{-1/2} N_{\mathrm{HI}} \frac{H(x, a)}{\sqrt{\pi}} \qquad (4.2)$$

(see [144]), where τ is the conventionally defined optical depth, T_4 is the temperature in 10,000 K, N_{HI} is the H I column density in cm^{-2}, and $H(x, a)$ is the Voigt profile that describes the absorption, as a function of normalized frequency x. Thus it is obvious that at $T = 10^4$ K, τ becomes 1 at H I column densities of 10^{13} cm^{-2}. Taking the Milky Way galaxy as an example, most randomly chosen sightlines, even above/below Galactic latitude of $+60/-60°$, have H I column densities above 10^{20} cm^{-2} [66]. Thus Lyα produced in the Milky Way would see upwards of $\sim 10^7$ optical depths of H I. Things may be similar or different in other galaxies, but we would need to go down to column densities far below observable thresholds in order for a medium to be optically thin to Lyα.

After absorption to the ^2P level, there is nothing the excited atom can do other than de-excite back to the ground state and release another Lyα photon. This marks the beginning of a radiation transfer problem of absorption and re-emission that will continue until either the photon is emitted from the galaxy, or absorbed by a dust grain. As the Lyα line forms at the wavelength where (measured) dust attenuation laws peak [11, 16], Lyα is also very susceptible to dust absorption. Thus with dust absorbing Lyα, and H I dictating the path taken and probability of interaction with dust, it is apparent that galaxies may not necessarily emit Lyα, even though a copious radiation field is produced by its H II regions.

From this radiation transfer problem we understand that it is particularly hard—even in the simplest geometrical constructions—to predict the emergent Lyα. Turning that statement around, it is equally difficult to estimate the intrinsic properties of a galaxy from an observed Lyα line.

4.1.1.3 The Solutions to the Problem: The Low-Redshift Universe

Above we discussed how Lyα offers a unique window on galaxy formation and into the high-z universe, and indeed this is the case. However, caution must be exercised until we know precisely how to interpret the observations. We need to know, in the simplest of cases, how much of the intrinsic Lyα is emitted—the '*escape fraction*'—in order to interpret photometric observations (fluxes and EWs). We need to know under what conditions—combinations of dust, H I distribution and kinematics—a galaxy becomes bright in Lyα, and what information is encoded in an emergent line profile. The Lyα escape fraction evolves positively by two orders of magnitude between $z = 0$ and $z = 6$ [47], and the underlying cause of this evolution has not been fully determined; however amounting to a factor of 100 it no doubt reflects a significant change in the properties of galaxies.

The relevant physics of absorption and scattering is simple enough that computer simulations could answer any of the above questions. The difficultly is knowing the realistic distributions of gas and dust, and how the ISM appears on small scales. It is currently not possible to capture such small scales—those of the star formation and fragmented molecular clouds themselves—while simultaneously retaining the volume of the galaxy halo; nor is this on the (self-consistent) computational horizon in the next generation of hydrodynamical simulations. Thus we must address the problem observationally by assembling all the relevant information about the H I, and dust content in galaxies, and do so with the highest possible spatial resolution. Naturally this is not possible in the populations of high-z Lyα-emitting galaxies, where a single ground-based seeing disk corresponds to 8 physical kpc. Indeed such observations—those necessary to interpret high-z galaxy surveys—can only be obtained in the nearby universe. In nearby star-forming galaxies one can observe H I using the 21 cm line in the radio, and simultaneously observe the distribution of dust at the resolution of HST. Indeed at low-z it is only the Lyα data themselves that is challenging to the observer. How we go about doing these observations, and what we have learned over the last ≈ 2 decades of doing so, makes the subject of these lectures.

4.1.2   Fundamental Expectations for Lyα in Star-Forming Galaxies

Here we address issues of the Lyα luminosity, expectations from stellar modeling, and nebular conditions.

4.1.2.1   Photoionization and the Rydberg Formula

While several processes may be responsible for producing Lyα in and around galaxies, we will discuss only the processes of photoionization, followed by recombinations and electron cascades. Collisional processes, and the 'cooling radiation' that is much discussed at high-z, will be completely ignored. In H II regions, Lyα is produced purely by the reprocessing of ionizing photons, and we begin with the Rydberg (1889) formula:

$$\frac{1}{\lambda} = R \left(\frac{1}{n_1^2} - \frac{1}{n_2^2} \right) \tag{4.3}$$

where λ is the vacuum wavelength, R is the Rydberg constant of $\approx 1.097 \times 10^{-3}$ Å^{-1}, and n_1 and n_2 are the lower and upper principal quantum numbers of the electronic energy levels. Inserting 1 and 2, respectively, for the quantum numbers gives the wavelength of Lyα, while 1 and infinity, respectively, gives 912 Å, the wavelength of the 'Lyman edge'. It is often useful to remember that the energy of a Lyα photon is precisely 3/4 the energy of a photon at the Lyman edge, which follows directly from this expression.

Under case B recombination, approximately $\approx 68\%$ of all recombination events include a transition to the ground-state through the Lyα channel ([26]; and see notes by M. Dijkstra in this series). Assuming every photon with energy $h\nu > 13.6$ eV produced by a stellar population ionizes a hydrogen atom, the above translates into roughly 2 of every 3 ionizing photons being 'converted' into Lyα. Were all ionizing photons emitted at the Lyman edge, about 50% of all the ionizing energy would be converted into the Lyα line. For more realistic ionizing spectra, this number will be somewhat lower ($\approx 30\%$), but regardless of the precise ionizing spectral shape it is clear that a very large amount of energy will be pumped into the narrow, discrete, Lyα emission feature. For a more detailed picture of how much Lyα is produced in star-forming galaxies, we need to review the necessary theory of stellar evolution and population synthesis.

4.1.2.2 Stellar Evolution

Stellar population synthesis (SPS) [10, 83], models are widely used throughout extragalactic astronomy to both predict and interpret observations. The principle is anchored in the fact that we know the spectrum and lifetime of stars of various mass and metallicity. Or, in terms of isochrones, that we have sufficiently well-sampled spectral libraries for stars over the appropriate mass range at each evolutionary point. These libraries may be determined observationally, by stellar evolutionary models with radiation transfer in the atmospheres, by interpolation, or some combination thereof. Post main-sequence evolution should be included as well as possible. Most importantly for questions relating to Lyα are the continuum level near $\lambda = 1216$ Å, and the ionizing spectrum of the most massive stars.

Standard Population Synthesis

Hydrogen-ionizing photons are produced mainly by O-stars, which have main-sequence lifetimes of ≈ 8 Myr. Thus for a constant SFR, the ionizing photon production rate, and therefore strength of recombination lines, will be constant after this timespan. The UV continuum, on the other hand, is produced not only by O stars, but has a significant contribution from later B and A spectral types; while obviously less luminous than O-stars, these are far more numerous and for the same constant SFR the UV continuum becomes constant after ≈ 100 Myr. Clearly the evolutionary stage of star formation rate is going to affect the intrinsic Lyα EW. In the limiting case of constant SFR the Lyα EW will begin high but converge to a constant value near 100 Å over a period of several tens of Myr. However in the case of an instantaneous burst of star formation (simple stellar population, SSP), the Lyα EW will also begin high but fade very rapidly to unobservable faint levels as O stars die, while the UV continuum may remain bright substantially afterwards.

This is illustrated in Fig. 4.1, using an example from Charlot and Fall [17]. The figure shows burst-like (exponentially decaying, with $\tau = 10$ Myr) and constant-SFR

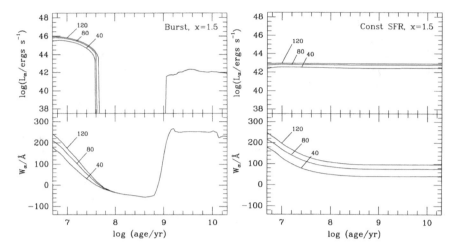

Fig. 4.1 The expected Lyα luminosity (*upper panels*) and EW (*lower panels*) as a function of time since the onset of star formation, for two different star formation histories. The *left* panel shows a very rapid burst after which the star formation ceases; the *right* panel shows star formation that proceeds at a constant rate. In both cases different population of the stellar initial mass function is shown, where the numbers 120, 80, and 40 show curves calculated for IMFs populated up to this maximum mass limit in $M_\odot$. *Figure is reproduced with permission from Charlot and Fall* [17], ©*AAS*

star-formation histories in the left and right panels respectively. Upper panels show the Lyα luminosity which decreases very rapidly when the O-stars leave the main sequence in the case of a burst, but is effectively constant in the right hand panel when star formation proceeds at a constant rate. The lower panels show the Lyα EW, which also decreases rapidly for the burst, but not as rapidly as the luminosity because the UV continuum flux is also decreasing. Of course both luminosity and EW will affect the detectability of a galaxy by Lyα: the object must be bright enough, and the line of sufficiently high contrast to observe above the continuum. In the right hand panel the EWs start high, and also decrease, but do so asymptotically as the UV continuum reaches its equilibrium value (UV-bright stars leave the MS at the same rate at which they are born, causing the luminosity vs. SFR to plateau), after ≈100 Myr. This plateau value is commonly taken to be 80–100 Å, a quotation that should usually be attributed to Charlot and Fall [17]. An increase in the Lyα flux and EW is visible at ages above ≈1 Gyr, which results from the emergence of planetary nebula nucleus stars, after leaving the extended horizontal branch but before becoming white dwarfs. Note, however, that the luminosity is three to four orders of magnitude less than that delivered by massive stars at early evolutionary times.

Beyond Standard Assumptions

Under 'normal' conditions, the largest uncertainty in calculations of the intrinsic Lyα luminosity and EW will be dominated by knowledge of the star formation history—indeed with O-stars so dominant at ionizing wavelengths and also so short-lived, the Lyα output will clearly be strong function of SFH. However the ionizing and non-ionizing flux depends strongly upon both the stellar metallicity and IMF: lower metallicity will decrease the stellar opacity in the outer envelopes and result in hotter surface temperatures, while flatter IMFs will increase the ionizing/non-ionizing UV flux ratio and over-produce Lyα. With regards to the properties of very low metallicity galaxies in the early universe, the inclusion of very metal-poor stars may significantly increase the Lyα EW [117, 128, 129], which could exceed 1000 Å for population III metallicities, and remain more than 500 Å down to $Z = 10^{-3} Z_\odot$. Much discussion pervades the astrophysical literature on whether the IMF is stochastically sampled at the high mass end, but any definition of stochasticity must occur relative to some reference upper mass limit. This is typically taken to be around 100 $M_\odot$, but in reality *every* IMF must become stochastically sampled at some stellar mass.

Recent literature has also addressed the importance of rapidly rotating stars and accreting binaries. Rapidly rotating stars have been included in the *Starburst99* population synthesis code [85]: rotating stars exhibit higher luminosities because of larger convective cores than their non-rotating counterparts. In the most extreme cases, the ionizing photon budget of a stellar population may be increased by a factor of a few. In addition less massive, and therefore longer lived, stars may also contribute to the ionizing luminosity, causing the phase over which a system exhibits a significant nebular spectrum to be extended beyond 10 Myr.

Binary stellar evolution has always been less well understood than evolution in isolation, but binary massive stars would also have a major impact on the ionizing photon production, as stars at a given age become hotter [30, 32]. Qualitatively the effects of binary evolution and rapid rotation may be somewhat similar, and may explain some of the higher ionization UV emission lines seen in the ultraviolet [31, 136].

It may be the case that these more exotic stellar populations are not dominant in the local universe, but that does not mean they do not occur. Evidence exists for particularly massive stars in the Large Magellanic Cloud [21] and IZw 18 [71, 72]. One must also keep in mind that extreme systems will always be identified as targets for followup spectroscopy in the ultraviolet, so catalogues of Lyα spectra may be skewed in this direction.

4.1.3   What We Would Like to Measure and How We Measure It

In physical terms one of the main quantities that we would like to measure is the SFR. Since this must be done through observable signatures, it may be better to cast the question of how rapidly the stellar population is producing ionizing photons (see arguments in Sects. 4.1.2.1 and 4.1.2.2). However, because we cannot observe the ionizing radiation directly, a more applicable quantity would be the rate at which ionizing photons are converted into nebular lines. This sequence represents increasing layers of abstraction, but also increasing possibility. Even measurements of the emission line spectrum will need some correction factor to be applied in order to go from the measured value to the intrinsic luminosity that correlates directly with SFR. Regarding Lyα, the radiation transfer effects mean the relationship between the observed and intrinsic Lyα are dependent upon a large number of additional quantities; understanding how the Lyα related observables—luminosity, EW, $f_{esc}^{Ly\alpha}$, and line profile shape—allow us to infer information about galaxies' physical properties is the main objective of this research field.

While we can never measure the intrinsic Lyα, we may estimate it from several other observable signatures (or combinations thereof). The entire hydrogen recombination line spectrum is the result of the same electronic cascade from higher energy levels that end in the ground state. All the lower energy levels are well separated with respect to the quantum mechanical uncertainty, and the likelihood of spontaneous de-excitation to a specific energy level is given purely by the Einstein coefficient A_{ul}. Consequently the ratios between the emergent flux in Lyα and strong emission lines that fall in the optical and NIR are set almost entirely by quantum mechanical properties with only minor sensitivity to macroscopic quantities such as temperature and density. Balmer lines, therefore, are vital components of the most robust studies of $f_{esc}^{Ly\alpha}$. However in certain environments optical emission lines may be unobservable or too optically thick to reliably interpret, and in such cases we must rely instead upon less direct tracers of the ionizing radiation. This inevitably is either direct observation of the ultraviolet continuum, or UV radiation that has been absorbed by dust, which then re-radiates the energy at FIR wavelengths. We now review some of the ways we estimate Lyα escape fractions, which we divide into direct observables of the nebular gas, and continuum-based methods.

4.1.3.1 Ionized Gas Observables

Here we cover observational methods that target discrete emission line feature only.

The Optical Domain—Balmer Emission Lines

Hα is the next strongest line in the hydrogen spectrum, and falls in the optical domain at $\lambda = 6563$ Å. Figure 4.2 shows the intrinsic Lyα/Hα ratio as a function of electron density n_e and temperature T_e. Obviously such ratios are valid only in the case of no dust or resonance scattering. Between 1000 and 30,000 K the ratio only ranges between around 8 and 10, and where the ISM is not subject to thermal instabilities (the plateau near 10^4 K) the line ratio is close to 9. Regarding dust attenuation, Hα is much closer to optically thin than Lyα, and becomes optically thick only at $E_{B-V} \approx 0.33$. It also remains visible from the ground out to $z \sim 2.5$, before it becomes lost in high thermal sky background in the L-band. Higher-order Balmer lines (e.g. Hβ) are in principle visible to somewhat higher redshift, but do not extend the redshift limit by much, and these lines are significantly weaker (e.g. $\approx$ Lyα/30 for Hβ). At lower optical depths, the Hα/Hβ ratio (if available) may be used to correct Hα for

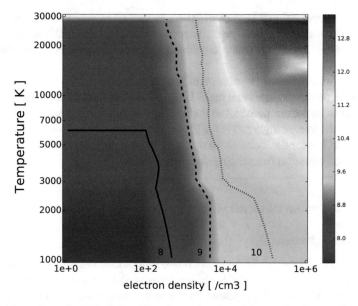

Fig. 4.2 Figure shows the intrinsic Lyα/Hα ratio as a function of electron density n_e and temperature T_e. Contours show constant values of 8, 9, and 10, to match the range of values typically adopted in publications. *Figure is produced using the PyNeb software* [87]

dust extinction which, at least so far, has been the most robust of the frequently used estimators of intrinsic Lyα.

The Near-Infrared—Paschen and Brackett Emission Lines

Near infrared lines in the H and K bands—e.g. Paα, Paβ, Brγ—suffer substantially less dust attenuation than Balmer lines, and are probably the best available tracer of the ionizing photon budget if observable. They remain optically thin even at $E_{B-V} \approx 2$–3, but have the disadvantage of being hundreds of times weaker than the Lyα line itself. Thus large telescopes are needed. Forming in the restframe NIR these lines are already born at wavelengths where the background is bright in a forest of night sky emission lines, and with the rising thermal background Pa and Br lines are not possible to observe at significant cosmological redshift.

Radio Recombination Lines (RRLs)

Spectra at radio wavelengths are awash with overlapping high-order hydrogen and helium transitions, such as H58α. These lines are entirely unaffected by dust obscuration, and some starburst galaxies in the local universe have been successfully observed in the RRLs (e.g. NGC 253 [74]). In principle high spatial and spectral resolutions could be obtained with interferometers like the *Very Large Array* (VLA). However, while the intrinsic morphology of H II regions may be revealed to RRL observations, the intrinsic quantum mechanical uncertainties on the energy levels (from Heisenberg's principle) mean that the transitions cannot be calibrated to factors better than

2–3. Further observational studies of RRLs is needed to determine if they can be measured in star-forming galaxies at distances where Lyα observations are possible, and it is noteworthy that no such combined Lyα and RRL observations currently exist.

The advantage of targeting observables of H II regions is that we are looking at transitions in the same atoms that make Lyα and are thus easy to compare. The main disadvantages, as just evaluated, are that they are significantly weaker that Lyα and difficult to observe at high-z. This will change in the coming years when JWST comes online, but for the moment we may rely instead upon continuum observations that also trace emission from the most massive stars.

4.1.3.2 Continuum Observables

Ultraviolet Continuum

As discussed in Sect. 4.1.2.2, the ultraviolet continuum is also dominated by O and B stars. Advantageously, the UV is observable over the same redshift range as Lyα, and to some extent must be captured by the same observation. Moreover the FUV continuum is (arguably) one of the most well known and exploited measurements of the SFR of a galaxy [73], and the comparison of Lyα- and UV-derived SFRs may provide a proxy for $f_{\mathrm{esc}}^{\mathrm{Ly}\alpha}$ as exploited, for example, as a function of redshift [47]. There is one significant disadvantage to comparing Lyα and UV radiations when estimating $f_{\mathrm{esc}}^{\mathrm{Ly}\alpha}$: that it is effectively an equivalent width measurement, and EW depends also upon evolutionary stage, metallicity, and IMF (see Sect. 4.1.2.2 and Fig. 4.1). For an individual galaxy where these quantities are not well constrained, then $f_{\mathrm{esc}}^{\mathrm{Ly}\alpha}$ measurements derived from SFR(Lyα)/SFR(UV) will be unreliable by a factor of a few, but in large samples of galaxies evolutionary effects are likely to average out.

Infrared Continuum Radiation

Like Lyα, the FUV continuum is also absorbed by dust and under some circumstances may require large corrections. This is accomplished by analysis of the spectral energy distribution (SED), but as argued above for Balmer line radiation, such corrections will only be accurate to optical depths of a few. If the target galaxy is an ultra-luminous infrared galaxy (ULIRG) or bright sub-mm galaxy at high-z, any SFR estimates based upon optical emission will be inadequate. In the case of the dustiest, and highest SFR galaxies, the only reliable estimate of the SFR comes from the FIR luminosity.

4.1.4   How Is Lyα Flux Measured: A Matter of Definition

It is never completely trivial to measure fluxes for hydrogen emission lines. Even Balmer lines are contaminated by stellar absorption that can be significant in populations dominated by stars in a given range of ages. The situation is worse for Lyα, where the resonant feature is scattered in both the atomic ISM and stellar atmospheres. Given that we may see both emission and absorption, what Lyα measurement we make depends upon what question we want to answer.

4.1.4.1   Interstellar Absorption

Since the Lyα feature may be present in both emission and absorption, one must ask the question of whether we want to isolate the emission-only part (produced purely in H II regions), or take the sum of emission and absorption. How well one can do the former will be dependent upon the spectral resolution. Figure 4.3 shows Lyα and continuum spectra of the nuclear regions of local star-forming galaxy IRAS 08339+6517. The black spectrum shows the high resolution spectrum obtained with STIS [95]: it shows the Lyα line to be rather narrow, with a sharp peak and steep rise on the blue side. It shows a deep, narrow, and well defined absorption on the blue side that goes to the zero flux level. The low resolution spectrum (green), however, naturally shows a much broader emission line, but the absorption trough on the blue side is partly filled

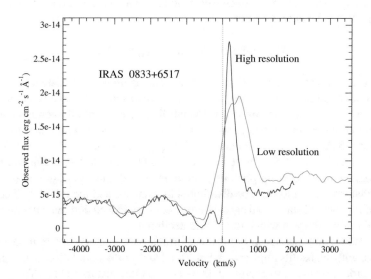

Fig. 4.3 Figure shows the Lyα spectrum of IRAS 08339+6517, obtained in high-resolution mode in black (≈40 km s⁻¹, with the STIS) and low-resolution mode in green, taken with the GHRS. The line profiles are clearly very different in each case. *Figure is reproduced with permission from Mas-Hesse et al. [95], ©AAS*

in. The absorption clearly appears deeper, and goes to zero at higher resolution. The peak is also shifted to the red in the lower resolution spectrum because of the intrinsic asymmetry of the line, which may also affect some velocity measurements. It is true that total flux and EW are invariant with spectral resolution (notes by Prochaska), but if we want to focus purely upon nebular recombination lines or ISM absorption properties, this will depend upon the resolution of the spectrograph.

In narrowband imaging we have no control over this, as the resolution even for the narrowest filters will be of the order $R = \lambda/\text{FWHM} \approx 100$ (resolution $\approx 3,000$ km s^{-1}). In this case we measure only the sum of the emitted nebular light minus interstellar absorption. If we want to focus purely upon the amount of nebular Lyα that escapes a galaxy we need high resolution, or to accept the fact that a measurement may be contaminated. If we want to make measurements that are comparable with those made at high redshift—e.g. to compare the EW distribution or determine where galaxies would fall on the luminosity function—we would more likely need to measure the sum.

4.1.4.2 The Stellar Absorption Feature

One of the difficulties with measuring hydrogen emission lines stems from the fact there is so much hydrogen in galaxies. For example, even Balmer lines that do not involve the $n = 1$ level suffer from absorption in the atmospheres of stars. This may amount to EWs of ≈ 10 Å in Hβ [35, 36] for A-star dominated stellar populations. For Lyα the situation may be much worse, and the EW in absorption has been shown from models to be described by:

$$W_{\text{Ly}\alpha} = -420 \, [\text{Å}] \exp(-T_{\text{eff}}/6100 \, [\text{K}]) \qquad (4.4)$$

over the $T = 30,000$–$10,000$ K range [142]. This is broadly consistent with observations of O and late B-type stars, although massive stars are also rare in the Milky Way and UV observations of stellar Lyman lines are potentially contaminated by foreground absorption in the galaxy. At $T_{\text{eff}} = 30,000$ K this amounts to an EW of about -3 Å which may be considered negligible, but the EW exceeds 35 Å in absorption at $T_{\text{eff}} = 15,000$ K. It may be argued that stars of these temperatures do not produce many ionizing photons to be reprocessed into Lyα, but one must also recall that star formation can be distributed in galaxies, and the total UV flux is the summation over all stars of all ages. B-type stars may indeed dominate the integrated UV light of a galaxy in which the SFR is declining.

A similar study has recently been performed with more complete model atmospheres, with the results implemented into the *Starburst99* population synthesis code [83] in order to study more representative star formation histories [111]. These results are broadly consistent with previous work, showing that when O-stars are dominant the absorption has negligible effect on the nebular Lyα. For a constant SFR, the absorption EW reaches ≈ 7 Å at an age of 10 Myr, but ≈ 14 Å for an instantaneous burst of the same age.

4.1.4.3   What Is the UV Continuum Level?

The ultraviolet continuum level is also not a constant, and changes shape rapidly with age and dust reddening. For example, in Fig. 4.3 the continuum level at $+3000$ km s^{-1} is about twice the value at -4000 km s^{-1}. Even if nebular Lyα may escape through preferential channels through the ISM, damping wings may also be visible in deep absorption spectra out to ± 6000 km s^{-1} [42, 77, 121]. Finally, there is a stellar wind P Cygni feature of N V at $\lambda\lambda = 1239, 1242$ Å, that may extend somewhat bluer because of blue-shifting in the wind—this is just 5500 km s^{-1} from Lyα and may be closer still if Lyα is redshifted. With a Lyα absorption wing that may extend close to N V stellar feature, defining the true continuum level near Lyα is not straightforward. This will not affect the determination of the Lyα flux, but it will affect the EW. The fact that Lyα may form upon a depressed continuum calls for the line flux and continuum levels to be measured completely independently in order not to very much under-estimate the nebular Lyα.

In the bulk of high-z galaxies, especially individual cases and faint LAEs, the continuum is undetected spectroscopically. EWs, therefore, tend to be estimated using photometric observations in broadband filters to determine the continuum level. Thus the observations sample the average flux over a wavelength range of a few hundred Å, and the effects of discrete features will be averaged out. This may not be a possibility for local galaxies unless FUV imaging has been obtained at a resolution that can be matched to the aperture of the spectrograph, and so for measurements to be truly comparable with high-z observations one may consider averaging the spectral continuum flux over a large wavelength window.

4.1.5   Constraints on Lyα Observation

The atmosphere does not transmit extra-terrestrial radiation bluewards of the UVA ozone cut-on at about 3200 Å. This sets the lower limit in redshift at which Lyα can be observed, although in practical terms it is somewhat higher redshifts (≈ 2) where the atmospheric transmission makes observations competitive. For all redshifts lower than this we need space-based observatories.

However simply getting the telescope above the absorbing layers of the atmosphere is not sufficient, and there are two more constraints that must be adhered to. Figure 4.4 shows a high resolution spectrum of BL Lacertae object, 1ES 1553+113, obtained from the HST Spectroscopic Legacy Archive.[1] These objects are bright active galaxies with a featureless, flat continua: all the features seen in the figure arise in the Milky Way and the outer layers of the Earth's atmosphere.

The *left* panel shows a zoom-in spectrum around Lyα, and two main features are visible. Firstly there is a strong absorption profile with damping wings visible out to at least ± 10 Å (2500 km s^{-1}), which arise from H I gas in the Milky Way. As discussed

---

[1] https://archive.stsci.edu/hst/spectral_legacy/.

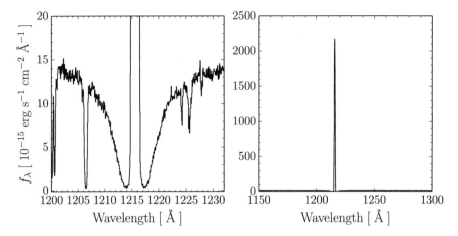

Fig. 4.4 The ultraviolet spectrum of BL Lac object 1ES1553+113, taken with the HST/COS. *Left* shows a zoom around the Lyα wavelength, and *right* shows an extended range in wavelength and flux. *These data were obtained from the HST Spectroscopic Legacy Archive (HSLA)*

in Sect. 4.1.1.2 this atomic halo gas has column densities above $\approx 10^{20}$ cm^{-2} on almost all sightlines [66], which will obviously absorb some/all of the Lyα emitted by an extragalactic source (Eq. 4.2).

The second feature is the emission line at $\lambda = 1216$ Å. This is Lyα emission but has nothing to do with the target: it arises in the outer atmosphere of the Earth (the 'geocorona') and is many times brighter than the Lyα emission from any extragalactic object. Should a Lyα line from a galaxy fall close in wavelength to the geocoronal feature it will be swamped by this emission. The *right* panel shows a larger range in wavelength and flux to illustrate just how strong the geocoronal Lyα line is.

The solution to this is straightforward: to only target galaxies that lie at sufficient cosmological redshift to well separate the target Lyα from the Galactic and geocoronal features. As above, to fully avoid the damping wing the recession velocity needs to exceed $cz \gtrsim 2500$ km s^{-1}. Unfortunately this has an important observational consequence: that it excludes all galaxies within a distance of about 40 Mpc. This limit is far outside the local volume, and excludes a large number of extremely interesting possible targets: the Magellanic Clouds, M 82, Arp 220, and many other local galaxies cannot be directly studied in Lyα.

4.2 Observational Facilities: The Contributors to Low-Redshift Lyα Science

Most ground-based optical telescopes are capable of observing Lyα at high-z in one capacity or another. However because of the expense and varied optimizations of satellite missions, our observations at low-z are assembled from a small number of

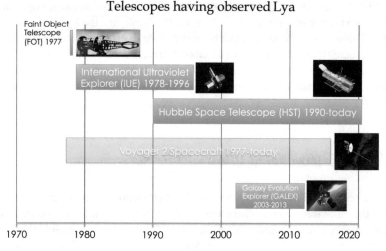

Fig. 4.5 Telescopes that have observed Lyα in the vacuum ultraviolet

platforms. Each of these has (or had) its own capabilities, and thus each is 'almost unique'. In this section we review the capabilities of each satellite in approximate chronological order of launch. This series of telescopes is illustrated in Fig. 4.5.

4.2.1 Faint Object Telescope

The far UV window was opened for the first time by the *Faint Object Telescope* (FOT). Operating on a handful of sounding rocket flights in 1977, the FOT was fed by a 40 cm primary mirror, and carried a medium resolution spectrophotometer, with a resolving power of $R = \lambda/\Delta\lambda \approx 140$, over the wavelength range of 1200–1700 Å. In its first flight the FOT returned the first ever spectrum of an extragalactic object in the vacuum UV: the brightest optical quasar in the sky, 3C 273, using an exposure time of just 235 s [24]. This observation is discussed in Sect. 4.3. The same telescope was used to observe local Seyfert galaxy NGC 4151 [23], providing the first non-QSO observation in Lyα, other FUV emission lines, and the ultraviolet continuum.

4.2.2 Ultraviolet Spectrometers on the Voyager Spacecraft

A pair of Ultraviolet Spectrographs (UVS) still fly onboard both of the *Voyager* spacecrafts, and thus formally have the longest operational lifespan: 1977–present. They operate a low dispersion slit spectrograph with wavelength coverage between 500 and 1700 Å, with a resolving power of $R = 50$.

The UVSs have one major advantage over all other ultraviolet spectrographs and imagers: the platforms have travelled far from the Earth, and the UV instrument does not suffer from the geocoronal background of Lyα. This completely mitigates one of the two observational constraints discussed in Sect. 4.1.5. Voyager/UVS has been able to send back raster-scanned spectra targeted at M 33 ([69] and see Sect. 4.3.2) and also mapped the diffuse Lyα emission from scattering inside the ISM of the Milky Way [81].

4.2.3   The International Ultraviolet Explorer

The *International Ultraviolet Explorer* (IUE [76]) was the first real workhorse instrument in the ultraviolet. The satellite was launched to a geosynchronous orbit in 1978, and the mission was nominally scheduled to last three years; 18 years later in 1996 the observatory was still operating, but was finally switched off. After almost two decades of operation, the IUE still represents a large legacy of UV spectra of many different astrophysical objects, including, for our purposes, large numbers of star-forming galaxies and AGN.

The IUE had an elliptical-like entrance aperture of 10×20 arcsec^2 that projected light onto a primary mirror of 45 cm in diameter. This fed two main spectrographs, the *Short Wavelength Prime* (SWP) and *Long Wavelength Prime* (LWP). The SWP covered a wavelength range of 1150–2000 Å, while the LWP operated between 1850 and 3300 Å. Thus it covered a spectral range from slightly bluer than Lyα to the wavelength at which atmospheric transmission permits ground-based observations in the U-band. Each spectrograph had both high and low resolution settings, although the latter, with resolving power, R between 250 and 400, was the only one used to image the faint (compared to the UV bright stars) galaxies discussed in these notes. Scientific results from the IUE are discussed mainly in Sect. 4.3.

4.2.4   The Hubble Space Telescope

Any moderate- or high-resolution extragalactic Lyα observations at $z \lesssim 2$ have been obtained by the *Hubble Space Telescope* (HST). HST has flown since 1990, and has been equipped with a diverse suite of instruments that are, or have been, capable of seeing Lyα. HST has returned Lyα images, and spectra at moderate resolution, using the following cameras:

The *Goddard High Resolution Spectrograph* (GHRS)

This spectrograph flew from the launch of Hubble, and had a 1.74×1.74 arcsec^2 entrance window (after COSTAR was mounted). The nominal resolving power was $R \sim 2000$ in the low resolution settings and up to 25,000 at mid resolution. GHRS was sensitive between $\lambda = 1150$ and 1800 Å, although only a narrow section of the wavelength range was observable at one time, and was tuned by the user. This provided contiguous wavelength coverage of ≈280 Å and ≈30–40 Å in the low and high resolution settings, respectively. Scientific results from the GHRS are discussed mainly in Sects. 4.5.1 and 4.5.2.

The *Space Telescope Imaging Spectrograph* (STIS)

STIS was mounted in 1997, during the second HST Servicing Mission, and is still available today. It is a particularly flexible instrument for ultraviolet spectroscopy and imaging, offering a wide array of modes. It offers slit (and slitless) spectroscopy in the far UV at wavelengths between 1150 and 1700 Å. The main advantage of STIS compared to other spectrographs is that it can provide true nominal spectral resolution of up to $R = 17,000$ in the medium resolution setting (a low resolution setting also provides $R \approx 1000$). The main disadvantage of the high R modes is that, like GHRS, only a small wavelength range of 55 Å can be obtained in a single observation. Thus, for observations of external galaxies, the low-R modes are more commonly used.

As a slit spectrograph the spectral resolution of STIS is dependent upon the size of the slit when observing extended sources: higher resolution is obtained with smaller slits. Thus unless the Lyα or UV surface brightness is particularly high, exposure times with STIS usually need to be longer than with other instruments, where the trade is on spectral resolution. Very high resolution settings also exist but it is unlikely that these could collect enough photons to be useful in observations of galaxies. Scientific results from the STIS spectrograph are discussed mainly in Sects. 4.5.1 and 4.5.2.

The STIS also has the capability to perform Lyα imaging, but as we will see in Sect. 4.6 and below, this is better accomplished with the Advanced Camera for Surveys (ACS).

The *Cosmic Origins Spectrograph* (COS)

The COS may be considered a GHRS on steroids, and was mounted in 2009 during HST Servicing Mission 4. It is an aperture spectrograph with a similar entrance window: 2.5 arcsec diameter circle, and also has a similar nominal resolution of $R \approx 20,000$. The major advantage of the COS over the other spectrographs is the large format FUV detectors, that provide a continuous wavelength coverage extending from 1150 to 1450 Å, with the G130M grating, almost without break. This means that as well as Lyα, many useful stellar and interstellar emission and absorption features can be obtained in a single observation without the need of additional settings. Other settings, centered around 1600 Å (G160M) and 1850 Å (G185M) probe longer wavelengths at similar resolution (although lower sensitivity), and the COS also has a low-resolution grating ($R \approx 1000$; G140L) that can capture the full 1100–1900 Å wavelength range continuously. Scientific results from the COS are discussed mainly in Sects. 4.5.2 and 4.5.3.

The *Advanced Camera for Surveys* (ACS)

ACS was mounted in 2002 during HST Servicing Mission 3B, and has two currently active channels. One of these—the *Solar Blind Channel* (SBC)—is sensitive in the

FUV between $\lambda \approx 1200$ and 1800 Å, using the same Multi-Anode Microchannel Array (MAMA) detectors as the STIS. The SBC is able to perform low-resolution prism (slitless) spectroscopy, but more importantly is loaded with a number of filters that transmit Lyα over various redshift windows between $z = 0$ and $z \approx 0.35$. Thus using narrowband synthesis techniques, ACS can provide Lyα images and the only fully two-dimensional morphological information of Lyα at low-z without employing raster scans with a slit spectrograph. Indeed it is the only instrument that has ever been able to accomplish this and, moreover, imaging over large apertures is the only way global Lyα fluxes can be measured. Details of narrowband synthesis can be found in Sect. 4.6.1, and scientific results from ACS Lyα imaging are discussed in Sects. 4.4 and 4.6.

The *Ultraviolet and Visible* (UVIS) channel of WFC3

For completeness we must also mention the UVIS channel of the *Wide Field Camera 3*. WFC3/UVIS contains a grism for slitless spectroscopic observations, for which the sensitivity peaks near 2400 Å. Because of the large field-of-view of UVIS, this setting can be used for large volume Lyα surveys at redshift $\sim$1. This is something of a 'honourable mention' in these notes because there are currently no UVIS study of Lyα from $z \sim 1$ galaxies in a refereed journal. In principle one could conduct a deeper survey than the existing $z \approx 1$ results with UVIS.

4.2.5 Galaxy Evolution Explorer

The *Galaxy Evolution Explorer* (GALEX [94]) flew between 2003 and 2013, and like the IUE exceeded its nominal mission lifespan by a large factor. Every telescope mentioned so far in this section has been a 'targeted' instrument, that must be pointed directly towards pre-selected objects. GALEX, in comparison, was the first real blind survey instrument in the UV.

GALEX flew a primary mirror with only a 45 cm diameter (the same as the IUE). However the main power of the GALEX satellite was its very large field-of-view, which covered a 1.2 degree diameter circular field. The efficiency of this is survey power, compared to HST, is highlighted in Fig. 4.6: the GALEX field-of-view is projected upon the M81 and M82 field in the left panel, while the right panel shows a zoom in to the orange box centred upon M82. The field-of-view of HST/STIS (33×33 arcsec^2; ACS/SBC is the same size and these are actual observations) is shown by the green squares. Each of these is 13,500 times smaller than the area covered by GALEX. Thus it is clearly illustrated why HST observations must adopt a pure pointing strategy, while GALEX was able to blindly survey the sky.

GALEX operated simultaneously with two bandpasses centered around 1500 Å (FUV) and 2200 Å (NUV). The main mission objective was to image the entire sky in the two bands, which in its own right has proven to be an enormously valuable resource for astrophysics in the local universe. Indeed many of the objects observed with HST, and discussed in the coming sections, were selected from the GALEX

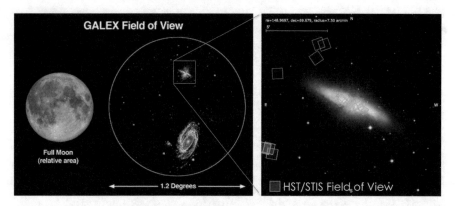

Fig. 4.6 A comparison of the GALEX field-of-view (*left*) with that of HST (*right*). The 1.2 degree circular field of GALEX encompasses the whole M81 and M82 complex, while the *right* figure zooms in on the orange box of M82. Each green box in the *right* panel shows the field-of-view of HST/STIS in the FUV. The ACS/SBC camera is the same size. In survey area the GALEX field area is 13,500 times larger

imaging catalogues. With direct relevance to Lyα, GALEX also operated a slitless spectrograph in both modes: while having only a low spectral resolution of $R \approx 100$–250, the field advantage enabled GALEX to survey large selected volumes in search of Lyα-emitting galaxies. In slitless mode the FUV and NUV channels were sensitive to Lyα emission at redshifts in the range $z \approx 0.19$–0.44 and $z \approx 0.7$–1.2, respectively. The combined field-of-view and large redshift range of slitless mode probes survey volumes of 5×10^5 cubic Mpc, which exceeds the volumes probed by typical single-pointing narrowband surveys at high redshift, and is comparable only with a Hyper-Suprime Cam pointing in volume. GALEX has discovered more than 100 Lyα-emitting galaxies in each of the redshift ranges; these results are discussed in detail in Sect. 4.4.

4.3   First Observations of Extragalactic Objects in Lyα

In the present climate of large samples and high resolution information (the HST and post-GALEX era) it is easy to be spoiled by the data quality and wealth of information. We also find ourselves embroiled in a discussion about how Lyα emission is a complicated, multi-parametric process in which we have not yet observationally determined the most important properties. It is often overlooked that this discussion began in the 1970s, driven by data of significantly lower quality than those with which we are familiar today.

The main points to take home from this section are:

- The first samples of local AGN and star-forming galaxies were observed in Lyα with the IUE between 1978 and 1996.
- Even given the long mission lifespan, aperture-matched UV and optical spectra of starburst galaxies numbered only about 20.
- Lyα/Balmer line ratios in both galaxies and AGN are significantly lower than the expectations based upon Case B recombination. In the majority of cases simple dust screen models do not reconcile the line ratios with expectation.
- Correlations between Lyα output (Lyα/Hβ, Lyα EW) and measure of dust and metal content ([O/H], E_{B-V}, UV colour) are present but the correlation is weak and scatter very large. Lyα output is no clear function of any single galaxy property.

4.3.1 Seyfert Galaxies and Quasars

The first vacuum UV observations of any extragalactic object were performed from a sounding rocket, carrying the Faint Object Telescope (FOT) [24]. Naturally targeting the sources expected to have the brightest UV flux, the FOT delivered a spectrum of 3C 273 between wavelengths of 1300 and 1700 Å, which includes the Lyα line redshifted from $z \approx 0.16$ (Fig. 4.7). This spectrum shows clearly detected continuum radiation on top of which a bright Lyα line is clearly visible with EW= 60 Å. The

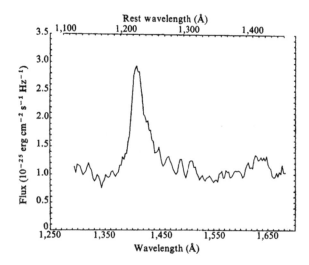

Fig. 4.7 The first Lyα spectrum obtained of an object in the local universe. This shows the spectrom of local quasar 3C 273 at $z \approx 0.16$, provided by the Faint Object Telescope [24]. The prominent Lyα line at 1216 Å in the restframe (upper x-axis) is blended with the N V 1240 Å feature, and appears asymmetric. *Figure is reproduced with permission from Davidsen et al.* [24] *and from Nature*

emission feature appears asymmetric, although this is because the Lyα line is blended with the N V 1240 Å line; this is often the case in AGN.

This spectrum immediately presented a puzzle because the Lyα line was around a factor of 10 weaker than it was predicted to be: Lyα/H$\beta \approx$ 40 was expected from recombination theory, but was observed to be just 4 in the brightest local quasar. The most obvious explanation—that reddening due to dust is responsible for the destruction of Lyα—seemed completely untenable in this case, because the Paschen and Balmer line ratios are all consistent with very little reddening (see Sect. 4.1.3.1). These authors concluded that the nebula must have a high optical depth to Lyα, and that fully understanding the output would lead to vastly enhanced understanding of QSO envelopes. The contemporary debate of what drives the Lyα emission from galaxies was effectively founded by this paper.

With the launch of the IUE in 1978, a new era of extragalactic science began, as it became possible to target larger number of objects and for longer exposure times. First observations of QSOs and Seyferts [103] verified and generalized the low Lyα/Hβ ratio observed in 3C 273 [24]: Lyα/Hβ (expected to be 40) was 2–5 without dust correction [103]. The first compilation of 16 AGN, complete with Lyα and Balmer fluxes was assembled by Wu et al. [150]. Again, dust corrections based upon Hα, Hβ, and Paα, fail to reconcile the Lyα fluxes with recombination theory in the vast majority of cases. A compilation of these results can be seen in Fig. 4.8. However, it is very noteworthy that in three systems [103, 150], the dust-corrected Lyα appears *too strong* for any known attenuation laws to explain. Again the issues of high gas densities, thick H I columns, high optical depths had to be invoked to explain the departure from recombination line ratios [80, 151].

The early IUE papers on AGN were also the first to report distributions of the Lyα EW, seen also in Fig. 4.8. While they were reporting low values of Lyα/Balmer lines, the EW (which is simply Lyα/UV continuum) remained quite high. For Seyfert 1 systems and radio galaxies, the EW distribution has a relatively high modal value of $\approx$150 Å, while for high-redshift quasars this number is $\approx$70 Å. Both of these numbers are high compared to the EWs expected for star formation. It is curious indeed that the Lyα/Balmer ratio is lower than expected, while the Lyα/FUV ratio is high: this implies that some of the processes that affect Lyα may also influence the adjacent continuum, or that the intrinsic Lyα EW is extremely high in these systems. As we will see in the coming sections, and by comparison with high-z galaxies (notes by M. Ouchi), this distribution is significantly flatter and skewed towards higher values than for star-forming galaxies. Even with a modest sample, some sources were found to have EWs that exceed the 240 Å predicted for star formation [17]. We are likely to be looking at systems in which the hard ionizing flux from accretion onto the supermassive black holes produces more ionizing photons per non-ionizing UV, as discussed in Sect. 4.1.2.2.

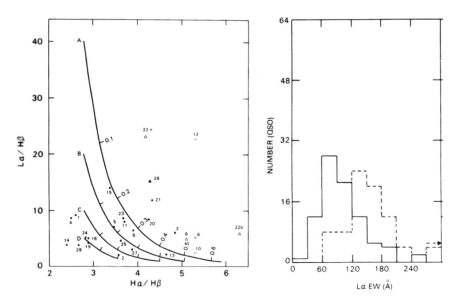

Fig. 4.8 A compilation of Lyα observations of Seyfert 1 galaxies and quasars [151]. The *left* panel shows the Lyα/Hβ ratio against Hα/Hβ. The black lines follow dust extinction laws assuming unreddened line ratios of Lyα/Hβ = 40, 20, 10, and 5 from top to bottom. Most systems lie below the Lyα/Hβ = 40 line but some fall above. The *right* panel shows the equivalent width distribution of Lyα for local Seyfert 1 and radio galaxies (dashed line) and high-z quasars (solid line). *Figures are reproduced with permission from Wu et al.* [151], ©*AAS*

4.3.2 Star-Forming Galaxies

Observations of star-forming galaxies were not far behind those of AGN. At the time, a handful of compact Lyα sources had been discovered in the immediate vicinity of bright quasars [27, 59] at high-redshift, but dedicated blind searches for high-z Lyα-emitting sources had still not located a single Lyα-emitter by 1994 [114]. This result was largely echoed also by pointed observations of known low-z starbursts; here observers had the distinct advantage of knowing that if IUE were pointed at a galaxy with strong Hα emission, Lyα photons must be produced within the spectroscopic aperture. We knew were to look, but still there was little to see.

The IUE was first pointed at three local starbursts [97]: in one case the object was likely too close to deblend Lyα from the geocoronal line; the other two are shown in Fig. 4.9. A Lyα feature is visible in Mrk 701 in the upper panel, but it is seen only in absorption, and there is no hint of a nebular emission feature. The third galaxy, however, C 1543+091 does indeed show a Lyα emission line. While the line is clearly detected, the Lyα/Hβ flux ratio was just 1/6 the expected strength based upon recombination theory. For the absorbing galaxies limits we place at Lyα/Hβ below 1. The first observations of starburst galaxies directly mirrored the results obtained in AGN (Sect. 4.3.1), and led to the conclusion that multiple resonance scatterings

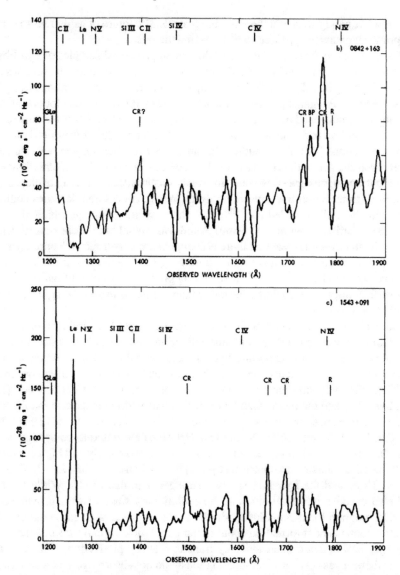

Fig. 4.9 The first star-forming galaxies observed in Lyα by the IUE. The *upper* panel shows Mrk 701, which shows only an absorption signature at Lyα. The *lower* panel shows C 1543+0907, in which a bright Lyα emission line is visible, redshifted to λ = 1250 Å. Note that both spectra appear to be rich with features, but most are labelled as suspected cosmic ray hits on the detector. *Figure is reproduced with permission from Meier and Terlevich [97], ©AAS*

of Lyα will quickly wash out the emission feature. The future of using Lyα to search for primeval galaxies appeared to be a pessimistic one.

The community proceeded to assemble several piecemeal samples of Lyα observations that had each been identified by their own selection method. The samples that followed were generally selected from objective prism surveys targeting strong Hα emission lines (H II galaxies), and compact galaxies with very blue colours that were identified by multi-band imaging (blue compact galaxies, BCGs). Most showed a mix of Lyα absorption and emission, but with Lyα fluxes an order of magnitude weaker than expectations [40]. Early authors began to contrast their targets with BCGs at lower redshifts, where the H I gas could be targeted directly by 21 cm emission; they noted that analogous galaxies were invariably enshrouded with H I envelopes that were 3–10 times larger than the stellar disks. Where H I column densities could be measured, these all exceeded 10^{20-21} cm^{-2} on the central pointings which, if homogeneously distributed within the beam, would correspond to millions optical depths in Lyα. In all cases it is clear that there is enough atomic hydrogen to suppress a Lyα line, even if the dust content is small. These were the first to present the idea that if H I envelopes are large, and the Lyα scattering process looks like diffusion, then the Lyα surface brightness could be far lower than predicted from H II regions, even if the photons do escape.

Interestingly in the very first observations, the single Lyα-emitting object was that with the lowest metallicity and dust extinction, and this lead to speculation that a trend between metal content and Lyα emission could be present. When samples of Lyα observations reached sizes of 12–14, weak trends began to emerge in the IUE data. Firstly it was seen that weak anti-correlations were present between {Lyα EW, Lyα/Hβ} and the metal abundance, which was verified in several publications [17, 41]. These are shown in the upper panels of Fig. 4.10, and indicate that at [O/H] $\lesssim -0.8$, Lyα EWs exceed 50 Å, and Lyα/Hβ ratios are found to be as high as 10. Note that 10 is still $\approx 1/3$ the recombination value, but is clearly still high compared to all the other galaxies. In contrast, at [O/H] ≈ -0.5 and upwards, Lyα EWs never exceed 10 Å, and the Lyα flux is never stronger than that of Hβ. While the Lyα EW also encodes information about the evolutionary stage of star formation, the Lyα/Hβ diagram looks remarkably similar to that of the EW. Because the Lyα/Hβ quantity contains no information about stellar populations, it indicates that the emergent Lyα EW is determined more by transfer than by production. Note also that metals themselves—at least free-floating ions—do not absorb Lyα and a correlation between Lyα/Hβ and [O/H] must be reflecting some other property of galaxies that makes low metallicity systems more amenable to Lyα *transfer*.

Using new data reduction tools, an IUE sample of 20 stabursts was finally reprocessed and analyzed [34], with aperture-matched Hα and Hβ obtained from the ground. Indeed measuring total fluxes—in 20 $\times$ 10 arcsec apertures that match the IUE, for galaxies at various recession velocities was, and still is, time consuming. With such improvements one would expect an improvement in any correlation analyses, and a tightening of trends; however Giavalisco et al. [34] showed that if anything trends became weaker in the homogenized dataset. The lower panels of Fig. 4.10 are from the re-analysis and should be compared with their corresponding upper panels.

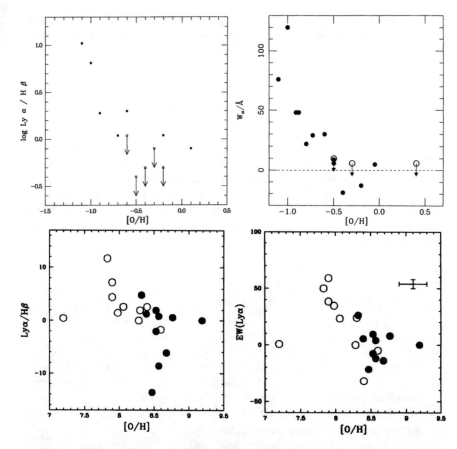

Fig. 4.10 Early trends of the Lyα output of galaxies compared with the gas-phase metallicity. *Left* panels show the Lyα/Hβ ratio which is a purely nebular measurement, while *right* panels show the Lyα EW which is also dependent upon the evolutionary stage of star formation. Anti-correlations are visible by eye in all cases, but clearly the dispersion is large with many outliers. *Figures are reproduced with permission from Hartmann et al. ([41]; upper left), Charlot and Fall ([17]; upper right), and Giavalisco et al. ([34]; lower), ©AAS*

In testing both Lyα EW and Lyα/Hβ against various dust and metal-related quantities, no significant correlations were found [34]. With dust playing an apparent minor role in Lyα transfer, atomic gas is left to explain why so few galaxies are strong emitters.

A worthy mention in this section is the Lyα spectrum of M 33, obtained with the UVS onboard *Voyager 2* by raster-scanning the disk with a long-slit [69]. This is the lowest redshift object with a recorded Lyα spectrum ($z = -0.0007$), and has by far highest apparent brightness, with a Lyα flux of 1.8×10^{-9} erg s^{-1} cm^{-2}. The EW of ≈ 100 Å is also particularly high for local systems. Interestingly, there is no enhancement of Lyα at the position of the bright H II region NGC 604, which contains $\approx$200 O-type and Wolf-Rayet stars in its central cluster. This suggests that the Lyα radiation from NGC 604 does not take a direct path out of M 33.

4.4 Systematic Surveys: The Global Picture

With the advent of the GALEX satellite and efficient, high-throughput cameras on HST, it became possible to really survey local galaxies in the Lyα line, both by conducing blind surveys (GALEX) and dedicated pointed observations (HST). Similarly to the last section, this section is concerned with information drawn from large-aperture, global Lyα spectrophotometry. As such the data-sets may be used as direct comparison against the studies at high-redshift, where only large-aperture, close-to-global photometry, is possible.

> Main points to take away from this section are:
> - GALEX discovered over 100 galaxies in each of two redshift windows around $z \approx 0.3$ and $z \approx 1$. HST samples with global photometry (from imaging data) currently total around 60 galaxies at $z \lesssim 0.2$.
> - Unbiased surveys with GALEX have allowed us to measure the Lyα luminosity function. LAEs at low-z are significantly less abundant than their counterparts at $z \gtrsim 2$. The volume-integrated Lyα luminosity density of the nearby universe is $\approx$100 times smaller than at $z \gtrsim 3$.
> - Lyα-emitting galaxies tend to be drawn from a population that is of lower mass, younger, more compact, bluer, more intensely star-forming, less dusty, of lower metallicity, more strongly ionizing, and of lower H I mass.

4.4.1 Approaches to Local Lyα Surveys

4.4.1.1 Blind Lyα and UV Surveys with GALEX

As discussed in Sect. 4.2.5, GALEX could survey a volume of 500,000 Mpc^3 for Lyα emission between $z = 0.195$ and 0.44 in a single observation, and was thus an extremely efficient observatory for low-z Lyα surveys. GALEX surveys were first conducted by Deharveng et al. using GALEX pipeline data [25], and later Cowie et al. using an independent data reduction and more fields [19, 20]. Both publications included catalogues of LAEs with Lyα EWs greater than 15–20 Å, which is a similar and very useful sensitivity limit because it closely matches the threshold of many high-redshift surveys making comparison relatively straightforward. Another significant advantage is that GALEX data also contain complete UV information for all the non-Lyα emitting galaxies and provide an unbiased control sample.

There are however some important differences between the GALEX-selected samples and high-z surveys. The majority of high-z LAEs are selected in narrowband imaging or blind spectroscopic surveys, whereas GALEX was a slitless spectrograph: while Lyα selection is very straightforward in narrowband data, it is not straightforward in most slitless observations. Slitless data usually require the detection of a source in an *undispersed* image first, and the trace is then identified (with or without emission lines) in the dispersed image. This detection image is usually dominated by continuum light, and therefore the GALEX LAEs (at least those published to date) are effectively continuum-selected first. The differing selection function must be kept in mind when comparing GALEX results to high-z, which has lead to a (admittedly minor) misconception.

Some more recent papers [7, 149] have begun to remedy this by leveraging the fact that GALEX slitless observations were obtained on many different position angles, and thus include many different orientations of the dispersion axis on the sky. Thus one can solve for the degeneracy between spatial and wavelength coordinates for a given pixel, and resample pixels from all dispersed images into datacubes that are similar to those delivered by integral field spectrographs. With datacubes one can then perform more conventional Lyα selection, and identify fainter sources, that may even lack continuum. Currently this has only been done for the NUV ($z \sim 1$) datasets [7, 149] but these results are currently promising, and are likely to significantly enhance the present status when finalized in all datasets.

4.4.1.2 Targeted Observations with Hubble

HST needs to be pointed at individual targets, so when interpreting any HST data one needs to keep in mind the selection function. Many HST datasets exist, performed by various groups, with either the motivation of studying Lyα directly, or acquiring UV observations of galaxies where the Lyα line is included. We list a few of these below, ordered approximately by sample size:

1. The *Lyman alpha Reference Sample* (LARS) [49, 106], which has obtained HST imaging and spectra of 42 galaxies. These are selected by UV luminosity (GALEX imaging) to be comparable to $z = 3$–6 LBGs, and Hα EW (SDSS spectroscopy) to guarantee current ongoing star formation.
2. *Lyman Break Analog* (LBA) samples [53, 54, 58] with UV spectroscopy from COS. These were selected from UV luminous galaxy samples (GALEX imaging) [58], and u-band compactness (SDSS imaging). The samples contain around 25 galaxies, of which three overlap with Sample 1.
3. GALEX LAEs [126]. 25 galaxies were observed with COS. These objects were selected from the blind Lyα surveys performed with GALEX [19, 20] in order to study the Lyα profiles at high spectral resolution.

4. COS Guaranteed Time star-forming targets [148]. 20 galaxies were observed with
   the COS, selected by Hα emission from the Kitt Peak International Spectroscopic
   Survey data release (KISSR sample [123]). Four of these targets overlap with
   Sample 1.
5. Green Peas [55, 61, 64, 152]. This mixed sample includes ≈20 galaxies, that were
   mostly observed with COS. These were initially selected from the SDSS Galaxy
   Zoo [15] by citizen scientists, to be extremely compact and green in color. Optical
   spectroscopy later revealed that this is because of extremely strong, redshifted
   [O III] and Hβ lines. One of these objects overlaps with Sample 1 and Sample
   2. Selection functions differ somewhat: some cut by UV luminosity to ease the
   spectroscopic target acquisition, and others also by high [O III]/[O II] ratios (see
   Sect. 4.4.3.3).
6. Ultra-luminous Infrared Galaxies (ULIRGs). 11 galaxies at $0.08 < z < 0.15$ were
   observed with COS [93], with the intent of studying Lyα output from galaxies of
   the highest bolometric luminosity ($L_{FIR} > 10^{12}L_\odot$). These were selected to be
   among the UV-brightest ULIRGs from the IRAS 2 Jy catalogue [138].

Figure 4.11 is a modified version of that from Cowie et al. [20], Barger et al. [7], and
Wold et al. [149] shows how some of the Lyα luminosities of local galaxies compare
with those of some high-z surveys. It is clear that at $z < 0.5$, the GALEX LAEs
and LARS objects are comparable in luminosity to some of the deeper observations
performed at high-z [13, 38, 46, 118]. In these volumes only the most extreme star-
forming galaxies (e.g. the green peas) [55] and specially selected individual targets
[50] are comparable to the brighter high-z LAEs. In contrast, the $z = 1$ GALEX
NUV samples exceed $L_{Lyα}$ of $10^{42.5}$ erg s^{-1}, and are comparable to the brighter end
of of the distribution of galaxies at high-z. The limit in survey sensitivity is close
to the characteristic luminosity, $L^\star$ at $z = 3$ [37, 108]. Such luminous Lyα-emitting

Fig. 4.11 The Lyα
luminosity of some samples
at low- and high-redshift.
Reference encoding is
R08 = Rauch et al. [118],
H10 = Hayes et al. [46],
B11 = Blanc et al. [9],
C11 = Cowie et al. [20],
G11 = Guaita et al. [38],
C12 = Cantalupo et al. [13],
H14 = Hayes et al. [49],
W14 = Wold et al. [149],
H15 = Henry et al. [55].
*Figure is expanded from
Hayes* [42]

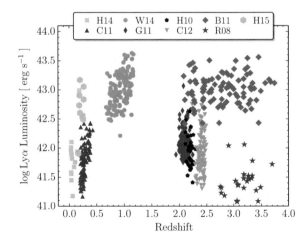

sources ($L_{Ly\alpha}> 10^{42.5}$ erg s^{-1}) do still exist in the local universe, but they appear to be extremely rare [20] and GALEX Lyα surveys did not cover the volume to find them.

4.4.2 Number Counts, the Luminosity Functions, and Cosmic Escape Fraction

With a statistical sample of LAEs at low-z, we will obviously want to examine their luminosity and EW distribution, and compare how they relate to the UV-selected local galaxy population, and the equivalent Lyα distributions at high-z.

Figure 4.12 shows the local FUV number counts of $z = 0.2$–0.4 LAEs, and compares them to the total FUV measurements in the same redshift range. The black line shows all the galaxies detected in the FUV by GALEX, and as usual there are few bright sources and many faint ones. The dashed and dotted lines show this function scaled down by factors of 20 and 100, respectively. The FUV counts of LAEs with EW above 20 and 45 Å are shown in blue and red points, respectively. These observed counts of LAEs are well-matched by the reduced total counts (dashed and dotted lines), which show that ∼5% of the local UV-selected galaxy population shows a bright Lyα line.

The data show that a few percent of local UV galaxies, with magnitudes of 20–22 in the AB system, show Lyα emission. This corresponds to absolute UV magnitudes of between −19 and −21, or between about 0.15 $L^{\star}$ and $L^{\star}$ at $z = 3$ [119]. At

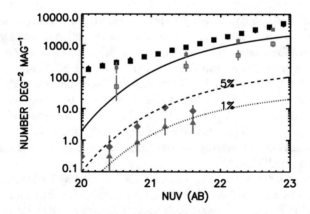

Fig. 4.12 NUV number counts. The black solid line shows a fit to the galaxy counts at $z = 0.2$–0.4 [1], which are shown by the magenta open squares. The solid magenta and black points show the counts for all galaxies in the GOODS-N field and all GALEX fields, respectively. The dashed and dotted black lines show the fit scaled down to 5 and 1% its total value. Blue and red points show the counts of LAEs with EW above 20 and 45 Å, respectively. *Figure is reproduced with permission from Cowie et al. [19], ©AAS*

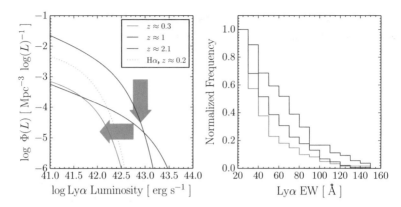

Fig. 4.13 *Left* shows the Lyα luminosity functions derived at $z \approx 0.3$ and $z \approx 1$ from GALEX FUV and NUV slitless spectroscopy, as well as at $z \approx 2$ from groundbased narrowband imaging. The *right* panel shows the EW distributions for the same samples

high-redshift there is some debate about precisely what this Lyα fraction ($X_{\mathrm{Ly}\alpha}$) is and how it evolves, but for $z = 3\text{–}4$ LBGs $X_{\mathrm{Ly}\alpha}$ is around 25% [131, 134], while it is nearer 50% by $z \approx 6$ [22, 130]. While $X_{\mathrm{Ly}\alpha}$ depends somewhat on the FUV luminosity, these fractions are clearly higher than the 5% found at $z \approx 0.3$.

Similar results are found when examining the Lyα luminosity function and its integral. Between $z = 0.3$ and 2, the star formation rate density of the universe increases by a factor of ≈ 5 [89], which is derived from integral of the galaxy LF—this is done with various selection functions and star formation tracers, but most methods give broadly consistent results. The luminosity function plots the number density (galaxies per comoving volume) at a differential luminosity, as a function of that luminosity. Thus, it can be used to contrast the evolution of that Lyα output of the universe as a function of z. The Lyα LF at $z \approx 0.3$, 1, and 2 is shown in Fig. 4.13, where measurements come from GALEX FUV, GALEX NUV, and ground-based narrowband imaging, respectively. Comparing the low-z universe with $z = 2$, there is clearly a large evolution: we do not attempt to place the evolution in terms of luminosity or number density, but note simply that the integral under the curve differs by a factor of 100 [47] between these two redshifts. Another way: the Lyα luminosity density of the universe is 100 times lower today than it was when the universe was around 3 Gyr old.

This factor of 100 decline in Lyα luminosity density may be compared with the change in the cosmic star formation rate density, which has also declined over the same epoch [89]. However this decline in SFRD amounts to only a factor of 5, which implies that as well as producing fewer Lyα photons per unit volume, galaxies in the modern universe also allow a smaller fraction of them to escape. The 'volumetric escape Lyα fraction', $f_{\mathrm{esc}}^{\mathrm{Ly}\alpha}$, also drops by a factor of around 10 over the same Δz [46, 47]. This is approximately in agreement with the decline in the fraction of UV-selected galaxies with strong Lyα emission ($X_{\mathrm{Ly}\alpha}$) that is discussed above.

We end this section by briefly discussing extended Lyα nebulae, dubbed 'Lyα blobs' (LABs). With GALEX pipeline products these are challenging to find at low-z: none were reported in the first papers that exploited the extragalactic survey data of GALEX in grism mode [19, 25]. However high-redshift results suggest that LABs live only in the most dense environments, but dedicated observations of two galaxy clusters at $z \approx 0.8$ also failed to find any spatially extended Lyα sources [70]. However this absence of LABs over significant survey areas enables a $z \approx 1$ constraint on the LAB density to be placed at $<0.5 \times 10^{-6}$ Mpc^{-3}.

Going beyond the pipeline data to the resampled datacubes described above [7], a single Lyα nebula was identified in the Chandra Deep Field South, with a projected diameter of 120 kpc. However this single blob does not alter the statistics by much: $z = 3.1$ narrowband observations of the SSA 22 protocluster found 10 extended nebulae in the same survey volume [96], and not accounting for luminosity differences the number density of low-z blobs remains around 1/10 that of $z = 3$. This $z = 0.9$ LAB would fall at the very lower end of the luminosity distribution at $z = 3$; while we do not have a sufficient number of LABs to constrain their density at low-z, it is clear that luminous systems have become rare.

4.4.3 Nebular Gas: Dust Content and Distribution, Metallicity, and Ionization Parameter

Naturally one of the reasons these GALEX samples attracted so much attention is that, unlike many sources at $z > 2$ they are easily amenable to followup observations. It is noteworthy that four followup papers had been published within one year of the first catalogue publication [25]. This allows much more precise determinations of the stellar and nebular properties of the galaxies than is typical in $z > 2$ Lyα-emitter samples. In this section we summarize some of the key results derived from optical emission lines that target the nebular regions in which Lyα radiation is formed. Stellar properties of the host galaxies are treated in Sect. 4.4.6.

4.4.3.1 Dust Content and Distribution

Hα and Hβ are among the strongest lines in the optical, and at $z < 0.4$ could be detected by spectroscopic observations on modest sized telescopes. The intrinsic Hα/Hβ ratio is set almost entirely by the physics of the hydrogen atom, and the dependency upon temperature and density is very weak [104]. Thus departures from an intrinsic Hα/Hβ ratio of around 2.86 most likely indicate that dust along the line-of-sight has absorbed the bluer Hβ line more than Hα, and increased the ratio. As discussed in Sect. 4.3.2, only very weak trends had been found between Lyα and dust reddening in the IUE samples [34], but recall that these comprise very inhomogeneously selected galaxy samples, with possibly differing aperture effects.

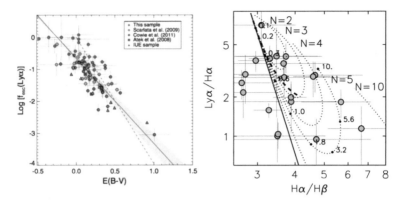

Fig. 4.14 The effect of dust on the Lyα output in LAE samples. *Left* shows $f_{esc}^{Ly\alpha}$ versus nebular E_{B-V}, and *right* shows the Lyα/Hα ratio versus Hα/Hβ. *Figures are reproduced with permission from Atek et al. ([4], left) and Scarlata et al. ([127], right), ©ESO and ©AAS*

With GALEX and homogenous preselection for HST, things could be very much improved.

Figure 4.14 shows $f_{esc}^{Ly\alpha}$ versus nebular E_{B-V} in the left panel [4] and Lyα/Hα versus Hα/Hβ in the right panel [127]. These axes are not identical, but are very similar: they can be regarded as physical properties on the left that have been derived from pure observables on the right. In contrast to the end of the IUE era, a clear anti-correlation between Lyα output and dust content is now visible. Dust obscuration is observed to reduce the Lyα output to some extent, with the highest $f_{esc}^{Ly\alpha}$ galaxies showing the lowest reddening. Note, of course, that the spread is large and many outliers remain. This spread most likely reflects the large number of other quantities that enter the transfer process, along with the viewing angle. Most importantly, these figures show only galaxies with Lyα EW greater than 20 Å, and these are just 1/20 of the UV-selected galaxy population (Sect. 4.4.2). If all galaxies were to be plotted it is likely that the region beneath the locus of points would be heavily populated, but little above. This likely represents the upper end of the distribution for all continuum-selected galaxies.

The dust extinction law derived from stars in the Milky Way [16] is indicated by red dashed line in the *left* plot and the black solid line in the *right* plot. This represents the idealized influence that dust would have if it were distributed as a homogenous screen in front of (and far from) the nebular gas, and goes through the origin: $f_{esc}^{Ly\alpha} = 1$ at $E_{B-V} = 0$. Compared to the dust reddening law, the locus of points have two main features: at low E_{B-V} (Hα/Hβ) there is a cloud of points below the line, while at higher Hα/Hβ (more reddened galaxies) most points lie above it. This flattens the slope of the effective attenuation, and the best linear fit to the data crosses $E_{B-V} = 0$ at $f_{esc}^{Ly\alpha} = 0.2$. Thus 80% of the Lyα radiation is still absorbed even in Lyα-selected galaxies with zero measured nebular dust extinction. Of course, this is most likely because H I gas resonantly scatters Lyα, increases the path length to escape, and therefore the probability of absorption.

The second main point of interest is that the observational data follow a locus that is much *flatter* than the Milky Way extinction law. This result is particularly curious, because we expect if Lyα were preferentially attenuated, and absorption were only enhanced by scattering, a steeper relationship would be derived. Indeed at all E_{B-V} there are galaxies where the inferred $f_{esc}^{Lyα}$ is up to 0.5 dex higher than expected. In short: at low E_{B-V} we see less Lyα than expected, and at high E_{B-V} we see more.

There are several explanations for this distribution of line ratios. Firstly, if dust is embedded only within dense H I clouds, then Lyα scattering on the surface of clouds may prevent Lyα from encountering dust grains, and preserve Lyα, while non-resonant lines must pass directly through the clouds and be reddened [39, 101]. This could preserve high Lyα/Hα ratios at high optical depths. However the difficulty with this interpretation is that this geometric construction would lead to these same high Lyα/Hα galaxies having higher Lyα EWs, and this is not observed. Moreover, three dimensional radiation transfer simulations show that this model can only work if the intra-clump medium is completely evacuated *and* the clumps are static [28, 82].

It is likely that the answer lies in departure from the pure dust screen model. The *right* panel of Fig. 4.14 [127] shows a series of curved dotted lines that extend upwards from the dust-screen attenuation curve, which are representative if the dust distribution is clumped. Each concentric loops represents an increasing number of clumps (labeled), and follows a curve of increasing τ. Should a number of clumps be distributed along the line-of-sight, then the effective attenuation becomes dependent upon the number of clumps and their optical depth, τ. At very low τ one expects little obscuration, just as for a screen, and the curves agree. At very high τ one expects effectively no reddening whatsoever because high τ clumps absorb all the light that impinges upon them, the emergent light is dominated by rays that do not pass through clumps. In this limit the emergent light is purely dependent upon the covering fraction, and not on τ, and again the curves agree. However at intermediate τ the clumpy medium makes the effective attenuation law appear much grayer than the conventional extinction law that governs the individual clumps; in such a scenario clumpy dust models can maintain UV/optical colours that are bluer than would be expected from colours measured at longer wavelength. This scenario cannot explain the sub-recombination Lyα fluxes, and resonance scattering/preferential absorption must still invoked for this.

All of the above discussion relates to Lyα-selected galaxies from the GALEX surveys, but when similar analyses are performed with HST in UV-selected samples, similar conclusions are also reached [49].

4.4.3.2 Nebular Metal Abundance

While correlations that emerged from the IUE samples were all weak, the strongest of the bunch was that between Lyα output (Lyα/Hα and Lyα EW, see Fig. 4.10), where more Lyα is emitted at lower nebular oxygen abundance [34, 41]. This key result is also borne out by later studies with GALEX and HST. By comparing the metallicity of GALEX LAE samples with non-Lyα-emitting counterparts of the same magnitude, it is clear that the metal distribution of the emitters is drawn from the low end of the parent distribution [19, 20]: the median value reported for the LAEs is $12 + \log(O/H) \approx 8.1$, while for the UV-selected population at the same magnitude is $12 + \log(O/H) \approx 8.6$.

The same conclusion is reached in the LARS sample, which was selected to span a range of UV luminosities with no a priori Lyα information. Galaxies with $f_{esc}^{Ly\alpha} > 0.1$ have [N II]6584/Hα ratios below 0.1 [49], which corresponds to $12 + \log(O/H) <$ 8.3. Green pea galaxies all have particularly bright Lyα lines—systematically the highest EW galaxies at low-z—and all have metallicities below $12 + \log(O/H) \approx$ 8.2 [55]. Again we must note that when considering Lyα EW, the intrinsic value may also be increased if low metallicity stars are formed, but still the high $f_{esc}^{Ly\alpha}$ values cannot be explained in this way without some unseen quantity affecting the escape of the Lyα that is produced. This, for example, could be a positive correlation between metal and dust abundances, but to date the causal mechanism has not been demonstrated.

4.4.3.3 Ionization State of the ISM

In the classical Strömgren sphere model, a star (or population thereof) embedded within a neutral medium will ionize its local volume. The ionization front will continue until the recombination rate balances the photoionization rate, and the volume of the Strömgren sphere is proportional to the ionizing luminosity, and inversely to square of the gas density. Ionizing radiation will propagate only to the edge of the Strömgren sphere—the ionization boundary—where one may expect an abrupt transition in the column density of H I, and consequently the optical depth to Lyα. In reality the gas will contain H, He, and various metals, and the continuum may have a spectrum that extends beyond the ionization edges of several more highly ionized species (e.g. oxygen, sulphur, etc.), and one expects each ionic species to be confined tightly to concentric embedded zones with the most highly ionized regions being more central.

If the ionizing flux is high enough, or the galaxy disk thin enough, it is possible that ionized bubbles, or fragments thereof, may extend to the 'edge' of a galaxy—the density boundary. This is known as a 'density bounded nebula' or a 'truncated Strömgren sphere'. In such a case, ionizing radiation ('Lyman continuum', LyC) could propagate into the local IGM. One may argue, therefore, that because the H I column density should not exceed 10^{17} cm^{-2}, that the transfer of Lyα photons would also be eased. As stated above, the zones of highly ionized metals should be smaller than those of lowly ionized species, and could act as a probe of the truncated Strömgren sphere. Some of the most interesting recent studies have formulated the theory necessary to use ratios of forbidden metal lines in the optical as an observational tracer of highly ionized channels through which Lyα and LyC may propagate [112]. Specifically the ratios of emission line fluxes from p^2 ions (with split ground state) to those of p^3 ions (with split first excited state) of the same atom, can be used as observational traces of possible density-bounded nebulae.

Ideally one would use transitions of the same element: e.g. [S III]/[S II] (doublets at 9071, 9533 Å and 6718, 6732 Å) or [O III]/[O II] (doublets at 4959, 5007 Å and 3726, 3729 Å), which are all relatively strong, and well positioned in wavelength to observe from ground-based telescopes in the optical. Note that the p^2 doublets of both elements have relative line strengths that are density-dependent, and so both lines really need to be measured; either of the p^3 ions is sufficient because these are independent of density. In practice this is much easier for the [O II] doublet, for which the split excited levels are closer in energy and the lines lie only $\approx$3 Å apart. It is also possible to use p^2 and p^3 transitions of different elements (e.g. [S II]/[O III]), so long as the ionization potentials significantly differ. Observationally the goal is to search for galaxies—or regions within a galaxy—where the more highly ionized line becomes significantly enhanced compared to the line of lower potential.

In starburst galaxies, the [S III]/[S II] ratio was first used to identify a highly ionized cone in NGC 5253, that extends from the nucleus along the minor axis to far beyond the star-forming regions [153]. The same experiment was later performed on a sample of six more local dwarf starbursts [154], although optically thin channels were only identified in one further target: NGC 3125. These two objects are therefore excellent candidates to emit LyC and Lyα radiation, although both galaxies are too nearby to actually test this.

A similar study has recently been performed in local luminous blue compact galaxy ESO 338–IG04 [8], as shown in Fig. 4.15. The [S II]/[O III] map is shown in the *right* panel, which drops substantially in two cones that extend in the north and south directions from the galaxy, and extend as far as we can trace the ionized gas. This is the only object in which ionization parameter mapping has been performed and HST Lyα images have also been obtained, and indeed the Lyα surface brightness is also enhanced in the direction of the ionization cones. At the same time, the deep integral field spectra reveal the presence of outflows in the same direction, as shown in the velocity field in the *left* panel of the figure, which also would ease the transmission of Lyα.

The same methods can be applied to galaxies on the global scale, to test whether the ISM may be density bounded on average. In this capacity some of the Green Pea galaxies with the highest [O III]/[O II] ratios have been studied using the SDSS spectroscopy and photoionization models [63]. Modeling results are not entirely conclusive, being somewhat degenerate with the properties of the ionizing continuum, gas geometry and filling factor, but CLOUDY modeling suggests that optically thin nebulae are a possibility. Note however, that this conclusion is not unique, and similar line ratios can also be reproduced with very low metallicities and shocks [135].

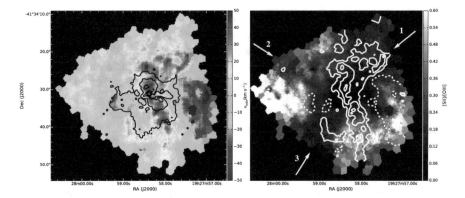

Fig. 4.15 *Left*: the Hα velocity field of ESO 338-IG04. Fast outflows can be seen emanating from above and below the galaxy in red. Lyα contours are overlaid in black solid lines, which appear to be enhanced in the same (north-south) direction. *Right*: the [S II]/[O III] ratio map, with overlaid contours of the velocity field. The Lyα is emitted in 'fans' that extend above and below the disk of the galaxy, possibly as a result of the highly ionized outflowing gas. *Figure is reproduced with permission from Bik et al.* [8], ©*ESO*

However COS observations in the Lyman continuum, obtained of a subsample of these Green Peas, support the optically thin scenario: in a very recent study that targeted five Green Peas with COS, all five were reported to be LyC leakers [60, 61]. While these are not identical to those modeled in detail [63], the rather homogenous optical spectra of Green Pea galaxies suggest that to some extent the predictions are being observationally confirmed.

As the Lyα line is significantly easier to observe than LyC radiation, Lyα tests lead over direct LyC experiments. Of the six galaxies studied with CLOUDY modeling by [63], five have been observed in Lyα with COS, and all show Lyα emission. Two of the galaxies—SDSS J1219+1526 and SDSS J0814+2156—have remarkably high Lyα EWs at 149 and 71 Å, respectively [55, 64], which ranks them as the highest in the sample, and among the highest EWs measured at low-z. Of eight additional Green Peas, six have EWs that exceed 40 Å [55]. Taken as a whole, the selection of galaxies based upon high equivalent widths in [O III] almost invariably selects galaxies with ISMs conducive to Lyα emission. More discussion of these galaxies is presented in Sect. 4.5.3.

4.4.4 Ionized Gas Kinematics

Regarding the pure transfer of Lyα, kinematic studies of ionized gas will in one sense be inferior to those of the atomic medium (Sects. 4.4.7 and 4.5) because they do not trace the motion of the envelope in which the Lyα radiation scatters. However integral field spectroscopy of strong emission lines such as Hα has several other major

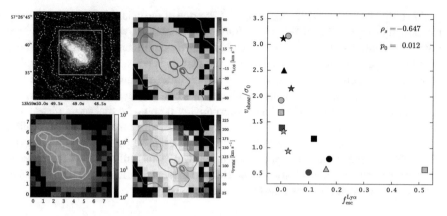

Fig. 4.16 The images in the *left* panels show an example of the IFU data for LARS 05. The left-most panels show the Hα intensity as derived from HST (upper) and PMAS (lower). The next column shows the line-of-sight Hα velocity field (upper) and velocity dispersion (lower). The figure to the *right* shows the Lyα escape fraction compared with the ratio of the shear velocity to the velocity dispersion ($v_{\text{shear}}/\sigma_0$), and indicates that $f_{\text{esc}}^{\text{Ly}\alpha}$ is higher in galaxies where the velocity field is more turbulent. *Figures are reproduced with permission from Herenz et al.* [56], ©*ESO*

advantages: firstly resolved spectroscopy of the H II regions will give the intrinsic distribution of Lyα radiation in both space and frequency, which enables simple aperture matching with a UV spectrograph, and later comparison with H I kinematics. Secondly, the Hα velocity field contains a wealth of information about how the ISM is supported: whether the galaxies are dominated by rotation or if the velocity field is disturbed because of violent mergers or feedback processes. Because the Hα line is so strong, and visible in the restframe R-band, modest observations with modern large-format IFUs can provide this information in relatively short integrations, and at a spatial resolution limited by seeing conditions.

Galaxies in the LARS sample have also been followed up systematically with the PMAS IFU at the Calar Alto 3.5 m telescope, in order to precisely measure the kinematics of the ionized gas [56]. A compilation of results is shown in Fig. 4.16, where Hα frames are shown in the left-most pair of images, and the line-of-sight velocity field, and velocity dispersion, are shown in the next pair of panels. This object—LARS 05 [29] or Mrk 1486—shows a relatively ordered velocity field, consistent with that of an inclined disk. However the amplitude of rotation is small, even for disk galaxies, with a shear velocity [$v_{\text{shear}} = 1/2(v_{\text{max}} - v_{\text{min}})$] of just 37 km s^{-1}. The velocity dispersion map, instead shows values above 200 km s^{-1} in individual spaxels, and when weighted by Hα intensity the total velocity dispersion (σ_0) is almost 50 km s^{-1}. The ratio of these two quantities, $v_{\text{shear}}/\sigma_0$, is an indicator of whether the galaxy is more supported by rotation or dispersion; LARS 05 is slightly dispersion dominated.

The right panel of Fig. 4.16 assembles this $v_{\text{shear}}/\sigma_0$ ratio for the full sample, and contrasts the result with the Lyα escape fraction. Here we see that $f_{\text{esc}}^{\text{Ly}\alpha}$ above 0.1

occurs only in galaxies where $v_{shear}/\sigma_0 \lesssim 1.2$: none of the galaxies that are obviously rotationally supported emit significant fractions of their Lyα, while all of those where unordered motion contributes significantly to the dynamics are Lyα-emitters. There is only one galaxy with a very turbulent ISM that does not show an escape fraction above 10%. Again this result may have multiple interpretations. It is possible that the disturbed Hα kinematics is the result of feedback, which in turn accelerates the atomic ISM allowing Lyα to escape. Alternatively it could be that more disordered galaxies produce their Lyα radiation with an intrinsically broader velocity profile, and the H I medium could see a higher fraction of the Lyα in the wings of the absorption profile where the optical depth is lower. This will also be an avenue of further research, and comparisons of the H II and H I velocity fields will likely resolve the issue when such high-resolution 21 cm data exist.

4.4.5 AGN Content

Whether the luminosity function should have AGN removed is a matter of definition, but in the LFs discussed in Sect. 4.4.5, best efforts had been made to remove galaxies where the nebular line spectrum could be dominated by non-thermal processes [7, 20]. It remains interesting, however, to examine the AGN fraction in galaxy samples at both low and high redshift, and to determine how much of the luminosity could arise from central black holes.

AGN are selected against by most HST surveys, so they are not present in the samples, but GALEX surveys are agnostic to the presence of AGN and provide the perfect sample in which to test. This was first undertaken in detail by Finkelstein et al. [33], who used five independent methods of determining the AGN content of GALEX LAEs: (1) the velocity width of permitted transitions of H I, He I, and He II; (2) the presence of high ionization lines such as He II, and [Ne V]; (3) line ratios that place galaxies in the AGN-dominated regions of the BPT diagnostic diagram [6, 68, 75]; (4) Mid IR (Spitzer/IRAC) colors that indicate a power-law dominated spectrum of hot dust [137]; (5) X-ray data that would indicated an accreting supermassive black hole.

In this first study, 10/23 LAEs were robustly classified as AGN using one or more of the methods described above, giving an AGN fraction of $\approx$40%. This is significantly higher than reported for all surveys at high-redshift, although many of these observations are not possible for high-z LAEs of similar luminosity because of their intrinsic faintness. It should also be noted that not all of this information (especially X-ray and Spitzer/IRAC data) exist for all the objects, and that some characteristics (e.g. some high ionization emission lines) may be mimicked by photoionized nebulae or shocks (see, for example, the photoionization modeling discussed in Sect. 4.4.3.3). Even if somewhat high compared to high-z surveys, and even other reports in the GALEX LAE samples [4, 20, 127], it nevertheless reflects that a significant fraction of the local LAE population harbors an accreting black hole in its nucleus. While the luminosity limits are different at low- and high-redshift and survey to survey (Fig. 4.11), it is certainly possible that the fraction of AGN is higher than reported.

At all redshifts the AGN fraction will be a function of luminosity, and information in Figs. 4.12 and 4.13 is important when interpreting the AGN fraction, and vice versa. For example the AGN fraction varies by a factor of 10 between $L_{Ly\alpha} = 10^{42.5}$ and luminosities ten times higher, where AGN almost certainly dominate the population [149].

4.4.6   Stellar Properties: Colors, Morphology, and Age

In this section we summarize some of the key results derived from optical and infrared continuum observations, which target the stellar populations of the galaxies.

4.4.6.1   Morphology and Size

Stellar morphologies of a small number of Lyα-emitting and non-emitting galaxies from the GALEX samples [20] are shown in Fig. 4.17. Two points are immediately apparent by eye: (1) there is a higher incidence of compact galaxies among the LAEs than non-LAEs; and (2) where spirals are found to be LAEs they tend to be orientated face-on. Similarly, when we look at the UV sizes of LARS galaxies, the galaxies with smaller 90% light radii have the highest Lyα escape fractions [49], as shown in the right panel of Fig. 4.17. As previously stated, Green Peas are among strongest Lyα emitting galaxy samples known, and are also among the most compact of star-forming galaxies in the low-z universe [55]; all are unresolved under typical ground-based seeing limitations in the SDSS, and some are barely resolved by HST. The strongest Lyα-emitters at low-z are preferentially compact star-forming galaxies.

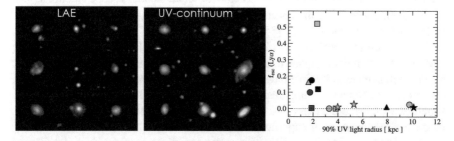

Fig. 4.17 The morphology of Lyα-emitting and non-emitting galaxies. The two larger images to the *left* show LAEs and non-LAEs, and the *right* image shows the Lyα escape fraction against the 90% light radius in the ultraviolet continuum. *Figures are reproduced with permission from Cowie et al.* [20] *and Hayes et al.* [49], ©*AAS*

The inferred Lyα output of galaxies has also been studied as a function of viewing angle using radiation transfer simulations [82, 145]. These simulations predict that line-of-sight (i.e. measured) escape fractions and EWs are higher when disks are viewed face-on, as Lyα radiation is more likely to follow paths of lower H I column density, and also directions in which the outflows are strongest (i.e. perpendicular to the disk). Thus the observation that Lyα-emitting disks tend to be face-on could be straightforward to explain.

The observation that stronger LAEs tend to be more compact is harder to explain. One may speculate that this phenomenon is related to the feedback from stellar winds and supernovae: the thermalization efficiency of SN ejecta is higher in denser environments, and more compact galaxies drive faster winds [51, 54]. If the generation of a hot expanding wind fluid is more efficient in compact galaxies, it could enhance the mechanical energy available to accelerate dense neutral material and instigate fluid instabilities that disrupt the medium—these points will be discussed in detail in Sects. 4.5.2 and 4.5.3, as well as in Sect. 4.7.2. Compactness may also be, to some extent, under the influence of the AGN fraction (Sect. 4.4.5). Ultimately this will be an avenue for future work for both Lyα observational studies and studies of feedback and galaxy superwinds.

4.4.6.2 Color and Age

Within the GALEX LAE sample it is difficult to see a clear trend between Lyα/Hα and the UV/optical colors alone. However when comparing the LAEs with their non-emitting control sample, the average ultraviolet continuum slopes differ significantly. UV color is conventionally described by the slope, β, where β describes the continuum slope when parameterized by a power-law in f_λ of the form $f_\lambda \propto \lambda^\beta$. LAEs show bluer slopes, with a median value of $\beta = -1.8$, while the non-emitting control samples show a median β of about -1.3 [20]. HST observations reveal that galaxies with $f_{\rm esc}^{\rm Ly\alpha}$ higher than 10% all have UV slopes bluer than $\beta = -1.7$, while redder galaxies do not emit Lyα, or do so with much smaller escape fractions [49]. Again, the highest $f_{\rm esc}^{\rm Ly\alpha}$ samples at low-z are the Green Pea galaxies, which are bluer still with $\beta \approx -1.9$ on average [55].

The UV continuum slope encodes two main quantities: the age of the star forming episode, and the reddening due to dust. It is also affected, to a less significant degree, by the stellar metallicity. We have already seen in Sect. 4.4.3.1 that Lyα-emitting objects tend to be less extinguished, and so on first sight it is not surprising that they should also appear bluer in the UV. However age should not be completely neglected, because large numbers of supernovae can only be supplied when the star-forming episode is very young: this is discussed at the end of the previous Sect. 4.4.6.1 and also in more detail in Sect. 4.5.2. It may be that the fact that LAEs are bluer in the UV is due to them having *both* younger stellar populations *and* less dust. Indeed if the dust is produced during the star formation episode then such a positive correlation between age and dust content may arise naturally.

4.4.7 Atomic Gas Envelopes

This section is concerned with targeted emission line observations of galaxies in the hyperfine structure transition of the H I ground-state, with a wavelength of $\lambda = 21$ cm. Comments on this subject, regarding specific objects, can also be found in Sect. 4.3.2 and more will follow in Sect. 4.5.1. Here we mainly discuss Lyα *samples* for which 21 cm-derived masses have been obtained and sometimes column densities can be derived.

Lyα observations require galaxies to lie at sufficient recession velocity to redshift the intrinsic Lyα feature away from atmospheric and Milky Way features (Sect. 4.1.5). Consequently the majority of samples reside at distances around ~150 Mpc (median of LARS) and ~1000 Mpc (average for GALEX and Green Peas). While many detailed H I studies have been performed in the local universe, these distances are very far by standards of 21 cm observations. Significant samples of Lyα and H I observations have therefore been hard to assemble, and only a handful of articles are currently in existence.

Figure 4.18 shows a compilation of some recent results. The upper row shows 21 cm observations of ESO 338–IG04 [12], which is a bright Lyα emitting galaxy [43]. The observations localize the H I to the star-forming galaxies, which are located towards the upper left and lower right of the figure. The small inset square, labeled 'HST FoV' is the same image as shown in Fig. 4.26, which contrasts directly the resolutions available with HST and radio telescopes working at 21 cm, where in this case the elongated synthesized beam has a size of 26×16 arcsec. These interferometric H I observations enable us to constrain the column density of H I, which peaks on top of ESO 338–IG04 at column densities above 10^{21} cm^{-2}. The fact that the galaxy shows strong Lyα emission indicates that this column is likely comprised of multiple clumps that are not resolved by the VLA beam, or that the gas is rapidly outflowing. Note also from the velocity map in the *upper right* panel that a coarse bulk rotation is also visible. Large amounts of H I are also visible in a stream that connect ESO 338–IG04 to the counterpart galaxy of a recent interaction, which is visible at the lower right of the frames.

The lower panels show the Lyα escape fraction measured in large apertures from HST imaging [49], contrasted with different mass estimates [109]. The *left* panel shows the estimated H I mass obtained from the total 21 cm flux integral converted into a luminosity: it clearly shows that the Lyα-emitting galaxies with escape fractions above 10% are drawn from the lower end of the H I-mass distribution. This relationship with H I could be directly causal, in the sense that these galaxies may have less H I available to trap the Lyα radiation in resonance. However in every case where the objects are resolved with an interferometer the column density exceeds 10^{21} cm^{-2} which is extremely optically thick to Lyα if the gas is homogeneously distributed. Another, less direct, possibility is that we are seeing a correlation because these galaxies are less massive in general and many properties of the galaxy population are correlated (see Sect. 4.7.2), and lower mass galaxies may have less dust or it may be easier for feedback to drive out gas.

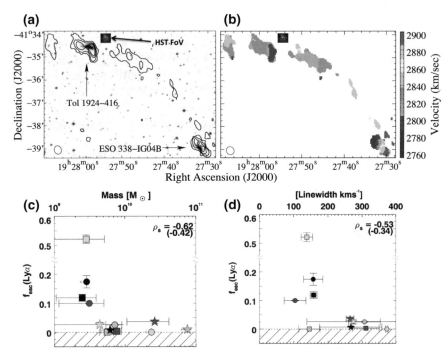

Fig. 4.18 The *upper* panels show the 21 cm observations of ESO 338–IG04 (labelled Tol 1924–416), obtained from VLA. The *left* panel shows the intensity image, and the *right* panel the velocity map. The main galaxy is towards the upper left, and a small inset RGB image shows the Lyα frame on approximately the same scale. This is the same image as shown in Fig. 4.26. Large quantities of H I have clearly been ejected in a stream, that continues almost unbroken to the interaction counterpart galaxy at the lower right of the frame. The small ellipse in the lower left of each image represents the synthesized beam. The *lower* panels shows H I observations of the LARS sample: the Lyα escape fraction (note the broken *y*-axis) from HST imaging is contrasted with the velocity width of the 21 cm lines, and the total inferred H I mass obtained from the flux integral. *Figures are reproduced with permission from Cannon et al.* [12] *and Pardy et al.* [109], ©*AAS*

4.5 High Resolution Spectroscopy: The Detailed Picture

This section deals exclusively with high resolution (better than ≈ 30 km s^{-1}) spectroscopy obtained with aperture spectrographs on the Hubble Space Telescope. These observations are typically obtained in apertures of size ≈ 2 arcsec, which is small compared to the scales probed by the IUE and GALEX, and thus we are mostly discussing Lyα and UV observations obtained on individual sightlines into a galaxy. Usually this will be towards the highest surface brightness regions in the ultraviolet continuum.

Main points to take home from this section are:

- The first galaxies observed with high dispersion spectrographs were blue compact dwarf galaxies. These tend to be *absorption dominated* systems, and fall far from the trends predicted by early IUE results. Most show very deep absorption profiles despite their metallicities being below 1/30 solar with almost negligible nebular dust extinction.
- Outflows in the atomic gas phase are usually probed by metal absorption lines and play a key role in promoting the emission of Lyα. However outflows alone are insufficient to guarantee a galaxy will become Lyα-emitter.
- Lyα escape fractions are found to be higher when the estimated atomic gas covering fraction falls significantly below 1. However the almost complete absence of Lyα emission at line centre, and the deep absorption seen in higher order Lyman series lines indicates that residual atomic gas is pervasive, and probably covers the star forming regions in even the most gas deficient galaxies.
- Substantial spectral re-shaping of the Lyα emission feature occurs within galaxies. This must be accounted for when using Lyα profiles and fluxes to constrain models of an increasingly neutral intergalactic medium at high redshift.

4.5.1   Blue Compact Dwarfs

As discussed in Sect. 4.3.2, the early studies with the IUE suggested that Lyα output may be anti-correlated with metallicity. Indeed the Meier and Terlevich paper [97] includes a statement about how it is unfortunate that IZw 18 is too near to observe with low-dispersion spectrographs because the line blends with the Milky Way geocoronal Lyα feature. With the arrival of high-R spectrographs on HST, much more nearby systems could be observed. Naturally IZw 18 was among the first starbursts to be targeted, and—contrary to expectation—was found to be a strong Lyα absorbing galaxy with an EW of ≈ -40 Å [77]. Shortly after this discovery, two further blue compact dwarfs—SBS 0335–052 and Tol 65—were also studied with the GHRS: these objects have comparable stellar masses and optical emission line spectra to IZw 18, and similar Lyα profiles were found—damped absorption with equivalent widths below -30 Å [141]. Thus these galaxies provide a remarkable departure from the preliminary trends of Lyα output versus metallicity derived with the IUE [17, 34]. On the relationships shown in Fig. 4.10 they would fall off the bottom of the x-axis in the upper two figures but with Lyα formally below zero; in the lower panel IZw 18 is the galaxy with $12 + \log(O/H) = 7.2$, where the Lyα has been set to zero.

The immediate interpretation of these results is that large slabs of atomic gas must overlie the star-forming regions, and cause Lyα radiation to undergo a large number of resonant scattering events. Ultimately the Lyα radiation would be absorbed by the small amounts of dust present in the ISM. Fitting Voigt profiles to the absorption feature shows that the H I column density is $\approx 5 \times 10^{21}$ cm^{-2}, and centered close to zero velocity [77, 95]. This is easily sufficient to produce an absorption profile with strong damping wings. Further modeling of the IZw 18 line profile confirms that radiation transfer effects can reproduce precisely this absorption shape with the available dust content [3]. More recent, and very deep, COS spectroscopy in fact reveals the hint of a weak emission line embodied within the absorption trough, although this amounts only to an escape fraction of around 1/1000 [42]. A similar very strong absorption profile, with a tiny bump of emission is also visible in the spectrum of another extremely metal-poor galaxy: SBS 1415+437 [62].

Two of the galaxies discussed above have been observed at 21 cm with the Very Large Array, and results indeed confirm that both are enshrouded in large envelopes of atomic hydrogen. Central column densities also exceed several 10^{21} cm^{-2}, which is consistent with the quantities measured by Voigt profile fitting (see also Sect. 4.4.7). Particularly in the case of IZw 18, this envelope is many times larger than the starburst region: H I is detected over ≈ 1 arcmin [143], which must be compared with a starburst of just a few arcsec when measured in UV and optical wavelengths.

These BCDs offer some very interesting insights into the processes that affect Lyα emission and absorption, but nevertheless they represent a very rare sub-population of local galaxies. For their magnitude, their low metallicity, high Hα equivalent width makes them less than 1/1000 of the local galaxy population, and one must exercise caution when using them as a representative description of local galaxies. However as we look back to the earliest times, such low-luminosity dwarf galaxies—that are likely the building blocks of more massive galaxies—become more and more abundant.

4.5.2 Outflows, Galaxy Winds, and Kinematics

Following the results of Sect. 4.5.1, GHRS was immediately turned to a larger sample of local starbursts. Most GHRS observations of Lyα also included observations of low ionization stage metal lines (e.g. O I λ1302 and Si II λ1304) that have ionization potentials such that they most likely form in the H I clouds that also absorb and scatter Lyα. The first HST spectra of a sample of local galaxies [78] included the two strongest absorption systems discussed in the previous section, as well as two additional Lyα-absorbing galaxies (II Zw 70 and Mrk 36). These spectra are shown in the left panel (lower four) and far right panel of Fig. 4.19. In these four galaxies the metallic absorption features were found to be close to static with respect to the system recession velocity of the galaxies, with all outflow velocities (ΔV_{out}) found to be less than 50 km s^{-1}. The implication of this is that large columns of H I exist in front of the UV-bright sources, and are sufficient to cause substantial resonant trapping of Lyα as mentioned in the previous section.

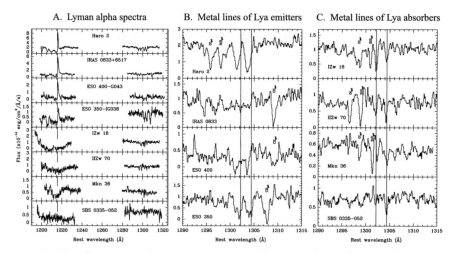

Fig. 4.19 Key results on local starburst galaxies. The *left* panel shows the Lyα spectra, ordered with the four emitting galaxies at the top and the four absorbing galaxies at the bottom. In the case of the emitters, the Lyα profile is clearly asymmetric with absorption on the blue side and emission on red. The *center* and *right* panels show the absorption profiles of O I λ1302 and Si II λ1304, which form in the atomic ISM; the vertical black lines show the wavelength at which these lines would be expected to fall if the atomic medium were static. The *centre* panel shows the profiles of the Lyα emitters, for which the metal lines are clearly blueshifted compared to the systemic velocity; the *right* panel shows the Lyα-absorbing galaxies for which the metal absorption lines are close to static. *Figure reproduced with permission from Kunth et al.* [78], *©ESO*

This same sample also included four Lyα-emitting galaxies—Haro 2 (also [86]), Haro 11 (a.k.a. ESO 350–IG38), IRAS 0833+65, and ESO 400–G043—which have Lyα equivalent widths in the range of 13–40 Å. These spectra are shown in the left panel (upper four) and central panel of Fig. 4.19. Recall again that these are measured in small apertures, and the fluxes are likely estimated with HST. Very importantly these Lyα profiles are not symmetric in frequency as may be predicted from the Balmer line emission, but are offset and skewed to the red. Indeed no Lyα emission is seen at zero velocity in this sample. Every example can be broken down into an absorption and emission component, with the absorption dominating on the blue side (and at zero velocity), and emission on the red. This is precisely what is expected in the case of an optically thick H I medium that is outflowing from the star forming region along the line-of-sight. Indeed as Fig. 4.19 shows, the ISM absorption lines are completely static in the case of the four Lyα-absorbing galaxies but are offset to the blue by velocities of 60–225 km s^{-1} for the Lyα-emitters. IRAS 0833+65 shows no apparent metal absorption lines in this data, but subsequent observations with COS [84] detect metal lines that are significantly blueshifted, and reveal a multicomponent outflow in this galaxy.

After this point, little further work was done on the comparison of Lyα emission with outflow signatures until COS was mounted on HST in 2009. The higher throughput, and much larger wavelength coverage of COS allowed Lyα and many

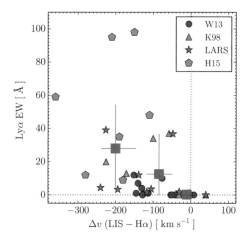

Fig. 4.20 The Lyα EW compared with the average outflow velocity of the atomic gas for a compilation of galaxies. References expand as: W13 = Wofford et al. [148]; K98 = Kunth et al. [78]; H15 = Henry et al. [55]; LARS = Rivera-Thorsen et al. [121]. The scatter is large, but the pink points with errorbars show averages of three bins when arranged by outflow velocity (68% ranges are shown). This paints a clear picture in which the average Lyα EW increases with increasing outflow velocity. *Figure is expanded from Hayes* [42]

UV absorption lines to all be observed in a single observation, which has led several teams to all present detailed studies of Lyα and atomic gas kinematics. Figure 4.20 paints the current picture of how Lyα emission compares with outflow velocity. The figure shows how the Lyα EW compares with the outflow velocity for a sample of 50 low-z galaxies, shown in gray [55], orange [121], green [78], and purple [148]. The large pink points with errorbars show the average values of galaxies, when divided into bins by ΔV_{out}, which contain around 17 objects each. The right-most point shows that for ΔV_{out} in the range -50 to $+50$ km s^{-1}, not a single galaxy exhibits Lyα in emission. Objects with $\Delta V_{out} \approx -100$ km s^{-1} (central bin) show an average Lyα EW of 12.5 Å, which increases to 28 Å for galaxies with $\Delta V_{out} \approx -200$ km s^{-1} (leftmost bin). These galaxies fulfill the canonical criterion to be considered a LAE (EW > 20 Å) and would conform to most narrowband selection criteria at high redshift; a large fraction of this bin is comprised by Green Pea galaxies, discussed in Sects. 4.4.3.3. There is large scatter in the diagram here as well, but every galaxy has a positive EW with some almost 100 Å. Selection effects may be at play in this compilation of galaxies from various samples, but on the surface it appears that strong Lyα-emitters have faster outflows.

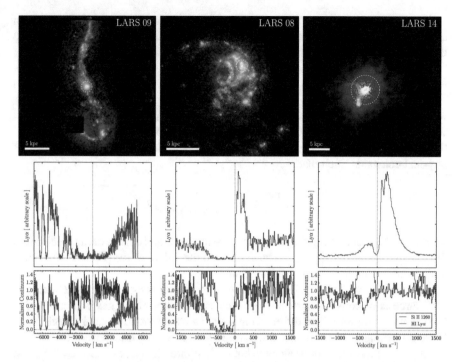

Fig. 4.21 Lyα and Si II spectra in velocity space for three LARS galaxies, that were selected to demonstrate the variety of Lyα profiles. Identifiers are labeled in the images in the *upper* row. In the spectra, the *lower* row shows the Si II lines in red and Lyα absorption component in blue, both normalized by the local continuum level. The *central* row shows the same Lyα spectra but renormalized to show the emission profile where present. Note that the velocity scale is different in the leftmost panel from the other two. *Figure is reproduced with permission from Hayes* [42]

4.5.3   Gas Covering Fraction and Column Density

The velocity centroid of an absorption profile is among the easiest things to measure, but far more can be done with the high-quality spectra that are returned by the COS, especially when combined with Lyα itself. Figure 4.21 shows a variety of Lyα and Si II absorption profiles that have been obtained as part of the LARS project. In the *left* panel we see a damped Lyα absorbing galaxy, where the Lyα absorption is far broader than the silicon line, which has zero discernible velocity shift. The *central* galaxy also displays a strong Si II line but one that is offset in velocity by around 250 km s^{-1}, and the absorption line does not reach zero at 0 km s^{-1}. A Lyα emission line is obvious in this galaxy, but it is redshifted with respect to the systemic velocity because the bulk of the atomic gas is shifted out of resonance with Lyα. The *right* panel also shows a galaxy with an outflow of ≈300 km s^{-1}, but the EW of the metal line is much smaller. In this galaxy the Lyα line is very strong (EW ≈ 40 Å) and a blue bump is also present.

When absorption occurs far from the continuum-emitting source (for example in a quasar absorption line spectrum; Lecture Notes in this series by J. X. Prochaska) the transferred intensity in the line is simply given by:

$$I = I_0 \exp(-\tau) \tag{4.5}$$

where I is the transmitted intensity, I_0 is the continuum level, and τ is the conventionally defined optical depth. For a given transition τ is simply a function of the column density (N), oscillator strength (f), wavelength (λ), and a number of physical constants, such that:

$$\tau = f\lambda \frac{\pi e^2}{m_e c} N = \frac{1}{3.768 \times 10^{14}} f\lambda N \tag{4.6}$$

where c is the speed of light and m_e and e are the mass and charge of the electron. In the second equality of Eq. 4.6 we have inserted the values of the physical constants, while f and λ are general for all transitions.

Equation 4.5 will also hold for an extended illuminating background source if the source is of constant surface brightness and the absorbing gas is homogenous (and again a long way from the emitting source so that scattering can be neglected). If the absorbing gas is optically thick, no radiation will be transferred. However it the absorbing gas is not homogenous but broken into clumps, as expected for a fast unstable outflow that is subject to fluid instabilities, the transferred flux will instead simply be the fraction of sightlines that are not covered, such that:

$$I = I_0(1 - f_c) \tag{4.7}$$

where f_c is the gas covering fraction. Finally, if a real outflow is composed of a clumpy neutral medium with a finite optical depth per clump, the throughput radiation will instead become

$$I = I_0[1 - f_c(1 - \exp(-\tau))]. \tag{4.8}$$

While I and I_0 are both known, τ and f_c are not and the solution will be degenerate for a single absorption line. However one of big advantages of the Si II ion is that it has many strong transitions from the ground state, all of which form in the UV: lines at $\lambda = 1190, 1193, 1260, 1304, 1526$ Å, are frequently observed in galaxy spectra, and at $z \sim 0$ the first four can be captured with a single observation with COS/G130M. Because these lines have differing oscillator strengths, the product $f\lambda$ varies by an order of magnitude among the transitions. Thus using more than two Si II absorption lines, we can solve for the average optical depth and covering fraction of the absorbing clouds. Moreover, if the absorption lines are spectrally resolved, as can be expected for the ≈ 15 km s^{-1} resolution of COS (nominal, for point sources), N_{SiII} and f_c may be recovered as a function of velocity and the structure of outflow may be kinematically decomposed. Such analyses have been employed several times at

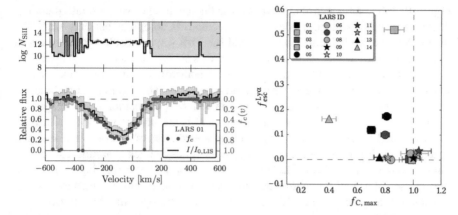

Fig. 4.22 The *left* panel shows how N_{SiII} (*upper*) and f_c (*lower*) are estimated as a function of velocity. The *right* panel shows the maximum covering fraction at any velocity, and how the Lyα escape fraction varies with this quantity. *Figure is reproduced with permission from Rivera-Thorsen et al.* [121], *©AAS*

high redshift [65, 113, 116, 125], and more recently for galaxies in the more nearby universe [121].

Figure 4.22 shows an example of how the covering fraction and column density of Si II-bearing clouds are measured in LARS 01. Using four lines of Si II, we solve for f_c and $N_{Si II}$ using Eq. 4.6 inserted into Eq. 4.8. The column density profile, shown in the left upper panel, shows that the clouds have column densities of $\approx 10^{12.5}$ cm^{-2} at all velocities, which is the column density at which the absorption lines become optically thick. Thus these numbers can be regarded as lower limits to $N_{Si II}$. For 'normal' metallicities this may correspond to $N_{H I} \gtrsim 10^{17-18}$ cm^{-2} in atomic hydrogen. The shape of the absorption profile is instead driven mainly by variations in the covering fraction (*left lower*), which shows that covering reaches a maximum value of ≈ 0.85 in this galaxy, and decreases significantly towards the higher and lower velocity wings of the absorption line.

The *right* panel of Fig. 4.22 shows how the Lyα escape fraction—determined by imaging observations in large apertures—varies with maximum f_c. There is a cloud of points for which f_c^{max} is consistent with complete covering at a single velocity: these galaxies have $f_{esc}^{Ly\alpha}$ below 3%, and most are dominated by Lyα absorbing systems. However at $f_c^{max} < 0.9$, anything can happen: there are galaxies here that have $f_{esc}^{Ly\alpha}$ consistent with zero, but also the five galaxies that are strong net emitters cluster around this part of the diagram. Here we see $f_{esc}^{Ly\alpha}$ in the range of 8–50%, and data are consistent with incomplete gas covering being in part responsible for the emission of Lyα.

However this story is also incomplete: if covering were truly < 1 at every velocity, then some Lyα would be emitted without interacting with H I at all, and there would be flux at $\Delta V_{out} = 0$. The full atlases of spectra that enter Fig. 4.20 contain almost no objects that show evidence of Lyα emission at line-centre. The simplest interpretation of this is that the covering of dense atomic clouds is complete, but only when the integration over all velocities is accounted for. This way we could expect non-zero covering *and* the need for radiation transfer effects in Lyα, but there are still important ramifications to be derived from the method: specifically that the clouds themselves must be discrete in both the x, y plane and in velocity. Thus there must be holes or channels between the densest clouds through which Lyα radiation can propagate.

Analyses based upon only metal lines are also not perfect when the aim is to study H I, and the reason is the very different abundances of hydrogen and metallic species. From Eq. 4.6 we see that Si II becomes optically thick at around 10^{13} cm^{-2}, but the same is true for H I in the Lyα resonance. For most star-forming galaxies we may not see Si II at all for column densities less than $\approx 10^{12}$ cm^{-2}, which for normal metallicities may be around $\sim 10^{17}$ cm^{-2} in H I and still optically thick to Lyα.

This has been shown directly using absorption line spectroscopy in higher order Lyman series lines in the shorter wavelength UV [55]. Observations of the Green Pea galaxies have been obtained that cover Lyγ and Lyβ, as well as Lyα and the metal lines between 1200 and 1400 Å. Since these Lyman transitions are not resonant, the nebular photons cannot scatter their way out of the H II regions (case B recombination) and are never seen in emission from star-forming galaxies. Since they also have high oscillator strengths (within a factor of 10 of Lyα) their absorption can trace absorbing gas to much lower column densities than metal lines can, and they therefore provide a more robust tracer of H I. In this study of ten galaxies, the Si II $\lambda 1260$ Å lines are only saturated in one case, and only one more shows an absorption line going to a residual intensity of $I/I_0 \approx 0.1$; the remainder are far weaker, and two galaxies show no clear evidence of Si II absorption at all. However the Lyβ or Lyγ absorption lines do saturate in every galaxy, which quite conclusively reveals H I column densities above $\approx 10^{16-17}$ with unity covering fraction. The Lyα output also indicates that the gas column must be relatively thin in the Green Peas: while still there is little or no Lyα emission at line centre, 9/10 galaxies in the sample exhibit strong blueshifted Lyα peaks. The former point is expected if some radiation transfer effects are at play, while the latter indicates that the column density in which this is happening must be low [146]. As we discussed in Sect. 4.4.3.3, these galaxies have already been cited as candidates for partially density bounded nebulae, based upon their optical emission line ratios.

The derivations outlined above were based upon the premise that the absorbing material lies sufficiently far on the front side of the UV continuum sources, and/or that the aperture of the spectrograph is able to well-isolate the source. However this need not be the case, and absorption spectroscopy may be complicated by the resonance scattering of metal lines that may partially re-fill the absorption line [115, 126]. This will act to make the different lines appear more similar, and could potentially cause intermediate column density gas to masquerade as gas with $f_c < 1$. The contribution of scattered light will also be a function of the distance along the line-of-sight and the size of the aperture. Fortunately more vital geometrical information can be gained from deep, high-resolution ultraviolet spectroscopy with a long wavelength baseline. The Si II lines discussed so far, as well as O I $\lambda 1302$ Å and C II $\lambda 1334$ Å all have

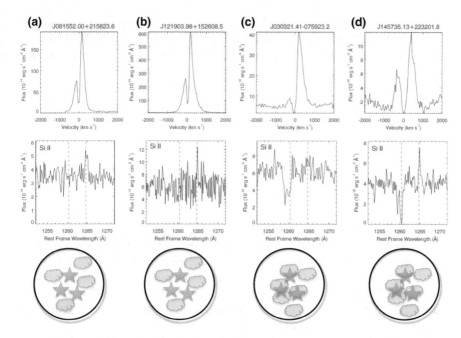

Fig. 4.23 The *upper* row shows the Lyα spectra of four Green Pea galaxies, the *left* two of which show very strong emission lines, while the *right* two are significantly weaker. The *central* row shows the associated Si II λ1260 Å absorption lines (left red dashed line) and the fluorescent emission line (right dashed line). The *lower* panel shows cartoons of the line-of-sight viewing geometry: the COS aperture is illustrated by the gray circle, the orange stars represent the UV continuum-emitting stars, and the blue clouds represent the absorbing and fluorescing atomic gas (bearing the Si II ions). *Figure is reproduced with permission from* [64], *©AAS; the cartoon apertures have been added*

ground-states that are split by fine structure. While resonant channels exist, there is a roughly equal probability that each absorption will be followed by a transition to a slightly elevated lower level, in which case a photon with slightly longer wavelength (a few Å) will be emitted [64, 126]. An illustration of this can be found in Fig. 4.23.

Figure 4.23 shows some illustrative spectra of Lyα-emitting Green Peas [64]. In the *top* row: the left-most two galaxies have Lyα EWs above 70 Å, and the rightmost two have EWs between 0 and 10 Å (including the absorption). The *central* row shows the wavelength region of the Si II λ1260 Å absorption feature and its corresponding fluorescent emission channel at λ ≈ 1265 Å. The leftmost two galaxies do not show a clear absorption feature, which indicates that the UV-bright clusters are not covered by a significant column of Si II-bearing cold gas. This geometry is illustrated in the *lower* panels where the black circle indicates the COS aperture: the absorbing clouds (blue) do not cover the stars (orange) so there is no strong absorption feature, and the Lyα radiation can propagate relatively freely. However an emission feature at 1265 Å is detected in both of these left-most cases, which indicates that there are clouds within the aperture that are absorbing in the 1260 Å line and fluorescing, but they are not directly in front of the clusters. The counterpoint is presented in the right-most pair of galaxies, which have much lower Lyα EWs. Here the Si II 1260 Å absorption is much stronger: broad and blueshifted in J0303 (suppressing the blue peak) and with a static component in J1457 that produces much broader, double-peaked Lyα. In these two galaxies the atomic clouds must cover the stellar sources, as illustrated by the cartoon diagram in the *lower* row. In this geometry the fluorescent line may or may not be visible, and indeed both cases are seen in this example.

4.5.4 Alternative Possibilities in the Optical

UV spectroscopy reveals a very detailed picture in the spectral direction, but because of the nature of UV spectrographs it is very hard to extract spatial information. This would require multiple pointings with an aperture spectrograph (e.g. COS) or long integrations with a slit (e.g. STIS), which would make the observations costly. Few comparable absorption lines exist in the optical, but one possible alternative is the Na ID $\lambda\lambda$5896, 5889 Å doublet, which has been used to measure atomic gas kinematics in various star-forming galaxies [18, 52, 92, 122].

Efforts have been made to use the Na ID doublet to asses the atomic gas kinematics of low-z galaxies with Lyα observations, to assess the importance of kinematics. Efforts have been made to detect the optical continuum of GALEX LAEs with the Palomar 5.1 m Hale telescope and 8.2 m VLT with FLAMES fibers, have shown the continuum to extremely faint, and the absorption lines are hard to find (Scarlata and Atek, respectively, private communication). The optical continuum is well detected in the Calar Alto/PMAS IFU spectra of the LARS sample [56], but the lines are also particularly weak in the majority of these galaxies. VLT/FLAMES IFU spectroscopy has also been obtained for ESO 338–IG04 and Haro 11 [124], with the intent of determining the line-of-sight H I velocity field, but even though deep observations were obtained (4 h integrations on an 8 m telescope) the absorption lines were extremely weak, and kinematic measurements are possible only towards three of the very highest surface brightness regions.

We must conclude that Na ID is not a good probe of H I kinematics in most Lyα-emitting galaxies, or UV-selected galaxies with blue continuum colors. This is most likely because little sodium exists in the neutral form in starburst galaxies. Na has an ionization potential of just 5.1 eV, which is much lower than the C II and Si II lines discussed in Sects. 4.5.2 and 4.5.3 and, most critically, much lower than that of H I. Thus Na I is ionized by radiation with wavelengths between the Lyman edge and $\lambda = 2430$ Å, and is not shielded by H I; instead Na I needs to be shielded from the UV by dust, and thus only probes the dustiest of H I clouds. Most Lyα-emitting, blue galaxies will not have sufficient dust optical depths to shield Na I from the UV continuum, and low-metallicity, faint galaxies will be very hard to study this way.

4.5.5 Is the Intergalactic Medium Important at High Redshift?

Another particularly interesting experiment that can be performed in local galaxies is to contrast the line shapes with high-z galaxies, where the effects of the IGM may (or may not) be important in reshaping the Lyα line. It has been argued that the asymmetry of the Lyα line could be an important diagnostic by which to separate LAEs from contaminant emission line galaxies at the highest redshifts because a partially neutral intergalactic medium will preferentially absorb the blue side of Lyα [67, 132]. The redshift evolution of the Lyα profile may be a possible diagnostic of a change in the neutral fraction of the universe, which would make Lyα an important probe of reionization, but for this to hold then we must be able to disentangle the IGM from radiation transfer effects in the ISM that already redshift and asymmetricize the line. To begin to test this we need a sample of Lyα profiles as free of IGM absorption as possible, which is only possible at low-z.

Several parameterizations of the asymmetry have been defined, that are based upon skewness statistics [67, 132], or comparison of the flux bluewards and redwards of the peak [120]. Taking the *'weighted skewness'* (S_W) as an example, this is based upon the standard statistical skewness (3rd moment of the probability density function):

$$S = \frac{1}{I\sigma^3} \sum_{i=1}^{n} (\lambda_i - \bar{\lambda})^3 f_i \tag{4.9}$$

where λ_i is wavelength and f_i is the flux in the ith spectral pixel, $\bar{\lambda}$ is the mean wavelength, σ is the standard deviation (2nd moment of the distribution) and I is the integrated line intensity. In the definition of S_W, only spectral pixels where the flux density is above 10% of the peak value are classified as Lyα, providing a wavelength range of $\lambda_{10,b}$ on the blue side to $\lambda_{10,r}$ on the red. Thus the calculation of S_W is made as:

$$S_W = S(\lambda_{10,r} - \lambda_{10,b}) \tag{4.10}$$

We have computed these asymmetries from high-resolution COS observations of the $\langle z \rangle \approx 0.3$ LAEs discovered in the GALEX surveys (Scarlata et al. in preparation). We begin by computing S_W in the COS spectra, the distribution of which we show in the *left* panel of Fig. 4.24 as the black solid histogram. The red hatched histogram in the same Figure shows the S_W distribution at $z \approx 6.5$ [67], for which S_W falls in precisely the same range. A K-sample Anderson-Darling test indicates that there is a $\approx$70% chance that the two distributions are drawn from the same parent sample.

We can also examine how different the fluxes would be if the GALEX LAEs were at high-z, in which a partially neutral IGM could depress the Lyα luminosity function. In this example we take the spectra and shift each to $z = 6.5$. We then convolve the redshifted spectra with the Madau (1995) prescription for Lyman series line blanketing in the IGM [88], which at $z = 6.5$ effectively sets all flux bluewards

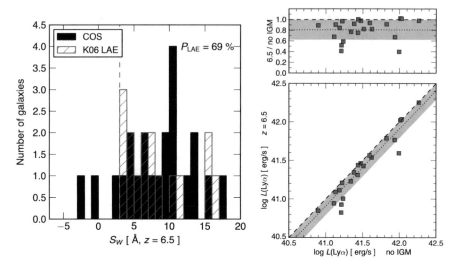

Fig. 4.24 *Left*: the distribution of the weighted skewness parameter S_W [132] for low-z GALEX-LAEs observed with COS (black histogram) and for $z = 6.5$ LAEs [67] observed with Subaru and Keck. *Right*: the comparison of the measured Lyα luminosity of the $z \approx 0.3$ LAEs (x-axis) with the same measurement after the spectrum is convolved with the IGM absorption profile appropriate for $z = 6.5$ (y-axis)

of restframe 1216 Å to zero. We then recompute the luminosity and compare this with the value measured in the case of no IGM. The results are shown in the *right* panel of Fig. 4.24. The Lyα luminosity is significantly underestimated in only a handful of galaxies and, on average, 81% of the total Lyα would be recovered if these galaxies resided in a $z = 6.5$ universe. Of course we cannot say that the IGM does not have an impact on Lyα emission and at higher redshift still (and it probably does), but it is clear that the ISM transfer of Lyα is sufficient to make a big difference to the line profile and one that must be accounted for if we are to use Lyα-emitter spectra to estimate the redshift onset of a neutral IGM.

4.6 Detailed Studies Using High Resolution Imaging

The *Hubble Space Telescope* has carried filters that transmit Lyα in the restframe since launch, meaning that in principle Lyα imaging has always been a possibility. However the particularly low quantum efficiencies of detectors on the *Wide Field and Planetary Camera 1* (WF/PC-1), and its successor WFPC2, meant that Lyα observations were never carried out. It took until the 2002 installation of the *Advanced Camera for Surveys* (ACS) that peak system efficiencies broke 1% at Lyα, and only then [79] did we begin to see how local galaxies appear in Lyα. This section discusses the Lyα

morphologies of galaxies in the local universe ($z \lesssim 0.2$), and what is learned from their shapes, sizes, and the comparison of local Lyα intensity with other properties.

Main points to take home from this section are:
- Lyα is most efficiently imaged with HST by using adjacent pairs of long-pass filters on the *Solar Blind Channel* (SBC) of ACS. This enables large-aperture fluxes to be measured, and makes imaging a vital complement to small-aperture spectroscopy.
- Special care is also needed to subtract the stellar continuum, which requires observations to be obtained in several additional bandpasses.
- The Lyα morphology of a galaxy can differ dramatically from the morphology of radiation that probes the hot stars and H II regions. Lyα can be seen in both emission and absorption, and may vary between the two on scales of a few tens of kpc. This points directly to the scattering of Lyα in atomic gas.
- When Lyα is seen in emission it is almost invariably accompanied by an extended envelope of low surface brightness emission that surrounds the star-forming regions, dubbed 'scattering halos'. On average these are twice the linear projected size of the Hα emission. A consequence of this is that apertures several times the optical size of a galaxy may be needed to capture the total Lyα flux.

4.6.1 How to Image Lyman Alpha with HST

No HST cameras have ever carried dedicated narrowband filters in the UV. Instead one has to rely upon either medium band filters such as F122M or long-pass filters such as F125LP. The most effective way to isolate Lyα is to use wavelength-adjacent pairs of long-pass filters [45], as shown in the left panel of Fig. 4.25. The long-pass filters on SBC are all described by a different cut-on wavelength in the blue, that rises very sharply from zero to the peak transmission. The red edge of each bandpass is defined by the decreasing quantum efficiency of the detector with wavelength, and this results in several inset triangular profiles. The subtraction of adjacent pairs yields a relatively narrow window in wavelength over which the throughput edges are steep, and reasonably well-defined. These are shown in the right panel of Fig. 4.25. Galaxies then must be selected to lie in the range of recession velocities that will redshift the Lyα line into the bandpass.

Even though the filters appear well-suited to isolate an emission line, they are still broad (≈ 100 Å at half-maximum; $R \sim 10$) compared to conventional optical narrowband filters (which may have $R \sim 100$). This reduces the line—continuum contrast compared to, for example, Hα narrowband imaging, or narrowband surveys

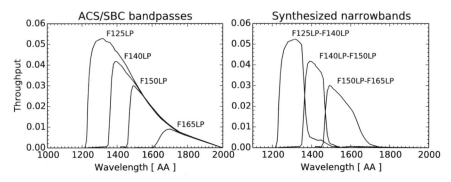

Fig. 4.25 *Left*: the bandpass profiles of the four main imaging filters on the Solar Blind Channel of HST/ACS. Each filter is labelled in the image. The F115LP filter is not shown because it transmits the bright geocoronal Lyα line and has never been used for Lyα imaging. *Right*: the synthetic narrow bandpasses that are released by the subtraction of filter pairs that are adjacent in wavelength. Again, each pair subtraction is labelled in the figure

from the ground that target high-z Lyα. Furthermore, there is a significant difference in wavelength between filters that capture the line and the adjacent continuum—the strategy cannot be like the convenient positioning of narrow and broadband filters for high-z surveys, and an extrapolation in wavelength is needed over a spectral region where the continuum slope evolves rapidly. A lot of attention therefore needs to be paid to how the continuum under Lyα is estimated.

These issues have been discussed at length in [43, 45, 106] and do not need to be restated in detail in these notes. A summary of the key points includes:

- The most massive stars form in very clumped environments [98, 102], and the level and shape of the UV continuum may vary dramatically on scales of just a few–10 pc. This scale, close to the core radius of star clusters, is easily resolved by HST and any continuum-subtraction method must account for spatial variations in the UV color.

- No complete set of analytic functions has been derived to describe the UV continuum colors between Lyα and offline filters (typically at $\lambda \approx 1350$ Å). The effects of stellar evolution, metallicity, and dust reddening are too complicated to boil down to a set of formulae. For example, while often parameterized as a power law,[2] characterizing the UV slope by β has insufficient fidelity to continuum subtract Lyα because the UV slope does not follow a sufficiently predictable shape between the near and far UV.

- The only currently published solution is to perform spatially resolved SED fitting, using stellar population synthesis models and prescriptions for dust attenuation. SED fitting in every pixel is computationally expensive, and carries the observational penalty that additional optical observations have to be obtained: it is necessary to sample at least the UV continuum slope and the 4000 Å break because

---

[2]Usually $f_\lambda \propto \lambda^\beta$, where f_λ is the flux density per wavelength interval, and β defines the logarithmic gradient of the continuum.

of well-known degeneracies between age, reddening, and metallicity [43]. Higher precision estimates can be obtained by adding redder broadband filters to allow for the contribution from more evolved stars, and by including narrowband observations of nebular lines to remove contamination to the 4000 Å break by the nebular Balmer discontinuity [45]. It is true that these methods also use an extrapolation to estimate the continuum at Lyα, but that is at least physically motivated by stellar evolution and dust reddening.

4.6.2   Lyα Morphologies and Scattering Halos

The first two local galaxies with calibrated Lyα images were ESO 338–IG04, and Haro 11 [43, 44], which were part of a pilot study of six local starbursts [79, 105]. Previous studies [79] had run into difficulties with the continuum-subtraction, as described in Sect. 4.6.1. Figure 4.26 shows ESO 338–IG04, which is the nearest galaxy that has been imaged, and that with the highest Lyα surface brightness. One of the most obvious points to materialize from these early studies—and very well demonstrated by this example—is that the Lyα morphology rarely resembles that traced by either the UV continuum or the Hα emission. In Fig. 4.26 the UV emission (green) is clearly confined to bright, compact star clusters, and many of these are also surrounded by H II regions (red). However the Lyα emission (blue) shows an entirely different structure: some H II regions have no cospatial Lyα emission, while Lyα is also found above and below the galaxy in extended fan-like structures that do not obviously correspond to H II regions. This example generalizes well over the first study of six starbursts [105], and throughout the LARS sample [48].

Superficially this result may be quite surprising because Lyα and Hα are produced in the same recombination nebulae, that in turn are ionized by the most massive UV-bright stars. Some local correspondence between Lyα and Hα or the UV could have been expected. Instead the central starbursting regions of a galaxy typically present as a complex mixture of emission and absorption, while the bulk of the Lyα flux (when net emission is found) is invariably located in a large-scale scattering halo that surrounds the starburst. Note that while the net Lyα measured in the central regions is formally negative, there is likely to be Lyα emission here too: the absorption arises because the UV continuum is bright, and even a modest EW in absorption can amount to a large absorbed flux, which dominates over the scattered light and produces negative values in the continuum subtraction. See also the examples in Fig. 4.21 where spectra are also shown, and arguments about spectral resolution in Sect. 4.1.4. In the extended halos, on the other hand, there is almost no continuum to subtract, and such uncertainties do not affect the halo measurements.

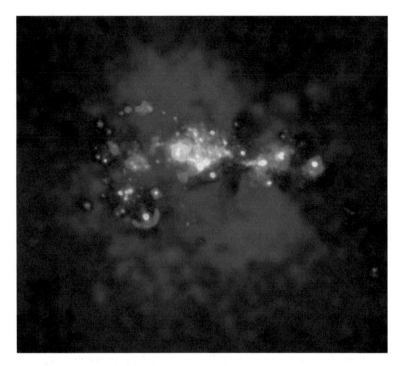

Fig. 4.26 Local starburst galaxy ESO 338–IG04. Hα is shown in red, UV continuum emission in green, and Lyα in blue. Clearly each radiation shows a very different morphology

4.6.2.1 Lyα Sizes Compared to Other Wavelengths

Scattered Lyα, being emitted in large halos that surround the starbursts, dominates the integrated light output. In some systems scattered halo emission accounts for 100% of the Lyα, but such estimates are generally quite hard to make because the central starbursting regions are often dominated by absorption (above). The sizes of galaxies measured in the Lyα line are often significantly larger than in the ultraviolet; the linear extension of Lyα/UV is a factor of ≈2 on average [48], as shown in Fig. 4.27. The extension of these scattering halos—their size in comparison to either the UV or Hα—is not an independent quantity, but is correlated with the total Lyα luminosity [48]. Moreover, bluer, less massive more strongly ionizing galaxies drive more extended halos than redder, more massive galaxies. It has not yet been possible to determine the causal mechanism for this phenomenon, but it is not reproducible by simply tuning the dust abundance in radiation transfer simulations, and may be related to evolution.

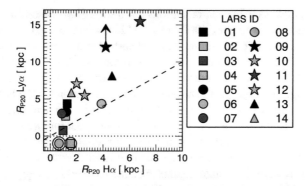

Fig. 4.27 The Lyα sizes of LARS galaxies compared to their Hα sizes. R_{P20} is the Petrosian radius for an η of 0.2. The dashed line shows the 1:1 relation; the majority of galaxies are larger in Lyα, and by a factor of ≈ 2 on average. Galaxies marked with a ring are Lyα absorbers, for which the size cannot be measured. *Figure is reproduced with permission from Hayes et al.* [48], *©AAS*

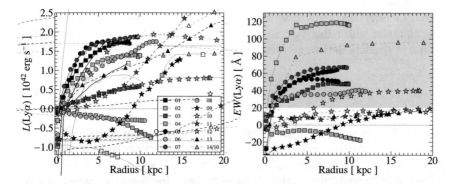

Fig. 4.28 Measurements of the Lyα luminosity (*left*) and EW (*right*) as a function or aperture radius. The gray shaded region indicates an EW in excess of 20 Å, the canonical definition of a Lyα-emitter. Every galaxy is a Lyα absorber at small radius, but several become net emitters when radii larger than 1 kpc are used. A handful of galaxies display ever-increasing EWs that come close to the 20 Å threshold but do not quite make it by the edge of the detector. Perhaps if very large aperture imaging were possible they would. *Figure is reproduced with permission from Hayes et al.* [48], *©AAS*

4.6.2.2 Aperture Effects and Implications for High-Redshift

Only with imaging data can we measure the surface brightness profiles and aperture curves of growth in Lyα. Figure 4.28 shows the recovered Lyα luminosity and EW measured in increasing apertures for the 14 galaxies in the LARS sample, including the Lyα absorbers. It is clear that on the smallest scales, most galaxies have negative measured Lyα. Six galaxies have Lyα aperture curves of growth that converge at radii less than ≈ 10 kpc but for another four targets, additional Lyα flux is still being captured in apertures that exceed 15 kpc. In fact, the ACS/SBC detector is not large enough to capture the full Lyα flux in some cases. LARS 09 is a particularly interesting example as in the central regions it is a strong Lyα absorbing galaxy, with damping wings visible to over 6000 km s^{-1} [121], which is also shown in the *left* panel of Fig. 4.21. Continually more flux is absorbed to an aperture of ≈4 kpc, after which scattered radiation dominates over absorption and the measured flux begins to increase; at 10 kpc it becomes a net emitter, and at 15 kpc its Lyα luminosity is still rapidly growing. The right panel shows the same calculation for the EW, where the shaded region represents EWs above 20 Å, which is the canonical selection EW for LAEs at high-z. Eight galaxies clearly enter the region, but four more are still rising at 20 kpc and may become LAEs if larger apertures could be measured.

This example suggest that there may be significant caveats to measurements of the Lyα output/fraction from high-z galaxies: a radius of 20 kpc suggests an aperture of 40 kpc would be needed to capture all the Lyα. This corresponds to ≈5 arcsec at $z = 3$, which is several times larger than the typical spectroscopic slit, or typical photometric aperture used in a narrowband observation. Indeed recent studies with the very high sensitivity IFU MUSE have put forward similar arguments in the high-z universe directly [147].

Surface brightness profiles are very different in Lyα from the UV continuum and, as is common with radial light distributions, Lyα halos have been fit with Sérsic profiles [49], using the intensity expression:

$$I(R) = I_0 \exp \left\{ - b_n \left[\left(\frac{R}{R_e} \right)^{1/n} - 1 \right] \right\} \tag{4.11}$$

where $I(R)$ and I_0 are the surface brightness at radius R and $R = 0$, respectively, and R_e is the effective radius which describes the extent of the light profile. The quantity n is the Sérsic index and describes the shape: when $n = 1$ the expression reduces to an exponential which is commonly found for disk galaxies; when $n = 4$ the equation describes the de Vaucouleurs profile which is commonly found for the stellar light distribution in elliptical galaxies and mergers.

This functional form provides an adequate fit to the Lyα data, but the Sérsic indices, n, are found to be very different in the Lyα images compared to the FUV continuum [49]. The average value for the sample in the FUV is $n = 4.5$, while in Lyα it is 1.4. The UV continuum which roughly traces the intrinsic ionizing photon production is very strongly centrally peaked, and similar to de Vaucouleurs profile. This central peaking, however, is almost entirely wiped out by the Lyα scattering process, which results in profiles that are closer to exponential. Whether this reflects the underlying distribution of H I remains to be determined, but if a correlation is ultimately found between Lyα surface brightness profiles and H I column density profiles, it will be very useful for interpreting the Lyα profiles derived for galaxies at high-z.

4.6.2.3   Surface Brightness Comparisons

Having images in Lyα, Hα, and the UV continuum enables us to contrast the local surface brightnesses on almost arbitrarily small scales, and down to the level of individual resolution elements. The comparison of fluxes in individual pixels can encode information about the extent of scattering, and departures from the canonical Case B recombination intensity.

The *left* panel of Fig. 4.29 shows the distribution of pixel surface brightness in the space of Lyα versus FUV. Diagonal red lines have unit slope, indicating a constant Lyα/FUV, and thus they encode the Lyα EW: 1, 10, and 100 Å are shown. These figures enable us to visualize how Lyα is emitted in comparison with the massive stellar population that produces the ionizing photons. A broad correlation is obviously visible, but the scatter in Lyα is several dex at intermediate FUV surface brightness (around 10^{-16} to 10^{-15} erg s^{-1} cm^{-2} Å^{-1}). The peak in the Lyα surface brightness distribution corresponds well to the brightest UV sources, indicating that some UV-bright regions are closely spatially correlated with bright Lyα emission; this would indicate little spatial decoupling in Lyα for at least some rays. The large spread is seen in directions of both enhanced and suppressed Lyα. For points dropping to lower EW, this may be the result of stellar evolution that could lower the intrinsic EW (Sect. 4.1.2.2), of H I scattering that removes some fraction of Lyα from the line-of-sight.

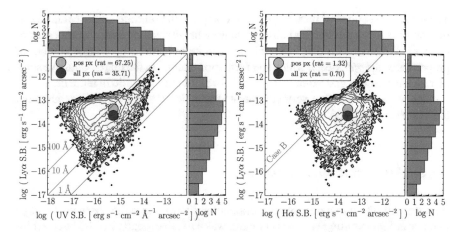

Fig. 4.29 The resolved 'pixel-by-pixel' surface brightness diagram for the Lyα-emitting regions of local luminous blue compact galaxy Haro 11. The contours describe the pixel density, and histograms above and to the side show the (logarithmic) net distribution. *Left* shows the Lyα surface brightness compared to the FUV measured in the corresponding pixel. Red diagonal lines have unit slope and represent constant equivalent width: EWs of 1, 10, and 100 Å are shown. *Right* shows the same for Lyα versus Hα, and red lines of unit slope represent fixed line ratios; here the case B value of 8.7 is indicated. In both figures the large circular points show the mean of all the points for which absorption is neglected (cyan) and included (magenta), so that the intensity-weighted average for the galaxy can be visualized. These ratios are quoted in the legend

The right plot in Fig. 4.29 shows the same as the left, but with Hα substituted for the FUV continuum on the x-axis. Lines of unit slope now follow constant Lyα/Hαratios, and the Case B value of 8.7 is shown. The morphology of the pixel contours is rather similar, and points fall both above and below the line. However the Lyα/Hα ratio, unlike the Lyα/FUV ratio discussed in the previous paragraph, is independent of stellar evolutionary stage. Points can only fall below the Case B line because of either dust reddening or scattering. Dust alone cannot alter the Lyα/FUV ratio, so the morphology of points must be driven mainly by radiation transfer effects.

Both figures show a tail of the distribution of points that extend to Lyα/FUV and Lyα/Hα ratios that greatly exceed the expectations from recombination theory and local reprocessing of ionizing radiation. These are the result of photons scattering into the line-of-sight, and are found mostly at lower FUV and Hα surface brightness. These points come from regions between stellar clusters and H II regions (at intermediate surface brightness) and from the scattering halos at the faint end of the distribution (see also Sect. 4.6.2). Were it not for masking of the images to include only pixels detected in the UV continuum, these tails would extend to very much higher ratios, that are orders of magnitude above expectation.

4.7 Putting It All Together: A Unifying Scenario for Lyman Alpha Emission and Absorption?

The previous sections of these notes have compiled the relevant empirical information on how Lyα emission and absorption may be affected by various properties of galaxies. We have seen that Lyα emission can be influenced by a large number of properties of the galaxies in which star formation occurs. In the final section we summarize these results and examine a possible evolutionary scenario in more detail. We discuss ways in which the very complicated situation may be simplified, and what the next steps may be.

4.7.1 An Evolutionary Scenario

First we will return to a possible evolutionary scenario. Since the absence of a large number of Lyα-emitting galaxies at high-z was a relatively early result [114], evolutionary scenarios and 'duty cycles' for Lyα have been long discussed in the literature [91, 95, 139, 140].

Hydrodynamical simulations of galaxies and star clusters have computed the phase and kinematic structure of the ISM [133]. As we have seen, Lyα transfer may be strongly affected by feedback and galaxy winds, and it is important to note that the massive stars are responsible for producing both the ionizing radiation to make Lyα, and the stellar winds and supernovae that heat the ISM. Simulations may produce a realistic description of how the atomic gas moves as a function of evolutionary stage, which we may then post-process with simplified sketch of how Lyα may appear, or indeed full 3-dimensional radiation transfer simulations.

At the earliest times, the atomic ISM may be uniformly distributed and largely static compared to the star-forming regions; indeed the densest H I gas will be the first to form the cold molecular clouds in which star formation is ignited, and we may expect the kinematics do be similar. H II regions will form when the first ionizing stars turn on, but at these times we expect there to be no velocity offset between the H II regions and local H I gas outside the Strömgren sphere. With complete covering, this would be sufficient to scatter all the Lyα, and also the adjacent continuum radiation in the Lyα resonance.

Stellar winds from O-stars may begin almost immediately as well, and start to create a hot central bubble. However these continuous sources of energy injection are shown to have less impact on the disruption of molecular clouds than the almost instantaneous injection that comes from SNe, even though SNe contribute only around half of the total mechanical energy. However we must wait an extra few Myr for the first supernovae to explode. Once an overpressurized bubble has been created it will expand and sweep up the ISM, producing a shell of outflowing atomic gas (as well as other phases). During this time we may expect some of the Lyα to escape through the shell, possibly with an asymmetric profile [139]. The ionizing photon production rate as a function of time is plotted in Fig. 4.30 against the rate of mechanical energy injection in the left panel, and the time-integrated amount of mechanical energy that has been injected in the right panel. Taking a simple stellar population—all stars form at a single time $t = 0$—as an example, some important evolutionary milestones include:

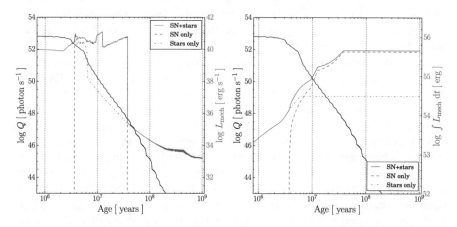

Fig. 4.30 Both panels show the rate at which hydrogen-ionizing photons are produced by massive stars in black. In the *left* panel, this shares an axis with the rate at which mechanical energy is produced by stellar winds and supernovae. The *right* panel shows the same with the integral of the mechanical luminosity to illustrate how much energy has been delivered to the ISM in total. Total energy, divided by subtype is shown separately in each

- $t \approx$ **0 Myr**: Stars form; ionizing photons produced; and O-star winds begin.
- $t \approx$ **4 Myr**: First supernova explosions occur. In fact this may be somewhat delayed if the upper mass limit for core-collapse SNe is reduced to ~ 15 $M_\odot$, in line with the current absence of more massive SN progenitors.
- $t \approx$ **5 Myr**: Ionizing luminosity decreased by factor of 10 compared with $t = 0$.
- $t \approx$ **6.5 Myr**: Last O-star explodes as SN.
- $t \approx$ **10 Myr**: Ionizing luminosity decreased by factor of 100 compared with $t = 0$.
- $t \approx$ **40 Myr**: Last supernova explodes; ionizing luminosity decreased by factor of $\approx 100{,}000$ compared with $t = 0$.

Important points to note from this sequence is that just 1 Myr after the first SN explodes, the ionizing luminosity has decreased by a factor of 10 compared to that of zero age. Thus if we must wait for the first supernovae to explode in order to see Lyα, we should never see EWs above ≈ 20–30 Å, and by 10 Myr, when only 30% of the total mechanical energy has been returned, the ionizing luminosity is less than 1% its value at $t = 0$. For a SSP, 70% of the mechanical energy is delivered when the stellar population is no longer an effective ionizing source, and the majority of the SNe are 'wasted'.

Of course we do not expect entire galaxies to behave as simple stellar populations. Local starburst galaxies do have UV light distributions that are very clumpy with a large fraction of the star formation being concentrated to compact star clusters. Perhaps star formation could better be described as a sequence of SSPs, in which even a narrow range of ages—star formation proceeding in different clusters over $\gtrsim 10$ Myr—could allow for some 'pre-disruption' of the ISM by the first generations, that initiate the winds. An extension of the above SSP scenario [95] shows how the range of spectral profiles observed in local starbursts can be described as a function of evolutionary phase: dwarf galaxies such as ɪZw 18 and SBS 0335–052 (Sect. 4.5.1) represent the earliest stages and indeed show Lyα absorption, while Lyα-emitting objects with P Cygni profiles (e.g. IRAS 08339+6517 and Haro 2) represent somewhat more evolved stages.

An important question would be over what timescales star formation can be synchronized in galaxies. The lowest mass objects discussed above have SFRs on the order of 1 $M_\odot$ yr^{-1}, while the more evolved galaxies have SFRs and stellar masses around 10–100 times higher; Green Peas [55] also have SFRs of 5–25 $M_\odot$ yr^{-1} but emission line spectra that look similar to the dwarfs. It is possible that the more evolved galaxies have blown out their expanding shells into the circumgalactic environment, which have become unstable and fragmented. This clumpy distribution of gas and dust is liable to explain the Lyα line profiles, UV absorption lines, and also broad range of Lyα/Balmer line ratios. Such a scenario is likely to lead to a wide range in combinations of H ɪ kinematic distributions/velocity ranges, covering fractions, and column densities. Feedback, coupled with dust obscuration still has the potential to explain the wide variety of Lyα outputs, together with spectral profiles, observed in star forming galaxies, but the key variable could be time.

4.7.2 How to Boil Down the Soup of Quantities

The literature, as well as these notes, contains many studies in which a dependent Lyα quantity—e.g. EW or $f_{esc}^{Ly\alpha}$—is presented on the y-axis of a graph, while some independent variable—e.g. reddening (Fig. 4.14) or outflow velocity (Fig. 4.20)—is presented on x. This way we test whether Lyα output depends upon certain quantities, assess the spread, and employ various statistical tests for correlation. After many observations, we arrive at the situation where Lyα has been claimed, probably correctly, to be stronger in younger galaxies, with lower stellar mass, more compact stellar continua, and higher sSFR. These are typically bluer and less dusty, with lower metallicities and higher ionization parameters. They also appear to drive faster outflows, with less rotationally dominated velocity fields, and have lower H I covering fractions and/or column densities. Given the large samples of observational data that are now available, it is certain that much more can be done. However firstly, we also need to be more clear about what question we want to answer.

At the Saas-Fee meeting we posed two questions: *what makes a high escape fraction?* and *what conditions make a galaxy a Lyα-emitter?* These two questions typically underlie the approaches taken in the field; they look the same but the questions are subtly different. The first question, regarding the escape fraction, is focussed upon the interstellar and circumgalactic medium in which Lyα radiation transfer occurs. It deals with elements such as dust and H I distribution, which are the properties of a galaxy that determine whether a Lyα photon will be emitted or absorbed. Regardless of what the input spectrum is, $f_{esc}^{Ly\alpha}$ is unchanged. Properties like metallicity or stellar mass *do not causally affect Lyα transfer* themselves as neither metals nor mass can destroy Lyα. However they may easily be correlated with $f_{esc}^{Ly\alpha}$, because they are in turn correlated with properties that do influence the radiation transfer.

Observationally, and particularly when comparing the Lyα-emitting and non emitting galaxy populations, we may be concerned with the second question of what makes a galaxy bright in Lyα. Obviously a high escape fraction (question 1) will help, as the Lyα is simply the product of $f_{esc}^{Ly\alpha}$ and the intrinsic spectrum. In this context one may consider a passive elliptical galaxy with no star-formation, and only a hot gaseous ISM: it may in principle transfer a very large fraction of any Lyα photons that were produced, but without an intrinsic Lyα production one may question how relevant this is. In making a galaxy bright in Lyα properties like SFR will obviously be a key factor, as well as anything that contributes to a higher production of ionizing photons, for example the initial mass function and stellar metallicity.

In the final lecture of the school, we attempted to segregate all the properties that we have discussed so far into those that influence the production of Lyα, and those that influence its escape. We were not entirely successful (as video evidence documents), but we summarize the results in Fig. 4.31, and list various properties one-by-one below with motivation for each.

- **Dust—escape**. No influence over Lyα production, but absorbs Lyα radiation when produced and influences only the escape. See Sect. 4.4.3.1.

PRODUCTION

		ESCAPE
Age	Compactness	Stellar mass
Specific SFR	UV colour	Turbulence
	Metallicity	Dust
	Ionization/Excitation	HI mass
		Outflow velocity
		HI column density
		HI covering fraction

Fig. 4.31 Division of physical properties into those that enter Q and those that enter $f_{\rm esc}^{\rm Ly\alpha}$. *Left* shows the blackboard at the end of the exercise, and *right* shows an expanded and tidied version

- **Stellar mass—escape**. Mass has no direct influence over radiation transfer, but could indirectly affect the transfer because feedback may be more efficient at low mass. On the other hand, feedback efficiency would be dependent on the total gravitational potential, which is normally dark matter dominated. Star formation itself is in fact *less* efficient at low mass. This caused an argument.
- **Age—production**. Influences primarily the production of Lyα because the ionizing photon production is so strongly dependent upon time (Sects. 4.1.2.2 and 4.7.1).
- **Compactness—production and escape**. I had intended to argue that more compact star formation leads to more efficient generation of winds, but it was pointed out that this may also alter the initial mass function and therefore also production. See Sect. 4.4.6.1 and Fig. 4.17. This also caused an argument.
- **Specific SFR or Hα EW—production**. This is broadly a consequence of the stellar population age and recent SFH, so affects mainly the intrinsic Lyα line.
- **Color—production and escape**. Bluer galaxies are likely to be younger and therefore produce more Lyα. Bluer galaxies may also be less dusty and therefore allow more to escape.
- **Metallicity—production and escape**. I had originally planned to file metal abundance under production: metal abundance alters the composition of stars and lower metallicity stars burn hotter and produce more ionizing radiation, increasing the intrinsic Lyα. However metals are required to make dust grains, and metallicity and dust content are definitely correlated.
- **Ionization parameter—production and escape**. Harder ionizing spectra are likely to produce more ionizing photons per non-ionizing UV (increasing the intrinsic Lyα EW) and may also more highly ionize the ISM reducing the H I column available to scatter Lyα. See Sect. 4.4.3.3.
- **Atomic gas mass—escape**. With every other quantity held constant, more atomic gas in a galaxy will increase scattering of Lyα and decrease the escape fraction. See Sect. 4.4.7
- **Outflows—escape**. Faster outflows may shift gas out of resonance allowing Lyα to escape. See Sect. 4.5.2.

- **Atomic gas covering fraction and column density—escape**. Both determine the optical depth a Lyα photon will see, and therefore the number of scatterings it will encounter. See Sect. 4.5.3.
- **Turbulent velocity fields—escape**. Disrupted velocity fields, whether because of feedback or interactions, will broaden the intrinsic Lyα line profile, allowing the wings to shift out of resonance more easily. If the ionized gas is more turbulent because of feedback then the atomic gas may also be more disrupted or outflowing. See Sect. 4.4.4.

While we have listed a dozen quantities above, each point in isolation assumes that all the other quantities are held constant. All are to some extent confirmed by observation and this enables us to build an intuitive picture, but many/all of these quantities will covary without question. Moreover most of these correlations are likely to be stronger than any of the ones with Lyα individually. It would therefore be helpful to combine some of these measured properties into groups, and deal with fewer parameters. For example, metallicity, SFR, and stellar mass are strongly related through the fundamental metallicity relation [90]. In addition sSFR, ionization parameter, and color are related to age, and outflow velocity, column density and covering fractions are all consequences of feedback. This in turn may be linked to the compactness of star formation because of the threshold of SFR surface density to drive global outflows [52].

This illustration is cartoon-like of course, but serves to illustrate how these quantities are strongly intertwined and many have been studied in large samples. The majority of Lyα studies, in contrast, have been restricted to small samples and treated only a handful of independent variables. At the current time, HST and GALEX have provided statistical samples of Lyα spectra—over 100 objects from each telescope—and large amounts of complementary multi-wavelength data have been obtained. These have not yet been combined into a single homogenous database of measurements that would allow more sophisticated statistical tools to be applied. Until the tests are performed we will not know whether the complex multi-parametric problem of Lyα escape can be reduced to a simpler set of equations with a smaller number of quantities. If such a 'fundamental relation of Lyα' would emerge it would be enormously valuable for the interpretation of large-scale high-z Lyα surveys that are now on the horizon, and likely vital for Lyα observations in the reionization epoch.

4.7.3   What More Do We Want to Know?

The final lecture closed with a discussion of what more we would like to know from the observational standpoint: *what key observations are required—but missing—in order to observationally figure out the 'Lyα problem'?* We close by summarizing some of the main themes.

4.7.3.1   The Spatial Distribution of Lyα-Scattering Gas

The main limiting factor to observational studies is in fact the same as it is in hydrodynamical simulations: we are limited by physical resolution. Lyα transfer is governed by the H I and dust distribution inside galaxies, and all the empirical studies
discussed in these notes are about how these quantities impact Lyα transfer. Specifically focusing upon the H I medium, the VLA can currently probe angular scales
of about 2.5 arcsec at $\lambda = 21$ cm using the A-configuration. This is the same size as
the entire aperture of HST/COS, which samples a size of just 0.5 kpc in the nearest
galaxy for which Lyα imaging is possible. For all other galaxies, larger physical
sizes will be probed. Obtaining 21 cm data in these high resolution configurations
also requires very long integration times and to date such high-resolution data have
not been obtained; the best current observations of galaxies with Lyα data is a small
number of galaxies observed in VLA configuration B and C (synthesized beams of
5–15 arcsec). This samples ≈3 kpc at 150 kpc (nearest systems with Lyα data) and
we can only resolve scales an order of magnitude larger than the scale height of the
warm neutral medium in the Milky Way. The cold neutral gas exists in clouds that
are significantly smaller, and it is this phase that makes the 'clumpy' H I medium
discussed throughout the manuscript.

Twenty-one centimeter observations in the highest resolution settings of the VLA
would really be essential to understand the distribution of H I in Lyα-emitting galaxies. This is an incremental process where the resolution is increased year-on-year, but
accepting the longest integration times is the only way the large-scale distribution of
H I can be measured and the column density mapped. The *Square Kilometer Array*
(SKA) will come online in the next decade, and will provide angular resolutions 100
times better at 1.4 GHz. This would in principle be able to resolve the issues, and
will provide significant improvements in many respects, but still the limiting column
densities will be on the order of $\approx 10^{19}$ cm^{-2} at these resolutions, or $\approx 10^{17}$ cm^{-2} at
resolutions currently possible with the VLA.

The alternative in obtaining high-resolution H I data is to target galaxies that are
much more nearby. This would dramatically improve the H I observations, but as
discussed in Sects. 4.1.5 and 4.4.7, Lyα observations require a significant recession
velocity so that the Lyα of the target does not blend with geocoronal emission or
get absorbed by Milky Way H I. A new ultraviolet satellite at L2 would mitigate
the geocoronal emission problem, but not the foreground H I. Nevertheless some
progress could still probably be made using high resolution spectrographs or the
F122M filter on ACS/SBC, at the cost of a somewhat increased integration time with
HST.

4.7.3.2   The Distribution of Dust

HST imaging can provide spatial resolution of ≈0.1 arcsec, which is orders of magnitude higher than 21 cm telescopes. It can therefore, with caveats, probe the distribution of dust on such scales. If dust exists only in dense molecular clouds, the HST

resolution may be sufficient to observe the dust distribution. This is usually done using either Hα/Hβ in narrowband filters [2] or modeling the attenuation on the stellar continuum [44]. However, any measurement of E_{B-V} or A_V from these methods has to assume a dust geometry, which is usually that dust is uniformly distributed in a screen that is in front of, and distant from, the gas and stars. Yet we know already from integrated measurement of GALEX LAEs that the dust screen approximation does not fully describe Lyα fluxes even before we consider resonance scattering ([4, 127], see Sect. 4.4.3.1) so it is not obvious that these dust measurements are sufficient.

Local Lyα studies have not employed more information than can be extracted from the Hα/Hβ ratio on either global or local scales, and we may benefit from improved understanding of the dust geometry. Studies of clumpy dust [11, 14, 100] have been performed [127] but rely upon the resonant Lyα line, whereas more robust estimates can be derived using additional Balmer lines and/or redder hydrogen lines from the Paschen or Brackett series. These may be able to constrain the average clump optical depth and covering without the use of the Lyα line itself, and the results could be tested against the Lyα output.

Finally these above measurements probe dust purely by the effect of its extinction, and the total amount of dust cannot be computed. Integrated dust masses are completely absent from local Lyα studies because the samples are observationally incomplete at FIR and sub-mm wavelengths. Piecemeal observations in the FIR do exist from some samples, and sub-mm observations in the 450 and 850 μm atmospheric windows could constrain the shape of the Rayleigh-Jeans tail. This would provide a total dust mass for a galaxy that could then be compared against the Lyα observation.

4.7.3.3   How Is the Lyα Spectral Profile Built Spatially?

Aside from a small number of slit-spectra obtained with STIS [3, 95, 107], we have almost no spatial information regarding the Lyα spectral morphology as a function of position. At low-z most resolved spectra come from COS, and have been obtained on one pointing per galaxy where they sample a smaller physical size than that over which Lyα emission is seen. Using the UV continuum for target acquisition reveals the Lyα profile where outflowing gas is likely to point in close to the direction of the observer. Consequently we have very limited information on how the line spectral profile of a Lyα halo appears.

Most observations at high-z also only give one positional measurement per galaxy, but the light will usually come from a larger region of the galaxy: 1 arcsec slit would sample 6–8 kpc at $z > 2$. Recent IFU observations in which the halo is recovered are certainly an exception to this, but still are seeing-limited. In these spectra we also do not know how the spectral morphology is built from compact central emission and diffuse scattering gas which may still begin to dominate on scales not resolved by the seeing. In interpreting the high-z sources (as well as those at low-z) we would like to know how the profile is built out of the 'down-the-barrel' profile that we find towards the brightest stellar clusters, and the halo emission. Ideally we would build an integral field spectrograph to fly on a UV satellite, but that does not appear to be on the horizon. The only way to go about such a project is to dissect a galaxy with multiple STIS slits, or obtain spectra on many different positions with the COS. In fact STIS has been used in this way in one target, IZw 18 [3], but without an emission line we do not learn much about how the emission profile is built, so clearly bright emitting sources need to be studied.

Acknowledgements Firstly it is my pleasure to thank the organizers of 46th Saas Fee school *Lyman-alpha as an astrophysical and cosmological tool* for this exciting invitation and their warmest hospitality: Anne Verhamme, Pierre North, Hakim Atek, Sebastiano Cantalupo, and Myriam Burgener Frick. While on the subject, I also need to thank them for their patience in waiting for this manuscript. Without exception, I would like to thank each attendee for their very active participation, interaction in during the sessions, stimulating discussions, and *La Pote* on Friday→Saturday. Warm thanks also need to be extended to Mark Dijkstra, X. Prochaska, and Masami Ouchi for their outstanding lectures, to which I very much wish I could have devoted more attention. I would like to thank the people whose materials have made it directly into these notes, many of whom are naturally close friends and collaborators: Hakim Atek, Arjan Bik, Joanna Bridge, John Cannon, Stéphane Charlot, Lennox Cowie, Alaina Henry, Anne Jaskot, Daniel Kunth, Miguel Mas-Hesse, David Meier, Sally Oey, Göran Östlin, Ivana Orlitová, Stephen Pardy, Thøger Rivera-Thorsen, Claudia Scarlata, Daniel Schaerer, Anne Verhamme. My apologies go to the probably long list of names missing from this skewed list. I gratefully acknowledge the support of the Swedish Research Council, Vetenskapsrådet and the Swedish National Space Board (SNSB), and am Fellow of the Knut and Alice Wallenberg Foundation.

References

1. Arnouts, S., Schiminovich, D., Ilbert, O., Tresse, L., Milliard, B., Treyer, M., Bardelli, S., Budavari, T., Wyder, T.K., Zucca, E., Le Fèvre, O., Martin, D.C., Vettolani, G., Adami, C., Arnaboldi, M., Barlow, T., Bianchi, L., Bolzonella, M., Bottini, D., Byun, Y.I., Cappi, A., Charlot, S., Contini, T., Donas, J., Forster, K., Foucaud, S., Franzetti, P., Friedman, P.G., Garilli, B., Gavignaud, I., Guzzo, L., Heckman, T.M., Hoopes, C., Iovino, A., Jelinsky, P., Le Brun, V., Lee, Y.W., Maccagni, D., Madore, B.F., Malina, R., Marano, B., Marinoni, C., McCracken, H.J., Mazure, A., Meneux, B., Merighi, R., Morrissey, P., Neff, S., Paltani, S., Pellò, R., Picat, J.P., Pollo, A., Pozzetti, L., Radovich, M., Rich, R.M., Scaramella, R., Scodeggio, M., Seibert, M., Siegmund, O., Small, T., Szalay, A.S., Welsh, B., Xu, C.K., Zamorani, G., Zanichelli, A.: The GALEX VIMOS-VLT deep survey measurement of the evolution of the 1500 Å luminosity function. ApJ **619**, L43–L46 (2005). https://doi.org/10.1086/426733, astro-ph/0411391
2. Atek, H., Kunth, D., Hayes, M., Östlin, G., Mas-Hesse, J.M.: On the detectability of Lyα emission in star forming galaxies. The role of dust. A&A **488**, 491–509 (2008). https://doi.org/10.1051/0004-6361:200809527, 0805.3501
3. Atek, H., Schaerer, D., Kunth, D.: Origin of Lyα absorption in nearby starbursts and implications for other galaxies. A&A **502**, 791–801 (2009). https://doi.org/10.1051/0004-6361/200911856, 0905.1329
4. Atek, H., Kunth, D., Schaerer, D., Mas-Hesse, J.M., Hayes, M., Östlin, G., Kneib, J.P.: Influence of physical galaxy properties on Lyα escape in star-forming galaxies. A&A **561**, A89 (2014). https://doi.org/10.1051/0004-6361/201321519, 1308.6577

5. Bacon, R., Accardo, M., Adjali, L., Anwand, H., Bauer, S., Biswas, I., Blaizot, J., Boudon, D., Brau-Nogue, S., Brinchmann, J., Caillier, P., Capoani, L., Carollo, C.M., Contini, T., Couderc, P., Daguisé, E., Deiries, S., Delabre, B., Dreizler, S., Dubois, J., Dupieux, M., Dupuy, C., Emsellem, E., Fechner, T., Fleischmann, A., François, M., Gallou, G., Gharsa, T., Glindemann, A., Gojak, D., Guiderdoni, B., Hansali, G., Hahn, T., Jarno, A., Kelz, A., Koehler, C., Kosmalski, J., Laurent, F., Le Floch, M., Lilly, S.J., Lizon, J.L., Loupias, M., Manescau, A., Monstein, C., Nicklas, H., Olaya, J.C., Pares, L., Pasquini, L., Pécontal-Rousset, A., Pelló, R., Petit, C., Popow, E., Reiss, R., Remillieux, A., Renault, E., Roth, M., Rupprecht, G., Serre, D., Schaye, J., Soucail, G., Steinmetz, M., Streicher, O., Stuik, R., Valentin, H., Vernet, J., Weilbacher, P., Wisotzki, L., Yerle, N.: The MUSE second-generation VLT instrument. In: Society of Photo-Optical Instrumentation Engineers (SPIE) Conference Series, vol. 7735 (2010), https://doi.org/10.1117/12.856027
6. Baldwin, J.A., Phillips, M.M., Terlevich, R.: Classification parameters for the emission-line spectra of extragalactic objects. PASP **93**, 5–19 (1981). https://doi.org/10.1086/130766
7. Barger, A.J., Cowie, L.L., Wold, I.G.B.: A flux-limited sample of $z \sim 1$ Lyα emitting galaxies in the chandra deep field south. ApJ **749**, 106 (2012). https://doi.org/10.1088/0004-637X/749/2/106, 1202.2865
8. Bik, A., Östlin, G., Hayes, M., Adamo, A., Melinder, J., Amram, P.: VLT/MUSE view of the highly ionized outflow cones in the nearby starburst ESO338-IG04. A&A **576**, L13 (2015). https://doi.org/10.1051/0004-6361/201525850, 1503.06626
9. Blanc, G.A., Adams, J.J., Gebhardt, K., Hill, G.J., Drory, N., Hao, L., Bender, R., Ciardullo, R., Finkelstein, S.L., Fry, A.B., Gawiser, E., Gronwall, C., Hopp, U., Jeong, D., Kelzenberg, R., Komatsu, E., MacQueen, P., Murphy, J.D., Roth, M.M., Schneider, D.P., Tufts, J.: The HETDEX pilot survey. II. The evolution of the Lyα escape fraction from the ultraviolet slope and luminosity function of $1.9 < z < 3.8$ LAEs. ApJ **736**, 31 (2011). https://doi.org/10.1088/0004-637X/736/1/31, 1011.0430
10. Bruzual, G., Charlot, S.: Stellar population synthesis at the resolution of 2003. MNRAS **344**, 1000–1028 (2003). https://doi.org/10.1046/j.1365-8711.2003.06897.x, astro-ph/0309134
11. Calzetti, D., Kinney, A.L., Storchi-Bergmann, T.: Dust extinction of the stellar continua in starburst galaxies: the ultraviolet and optical extinction law. ApJ **429**, 582–601 (1994). https://doi.org/10.1086/174346
12. Cannon, J.M., Skillman, E.D., Kunth, D., Leitherer, C., Mas-Hesse, M., Östlin, G., Petrosian, A.: Extended tidal structure in two Lyα-emitting starburst galaxies. ApJ **608**, 768–771 (2004). https://doi.org/10.1086/420868. arXiv:astro-ph/0403191
13. Cantalupo, S., Lilly, S.J., Haehnelt, M.G.: Detection of dark galaxies and circum-galactic filaments fluorescently illuminated by a quasar at $z = 2.4$. MNRAS **425**, 1992–2014 (2012). https://doi.org/10.1111/j.1365-2966.2012.21529.x, 1204.5753
14. Caplan, J., Deharveng, L.: Extinction and reddening of H II regions in the Large Magellanic Cloud. A&A **155**, 297–313 (1986)
15. Cardamone, C., Schawinski, K., Sarzi, M., Bamford, S.P., Bennert, N., Urry, C.M., Lintott, C., Keel, W.C., Parejko, J., Nichol, R.C., Thomas, D., Andreescu, D., Murray, P., Raddick, M.J., Slosar, A., Szalay, A., Vandenberg, J.: Galaxy Zoo Green Peas: discovery of a class of compact extremely star-forming galaxies. MNRAS **399**, 1191–1205 (2009). https://doi.org/10.1111/j.1365-2966.2009.15383.x, 0907.4155
16. Cardelli, J.A., Clayton, G.C., Mathis, J.S.: The relationship between infrared, optical, and ultraviolet extinction. ApJ **345**, 245–256 (1989). https://doi.org/10.1086/167900
17. Charlot, S., Fall, S.M.: Lyman-alpha emission from galaxies. ApJ **415**, 580 (1993). https://doi.org/10.1086/173187
18. Chen, Y.M., Tremonti, C.A., Heckman, T.M., Kauffmann, G., Weiner, B.J., Brinchmann, J., Wang, J.: Absorption-line probes of the prevalence and properties of outflows in present-day star-forming galaxies. AJ **140**, 445–461 (2010). https://doi.org/10.1088/0004-6256/140/2/445, 1003.5425
19. Cowie, L.L., Barger, A.J., Hu, E.M.: Low-redshift Lyα selected galaxies from GALEX spectroscopy: a comparison with both UV-continuum selected galaxies and high-redshift Lyα emitters. ApJ **711**, 928–958 (2010). https://doi.org/10.1088/0004-637X/711/2/928, 0909.0031

20. Cowie, L.L., Barger, A.J., Hu, E.M.: Lyα emitting galaxies as early stages in galaxy formation. ApJ **738**, 136 (2011). https://doi.org/10.1088/0004-637X/738/2/136, 1106.0496

21. Crowther, P.A., Caballero-Nieves, S.M., Bostroem, K.A., Maíz Apellániz, J., Schneider, F.R.N., Walborn, N.R., Angus, C.R., Brott, I., Bonanos, A., de Koter, A., de Mink, S.E., Evans, C.J., Gräfener, G., Herrero, A., Howarth, I.D., Langer, N., Lennon, D.J., Puls, J., Sana, H., Vink, J.S.: The R136 star cluster dissected with Hubble Space Telescope/STIS. I. Far-ultraviolet spectroscopic census and the origin of He II λ1640 in young star clusters. MNRAS **458**, 624–659 (2016). https://doi.org/10.1093/mnras/stw273, 1603.04994

22. Curtis-Lake, E., McLure, R.J., Pearce, H.J., Dunlop, J.S., Cirasuolo, M., Stark, D.P., Almaini, O., Bradshaw, E.J., Chuter, R., Foucaud, S., Hartley, W.G.: A remarkably high fraction of strong Lyα emitters amongst luminous redshift $6.0<z<6.5$ Lyman-break galaxies in the UKIDSS ultra-deep survey. MNRAS **422**, 1425–1435 (2012). https://doi.org/10.1111/j.1365-2966.2012.20720.x, 1110.1722

23. Davidsen, A.F., Hartig, G.F. (eds.): Far-Ultraviolet Spectrum of the Seyfert Galaxy NGC 4151 (1978)

24. Davidsen, A.F., Hartig, G.F., Fastie, W.G.: Ultraviolet spectrum of quasi-stellar object 3C273. Nature **269**, 203–206 (1977). https://doi.org/10.1038/269203a0

25. Deharveng, J.M., Small, T., Barlow, T.A., Péroux, C., Milliard, B., Friedman, P.G., Martin, D.C., Morrissey, P., Schiminovich, D., Forster, K., Seibert, M., Wyder, T.K., Bianchi, L., Donas, J., Heckman, T.M., Lee, Y.W., Madore, B.F., Neff, S.G., Rich, R.M., Szalay, A.S., Welsh, B.Y., Yi, S.K.: Lyα-emitting galaxies at $0.2<z<0.35$ from GALEX spectroscopy. ApJ **680**, 1072–1082 (2008). https://doi.org/10.1086/587953, 0803.1924

26. Dijkstra, M.: Lyα emitting galaxies as a probe of reionisation. PASA **31**, e040 (2014). https://doi.org/10.1017/pasa.2014.33, 1406.7292

27. Djorgovski, S., Spinrad, H., McCarthy, P., Strauss, M.A.: Discovery of a probable galaxy with a redshift of 3.218. ApJ **299**, L1–L5 (1985). https://doi.org/10.1086/184569

28. Duval, F., Schaerer, D., Östlin, G., Laursen, P.: Lyman α line and continuum radiative transfer in a clumpy interstellar medium. A&A **562**, A52 (2014). https://doi.org/10.1051/0004-6361/201220455, 1302.7042

29. Duval, F., Östlin, G., Hayes, M., Zackrisson, E., Verhamme, A., Orlitova, I., Adamo, A., Guaita, L., Melinder, J., Cannon, J.M., Laursen, P., Rivera-Thorsen, T., Herenz, E.C., Gruyters, P., Mas-Hesse, J.M., Kunth, D., Sandberg, A., Schaerer, D., Månsson, T.: The Lyman alpha reference sample. VI. Lyman alpha escape from the edge-on disk galaxy Mrk 1486. A&A **587**, A77 (2016). https://doi.org/10.1051/0004-6361/201526876, 1512.00860

30. Eldridge, J.J., Stanway, E.R.: Spectral population synthesis including massive binaries. MNRAS **400**, 1019–1028 (2009). https://doi.org/10.1111/j.1365-2966.2009.15514.x, 0908.1386

31. Eldridge, J.J., Stanway, E.R.: The effect of stellar evolution uncertainties on the rest-frame ultraviolet stellar lines of C IV and He II in high-redshift Lyman-break galaxies. MNRAS **419**, 479–489 (2012). https://doi.org/10.1111/j.1365-2966.2011.19713.x, 1109.0288

32. Eldridge, J.J., Stanway, E.R.: BPASS predictions for Binary Black-Hole Mergers (2016). arXiv:1602.03790

33. Finkelstein, S.L., Cohen, S.H., Malhotra, S., Rhoads, J.E., Papovich, C., Zheng, Z.Y., Wang, J.X.: A plethora of active galactic nuclei among Lyα galaxies at low redshift. ApJ **703**, L162–L166 (2009). https://doi.org/10.1088/0004-637X/703/2/L162, 0906.4554

34. Giavalisco, M., Koratkar, A., Calzetti, D.: Obscuration of LY alpha photons in star-forming galaxies. ApJ **466**, 831 (1996). https://doi.org/10.1086/177557

35. González Delgado, R.M., Leitherer, C., Heckman, T.M.: Synthetic spectra of H Balmer and He I absorption lines. II. Evolutionary synthesis models for starburst and poststarburst galaxies. ApJS **125**, 489–509 (1999). https://doi.org/10.1086/313285, arXiv:astro-ph/9907116

36. González Delgado, R.M., Cerviño, M., Martins, L.P., Leitherer, C., Hauschildt, P.H.: Evolutionary stellar population synthesis at high spectral resolution: optical wavelengths. MNRAS **357**, 945–960 (2005). https://doi.org/10.1111/j.1365-2966.2005.08692.x, astro-ph/0501204

37. Gronwall, C., Ciardullo, R., Hickey, T., Gawiser, E., Feldmeier, J.J., van Dokkum, P.G., Urry, C.M., Herrera, D., Lehmer, B.D., Infante, L., Orsi, A., Marchesini, D., Blanc, G.A., Francke, H., Lira, P., Treister, E.: Lyα emission-line galaxies at $z = 3.1$ in the extended Chandra Deep Field-South. ApJ **667**, 79–91 (2007). https://doi.org/10.1086/520324, 0705.3917

38. Guaita, L., Acquaviva, V., Padilla, N., Gawiser, E., Bond, N.A., Ciardullo, R., Treister, E., Kurczynski, P., Gronwall, C., Lira, P., Schawinski, K.: Lyα-emitting galaxies at $z = 2.1$: stellar masses, dust, and star formation histories from spectral energy distribution fitting. ApJ **733**, 114 (2011). https://doi.org/10.1088/0004-637X/733/2/114, 1101.3017

39. Hansen, M., Oh, S.P.: Lyman α radiative transfer in a multiphase medium. MNRAS **367**, 979–1002 (2006). https://doi.org/10.1111/j.1365-2966.2005.09870.x. arXiv:astro-ph/0507586

40. Hartmann, L.W., Huchra, J.P., Geller, M.J.: How to find galaxies at high redshift. ApJ **287**, 487–491 (1984). https://doi.org/10.1086/162707

41. Hartmann, L.W., Huchra, J.P., Geller, M.J., O'Brien, P., Wilson, R.: Lyman-alpha emission in star-forming galaxies. ApJ **326**, 101–109 (1988). https://doi.org/10.1086/166072

42. Hayes, M.: Lyman alpha emitting galaxies in the nearby universe. PASA **32**, e027 (2015). https://doi.org/10.1017/pasa.2015.25, 1505.07483

43. Hayes, M., Östlin, G., Mas-Hesse, J.M., Kunth, D., Leitherer, C., Petrosian, A.: HST/ACS Lyman α imaging of the nearby starburst ESO 338-IG04. A&A **438**, 71–85 (2005). https://doi.org/10.1051/0004-6361:20052702. arXiv:astro-ph/0503320

44. Hayes, M., Östlin, G., Atek, H., Kunth, D., Mas-Hesse, J.M., Leitherer, C., Jiménez-Bailón, E., Adamo, A.: The escape of Lyman photons from a young starburst: the case of Haro 11. MNRAS **382**, 1465–1480 (2007). https://doi.org/10.1111/j.1365-2966.2007.12482.x, 0710.2622

45. Hayes, M., Östlin, G., Mas-Hesse, J.M., Kunth, D.: Continuum subtracting Lyman-alpha images: low-redshift studies using the solar blind channel of HST/ACS. AJ **138**, 911–922 (2009). https://doi.org/10.1088/0004-6256/138/3/911, 0803.1176

46. Hayes, M., Östlin, G., Schaerer, D., Mas-Hesse, J.M., Leitherer, C., Atek, H., Kunth, D., Verhamme, A., de Barros, S., Melinder, J.: Escape of about five per cent of Lyman-α photons from high-redshift star-forming galaxies. Nature **464**, 562–565 (2010). https://doi.org/10.1038/nature08881, 1002.4876

47. Hayes, M., Schaerer, D., Östlin, G., Mas-Hesse, J.M., Atek, H., Kunth, D.: On the redshift evolution of the Lyα escape fraction and the dust content of galaxies. ApJ **730**, 8 (2011). https://doi.org/10.1088/0004-637X/730/1/8, 1010.4796

48. Hayes, M., Östlin, G., Schaerer, D., Verhamme, A., Mas-Hesse, J.M., Adamo, A., Atek, H., Cannon, J.M., Duval, F., Guaita, L., Herenz, E.C., Kunth, D., Laursen, P., Melinder, J., Orlitová, I., Otí-Floranes, H., Sandberg, A.: The Lyman alpha reference sample: extended Lyman alpha halos produced at low dust content. ApJ **765**, L27 (2013). https://doi.org/10.1088/2041-8205/765/2/L27, 1303.0006

49. Hayes, M., Östlin, G., Duval, F., Sandberg, A., Guaita, L., Melinder, J., Adamo, A., Schaerer, D., Verhamme, A., Orlitová, I., Mas-Hesse, J.M., Cannon, J.M., Atek, H., Kunth, D., Laursen, P., Otí-Floranes, H., Pardy, S., Rivera-Thorsen, T., Herenz, E.C.: The Lyman alpha reference sample. II. Hubble Space Telescope imaging results, integrated properties, and trends. ApJ **782**, 6 (2014). https://doi.org/10.1088/0004-637X/782/1/6, 1308.6578

50. Hayes, M., Melinder, J., Östlin, G., Scarlata, C., Lehnert, M.D., Mannerström-Jansson, G.: O VI emission imaging of a galaxy with the Hubble Space Telescope: a warm gas halo surrounding the intense starburst SDSS J115630.63+500822.1. ApJ **828**, 49 (2016). https://doi.org/10.3847/0004-637X/828/1/49, 1606.04536

51. Heckman, T.M.: Galactic superwinds circa 2001. In: Mulchaey, J.S., Stocke, J.T. (eds.) Extragalactic Gas at Low Redshift. Astronomical Society of the Pacific Conference Series, vol. 254, p. 292 (2002). astro-ph/0107438

52. Heckman, T.M., Lehnert, M.D., Strickland, D.K., Armus, L.: Absorption-line probes of gas and dust in galactic superwinds. ApJS **129**, 493–516 (2000). https://doi.org/10.1086/313421. arXiv:astro-ph/0002526

53. Heckman, T.M., Borthakur, S., Overzier, R., Kauffmann, G., Basu-Zych, A., Leitherer, C., Sembach, K., Martin, D.C., Rich, R.M., Schiminovich, D., Seibert, M.: Extreme feedback

and the epoch of reionization: clues in the local universe. ApJ **730**, 5 (2011). https://doi.org/
10.1088/0004-637X/730/1/5, 1101.4219

54. Heckman, T.M., Alexandroff, R.M., Borthakur, S., Overzier, R., Leitherer, C.: The systematic
    properties of the warm phase of starburst-driven galactic winds. ApJ **809**, 147 (2015). https://
    doi.org/10.1088/0004-637X/809/2/147, 1507.05622
55. Henry, A., Scarlata, C., Martin, C.L., Erb, D.: Lyα emission from green peas: the role of
    circumgalactic gas density, covering, and kinematics. ApJ **809**, 19 (2015). https://doi.org/10.
    1088/0004-637X/809/1/19, 1505.05149
56. Herenz, E.C., Gruyters, P., Orlitova, I., Hayes, M., Östlin, G., Cannon, J.M., Roth, M.M., Bik,
    A., Pardy, S., Otí-Floranes, H., Mas-Hesse, J.M., Adamo, A., Atek, H., Duval, F., Guaita,
    L., Kunth, D., Laursen, P., Melinder, J., Puschnig, J., Rivera-Thorsen, T.E., Schaerer, D.,
    Verhamme, A.: The Lyman alpha reference sample. VII. Spatially resolved Hα kinematics.
    A&A **587**, A78 (2016), https://doi.org/10.1051/0004-6361/201527373, 1511.05406
57. Hill, G.J., Gebhardt, K., Komatsu, E., Drory, N., MacQueen, P.J., Adams, J., Blanc, G.A.,
    Koehler, R., Rafal, M., Roth, M.M., Kelz, A., Gronwall, C., Ciardullo, R., Schneider, D.P.:
    The Hobby-Eberly Telescope Dark Energy Experiment (HETDEX): description and early
    pilot survey results. In: Kodama, T., Yamada, T., Aoki, K. (eds.) Panoramic Views of Galaxy
    Formation and Evolution. Astronomical Society of the Pacific Conference Series, vol. 399,
    p. 115 (2008). 0806.0183
58. Hoopes, C.G., Heckman, T.M., Salim, S., Seibert, M., Tremonti, C.A., Schiminovich, D.,
    Rich, R.M., Martin, D.C., Charlot, S., Kauffmann, G., Forster, K., Friedman, P.G., Morrissey,
    P., Neff, S.G., Small, T., Wyder, T.K., Bianchi, L., Donas, J., Lee, Y.W., Madore, B.F., Milliard,
    B., Szalay, A.S., Welsh, B.Y., Yi, S.K.: The diverse properties of the most ultraviolet-luminous
    galaxies discovered by GALEX. ApJS **173**, 441–456 (2007). https://doi.org/10.1086/516644.
    arXiv:astro-ph/0609415
59. Hu, E.M., Cowie, L.L.: The distribution of gas and galaxies around the distant quasar PKS
    1614 + 051. ApJ **317**, L7–L12 (1987). https://doi.org/10.1086/184902
60. Izotov, Y.I., Orlitová, I., Schaerer, D., Thuan, T.X., Verhamme, A., Guseva, N.G., Worseck,
    G.: Eight per cent leakage of Lyman continuum photons from a compact, star-forming dwarf
    galaxy. Nature **529**, 178–180 (2016a). https://doi.org/10.1038/nature16456
61. Izotov, Y.I., Schaerer, D., Thuan, T.X., Worseck, G., Guseva, N.G., Orlitová, I., Verhamme,
    A.: Detection of high Lyman continuum leakage from four low-redshift compact star-
    forming galaxies. MNRAS **461**, 3683–3701 (2016). https://doi.org/10.1093/mnras/stw1205,
    1605.05160
62. James, B.L., Aloisi, A., Heckman, T., Sohn, S.T., Wolfe, M.A.: Investigating nearby star-
    forming galaxies in the ultraviolet with HST/COS spectroscopy. I. Spectral analysis and
    interstellar abundance determinations. ApJ **795**, 109 (2014). https://doi.org/10.1088/0004-
    637X/795/2/109, 1408.4420
63. Jaskot, A.E., Oey, M.S.: The origin and optical depth of ionizing radiation in the "green pea"
    galaxies. ApJ **766**, 91 (2013). https://doi.org/10.1088/0004-637X/766/2/91, 1301.0530
64. Jaskot, A.E., Oey, M.S.: Linking Lyα and low-ionization transitions at low optical depth. ApJ
    791, L19 (2014). https://doi.org/10.1088/2041-8205/791/2/L19, 1406.4413
65. Jones, T.A., Ellis, R.S., Schenker, M.A., Stark, D.P.: Keck spectroscopy of gravitationally
    lensed $z \simeq 4$ galaxies: improved constraints on the escape fraction of ionizing photons. ApJ
    779, 52 (2013). https://doi.org/10.1088/0004-637X/779/1/52, 1304.7015
66. Kalberla, P.M.W., Kerp, J.: The Hi distribution of the Milky Way. ARA&A **47**, 27–61 (2009).
    https://doi.org/10.1146/annurev-astro-082708-101823
67. Kashikawa, N., Shimasaku, K., Malkan, M.A., Doi, M., Matsuda, Y., Ouchi, M., Taniguchi,
    Y., Ly, C., Nagao, T., Iye, M., Motohara, K., Murayama, T., Murozono, K., Nariai, K., Ohta,
    K., Okamura, S., Sasaki, T., Shioya, Y., Umemura, M.: The end of the reionization epoch
    probed by Lyα emitters at $z = 6.5$ in the Subaru Deep Field. ApJ **648**, 7–22 (2006). https://
    doi.org/10.1086/504966, arXiv:astro-ph/0604149
68. Kauffmann, G., Heckman, T.M., Tremonti, C., Brinchmann, J., Charlot, S., White, S.D.M.,
    Ridgway, S.E., Brinkmann, J., Fukugita, M., Hall, P.B., Ivezić, Ž., Richards, G.T., Schneider,

D.P.: The host galaxies of active galactic nuclei. MNRAS **346**, 1055–1077 (2003). https://doi.org/10.1111/j.1365-2966.2003.07154.x, astro-ph/0304239

69. Keel, W.C.: A nearby galaxy in the deep-ultraviolet: Voyager 2 observations of M33 from Lyα to the Lyman limit. ApJ **506**, 712–720 (1998). https://doi.org/10.1086/306277

70. Keel, W.C., White III, R.E., Chapman, S., Windhorst, R.A.: The disappearance of Lyα blobs: a GALEX search at $z = 0.8$. AJ **138**, 986–990 (2009). https://doi.org/10.1088/0004-6256/138/3/986, 0907.2201

71. Kehrig, C., Vílchez, J.M., Pérez-Montero, E., Iglesias-Páramo, J., Brinchmann, J., Kunth, D., Durret, F., Bayo, F.M.: The extended He II λ4686-emitting region in IZw 18 unveiled: clues for peculiar ionizing sources. ApJ **801**, L28 (2015). https://doi.org/10.1088/2041-8205/801/2/L28, 1502.00522

72. Kehrig, C., Vílchez, J.M., Pérez-Montero, E., Iglesias-Páramo, J., Hernández-Fernández, J.D., Duarte Puertas, S., Brinchmann, J., Durret, F., Kunth, D.: Spatially resolved integral field spectroscopy of the ionized gas in IZw18. MNRAS **459**, 2992–3004 (2016). https://doi.org/10.1093/mnras/stw806, 1604.08555

73. Kennicutt Jr., R.C.: Star formation in galaxies along the Hubble sequence. ARA&A **36**, 189–232 (1998). https://doi.org/10.1146/annurev.astro.36.1.189. arXiv:astro-ph/9807187

74. Kepley, A.A., Chomiuk, L., Johnson, K.E., Goss, W.M., Balser, D.S., Pisano, D.J.: Unveiling extragalactic star formation using radio recombination lines: an expanded very large array pilot study with NGC 253. ApJ **739**, L24 (2011). https://doi.org/10.1088/2041-8205/739/1/L24, 1106.4818

75. Kewley, L.J., Dopita, M.A., Sutherland, R.S., Heisler, C.A., Trevena, J.: Theoretical modeling of starburst galaxies. ApJ **556**, 121–140 (2001). https://doi.org/10.1086/321545, astro-ph/0106324

76. Kondo, Y. (ed.): Exploring the Universe with the IUE Satellite. Astrophysics and Space Science Library, vol. 129 (1987). https://doi.org/10.1007/978-94-009-3753-6

77. Kunth, D., Lequeux, J., Sargent, W.L.W., Viallefond, F.: Is there primordial gas in IZw 18? A&A **282**, 709–716 (1994)

78. Kunth, D., Mas-Hesse, J.M., Terlevich, E., Terlevich, R., Lequeux, J., Fall, S.M.: HST study of Lyman-alpha emission in star-forming galaxies: the effect of neutral gas flows. A&A **334**, 11–20 (1998). arXiv:astro-ph/9802253

79. Kunth, D., Leitherer, C., Mas-Hesse, J.M., Östlin, G., Petrosian, A.: The first deep advanced camera for surveys Lyα images of local starburst galaxies. ApJ **597**, 263–268 (2003). https://doi.org/10.1086/378396. arXiv:astro-ph/0307555

80. Lacy, J.H., Malkan, M., Becklin, E.E., Soifer, B.T., Neugebauer, G., Matthews, K., Wu, C.C., Boggess, A., Gull, T.R.: Infrared, optical, and ultraviolet observations of hydrogen line emission from Seyfert galaxies. ApJ **256**, 75–82 (1982). https://doi.org/10.1086/159884

81. Lallement, R., Quémerais, E., Bertaux, J.L., Sandel, B.R., Izmodenov, V.: Voyager measurements of hydrogen Lyman-α diffuse emission from the Milky Way. Science **334**, 1665 (2011). https://doi.org/10.1126/science.1197340

82. Laursen, P., Duval, F., Östlin, G.: On the (non-)enhancement of the Lyα equivalent width by a multiphase interstellar medium. ApJ **766**, 124 (2013). https://doi.org/10.1088/0004-637X/766/2/124, 1211.2833

83. Leitherer, C., Schaerer, D., Goldader, J.D., González Delgado, R.M., Robert, C., Kune, D.F., de Mello, D.F., Devost, D., Heckman, T.M.: Starburst99: synthesis models for galaxies with active star formation. ApJS **123**, 3–40 (1999). https://doi.org/10.1086/313233. arXiv:astro-ph/9902334

84. Leitherer, C., Chandar, R., Tremonti, C.A., Wofford, A., Schaerer, D.: Far-ultraviolet observations of outflows from infrared-luminous galaxies. ApJ **772**, 120 (2013). https://doi.org/10.1088/0004-637X/772/2/120, 1306.0419

85. Leitherer, C., Ekström, S., Meynet, G., Schaerer, D., Agienko, K.B., Levesque, E.M.: The effects of stellar rotation. II. A comprehensive set of Starburst99 models. ApJS **212**, 14 (2014). https://doi.org/10.1088/0067-0049/212/1/14, 1403.5444

86. Lequeux, J., Kunth, D., Mas-Hesse, J.M., Sargent, W.L.W.: Galactic wind and Lyman α emission in the blue compact galaxy Haro 2 = MKN 33. A&A **301**, 18 (1995)
87. Luridiana, V., Morisset, C., Shaw, R.A.: PyNeb: a new tool for analyzing emission lines. I. Code description and validation of results. A&A **573**, A42 (2015). https://doi.org/10.1051/0004-6361/201323152, 1410.6662
88. Madau, P.: Radiative transfer in a clumpy universe: the colors of high-redshift galaxies. ApJ **441**, 18–27 (1995). https://doi.org/10.1086/175332
89. Madau, P., Dickinson, M.: Annu. Rev. Astron. Astrophys. **52**, 415–486 (2014). https://doi.org/10.1146/annurev-astro-081811-125615, 1403.0007
90. Mannucci, F., Cresci, G., Maiolino, R., Marconi, A., Gnerucci, A.: A fundamental relation between mass, star formation rate and metallicity in local and high-redshift galaxies. MNRAS **408**, 2115–2127 (2010). https://doi.org/10.1111/j.1365-2966.2010.17291.x, 1005.0006
91. Mao, J., Lapi, A., Granato, G.L., de Zotti, G., Danese, L.: The role of the dust in primeval galaxies: a simple physical model for Lyman break galaxies and Lyα emitters. ApJ **667**, 655–666 (2007). https://doi.org/10.1086/521069, astro-ph/0611799
92. Martin, C.L.: Mapping large-scale gaseous outflows in ultraluminous galaxies with Keck II ESI spectra: variations in outflow velocity with galactic mass. ApJ **621**, 227–245 (2005). https://doi.org/10.1086/427277, astro-ph/0410247
93. Martin, C.L., Dijkstra, M., Henry, A., Soto, K.T., Danforth, C.W., Wong, J.: The Lyα line profiles of ultraluminous infrared galaxies: fast winds and Lyman continuum leakage. ApJ **803**, 6 (2015). https://doi.org/10.1088/0004-637X/803/1/6, 1501.05946
94. Martin, D.C., Fanson, J., Schiminovich, D., Morrissey, P., Friedman, P.G., Barlow, T.A., Conrow, T., Grange, R., Jelinsky, P.N., Milliard, B., Siegmund, O.H.W., Bianchi, L., Byun, Y.I., Donas, J., Forster, K., Heckman, T.M., Lee, Y.W., Madore, B.F., Malina, R.F., Neff, S.G., Rich, R.M., Small, T., Surber, F., Szalay, A.S., Welsh, B., Wyder, T.K.: The galaxy evolution explorer: a space ultraviolet survey mission. ApJ **619**, L1–L6 (2005). https://doi.org/10.1086/426387, astro-ph/0411302
95. Mas-Hesse, J.M., Kunth, D., Tenorio-Tagle, G., Leitherer, C., Terlevich, R.J., Terlevich, E.: Lyα emission in starbursts: implications for galaxies at high redshift. ApJ **598**, 858–877 (2003). https://doi.org/10.1086/379116. arXiv:astro-ph/0309396
96. Matsuda, Y., Yamada, T., Hayashino, T., Yamauchi, R., Nakamura, Y., Morimoto, N., Ouchi, M., Ono, Y., Kousai, K., Nakamura, E., Horie, M., Fujii, T., Umemura, M., Mori, M.: The Subaru Lyα blob survey: a sample of 100-kpc Lyα blobs at $z = 3$. MNRAS **410**, L13–L17 (2011). https://doi.org/10.1111/j.1745-3933.2010.00969.x, 1010.2877
97. Meier, D.L., Terlevich, R.: Extragalactic H II regions in the UV—implications for primeval galaxies. ApJ **246**, L109–L113 (1981). https://doi.org/10.1086/183565
98. Meurer, G.R., Heckman, T.M., Leitherer, C., Kinney, A., Robert, C., Garnett, D.R.: Starbursts and star clusters in the ultraviolet. AJ **110**, 2665 (1995). https://doi.org/10.1086/117721, astro-ph/9509038
99. Miyazaki, S., Komiyama, Y., Nakaya, H., Kamata, Y., Doi, Y., Hamana, T., Karoji, H., Furusawa, H., Kawanomoto, S., Morokuma, T., Ishizuka, Y., Nariai, K., Tanaka, Y., Uraguchi, F., Utsumi, Y., Obuchi, Y., Okura, Y., Oguri, M., Takata, T., Tomono, D., Kurakami, T., Namikawa, K., Usuda, T., Yamanoi, H., Terai, T., Uekiyo, H., Yamada, Y., Koike, M., Aihara, H., Fujimori, Y., Mineo, S., Miyatake, H., Yasuda, N., Nishizawa, J., Saito, T., Tanaka, M., Uchida, T., Katayama, N., Wang, S.Y., Chen, H.Y., Lupton, R., Loomis, C., Bickerton, S., Price, P., Gunn, J., Suzuki, H., Miyazaki, Y., Muramatsu, M., Yamamoto, K., Endo, M., Ezaki, Y., Itoh, N., Miwa, Y., Yokota, H., Matsuda, T., Ebinuma, R., Takeshi, K.: Hyper Suprime-Cam. In: Ground-based and Airborne Instrumentation for Astronomy IV. Proc. SPIE, vol. 8446, p. 84460Z (2012). https://doi.org/10.1117/12.926844
100. Natta, A., Panagia, N.: Extinction in inhomogeneous clouds. ApJ **287**, 228–237 (1984). https://doi.org/10.1086/162681
101. Neufeld, D.A.: The escape of Lyman-alpha radiation from a multiphase interstellar medium. ApJ **370**, L85–L88 (1991). https://doi.org/10.1086/185983

102. Oey, M.S., King, N.L., Parker, J.W.: Massive field stars and the stellar clustering law. AJ **127**, 1632–1643 (2004). https://doi.org/10.1086/381926, astro-ph/0312051
103. Oke, J.B., Zimmerman, B.: IUE and visual spectrophotometry of 3C 120 and Markarian 79. ApJ **231**, L13–L17 (1979). https://doi.org/10.1086/182996
104. Osterbrock, D.E.: Astrophysics of gaseous nebulae and active galactic nuclei (1989)
105. Östlin, G., Hayes, M., Kunth, D., Mas-Hesse, J.M., Leitherer, C., Petrosian, A., Atek, H.: The Lyman alpha morphology of local starburst galaxies: release of calibrated images. AJ **138**, 923–940 (2009). https://doi.org/10.1088/0004-6256/138/3/923, 0803.1174
106. Östlin, G., Hayes, M., Duval, F., Sandberg, A., Rivera-Thorsen, T., Marquart, T., Orlitová, I., Adamo, A., Melinder, J., Guaita, L., Atek, H., Cannon, J.M., Gruyters, P., Herenz, E.C., Kunth, D., Laursen, P., Mas-Hesse, J.M., Micheva, G., Otí-Floranes, H., Pardy, S.A., Roth, M.M., Schaerer, D., Verhamme, A.: The Lyα reference sample. I. Survey outline and first results for Markarian 259. ApJ **797**, 11 (2014). https://doi.org/10.1088/0004-637X/797/1/11, 1409.8347
107. Otí-Floranes, H., Mas-Hesse, J.M., Jiménez-Bailón, E., Schaerer, D., Hayes, M., Östlin, G., Atek, H., Kunth, D.: Physical properties and evolutionary state of the Lyman alpha emitting starburst galaxy IRAS 08339+6517. A&A **566**, A38 (2014). https://doi.org/10.1051/0004-6361/201323069, 1403.7687
108. Ouchi, M., Shimasaku, K., Akiyama, M., Simpson, C., Saito, T., Ueda, Y., Furusawa, H., Sekiguchi, K., Yamada, T., Kodama, T., Kashikawa, N., Okamura, S., Iye, M., Takata, T., Yoshida, M., Yoshida, M.: The Subaru/XMM-Newton Deep Survey (SXDS). IV. Evolution of Lyα emitters from $z = 3.1$ to 5.7 in the 1 deg^2 field: luminosity functions and AGN. ApJS **176**, 301–330 (2008). https://doi.org/10.1086/527673, 0707.3161
109. Pardy, S.A., Cannon, J.M., Östlin, G., Hayes, M., Rivera-Thorsen, T., Sandberg, A., Adamo, A., Freeland, E., Herenz, E.C., Guaita, L., Kunth, D., Laursen, P., Mas-Hesse, J.M., Melinder, J., Orlitová, I., Otí-Floranes, H., Puschnig, J., Schaerer, D., Verhamme, A.: The Lyman alpha reference sample. III. Properties of the neutral ISM from GBT and VLA observations. ApJ **794**, 101 (2014). https://doi.org/10.1088/0004-637X/794/2/101, 1408.6275
110. Partridge, R.B., Peebles, P.J.E.: Are young galaxies visible? ApJ **147**, 868 (1967). https://doi.org/10.1086/149079
111. Peña-Guerrero, M.A., Leitherer, C.: H I Lyman-alpha equivalent widths of stellar populations. AJ **146**, 158 (2013). https://doi.org/10.1088/0004-6256/146/6/158, 1310.1155
112. Pellegrini, E.W., Oey, M.S., Winkler, P.F., Points, S.D., Smith, R.C., Jaskot, A.E., Zastrow, J.: The optical depth of H II regions in the Magellanic Clouds. ApJ **755**, 40 (2012). https://doi.org/10.1088/0004-637X/755/1/40, 1202.3334
113. Pettini, M., Rix, S.A., Steidel, C.C., Adelberger, K.L., Hunt, M.P., Shapley, A.E.: New observations of the interstellar medium in the Lyman break galaxy MS 1512-cB58. ApJ **569**, 742–757 (2002). https://doi.org/10.1086/339355. arXiv:astro-ph/0110637
114. Pritchet, C.J.: The search for primeval galaxies. PASP **106**, 1052–1067 (1994). https://doi.org/10.1086/133479
115. Prochaska, J.X., Kasen, D., Rubin, K.: Simple models of metal-line absorption and emission from cool gas outflows. ApJ **734**, 24 (2011). https://doi.org/10.1088/0004-637X/734/1/24, 1102.3444
116. Quider, A.M., Pettini, M., Shapley, A.E., Steidel, C.C.: The ultraviolet spectrum of the gravitationally lensed galaxy 'the Cosmic Horseshoe': a close-up of a star-forming galaxy at $z \sim 2$. MNRAS **398**, 1263–1278 (2009). https://doi.org/10.1111/j.1365-2966.2009.15234.x, 0906.2412
117. Raiter, A., Schaerer, D., Fosbury, R.A.E.: Predicted UV properties of very metal-poor starburst galaxies. A&A **523**, A64 (2010). https://doi.org/10.1051/0004-6361/201015236, 1008.2114
118. Rauch, M., Haehnelt, M., Bunker, A., Becker, G., Marleau, F., Graham, J., Cristiani, S., Jarvis, M., Lacey, C., Morris, S., Peroux, C., Röttgering, H., Theuns, T.: A population of faint extended line emitters and the host galaxies of optically thick QSO absorption systems. ApJ **681**, 856–880 (2008). https://doi.org/10.1086/525846, 0711.1354

119. Reddy, N.A., Steidel, C.C.: A steep faint-end slope of the UV luminosity function at $z \sim 2$–3: implications for the global stellar mass density and star formation in low-mass halos. ApJ **692**, 778–803 (2009). https://doi.org/10.1088/0004-637X/692/1/778, 0810.2788

120. Rhoads, J.E., Dey, A., Malhotra, S., Stern, D., Spinrad, H., Jannuzi, B.T., Dawson, S., Brown, M.J.I., Landes, E.: Spectroscopic confirmation of three redshift $z \sim 5.7$ Lyα emitters from the large-area Lyman alpha survey. AJ **125**, 1006–1013 (2003). https://doi.org/10.1086/346272, arXiv:astro-ph/0209544

121. Rivera-Thorsen, T.E., Hayes, M., Östlin, G., Duval, F., Orlitová, I., Verhamme, A., Mas-Hesse, J.M., Schaerer, D., Cannon, J.M., Otí-Floranes, H., Sandberg, A., Guaita, L., Adamo, A., Atek, H., Herenz, E.C., Kunth, D., Laursen, P., Melinder, J.: The Lyman alpha reference sample. V. The impact of neutral ISM kinematics and geometry on Lyα escape. ApJ **805**, 14 (2015). https://doi.org/10.1088/0004-637X/805/1/14, 1503.01157

122. Rupke, D.S., Veilleux, S., Sanders, D.B.: Keck absorption-line spectroscopy of galactic winds in ultraluminous infrared galaxies. ApJ **570**, 588–609 (2002). https://doi.org/10.1086/339789, astro-ph/0201371

123. Salzer, J.J., Gronwall, C., Lipovetsky, V.A., Kniazev, A., Moody, J.W., Boroson, T.A., Thuan, T.X., Izotov, Y.I., Herrero, J.L., Frattare, L.M.: The KPNO international spectroscopic survey. II. Hα-selected survey list 1. AJ **121**, 66–79 (2001). https://doi.org/10.1086/318040, astro-ph/0010406

124. Sandberg, A., Östlin, G., Hayes, M., Fathi, K., Schaerer, D., Mas-Hesse, J.M., Rivera-Thorsen, T.: Neutral gas in Lyman-alpha emitting galaxies Haro 11 and ESO 338-IG04 measured through sodium absorption. A&A **552**, A95 (2013). https://doi.org/10.1051/0004-6361/201220702, 1303.2011

125. Savage, B.D., Sembach, K.R.: The analysis of apparent optical depth profiles for interstellar absorption lines. ApJ **379**, 245–259 (1991). https://doi.org/10.1086/170498

126. Scarlata, C., Panagia, N.: A Semi-analytical line transfer model to interpret the spectra of galaxy outflows. ApJ **801**, 43 (2015). https://doi.org/10.1088/0004-637X/801/1/43, 1501.07282

127. Scarlata, C., Colbert, J., Teplitz, H.I., Panagia, N., Hayes, M., Siana, B., Rau, A., Francis, P., Caon, A., Pizzella, A., Bridge, C.: The effect of dust geometry on the Lyα output of galaxies. ApJ **704**, L98–L102 (2009). https://doi.org/10.1088/0004-637X/704/2/L98, 0909.3847

128. Schaerer, D.: On the properties of massive Population III stars and metal-free stellar populations. A&A **382**, 28–42 (2002). https://doi.org/10.1051/0004-6361:20011619. arXiv:astro-ph/0110697

129. Schaerer, D.: The transition from Population III to normal galaxies: Lyα and He II emission and the ionising properties of high redshift starburst galaxies. A&A **397**, 527–538 (2003). https://doi.org/10.1051/0004-6361:20021525. arXiv:astro-ph/0210462

130. Schenker, M.A., Ellis, R.S., Konidaris, N.P., Stark, D.P.: Line-emitting galaxies beyond a redshift of 7: an improved method for estimating the evolving neutrality of the intergalactic medium. ApJ **795**, 20 (2014). https://doi.org/10.1088/0004-637X/795/1/20, 1404.4632

131. Shapley, A.E., Steidel, C.C., Pettini, M., Adelberger, K.L.: Rest-frame ultraviolet spectra of $z \sim 3$ Lyman break galaxies. ApJ **588**, 65–89 (2003). https://doi.org/10.1086/373922. arXiv:astro-ph/0301230

132. Shimasaku, K., Kashikawa, N., Doi, M., Ly, C., Malkan, M.A., Matsuda, Y., Ouchi, M., Hayashino, T., Iye, M., Motohara, K., Murayama, T., Nagao, T., Ohta, K., Okamura, S., Sasaki, T., Shioya, Y., Taniguchi, Y.: Lyα emitters at $z = 5.7$ in the Subaru Deep Field. PASJ **58**, 313–334 (2006). arXiv:astro-ph/0602614

133. Silich, S.A., Tenorio-Tagle, G.: On the fate of processed matter in dwarf galaxies. MNRAS **299**, 249–266 (1998). https://doi.org/10.1046/j.1365-8711.1998.01765.x, astro-ph/9805370

134. Stark, D.P., Ellis, R.S., Chiu, K., Ouchi, M., Bunker, A.: Keck spectroscopy of faint $3 < z < 7$ Lyman break galaxies—I. New constraints on cosmic reionization from the luminosity and redshift-dependent fraction of Lyman α emission. MNRAS **408**, 1628–1648 (2010). https://doi.org/10.1111/j.1365-2966.2010.17227.x, 1003.5244

135. Stasińska, G., Izotov, Y., Morisset, C., Guseva, N.: Excitation properties of galaxies with the highest [O III]/[O II] ratios. No evidence for massive escape of ionizing photons. A&A **576**, A83 (2015). https://doi.org/10.1051/0004-6361/201425389, 1503.00320
136. Steidel, C.C., Strom, A.L., Pettini, M., Rudie, G.C., Reddy, N.A., Trainor, R.F.: Reconciling the stellar and nebular spectra of high-redshift galaxies. ApJ **826**, 159 (2016). https://doi.org/10.3847/0004-637X/826/2/159, 1605.07186
137. Stern, D., Eisenhardt, P., Gorjian, V., Kochanek, C.S., Caldwell, N., Eisenstein, D., Brodwin, M., Brown, M.J.I., Cool, R., Dey, A., Green, P., Jannuzi, B.T., Murray, S.S., Pahre, M.A., Willner, S.P.: Mid-infrared selection of active galaxies. ApJ **631**, 163–168 (2005). https://doi.org/10.1086/432523, astro-ph/0410523
138. Strauss, M.A., Huchra, J.P., Davis, M., Yahil, A., Fisher, K.B., Tonry, J.: A redshift survey of IRAS galaxies. VII—The infrared and redshift data for the 1.936 Jansky sample. ApJS **83**, 29–63 (1992). https://doi.org/10.1086/191730
139. Tenorio-Tagle, G., Silich, S.A., Kunth, D., Terlevich, E., Terlevich, R.: The evolution of superbubbles and the detection of Lyα in star-forming galaxies. MNRAS **309**, 332–342 (1999). https://doi.org/10.1046/j.1365-8711.1999.02809.x. arXiv:astro-ph/9905324
140. Thommes, E., Meisenheimer, K.: The expected abundance of Lyman-α emitting primeval galaxies. I. General model predictions. A&A **430**, 877–891 (2005). https://doi.org/10.1051/0004-6361:20035863
141. Thuan, T.X., Izotov, Y.I.: Nearby young dwarf galaxies: primordial gas and Ly alpha emission. ApJ **489**, 623 (1997). https://doi.org/10.1086/304826
142. Valls-Gabaud, D.: On the Lyman-alpha emission of starburst galaxies. ApJ **419**, 7 (1993). https://doi.org/10.1086/173454. arXiv:astro-ph/9306008
143. van Zee, L., Westpfahl, D., Haynes, M.P., Salzer, J.J.: The complex kinematics of the neutral hydrogen associated with I ZW 18. AJ **115**, 1000–1015 (1998). https://doi.org/10.1086/300251. arXiv:astro-ph/9712070
144. Verhamme, A., Schaerer, D., Maselli, A.: 3D Lyα radiation transfer. I. Understanding Lyα line profile morphologies. A&A **460**, 397–413 (2006). https://doi.org/10.1051/0004-6361:20065554, arXiv:astro-ph/0608075
145. Verhamme, A., Dubois, Y., Blaizot, J., Garel, T., Bacon, R., Devriendt, J., Guiderdoni, B., Slyz, A.: Lyman-α emission properties of simulated galaxies: interstellar medium structure and inclination effects. A&A **546**, A111 (2012). https://doi.org/10.1051/0004-6361/201218783, 1208.4781
146. Verhamme, A., Orlitová, I., Schaerer, D., Hayes, M.: Using Lyman-α to detect galaxies that leak Lyman continuum. A&A **578**, A7 (2015). https://doi.org/10.1051/0004-6361/201423978, 1404.2958
147. Wisotzki, L., Bacon, R., Blaizot, J., Brinchmann, J., Herenz, E.C., Schaye, J., Bouché, N., Cantalupo, S., Contini, T., Carollo, C.M., Caruana, J., Courbot, J.B., Emsellem, E., Kamann, S., Kerutt, J., Leclercq, F., Lilly, S.J., Patrício, V., Sandin, C., Steinmetz, M., Straka, L.A., Urrutia, T., Verhamme, A., Weilbacher, P.M., Wendt, M.: Extended Lyman α haloes around individual high-redshift galaxies revealed by MUSE. A&A **587**, A98 (2016). https://doi.org/10.1051/0004-6361/201527384, 1509.05143
148. Wofford, A., Leitherer, C., Salzer, J.: Lyα escape from $z \sim 0.03$ star-forming galaxies: the dominant role of outflows. ApJ **765**, 118 (2013). https://doi.org/10.1088/0004-637X/765/2/118, 1301.7285
149. Wold, I.G.B., Barger, A.J., Cowie, L.L.: $z \sim 1$ Lyα emitters. I. The luminosity function. ApJ **783**, 119 (2014). https://doi.org/10.1088/0004-637X/783/2/119, 1401.6201
150. Wu, C.C., Boggess, A., Gull, T.R.: Lyman alpha fluxes of Seyfert galaxies and low-redshift quasars. ApJ **242**, 14–17 (1980). https://doi.org/10.1086/158439
151. Wu, C.C., Boggess, A., Gull, T.R.: Prominent ultraviolet emission lines from Type 1 Seyfert galaxies. ApJ **266**, 28–40 (1983). https://doi.org/10.1086/160756
152. Yang, H., Malhotra, S., Gronke, M., Rhoads, J.E., Dijkstra, M., Jaskot, A., Zheng, Z., Wang, J.: Green Pea galaxies reveal secrets of Lyα escape. ApJ **820**, 130 (2016). https://doi.org/10.3847/0004-637X/820/2/130, 1506.02885

153. Zastrow, J., Oey, M.S., Veilleux, S., McDonald, M., Martin, C.L.: An ionization cone in the dwarf starburst galaxy NGC 5253. ApJ **741**, L17 (2011). https://doi.org/10.1088/2041-8205/741/1/L17, 1109.6360
154. Zastrow, J., Oey, M.S., Veilleux, S., McDonald, M.: New constraints on the escape of ionizing photons from starburst galaxies using ionization-parameter mapping. ApJ **779**, 76 (2013). https://doi.org/10.1088/0004-637X/779/1/76, 1311.2227

Correction to: Physics of Lyα Radiative Transfer

Mark Dijkstra

Correction to:
Chapter 1 in: M. Dijkstra et al., *Lyman-alpha as an Astrophysical and Cosmological Tool*, Saas-Fee Advanced Course 46, https://doi.org/10.1007/978-3-662-59623-4_1

In the original version of this book, the following belated corrections have been incorporated:

In Chapter 1, p. 52, middle of the page (3 lines above Eq. 1.85) the following typo was corrected:

$$\frac{\partial^2 \phi}{\partial \phi^2} = \frac{6\phi}{x^2}$$

was corrected to

$$\frac{\partial^2 \phi}{\partial x^2} = \frac{6\phi}{x^2}.$$

The updated version of this chapter can be found at
https://doi.org/10.1007/978-3-662-59623-4_1

© Springer-Verlag GmbH Germany, part of Springer Nature 2020
M. Dijkstra et al., *Lyman-alpha as an Astrophysical and Cosmological Tool*,
Saas-Fee Advanced Course 46, https://doi.org/10.1007/978-3-662-59623-4_5

Figure 1.35 has been replaced with the revised figure.

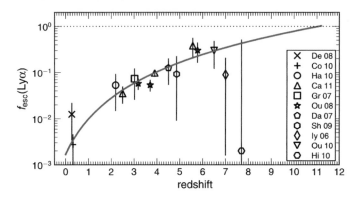

Fig. 1.35 Observational constraints on the redshift-dependence of the volume averaged 'effective' escape fraction, f_{esc}^{eff}, which contains constraints on the true escape *fraction* f_{esc}^{α} (*Credit from Fig. 1 of* [121] *©AAS. Reproduced with permission*)

The erratum chapter and the book have been updated with the changes.

Index

© Springer-Verlag GmbH Germany, part of Springer Nature 2019
M. Dijkstra et al., *Lyman-alpha as an Astrophysical and Cosmological Tool*,
Saas-Fee Advanced Course 46, https://doi.org/10.1007/978-3-662-59623-4

Printed in the United States
By Bookmasters